THE CULTURAL HISTORY OF
PLANTS

THE CULTURAL HISTORY OF
PLANTS

SIR GHILLEAN PRANCE
CONSULTING EDITOR

MARK NESBITT
SCIENTIFIC EDITOR

Routledge
New York • London

Published in 2005 by
Routledge
270 Madison Avenue
New York, NY 10016
www.routledge-ny.com

Published in Great Britain by
Routledge
2 Park Square
Milton Park, Abingdon
Oxon OX14 4RN U.K.

10 9 8 7 6 5 4 3 2 1

Library of Congress Cataloging-in-Publication Data

Cultural history of plants / edited by Ghillean Prance.
      p. cm.
ISBN 0-415-92746-3 (Hardcover : alk. paper)
1. Crops–History. I. Prance, Ghillean T., 1937-
SB71.C86 2004
630'.9–dc21   2002012820

Contents

Part 3   Today and Tomorrow

List of Contributors

Peter Barnes
Freelance horticultural botanist and writer,
UK

Hans T. Beck
Department of Biological Sciences, Northern
Illinois University, DeKalb, Illinois, USA

Paul Bremner
Centre for Pharmacognosy & Phytotherapy,
School of Pharmacy, University of London,
UK

Charles R. Clement
Instituto Nacional de Pesquisas da
Amazônia—INPA, Manaus, AM, Brazil

Ivan Crowe
Independent Scholar, UK

David R. Given
Botanical Services Curator, Christchurch City
Council, New Zealand

Daphne Hakuno
New York Botanic Gardens, New York, USA

David R. Harris
Institute of Archaeology, University College
London, UK

Michael Heinrich
Centre for Pharmacognosy & Phytotherapy,
School of Pharmacy, University of London,
UK

Vernon Heywood
Professor Emeritus, School of Plant Sciences,
The University of Reading, UK

Andrew Jacobson
Former Curator/Archivist, and Director of
Collections, New Jersey Museum of
Agriculture, USA
Archivist, AIG: American International
Group, Inc., New York, NY, USA

Nigel Maxted
School of Biosciences, University of
Birmingham, Birmingham, UK

Sue Minter
Horticultural Director, Eden Project,
Cornwall, UK

Mark Nesbitt
Centre for Economic Botany, Royal Botanic
Gardens, Kew, UK

Georgina Pearman
Researcher, Eden Project, Cornwall, UK

Barbara Pickersgill
School of Plant Sciences, University of
Reading, UK

Andrea Pieroni
School of Life Sciences, University of
Bradford, UK and Department of Social
Sciences, Wageningen University,
Netherlands

Ghillean Prance
Former Director of the Royal Botanic
Gardens at Kew, UK, and Scientific Director of
the Eden Project in Cornwall, UK

Jane Renfrew
Lucy Cavendish College, University of
Cambridge, UK

George A.F. Roberts
Emeritus Professor of Textile Science,
Nottingham Trent University, UK

Richard Rudgley
Institute of Social and Cultural
Anthropology, University of Oxford, UK

Tony Russell
Writer, Broadcaster and former Head
Forester of Westonbirt Arboretum,
Gloucestershire, UK

Helen Sanderson
Centre for Economic Botany, Royal Botanic
Gardens, Kew, UK

Frances A. Wood
Nottingham Trent School of Art & Design,
Nottingham Trent University, UK

Part I
The Seeds of Time

GHILLEAN T. PRANCE

All animals are dependent upon plants, since plants are the organisms at the base of the food chain, because of their capacity to photosynthesize—that is, to turn water and carbon dioxide into oxygen and sugars, in the presence of sunlight. As life on Earth gradually evolved from simple unicellular organisms to the variety of organisms we know today, the complexity of interactions between plants and animals increased, but plants remained the basis for life on Earth. Plants support all animal life. Humans are no exception to this rule, and we are just as dependent upon plants as any other animal. We depend on plants not only for their role in producing the oxygen we breathe, but also for food, shelter, medicines, clothing, and countless other uses.

The first chapter in this book describes how primates gradually developed into the hominids and eventually into our species, *Homo sapiens*. With the advent of *Homo*, an intelligent being, more than a basic subsistence from plants developed and a cultural relationship between plants and people began to evolve. The managed use of fire began at a very early stage, perhaps even by our ancestor *Homo erectus*, who began to use fire to flush out game from the vegetation. Later our species developed cooking, thus enabling the use of so many previously inedible species of plants. Because of this gradual evolution of our species, amongst many other animal species and with complete dependence on the plants around them, humans seem to have an inborn love of nature. This concept was termed "biophilia" by the great Harvard biologist Edward O. Wilson. It was therefore a natural reaction for humans to develop a close relationship with plants and with the landscape around them. Early in the evolutionary sequence of our australopithecine ancestors, dependence on plants was confirmed by the need for vitamin C in their diet. Unlike almost all other mammals, their bodies were unable to manufacture vitamin C. This meant that from an early stage plants were an essential part of their diet.

Today we still have a few glimpses of how a hunter-gatherer society works from studies of the indigenous peoples of the Amazon, New Guinea, and a few other places. These Stone-Age societies are very much plant-based cultures, and it is amazing how many uses for plants they have developed. They are much more in touch with plants than most people are today. The culture of indigenous peoples very much depends upon which plants they put to use and how they use them: to eat, as materials from which to build their houses, as medicines to heal, as tools for hunting and other tasks, in rituals to commune with their spirits through narcotics, and as materials to make their clothing. Studies by ethnoarcheologists and ethnobotanists show that the cultural history of plants began long before history began to be documented.

Between ten and twelve thousand years ago, a major change occurred that completely revolutionized human culture and its relationship to the environment. This was the invention of agriculture. It is interesting that this took place independently in various parts of the world, based on the local plant resources. In eastern Asia, rice was the basis of agriculture, whereas in the Middle East or western Asia wheat and barley became the staple crops. In Central and South America, maize was the cereal that enabled agriculture to prosper. The second chapter gives some of the fascinating details of the multiple origins of agriculture. Cultivation of plants led to a major cultural change because people no longer wandered from place to place as hunter-gatherers, but became settled in towns and villages near their agricultural fields. The greater availability of food led to population growth and consequently to greater destruction of the environment as demand increased for building materials and other resources from the natural ecosystems. The use of agriculture allowed people the spare time to develop in other ways, and so the great civilizations of the Incas, the Maya, China, Egypt, Greece, Rome, and the Middle East all gradually developed. At the same time many new uses for plants were developed (including new building materials and systems of medicine), and a greater variety of food crops were needed to sustain the growing population. The inborn biophilia in humans also led to the use of plants for ornamentation, to which the legendary hanging gardens of Babylon and the intricate Egyptian gardens attested. People began to use plants to flavor their food with spices and to produce pleasant odors with perfumes, incenses, and embalming. These first two chapters will take us back to the roots of the cultural history of plants and of human relationships to plants.

1

The Hunter-Gatherers

IVAN CROWE

Introduction

Living in a global economy based on agriculture, we tend to forget that wild plant foods previously played a pivotal role in the evolution of the primates, including humans. Wild resources also continue to sustain some of the few surviving hunter-gatherer societies. Over the past millions of years, since the appearance of the first humans, hunter-gatherers have occupied a vast array of different climatic zones and habitats, learning to survive by utilizing a staggering variety of flora and fauna. The means by which they exploited natural resources influenced the forms of agriculture and animal husbandry that have emerged in different locations throughout the world. This chapter surveys the role plant foods have played in human evolution and culture from the appearance of the first primates to the beginnings of cultivation.

Primate Diets

It was the spread of the flowering plants that provided the springboard for primate evolution. By 65 million years ago, toward the end of the Cretaceous period, the Angiosperms (flowering plants) had already become well established, and broad-leafed, fruit-bearing trees began to dominate the vast forests that eventually covered much of the Earth.

Fossil fruits and seeds indicate that the inland forests seem to have been dominated by species related to today's sweet-sop, *Annona squamosa,* and sour-sop, *Annona muricata*, with mangrove and swamp palms in coastal regions. Early forms of pistachio, walnuts, and mango appear to have been present. Trees such as bay, cinnamon, magnolias, and black gum trees grew alongside palms, *Sequoia* conifers and climbing plants such as vines and lianas.

The birds had already adapted to this change by feeding on fruit and nectar from the flowering plants. The new plants meant that a wider range of food became available, and in greater abundance. It was a mutually beneficial relationship, in which the birds ate the fruit and thereby helped to distribute seeds on their bills and feet and by defecation. Insects already played an active part in this relationship by transmitting pollen from plant to plant in their search for nectar.

The primates were able to exploit this ecology to great advantage. Their immediate ancestors were in all probability insectivores and it may well have been the presence of insects that initially led them to adapt to a life in the trees. Birds' eggs too could have provided an additional

Inuit berry pickers between 1900 and ca. 1930. Library of Congress, Prints & Photographs Division.

source of valuable nutrients. The earliest primates, being small, most probably had a predominately insectivorous diet. Small mammals lose body heat more quickly than larger creatures, so they need a mainly carnivorous diet in order to maintain the higher metabolic rate required to compensate for this heat loss. Plant foods generally take longer to digest. Thus a mainly plant-based diet was only possible for primates who evolved to a size that limited their heat loss and thus reduced their metabolic rate.

Initially, while continuing to obtain most of the protein they needed from insects, some primate species increasingly derived many of their energy requirements from plant resources such as nectar, gum, and fruit. Seeds and nuts provided an alternative source of proteins and lipids; eventually insects would play a less important role in the primates' diet as they exploited the plant foods available in the forest.

One peculiar aspect of the primate diet that was most probably acquired during this early period of evolution is the need to regularly include a source of vitamin C in what is eaten. Vitamin C is not a necessary component of the diet for most creatures, including some other mammals. It is probably safe to assume that the primates lost the ability to synthesize vitamin C because their diet was one that always included plenty of plants and fruit, which ensured the inclusion of vitamin C in most of what was being eaten. Color vision, a characteristic shared with the birds, probably also evolved during this period to enable the primates to locate and discriminate between poisonous and edible fruit (Crowe 2000, 18).

We know from today's primates that diet is closely linked with body size (Kay and Covert 1984), as explained previously. Small animals, because of their immediate energy demands, cannot tolerate the delay between eating the leaves and deriving energy from them. Hence smaller primates whose diet does include large quantities of leaves also have to eat fruit to obtain energy, as leaves contain fewer sugars that can be easily assimilated. Some primatologists have concluded that any species that came to rely mainly upon leaves for its survival must at some time have gone through an intermediate frugivore stage during the course of its evolution.

The largest of the living primates, the gorilla, has a largely folivorous diet. But leaves are a low-grade food. Depending on the species of plant, bacterial fermentation has to occur in either the stomach or in the intestines of folivorous primates in order to process the leaves before any nutritive value can be extracted. Therefore, the evolutionary increase in body weight seen in folivores was a necessary adaptation to accommodate modifications in the digestive tract. This adaptation is effectively a cul-de-sac as any radical changes in habitat resources can lead to the extinction of a species. Even when favored fruit is seasonally available the gorilla must continue to consume leaves, simply in order to maintain the gut micro-organisms it normally requires for digesting the plants that form the bulk of its diet during the rest of the year (Tutin 1992). This may well have been a factor that, much later in hominoid evolution, contributed to the eventual demise of the robust australopithecine (the upright ape preceding and probably ancestral to humans) during the early Pleistocene period (Crowe 2000, 18), as they were believed to be dependent upon similar resources (Foley 1989).

In contrast, frugivory (fruit eating) gives primates a flexibility of diet that allows them to avoid the specialization of either eating predominantly meat or predominantly leaves that is normally characteristic of most other mammals. It also enabled different primate species to adapt in varying degrees to their habitat, and to supplement their basic diet of fruit with insects, small mammals, or leaves and other similar vegetable matter.

One notable disadvantage of eating nothing but plant foods is that no single plant can provide all the amino acids required by the body as the building blocks to produce animal protein. A wide variety of plants must always be included in the diet to ensure that all the essential amino acids are present. This is reflected in the behavior of chimpanzees, whose dietary needs often lead them to engage in opportunistic hunting activities (Lawick and Goodall 1971, 182) to supplement their diet by eating meat. This is because the meat of all animals, unlike plant material, contains all the amino acids any other creature needs to synthesis their own body tissue. The diet of chimpanzees as a result is even more diverse than that of humans and this severely limits the habitats in which they can survive. In fact it is one of the main reasons why they are so endangered as a species. Our own ancestors, the slender australopithecines, on the other hand, probably owed their survival, after the demise of their robust cousins, to the fact that they were scavengers and possibly opportunistic hunters of small animals while inhabiting the fringes of the African savanna (Foley 1989).

The underlying factor here is that an exclusively, or predominantly, vegetarian diet can place a huge burden on animals whose habitat and particular digestive system limit the edible plant foods available to them. Climate and seasonality can both compound the problem. Once the early primates migrated to more temperate climes, suitable plants for primates' diets were both in short supply and often widely dispersed—this was especially so at certain times of the year.

The Value of Fire

Plants contain a wide range of structural tissues, such as cellulose, and contain chemical compounds that ward off predators. As a result, many plant tissues are inedible—and sometimes even poisonous—in their raw state. Nevertheless plants are an essential part of our diet and an important source of energy as well as of nutrients, vitamins, and essential trace elements. The effective exploitation of these diverse resources is therefore essential to our survival.

What made human survival possible in many regions was acquiring the use of fire. The control of fire was possibly first achieved by *Homo erectus*, the direct descendants of the australopithecines, maybe as long ago as two million years ago, while they were still confined to the African continent. Fire may not have originally been used to cook food but employed to keep dangerous animals at bay and to keep warm. The effects of fire on animal flesh and plants must have been observed in the aftermath of the fires that often swept across the savanna after lightening strikes during the routine thunderstorms.

When considering the exploitation of food plants alone, the control of fire was absolutely pivotal to our success as a species. Many otherwise inedible plants are made more palatable and more nutritious, and rendered free of toxins, by cooking. This means that, once our ancestors began to employ fire to cook their food, many plants that would previously have been inedible could be included in the diet; this vastly increased the potential resource base (Hillman 1999, personal communication). Cooking also helps to preserve most foods.

There is another important side effect of cooking; the process of cooking roots—and some other parts of plants—has the effect of bursting the cells, thereby releasing the nutrients stored by the plant to aid its growth when spring arrives. Therefore the advent of cooking not only made more plant resources available as food, but the nutritional value of those plants was also increased. Mastication of cooked material was also easier than that of the raw resource and this benefited the youngest and eldest alike and particularly those without a full head of teeth. Well-cooked vegetable matter can also be used as a weaning food. All these things must have aided the survival rate among those hominid populations possessing fire. Wrangham et al. (1999) has also suggested that access to additional nutrients, from root foods especially, could have helped fuel the evolutionary development of a larger hominid brain (Wandsnider 1997; McKie 2000, 110). Archaeological evidence for use of fire by *Homo erectus* is still controversial, but appears well established from at least 700,000 years ago.

Fats and Carbohydrates

Whilst relying heavily on scavenging, *Homo erectus* were also foragers, as were all humans until the advent of farming. Plants are particularly important as a source of carbohydrate and to a lesser extent fats in primate diets. This is especially so when the animals being taken in hunting are suffering from nutritional stress and themselves have depleted fat reserves (Speth 1990). On the African savanna this would have been a familiar scenario nearly every year during the dry season among the herbivores being hunted or scavenged, when arid conditions adversely affected the vegetation.

The importance of fats and carbohydrates in the human diet can best be explained by relating what happens when none are available. Proteins cannot be properly absorbed by the human body without the regular consumption of either fats or carbohydrates, which are needed to aid the metabolization of protein. When both of these nutrients are missing from the diet individuals may begin to show signs of protein toxemia. There have been extreme instances of people who have had nothing but protein-rich foods to eat over a period of several months becoming so disoriented that they seem to be suffering from a form of dementia; and yet once fats or carbohydrates are reintroduced into their diet they make a rapid and complete recovery within a matter of days if not hours. In the longer term death can result (Speth 1990). It is this kind of problem that our hominid ancestors would have encountered first as scavengers and then later as hunters on the African savanna. It is a problem that some hunter-gatherers still experience today.

The range of nutrients that can be obtained from the foods available is obviously crucial to survival. At times people have to go to extraordinary lengths when processing their food to make up for the deficits in particular food groups. Sometimes, though, there seems to be little or no ergonomic advantage, with far more energy being used in the processing procedure than the amount of energy eventually gained from the food. But the importance of these processing procedures can sometimes be better explained by the need to maintain a balanced diet. This is because, as has been illustrated with reference to fats and carbohydrates, if one particular food group is under-represented or, worse, missing entirely, the results can be catastrophic.

One of the main dietary constraints in certain parts of the South American rain forests, as in other similar habitats, is a scarcity of food plants that can provide sufficient carbohydrates for human foragers. Katherine Milton has suggested that a number of indigenous species of roots from families such as Araceae and Marantaceae (see Roots and Tubers) could have been important sources of carbohydrate in the past, before the introduction of cultigens such as manioc, as well as the nuts of the babacu palm, which are available throughout the year, as are the seeds of the banana brava plant, *Phenakospermum guyannense*. Various species of wild figs, which are highly nutritious, might have also made an important contribution to the diet in this respect (Milton 1992). Nuts like cashews and Brazil nuts would also have been available as a valuable addition to the diet in some areas. Depending on the region, fruit trees native to the American tropics, such as *Pourouma cecropiifolia* (uvilla), the pawpaw *Carica papaya*, and the avocado pear *Persea americana,* which has a particularly high content of both protein and fat, would have provided further useful food resources. In addition, some vines, such as the ceriman, *Monstera deliciosa*, also produced fruit.

Many wild fruits in this kind of habitat, on the other hand, are very small and require a great deal of time and effort to gather enough to meet dietary needs. The amount of energy gained from the food collected may be little more than the energy used to acquire the food. If foraging has been unduly prolonged or done too far afield, the problem is compounded. Situations such as this may occur periodically if trees previously relied upon for fruit are found to be barren, when, for example, weather has been unseasonable, as when rain is experienced during the dry season.

Antecedents to Cultivation: Some Case Studies

Tropical Rainforest

In the rain forest the treetops form a dense unbroken canopy, cutting out most of the sunlight at ground level, thus preventing the growth of many plants on the forest floor, but every now and then, one of the great trees crashes down, creating a gap in the canopy that allows the sunlight to flood in. Lying dormant, awaiting just such an event, is a wide variety of plant seeds and seedlings ready to take advantage of the situation. Sunlight causes them to germinate and spring into life to fill the space made briefly available. Such clearings would at times allow some of the food plants sought by humans to multiply as they colonized the open space, providing a natural forest garden for the inhabitants to exploit (Crowe 2000, 135).

Sometimes people may have constructed temporary shelters near these clearings. Wild plants would have also naturally colonized the open areas around dwellings, and this process would have been encouraged by the increased nitrogen content of the soil, enhanced by human waste that would also have contained the seeds from the food plants eaten. This and any other discarded plant material left over after eating, such as fruit stones or cores deposited nearby, would have also provided people with an excellent opportunity to observe just how some plants were propagated. Human waste and discarded plant debris would naturally have often contained viable seeds of the very plants most favored by the people inhabiting the site (Hawkes 1989, 481).

Given sufficient incentive, deliberate cultivation would have logically been the next step (Harris 1989). Initially human intervention may have simply been a matter of creating the right conditions to encourage the growth of favored plants by clearing areas of forest of vegetation that competed with these plants for the light and nutrients. Cutting back the young trees to maintain the open area created by a falling tree may have led to forest people producing temporary clearings themselves to create forest gardens. Later the deliberate propagation of particular species meant humans were taking the first steps toward cultivation and the full domestication of certain plants, which would have occurred as the direct result of human selection. Some of the earliest evidence we have for this kind of intervention is in New Guinea (Groube 1989, 298–301). Climate change may have also had an influence. In the Amazon, for example, a warmer, more seasonal climate following the last Ice Age would have had an effect upon the food resources available and the survival strategies adopted by humans inhabiting the rainforest (Colinvaux 1989). This scenario may well have led to humans exerting a more conscious control over the selection of the flora they relied upon for food, resulting in a more diverse food economy in which cultigens were eventually included.

The early inhabitants of the forest probably learned to use the same slash and burn techniques that present-day Amazonian Indian populations use. This involves the men of the group cutting down the trees and undergrowth in a small section of the forest and then setting fire to the debris. The burning is done near the end of the dry season, when the vegetation has dried out and is easier to ignite. The resulting ash helps to enrich the soil, which in the rain forest is otherwise very poor in nutrients. The forest soil is naturally low in plant nutrients and soon becomes exhausted. After two or three years the soil cannot support any more crops, so the gardens are abandoned and new ones begun. At any one time, a family may have several gardens at various stages under cultivation (Harris 1973).

Deep in the Amazon rain forest particular food resources are often restricted; however, on the flood plains adjacent to the main watercourses there are large areas of rich alluvial soil. Tribes living close to the rivers would have had a much wider variety of wild plant foods, together with fish, mollusks, and aquatic mammals and perhaps freshwater turtles. The food resources offered by a river and adjacent habitat would have therefore made it far easier to support a settled community. Human waste and discarded vegetable matter would have encouraged the germination of commonly used food plants near dwellings, which could have formed the basis of "doorstep gardens."

Food Processing in Australia

Long before cultivation began, grinding stones were being used by some foragers to process seeds and other plant foods. Seed gathering is of primary importance for survival in the desert regions of Australia. For this reason, grinding stones were probably essential for desert people. Seeds were often dehusked by rubbing them between the heel of one hand and the palm of the other. They were then dropped into a wooden dish, allowing the wind to blow away some of the chaff, or the chaff was removed using a technique called yanding. This process involves agitating the contents of a dish to separate the seeds, which were then ground into a flour and often mixed with water to be formed into cakes or dampers that were cooked in the hot ashes of a fire (Cane 1989). A wide range of sedges, edible grasses, and the seeds from several kinds of shrub were gathered as well as the seeds of several nontoxic varieties of acacia. Apart from making certain foods more palatable and aiding mastication, grinding had the added advantage of making foods easier to digest, thereby allowing the release of more nutrients. The processing of potentially edible plants is not only a means of extending the range of resources that can be exploited for food in a given area; it also has the advantage of ensuring that the nutritional value of the resources available is being maximized.

When the nomadic tribes would move to another area, they would leave behind their grinding stones. The stones were not abandoned, however. Similar to other nomadic people, the aborigines of Australia periodically return to the same locations on a regular basis. They may cover a vast area during their wanderings and sometimes, in more arid regions, it may be decades before they revisit an area. This constant movement is necessary to allow depleted resources to recover. Any given area can only support a finite number of people and in these arid environments nomad groups are normally quite small, between thirty and fifty individuals, according to Yellen (Renfrew and Bahn 1991, 173).

Exact numbers are partly dependent upon the resources available and how far it is necessary to travel when hunting and gathering plant foods.

The aborigines generally preferred roots and fruits, when they could be found, as these usually involved little or no preparation, unlike nuts and seeds. In addition, aboriginal people living in desert regions also ate succulents. Some species of bushes in Australia retain their berries even after they have matured and dried out; these represented another important resource particularly during the most arid periods when little else was available.

In common with many hunter-gatherer people, Aborigine women played a key role in acquiring food for their group and were the main collectors of plant foods. As today, yams were found by identifying the leaves and then tracing the tendrils of each plant, entwined among the branches of nearby bushes, back to their source. A digging stick was then used to excavate the yams by following the tendrils underground until the main body of the plant was located. Great care was taken not to remove the whole plant when foraging, so that some was left behind to ensure vegetative regeneration. The bitter tasting *Dioscorea bulbifera* (bitter yam, air potato) was cooked in an earthen oven (Jones and Meeham 1989, 124). Snail shells with holes cut in them were then used to grate the tubers. The prepared material was afterwards left to soak overnight to detoxify it. Other yams, such as the long yam *D. transversa,* required less stringent preparation.

Although systematic plant cultivation was never adopted on the Australian continent as a means of ensuring a sustainable food resource, according to one early explorer, Sir George Grey, there were some areas where tribes extensively harvested and deliberately propagated the yam *D. hastifolia* (Hallam 1989). For many aboriginal people such plant foods traditionally provided the staple diet.

Wetland resources were equally important in some areas. During the rainy season the Gidjingali Aborigine women and girls collected water lilies (*Nymphaea* spp.), an important source of carbohydrate. The stalks were eaten raw, and the small black seeds were ground into flour to make unleavened cakes. In the dry season, as swamps began to shrink, women dug out water chestnuts, the corms of the spike rush *Eleocharis dulcis*. At other times of the year cycad nuts (*Cycas* and *Macrozamia* spp.) were exploited as the staple food resource (Jones and Meeham 1989).

In several places in Australia, the Philippines, and throughout Indonesia, the highly toxic nut of the cycad palm is used for food. The nuts contain a dangerous neurotoxin, so great care has to be taken at all stages of preparation, including the avoiding even touching the mouth while handling the nuts. The nuts are first dehusked before being allowed to dry in the sun for a few days. A stone pestle and mortar are then used to crush the nuts into a pulp that is then put into woven bags (Jones and Meeham 1989). These are put into pits filled with water and left immersed for a further few days. Fermentation takes place, and a foul-smelling froth forms on the water surface as most of the toxins gradually leech out. After further grinding, the resulting paste was formed into loaves, wrapped in pandanus leaves, and baked in an earth oven. With some species of cycad in the Philippines, it has been found that despite all precautions, the toxins can still cause paralysis and severe, irreversible mental degeneration in later life, as the effects of even small amounts of the toxin are accumulative. One great advantage of this food however is that the loaves can be kept for several months.

Food Preservation in the Artic

The use of preserved foods may not be as prevalent among nomadic hunter-gatherers as sedentary people simply due to the transport problems imposed by a nomadic lifestyle. Nevertheless such foods often play an important part in their long-term survival strategy.

In colder latitudes, there is a general tendency for people to include a smaller proportion of plant foods in what they eat (Lee 1968). The food of people living near the Arctic Circle used to consist mainly of flesh foods but plant foods were still an essential element in their diet. Plant material from the stomachs of both terrestrial and marine mammals may have once played some part in people's diet and seaweed was also consumed. In such regions the summers are very brief and there is therefore only a short period when the gathering of most other plant foods is possible. In high latitudes today bushes belonging to the heather family such as crowberries, bilberries, and cowberries provide edible fruits that represent an important source of vitamin C, as does the creeping willow, *Salix arctica*. The leaves of this plant, which contain ascorbic acid, were traditionally plucked by some Arctic people and dropped into boiling water to extract the vitamin. This was then poured into a hole that had been excavated in the ice, where the mixture very quickly froze solid, preserving much of the vitamin C present that might otherwise have been destroyed by the boiling. In its frozen state, this concoction provided a source of vitamin C throughout the winter.

Root vegetables, where they were available, were excavated from the ground, usually with a digging stick. They were also obtained, for example among the Nabesna people of Alaska, by taking them from the caches of muskrats where the animals stored food ready for the winter. Berries were sometimes preserved in oil but usually were dried simply by laying them out on racks in the sun. Reducing the water content prevents or delays bacterial growth and the action of the enzymes naturally present in the tissue. During the winter, or late autumn when it often rained, food was preserved by being smoked, usually within the family dwelling. Smoking dries out the food and coats it with chemicals that inhibit the invasion of microorganisms that could cause it to go bad. Sometimes meat was also ground up and mixed with grease and berries to help preserve it to produce pemmican, which the buffalo hunters of the plains also often depended on during the winter months. Drying or smoking can preserve many kinds of plant foods, fruits, and vegetables. Many aspects of the kind of food economy seen in northern latitudes in recent times might well be equally applicable to the cultures that existed during the last Ice Age in both Europe and North America as well as in parts of Asia.

Seasonal Markers

In prehistory, before the advent of formal calendars, the natural world provided information by which people could assess the passage of time throughout the year. The migration of certain animals and the appearance and particularly the growth stages reached by various plants ("calendar plants"), allowed hunter-gatherers to judge the passing of the seasons. They would be able to tell from their observations which food resources were likely to be available at that time or in the immediate future. For example, it has been suggested that some of the artifacts from the Upper Paleolithic in Europe depict such seasonal markers. This sort of information was especially important when trying to determine when certain resources would become available in remote locations if fruitless expeditions, which would waste precious time and energy, were to be avoided.

Incentives to Settle

Not all hunter-gatherers were nomadic. There is abundant archaeological and ethnographic evidence to show that hunter-gatherers have had permanent settlements in areas rich in plant or animal resources. Today's sample of hunter-gatherer cultures is atypical because foragers have been displaced from richer environments by farmers over most of the world, and only survive in areas unsuitable for farming.

Where there are aquatic resources, settlement is far more likely. Rivers, lakes, and seas often provided most of the animal food resources required, and usually in freshwater locations a range of seasonal plant foods too. Tribes such as the Menominee, who lived alongside the Great Lakes in regions of what is now Wisconsin and Michigan, relied heavily upon wild rice (*Zizania palustris*) as a major constituent of their diet (Taylor 1991, 236). Not a true rice, this plant is a long-stemmed wild grass that grew on lake margins and especially in the marshes around the Great Lakes. It was usually collected by the women of the tribe in their canoes; they bent the tall grasses over the side of the canoe and struck the seed heads with their paddle so that the grains fell inside. In these circumstances there is a strong incentive to remain in one place and find ways of utilizing the local flora in a sustainable manner. This must have been one of the factors that led to the next stage in plant exploitation—cultivation.

Few if any of the techniques referred to in this chapter were confined to the cultures described: they appear and reappear in different guises throughout the world and were probably utilized in various forms throughout human history. Nor has this short appraisal done justice to what is potentially an exhaustive study. Plant resources were used by hunter-gatherers for weapons, clothing, building shelters, and cordage and to fashion many artifacts, as well as for food. Plants were also a source of medicine, dyes, poisons, and hallucinogenics. The people of prehistory had to rely upon an intimate knowledge of the plants in their habitat to effectively exploit the flora available. One thing of which we can be sure is that more of this knowledge is lost than is now known.

References and Further Reading

Cane, S. 1989. Australian aboriginal seed grinding and its archaeological record. In *Foraging and Farming: The Evolution of Plant Exploitation,* edited by D.R. Harris and G.C. Hillman. London: Unwin Hyman, 99–129.

Chivers, D., Wood, B.A., and Bilsborough, A. 1984. *Food Acquisition and Processing in Primates.* New York: Plenum Press.

Colinvaux, P.A. 1989. Past and Future Amazon. *Scientific American,* 260(5): 102–8.

Crowe, I. 2000. *The Quest for Food.* Stroud: Tempus Publishing.

Foley, R. 1989. *Another Unique Species: Patterns in Human Evolutionary Ecology.* Harlow: Longman and New York: Wiley.

Groube, L. 1989. The taming of the rainforest: A model for Late Pleistocene forest exploitation in New Guinea. In *Foraging and Farming: The Evolution of Plant Exploitation,* edited by D.R. Harris and G.C. Hillman. London: Unwin Hyman, 292–304.

Hallam, S. 1989. Plant usage and management in SW Australian Aboriginal societies. In *Foraging and Farming: The Evolution of Plant Exploitation,* edited by D.R. Harris and G.C. Hillman. London: Unwin Hyman, 136–155.

Harris, D.G. 1973. The prehistory of tropical agriculture. In *The Explanation of Culture Change: Models in Prehistory,* edited by C. Renfrew. London: Duckworth, 391–417.

Harris, D.G. 1989. An evolutionary continuum of people-plant interaction. In *Foraging and Farming: The Evolution of Plant Exploitation,* edited by D.R. Harris and G.C. Hillman. London: Unwin Hyman, 11–26.

Harris, D.G., and Hillman, G.C. (editors) 1989. *Foraging and Farming: The Evolution of Plant Exploitation.* London: Unwin Hyman.

Hawkes, J.G. 1989. The domestication of roots and tubers in the American tropics. In *Foraging and Farming: The Evolution of Plant Exploitation,* edited by D.R. Harris and G.C. Hillman. London: Unwin Hyman, 481–503.

Jones, R., and Meehan, B. 1989. Food plants of the Gidjingali: ethnographic and archaeological perspectives. In *Foraging and Farming: The Evolution of Plant Exploitation,* edited by D.R. Harris and G.C. Hillman. London: Unwin Hyman, 120–135.

Kay, R.F., and Covert, H.H. 1984. Anatomy and behaviour of extinct primates. In *Food Acquisition and Processing in Primates,* edited by D. Chivers, B.A. Wood, and A. Bilsborough. New York: Plenum Press, 467–508.

Lawick, H., and Goodall, J. 1971. *In the Shadow of Man.* Boston: Houghton Mifflin and London: Collins.

Lee, R.B. 1968. What hunters do for a living, or, how to make out on scarce resources. In *Man the Hunter,* edited by R.B. Lee and I. DeVore. Chicago: Aldine, 30–48.

McKie, R. 2000. *Ape Man: The Story of Human Evolution.* London: BBC publications.

Milton, K. 1992. Comparative aspects of diet in Amazonian forest-dwellers. In *Foraging Strategies and Natural Diet of Monkeys, Apes and Humans,* edited by E.M. Widdowson and A. Whiten. Oxford: Clarendon Press, 253–63.

Ostwalt, W. 1973. *Habitat and Technology: The Evolution of Hunting.* New York: Holt, Rinehart and Winston.

Ostwalt, W. 1976. *An Anthropological Analysis of Food Getting Technology.* New York: Wiley.

Renfrew, C., and Bahn, P. 1991. *Archaeology Theories, Methods and Practice.* London: Thames and Hudson.

Speth, J. 1990. Seasonality, resource stress and food sharing, in so-called `egalitarian' foraging societies. *Journal of Anthropological Archaeology, 9,* 148–188.

Taylor, C.F. 1991. *The Native Americans.* London: Salamander.

Tutin, C. 1992. Foraging profiles of sympatric lowland gorillas and chimpanzees in the Lopé game reserve, Gabon. In *Foraging Strategies and Natural Diet of Monkeys, Apes and Humans,* edited by E.M. Widdowson and A. Whiten. Oxford: Clarendon Press, 19–26.

Ucko, P.J., and Dimbleby, G.W. (editors) 1969. *The Domestication and Exploitation of Plants and Animals.* London: Duckworth.

Wandsnider, L. 1997. The roasted and the boiled: food composition and heat treatment with special emphasis on pit-hearth cooking. Journal of Anthropological Archaeology 16(1):1–48.

Widdowson, E.M., and Whiten, A. 1992. *Foraging Strategies and Natural Diet of Monkeys, Apes and Humans.* Oxford: Clarendon Press.

Wrangham, R.W., Jones, J.H., Laden, G., Pilbeam, D. and Conklin-Brittain, N.L. 1999. The raw and the stolen: cooking and the ecology of human origins. Current Anthropology 40: 567–594.

2
Origins and Spread of Agriculture

DAVID R. HARRIS

In today's world most people depend on the products of agriculture for their daily sustenance, yet this is a recent development in the evolution of humanity. Modern humans (*Homo sapiens*) had emerged in Africa by 100,000 years ago and during the following 50,000 years they spread, as foraging hunter-gatherers, through most of Eurasia. But it was not until about 10,000 radiocarbon (^{14}C) years ago that some groups in Southwest Asia began to cultivate cereals and herbaceous legumes and thus became the world's first farmers. The transition from foraging to farming radically changed the relationship of humans to their environment, and because it allowed more people to be supported per unit area of cultivable land, it paved the way for settled village life, and ultimately for urban civilization.

By 1500 AD, when Europeans were beginning to colonize other continents, most of the world's population (estimated at 350 million) depended for their staple food on crops raised in a variety of agricultural systems in all the habitable continents except Australia. Archaeological and biological evidence suggests that this worldwide distribution of agriculture was mainly the result of expansion from a few core regions where independent transitions from foraging to farming took place, at different times, between about 10,000 and 3500 ^{14}C years ago. Why these transitions occurred, and where and when they did, remains a puzzling and controversial question. Many factors, singly or in combination, have been suggested to explain the process. These include climatic and other environmental changes; differences in the availability of wild plants and animals; population growth; technological innovation; and competition between, and wealth accumulation by, hunter-gatherer groups. The role of such factors in transitions to agriculture may have varied from region to region, and at present there is insufficient evidence to test alternative explanatory hypotheses. However, new data on climatic change and associated changes in vegetation at the end of the Pleistocene period about 11,000 ^{14}C years ago now point to environmental change as a major factor in some at least of the initial transitions from foraging to farming.

In recent years, great advances have been made in unraveling the origins of agriculture by applying new analytical techniques and integrating archaeological and biological data. For example, the development of radiocarbon dating by accelerator mass spectrometry (^{14}C AMS), which allows samples as small as individual cereal grains to be dated directly, has had a major impact, particularly on investigation of the beginnings of agriculture in the Americas (Kaplan and Lynch 1999;

Long et al. 1989; Piperno and Flannery 2001; B.D. Smith 1997; 2001), and new data from molecular genetics, especially from modern and ancient DNA, is beginning to revolutionize our understanding of plant and animal domestications (Jones and Brown 2000).

The earliest evidence for transitions from foraging to farming comes from two regions of the world—Southwest Asia and China—when particular juxtapositions of environmental and cultural conditions caused some groups of foragers to start cultivating and domesticating a limited range of plants. At present there is only sufficient evidence from one of these regions, southwest Asia, to draw fairly firm conclusions about how and why the transition to agriculture occurred, but recent archaeological investigations in East Asia, particularly in China, are now beginning to clarify the process there. Other regions where there is evidence that foragers independently developed forms of agriculture include northern tropical Africa south of the Sahara, peninsular India, possibly New Guinea, and, in the Americas, Mexico and adjacent areas (Mesoamerica), eastern North America, the central Andean highlands, and Amazonia; but all these transitions appear to have occurred later than those in Southwest Asia and China, probably between about 7000 and 3500 years ago.

In each of these regions shifts took place from the harvesting of wild plants, particularly the seeds of grasses and legumes, to their cultivation and domestication. The human populations gradually became more dependent for their food supply on a small selection of grain crops, and in some regions root and tuber crops. In most regions one or more domesticated cereal became a major staple: barley and wheats in Southwest Asia, rice in China, maize in North America, and sorghum in sub-Saharan Africa. Herbaceous legumes too were domesticated in most of the regions (New Guinea, Amazonia, and eastern North America excepted). These crops, known as pulses, complemented the cereals nutritionally by providing oils and essential amino acids such as lysine that the cereals lacked, as well as adding to the dietary supply of carbohydrate and protein. Thus the cereals were complemented in Southwest Asia by lentil, pea, chickpea, and other pulses; in China by soybean; in Mesoamerica by common bean; and in west Africa south of the Sahara by cowpea and groundnuts (Harris 1981).

Although this book is not concerned with domestic animals, their role in the origins and early spread of agriculture deserves brief mention here because their presence or absence strongly influenced how agriculture developed in each of the core regions. Thus in Southwest Asia several herd animals were domesticated and incorporated into a system of agro-pastoral production that gradually, between about 10,500 and 7500 ^{14}C years ago, integrated cereal and pulse cultivation with the raising of goats, sheep, pigs, and cattle. This system of mixed grain-livestock farming spread in later prehistoric times west into Europe and North Africa and east into central and south Asia, but it was not paralleled elsewhere. In China and parts of south and Southeast Asia domestic pigs, chickens, and water buffaloes became associated early on with rice cultivation; no indigenous herd animals were domesticated in sub-Saharan Africa but domestic sheep, goats, and cattle were introduced from Southwest Asia; and no domestic animals were integrated into systems of crop cultivation in the Americas before the arrival of Europeans in the 16th century (although turkeys, Muscovy ducks, llamas, alpacas, and guinea pigs were domesticated in parts of North and South America in pre-European times).

Before reviewing what is now known about the origins and early development of agriculture in the regions where independent transitions from foraging to farming appear to have taken place, it is first necessary to clarify the meaning of the terms *cultivation, domestication,* and *agriculture* because they are often used loosely in the voluminous literature on "agricultural origins." In this chapter they are defined as follows: *cultivation* refers to the sowing and planting, tending, and harvesting of useful wild or domestic plants, with or without tillage of the soil; *domestication* means that plants have been changed genetically and/or morphologically as a result of human (inadvertent or deliberate) selection and have become dependent on people for their long-term survival; *agriculture* is defined as the growing of crops (i.e., domesticated plants) in systems of cultivation that normally involve systematic tillage of the soil (Harris 1989, 17–22; 1996a, 444–56).

The distinction between cultivation and agriculture is particularly important because it focuses attention on the regions where there is evidence of the indigenous development of agriculture (as defined above), as opposed to the many parts of the world where foragers practiced various techniques of cultivation but did not domesticate any crops (for some historical examples of the cultivation of wild plants by Australian and North American foragers see Hallam 1989; Harris 1984). It also allows use of the term *pre-domestication cultivation*, which helps us to understand how agriculture originated.

Although it is helpful to make a clear distinction between cultivation and agriculture it is often difficult to do so on the basis of archaeological evidence. Nevertheless, many crops can be distinguished from their wild progenitors by identifying the morphological changes that occurred as a result of their domestication. For example, domesticated barley, wheats, and rice can be identified—provided that sufficiently well preserved (usually charred) archaeological samples are available—by the presence of rough scars on the spikelet forks, which are evidence of the replacement, under domestication, of the brittle rachis (the "spine" of the ear) of the wild grasses by the tough rachis of the cereals. Domesticated maize is much easier to identify because the morphology of the seed head is conspicuously different, even in small primitive varieties, from that of its wild progenitor. Distinguishing archaeologically between most pulses and their wild progenitors or other close relatives is however very difficult (Butler 1992), particularly because the seeds show few changes under domestication other than an increase in average size. Likewise, the remains of root and tuber crops are very difficult to identify in archaeological deposits because their soft tissues tend not to be well preserved and therefore morphological differences between domesticated and wild forms, such as increase in tuber size and reduction of roughness and spininess of the tubers, can seldom be recognized. Hather (1991, 1994) describes a new approach to this problem.

Bearing in mind the above definitions of cultivation, domestication, and agriculture, and with an awareness of how difficult it often is to distinguish between the remains of domestic and wild plants—and hence to establish when, in any given region, agriculture can be said to have begun—we can now briefly examine current evidence for the origins of agriculture in Asia, Africa, and the Americas. It is appropriate to consider Southwest Asia first because it is for that region that we have the most comprehensive, and earliest, archaeological evidence of a transition from foraging to farming.

Southwest Asia

The sites that have yielded the earliest archaeobotanical evidence of agriculture are concentrated in an arc (often referred to as the Fertile Crescent) around the Mesopotamian lowland from the southern Levant to the southern foothills of the Zagros Mountains (Figure 1) (Harris 1998b; B.D. Smith 1998, 48–89) One site on the middle Euphrates River in Syria (Figure 1)—Tell Abu Hureyra—has provided very early evidence for the beginnings of cereal cultivation. There, some 12,000 ^{14}C years ago, at the end of the Paleolithic period (locally referred to as the Natufian), the inhabitants of the site, whose food supply included a wide range of wild plants, evidently began to cultivate some of the native cereal grasses and herbaceous legumes that they were already accustomed to harvesting as staple foods. The cultivation of cereals and legumes probably began in response to these foods becoming less abundant in the wild as a result of a sudden change to colder and drier conditions between about 11,000 and about 10,000 ^{14}C years ago—a climatic phase known as the Younger Dryas stadial (Harris 2003; Hillman et al. 2001; Moore and Hillman 1992).

This interpretation is based on Hillman's work over many years on the plant remains from Abu Hureyra. He has found evidence of a decline in abundance from the least to the most drought-tolerant species, as, under the impact of the Younger Dryas, the climate became more arid and the vegetation more desert-like. In the archaeobotanical record from Abu Hureyra the seeds of open-woodland species decline first, followed in sequence by those of wild lentils (*Lens* spp.) and other large-seeded legumes, of wild wheats and ryes (*Triticum* and *Secale* spp.), of feather grasses and club rushes (*Stipa*, *Stipagrostis*, and *Scirpus* spp.), and finally of shrubs belonging to the family Chenopodiaceae. Also, towards the end of the Younger Dryas domesticated cereals and pulses increase, as do weeds typical of dryland cultivation (Hillman 2000, 376–93). The evidence thus suggests that it was a progressive decline in the availability of wild plant foods that were already dietary staples that prompted the people living at Abu Hureyra to try to sustain their food supply by cultivating some of the more productive grasses and legumes. They probably did so by sowing grain retained from the previous year's harvest on patches of relatively moist alluvial soil in drainage channels and small depressions—a cumulative process of seed selection and annual re-seeding that led to the emergence of the domesticated cereals and pulses.

It is likely, but cannot be verified for lack of adequate archaeobotanical evidence, that the inhabitants of other large Natufian sites, such as Mureybet, Ain Mallaha, Hayonim, El-Wad, and Wadi Hammeh in the southern Levant (Figure 1), responded in a similar way to reductions in the availability of wild plant foods induced by the Younger Dryas, a supposition that gains some support from the presence there of large quantities of stone sickle blades as well as pestles and mortars used for grinding grains.

During the subsequent earliest Neolithic period, the Pre-Pottery Neolithic A (PPNA), which lasted in the Levant from about 10,300 to about 9500 ^{14}C years ago, the climate became warmer and wetter, facilitating the expansion of grain cultivation on alluvial soils. There is very little conclusive evidence for crops in the PPNA, although domesticated barley (*Hordeum vulgare*) and (probably emmer) wheat (*Triticum turgidum* ssp. *dicoccum*) have been found at Iraq ed-Dubb and Tell Aswad (Figure 1) (Colledge 2001, 143; van Zeist and Bakker-Heeres 1982, 185–90), and at these and other Levantine sites such as Netiv Hagdud, Mureybet, and Jerf el-Ahmar (Figure 1) remains of wild (and indeterminate wild/domestic) cereals and legumes have been recovered. The evidence as a whole suggests that during the PPNA cultivation expanded only gradually, with the proportion of domesticated grains only slowly increasing in harvests of mainly wild cereals and legumes, and that people also continued to depend for much of their food on such wild resources as nuts, fruits, gazelle, fish, small mammals, and birds.

There is no conclusive evidence of domestic animals (other than the dog) during the PPNA, and it is only in the following PPNB period, which lasted from about 9500 to about 7500 ^{14}C years ago, that there is widespread evidence both for grain cultivation and for the herding of domestic goats

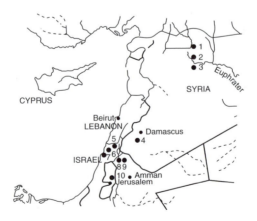

Figure 1. The southern Levant and middle Euphrates valley, showing the location of archaeological sites mentioned in the text. (1) Jerf el-Ahmar; (2) Mureybet; (3) Abu Hureyra; (4) Aswad; (5) Ain Mallaha; (6) Hayonim; (7) El-Wad; (8) Wadi Hammeh; (9) Iraq ed-Dubb; (10) Netiv Hagdud.

and sheep. The archaeobotanical record indicates a gradual increase through the PPNB in the geographical distribution of domesticated cereals and pulses (Garrard 1999). By the Middle PPNB there is evidence for all the "founder crops" (Zohary 1996): two-row and six-row barley, einkorn (*Triticum monococcum*), emmer and free-threshing bread wheat (*Triticum aestivum*), lentil (*Lens culinaris*), pea (*Pisum sativum*), chickpea (*Cicer arietinum*), bitter vetch (*Vicia ervilia*), and flax (*Linum usitatissimum*), and by the end of the PPNB agro-pastoralism combining grain cultivation with herding was being practiced widely throughout western Southwest Asia, where it had come to support most of the human population. By that time too, the new agro-pastoral way of life had begun to spread west into Cyprus and across Anatolia towards Europe, southwest into Egypt, and east towards central and southern Asia (Bar-Yosef and Meadow 1995, 73–93; Harris 1996b, 554–64; Harris 1998c; Meadow 1998; Wetterstrom 1993, 199–202).

East Asia

Whereas archaeological research on the origins of agriculture in Southwest Asia has been actively pursued since the pioneer excavations in the 1950s at the early Neolithic sites of Jarmo and Jericho (Figure 1), equivalent research in East Asia is much more recent. Since the 1980s, however, great advances have been made in the archaeological investigation of early agriculture in China, most of which has focused on the beginnings of rice cultivation (see, for example, references in Cohen 1998 and the special section on rice domestication published in *Antiquity*, 1998, 72, 855–907).

From Neolithic sites in east-central China, mainly in the middle and lower Yangtze valley, there is now strong evidence that rice (*Oryza sativa*) became a staple food crop between 8000 and 6000 ^{14}C years ago. The earliest grains of rice found in archaeological deposits are preserved in pottery and come from the site of Pengtoushan in the middle Yangtze valley (Figure 2) (Chen and Jiang 1997; Crawford and Shen 1998, 861). They have been dated (by the AMS radiocarbon method) to approximately 7800 ^{14}C years ago (equivalent to the early Pottery Neolithic period in the Levant), and by 6000 ^{14}C years ago there is plentiful evidence of domesticated rice, associated with the remains of pile dwellings, spade-like implements made of bone and wood, and abundant pottery (Glover and Higham 1996, 426–9). The oldest Neolithic sites in the Yangtze valley have also yielded the remains of domesticated dog, pig, chicken, and water buffalo (Yan 1993). Farther north, in the middle Huanghe (Yellow River) valley and on the associated loess plateau, there is evidence by 7000 ^{14}C years ago for villages supported by a mixed economy of hunting, fishing, cultivation of domesticated foxtail, proso, and Japanese millet (*Setaria italica, Panicum miliaceum* and *Echinochloa crusgalli*), and the raising of domestic dogs, pigs, and probably chickens (Chang 1986, 87–95; Crawford 1992, 13–14; Lu 1999; Underhill 1997, 117–25).

As a whole, this evidence indicates that by 7000 ^{14}C years ago, in the Chinese Early Neolithic period, substantial settlements were well established in east-central China at such sites as Hemudu and Luojiajiao southeast of the lower Yangtze, Jiahu north of the Yangtze, and Cishan and Peiligang in the region of the middle Huanghe valley (Figure 2). These settlements were supported by grain agriculture based on rice and millets, associated with the raising of domesticated pigs, chickens, and water buffaloes. A variety of indigenous vegetables, fruits, and possibly a pulse (soybean, *Glycine max*) may also have been cultivated in the Neolithic period, as they certainly were by early historical times, but this has not been confirmed by archaeobotanical data.

Very few Late Paleolithic occupation sites are known and none has yielded a sequence of well identified and dated plant remains that throw light on the beginnings of cultivation and domestication, as Abu Hureyra does in Southwest Asia. There is however a cave site, Diaotonghuan, in the Dayuan basin south of the middle Yangtze (Figure 2), that has been extensively excavated and has yielded possible evidence of very early rice cultivation in the form of rice phytoliths (silicified particles of plant epidermal tissues). There is uncertainty about the accuracy of the radiocarbon dating

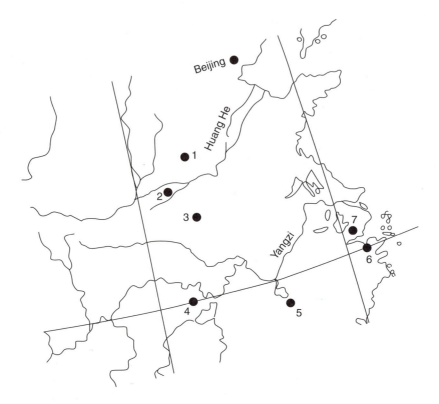

Figure 2. East-central China, showing the location of archaeological sites mentioned in the text. (1) Cishan; (2) Peligang; (3) Jiahu; (4) Pengtoushan; (5) Diaotonghuan; (6) Hemudu; (7) Luojiajiao.

of the sequence of occupation of the cave, but Zhao (1998), who recovered the rice phytoliths from the cave deposits, argues that the lowest level in which they occur (Zone G) probably dates to the Late Paleolithic, between 12,000 and 11,000 ^{14}C years ago. He suggests that this indicates the beginning of wild rice exploitation at the site, and that a marked reduction (in Zone F) in the abundance of rice phytoliths and a subsequent return to high counts (in Zones E and D) reflects the impact of, and the subsequent recovery after, the cold, dry climate of the Younger Dryas stadial. He further argues that the phytoliths in Zones E and D, which he equates with the beginning of the Neolithic, derive from domesticated rice, and that by Zone B, later in the Early Neolithic, between 7500 and 7000 ^{14}C years ago, the transition to rice agriculture had occurred.

Zhao's interpretation remains speculative, particularly because the dating of the site is problematic, but it corresponds remarkably closely to the model (presented in the previous section) for the origins of cereal cultivation in Southwest Asia derived from analysis of the plant remains recovered at the site of Abu Hureyra. However, the occurrence and nature of a Younger Dryas effect in East Asia is not well established and until more conclusive paleoenvironmental and archaeological evidence can be obtained the hypothesis that the Younger Dryas initiated the transition in China to (rice and perhaps millet) agriculture must be regarded as tentative.

There is much firmer evidence for the spread of rice north and south from the Yangtze valley during and after the Neolithic. The spread of rice agriculture depended on the selection of varieties adapted to different climatic and day-length regimes and was evidently a very slow process. The earliest rice recovered archaeologically in Korea dates to about 3200 ^{14}C years ago (Choe 1982, 520) and it did not become a staple crop in Japan until the 4th century BC (Imamura 1996, 453–7). Its southerly spread is poorly documented, but there is evidence of it in the Ganga (Ganges) valley in north India by about 4500 years ago, where, however, it may have been independently domesticated (Fuller 2002, 299–300; Glover and Higham 1996, 416–9). From there it spread west to the Indus valley and south into peninsular India where it was a late addition to an assemblage of indigenous millets and pulses—principally browntop millet (*Brachiaria ramosa*), bristly foxtail millet (*Setaria verticillata*), mung bean (*Vigna radiata*), and horsegram (*Macrotyloma uniflorum*)—that were staple crops of the Neolithic period in southern India between about 4800 and 3200 ^{14}C years ago (Fuller, Korisettar, and Venkatasubbiah 2001).

The introduction of rice into mainland Southeast Asia seems to have been delayed until after 5000 ^{14}C years ago, and its further spread in island Southeast Asia was eventually checked in the equatorial zone of eastern Indonesia (Glover and Higham 1996, 419–26). From there agriculture—still with pigs and chickens but with such root crops as taro (*Colocasia esculenta*) and greater and lesser yams (*Dioscorea alata* and *D. esculenta*) replacing rice as staple crops—spread across the Pacific by a rapid process of maritime colonization beginnng about 3500 years ago (Bellwood 1989; Spriggs 1996).

Whether this process of expansion was also responsible for the introduction of agriculture into New Guinea, or whether agriculture developed independently there, has long been debated. Until recently it was generally believed that the staple root crops cultivated in the highlands today had been domesticated outside New Guinea—taro and greater and lesser yams in mainland Southeast Asia and sweet potato (*Ipomoea batatas*) in tropical America—and that agriculture only became established in the highlands after these crops had been introduced, yams in prehistoric and sweet potato in early historic times. But, as a result of archaeological and paleoenvironmental research in the highlands since the 1960s, notably at Kuk swamp and other sites in the Waghi valley (Hope and Golson 1995; Golson 1989; Golson 1977) that view must now be revised (Denham et al. 2003; Yen 1991; Yen 1995). Also, recent biomolecular studies indicate that taro and greater yam were independently domesticated in New Guinea (as well as in mainland Southeast Asia), in addition to bananas (*Musa* spp.), sugarcane (*Saccharum officinarum*), and breadfruit (*Artocarpus altilis*) (Lebot 1999). The archaeological evidence at the Kuk site suggests that small-scale cultivation began there as early as 9000 ^{14}C years ago; that by about 4000 ^{14}C years ago taro, which was probably domesticated at lower elevations in New Guinea, was being extensively cultivated by a system of ditching and mounding; and that the sweet potato, of South American origin, was introduced as recently as 300 ^{14}C years ago, where it has since become the dominant crop in the highland valleys (Bayliss-Smith 1996; Denham et al. 2003).

It is now clear that cultivation began very early in New Guinea, that it focused almost exclusively on root and tree crops, and that the island was not one of the core regions from which agriculture spread—an observation that correlates with the total lack of agriculture in Australia prior to the beginnings of European settlement in the 18th century (Harris 1995).

Northern tropical Africa south of the Sahara

For many years archaeologists disregarded the possibility that agriculture might have originated independently in tropical Africa and assumed that both cereal cultivation and livestock raising were introduced from Southwest Asia via the Nile valley and the Sahara. But as early as 1950 the French botanist Portères suggested that there had been four "berceaux agricoles primaires" (cradles of primary agriculture) south of the Sahara, and in 1959 the American anthropologist Murdock proposed, on botanical and linguistic grounds, that agriculture had developed independently in West Africa possibly as early as 7000 years ago, and spread eastward through the Sudanic zone south of the Sahara (Harris 1976; Harris 1998a; Murdock 1959, 64–70; Portères 1950).

The crops indigenous to the Sudanic zone include cereals, of which sorghum (*Sorghum bicolor*) and pearl millet (*Pennisetum glaucum*) are most widely cultivated as staples; pulses, particularly cowpea (*Vigna unguiculata*) and, in West Africa, two types of groundnut (*Macrotyloma geocarpus* and *Vigna subterranea*); and the oil-yielding shea butter tree (*Vitellaria paradoxa*). To the south, in the forest zone of West Africa, yams (*Dioscorea cayenensis* and *D. rotundata*) have long been staple crops, and, west of the Bandama River, also African rice (*Oryza glaberrima*); and throughout the forest zone the oil palm (*Elaeis guineensis*) has made an important contribution to the diet.

Until very recently there has been almost no direct archaeobotanical evidence of early agriculture in northern tropical Africa, and such data are still extremely sparse. There is some evidence from pollen and seed remains for the exploitation, by about 5000 years ago, of two oil-yielding tree crops (oil palm and *Canarium schweinfurthii*) and also *Celtis* sp. fruits at sites in Ghana close to the present forest-savanna boundary, as well as at later sites to the west and east in Liberia, Cameroon, and Zaïre (Stahl 1993, 263). This suggests a long history of tree-crop management, very probably associated with yam cultivation (as in recent times) along the forest/humid savanna margins of western and central Africa. Farther north, in the drier savannas of the Sudanic zone, cultivation of the indigenous cereals and pulses evidently led to the establishment of grain agriculture by about 3500 years ago. This is attested by evidence of domesticated pearl millet at Tichitt in southern Mauritania (Amblard 1996, 425; Holl 1985, 159; Munson, 1976), at Birimi in northern Ghana (D'Andrea, Klee, and Casey 2001) and at Ti-n-Akof in northern Burkina Faso and Kursakata in northeastern Nigeria (Neumann 1999, 75–7; Neumann, Ballouche, and Klee 1996, 443) (Figure 3).

Thus, despite the sparsity of the archaeobotanical record from northern tropical Africa, such data as are available suggest that the zonal contrast between the southern emphasis on root crops and oil palm along the forest/savanna boundary and the northern focus on grains (millets and pulses) in the drier Sudanic zone has a time depth of at least 3500 years. The grain-crop system was capable of supporting large populations and it subsequently spread into eastern and southern Africa (together with pastoral systems based on domestic sheep, goats, and cattle introduced from Southwest Asia).

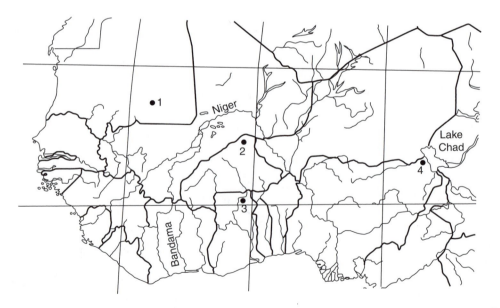

Figure 3. West Africa, showing the location of archaeological sites mentioned in the text. (1) Tichitt; (2) Ti-n-Akof; (3) Birimi; (4) Kursakata.

North America

Mesoamerica, and in particular southern Mexico, is usually regarded as both the principal and the earliest region of agricultural origins in the Americas. The three main crops of indigenous Mesoamerican agriculture—maize or corn (*Zea mays*), common bean (*Phaseolus vulgaris*), and pepo squash (*Cucurbita pepo*)—were domesticated in the region, as were several other important food plants such as avocado (*Persea americana*) and chili pepper (*Capsicum* spp.). These staple crops, together with tree fruits such as papaya (*Carica papaya*), guava (*Psidium guajava*), sapote blanco (*Casimiroa edulis*), sapote negro (*Diospyros digyna*) and ciruela (*Spondias mombin*), combined to provide (in the absence of significant quantities of protein and fat from domesticated animals) a well-balanced vegetarian diet that was the mainstay of the inhabitants of Mesoamerica until Europeans began to introduce Eurasian crops and livestock in the 16th century.

In Mesoamerica, as in Southwest Asia, archaeological investigation of the beginnings of agriculture began in the 1950s. The first field projects with that aim were undertaken by MacNeish in the state of Tamaulipas in eastern Mexico. Then in the early 1960s he shifted his attention to the Tehuacán valley southeast of Mexico City (Figure 4) where he excavated a series of cave and open sites that yielded well-preserved remains of maize, beans, squash, chili pepper, avocado, and bottle gourd (*Lagenaria siceraria*) (C.E. Smith 1967). Also in the 1960s Flannery recovered domesticated maize, pepo squash, and bottle gourd from the cave site of Guilá Naquitz in the Oaxaca valley (Figure 4) (C.E. Smith 1986), but since then the pace of archaeobotanical field research in Mexico on the origins of agriculture has slackened. Indeed Bruce Smith (2001, 1325–26) has pointed out

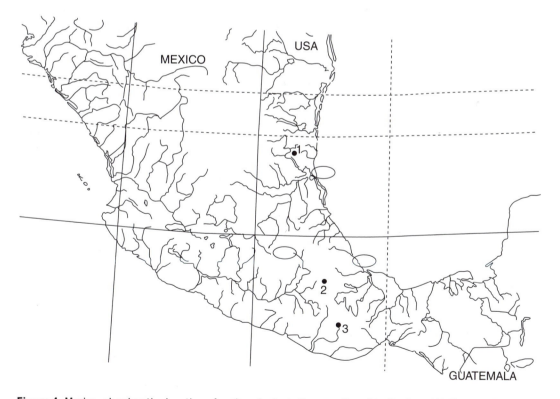

Figure 4. Mexico, showing the location of archaeological sites mentioned in the text. (1) Sierra de Tamaulipas sites; (2) Coxcatlán and other sites, Tehuacán valley; (3) Guilá Naquitz, Oaxaca valley.

that most of what is known archaeologically about the early history of maize, beans, and squash comes from only five dry caves excavated in the 1950s and 1960s (in Tamaulipas, Tehuacán, and Oaxaca). Nevertheless, a coherent if tentative interpretation of the domestication and spread of the three staple crops is now possible, largely because a small number of well-preserved and identified specimens have recently been radiocarbon dated by the AMS method.

These new results suggest that large time gaps separate the earliest evidence for the three crops, with domesticated pepo squash attested at Guilá Naquitz as long as 10,000 years ago (B.D. Smith 1997); maize cobs, also from Guilá Naquitz, about 6300 years ago (Piperno and Flannery 2001); and common beans, from Coxcatlán cave in the Tehuacán valley, no earlier than 2300 years ago (Kaplan and Lynch 1999). Despite the fact that archaeobotanical evidence of early crops is lacking for large areas of Mexico, these results strongly suggest that several millennia (perhaps 6000 or even 7000 years) elapsed after the initial domestication of pepo squash before the three crops began to be cultivated together in the productive agricultural system described in historical and ethnographic accounts and still widely practiced today.

In his overview of the new evidence Bruce Smith (2001) shows how all three crops dispersed northward from their south- and west-Mexican areas of origin, but that they did so at different periods and rates, with maize moving faster than squash and both reaching the southwestern United States about 3300 ^{14}C years ago, well before the common bean. They continued to spread north in later prehistoric times in eastern North America, where they were gradually incorporated into pre-existing systems of cultivation based on small-seeded plants native to the eastern woodlands, such as goosefoot (*Chenopodium berlandieri*) and marsh elder (*Iva annua*). These plants had evidently been locally domesticated by about 4000 ^{14}C years ago, and pepo squash appears also to have been domesticated from a local wild progenitor independently of its earlier domestication in Mexico; but maize and common bean did not reach the eastern woodland zone and become part of the slowly developing agricultural economy until over 2000 years later: maize by about AD 200 and common bean not until AD 1000 to AD 1200 (B.D. Smith 1992; B.D. Smith 1998, 184–200; Watson 1989).

The gradual dispersal of individual domesticates within and north from Mexico contrasts with the more rapid latitudinal spread in Eurasia of the founder crops of Southwest Asian agriculture. It implies that in North America as a whole, dependence on agriculture developed much more gradually, from the earliest cultivation of a few useful plants such as pepo squash and bottle gourd by foragers some 9000 ^{14}C years ago, to the gradual establishment of maize-based agriculture from perhaps 6000 ^{14}C years ago in Mexico and its subsequent spread and elaboration as more Mesoamerican plants were domesticated.

There is very little evidence in the form of seeds or other macroscopic plant remains of the beginnings of agriculture in Central America but microscopic evidence from phytoliths, pollen, and starch grains recovered at rockshelter sites in the Pacific catchment of the Rio Santa Maria in central Panama attests to the presence and possible cultivation, prior to 7000 ^{14}C years ago, of four root crops: manioc (*Manihot esculenta*), yam (*Dioscorea* spp.), arrowroot (*Maranta arundinacea*), and leren (*Calathea allouia*), a species of *Cucurbita* and a primitive form of maize (Piperno et al. 2000; Piperno and Holst 1998; Piperno and Pearsall 1998: 209–27).

South America

The Andean highlands, especially the central part of the cordillera in present-day Peru, constitute another core region of early agriculture in the Americas. Here an assemblage of indigenous plants and animals was domesticated: grain crops, including two species of *Chenopodium*—quinoa (*C. quinoa*) and cañihua (*C. pallidicaule*)—and pulses such as common bean and lima bean (*Phaseolus lunatus*); tubers, including the potato (*Solanum tuberosum*) and several minor root crops; and herd animals, including two camelids (llama and alpaca) and one small mammal (the guinea pig). Archaeological evidence of these domesticates is, however, fragmentary. Domesticated potato, common bean, lima bean, and chili pepper (*Capsicum chinense*) have been found at the central Peruvian mid-altitude cave sites of Tres Ventanas and Guitarrero, but the claimed age of 10,000-9500 ^{14}C years is not supported by direct dating of the plant remains (Ugent, Pozorski, and Pozorski 1982; Kaplan 1980; C.E. Smith 1980). Now a common bean from Guitarrero Cave has been dated to 4000 ^{14}C years ago, while the earliest lima bean, from Chilca Canyon near Peru, has been directly dated to 4800 ^{14}C years ago (Kaplan and Lynch 1999). Overall, there is evidence from these sites, and also from the Ayacucho valley in the central Peruvian highlands, that these crops and quinoa, squash, and bottle gourd were being cultivated by about 5000 ^{14}C years ago (Pearsall 1992). The patchy archaeobotanical evidence in the region as a whole points to crops having been domesticated at separate locations over the course of several millennia, and the sparse data on the camelids suggest that they were domesticated in the high Andes (Wing 1977), perhaps in association with cultivation of the chenopods and tubers (Pearsall 1989).

Although the early Andean agricultural assemblage included grains, tubers, and herd animals and was capable of sustaining large human populations, it did not develop into an integrated and expansive system of grain-livestock production on the Southwest Asian model. The camelids were an important source of meat for high-altitude populations, but they were valued as much or more as pack animals (llama) and wool producers (alpaca) and were neither milked nor used as draft animals in the cultivation of the grain and root crops (Gade 1969; Murra 1965). Individual crops spread within and beyond the Andean highlands, especially to the desert zone along the Pacific coast, but the highland agricultural system as a whole did not expand extensively from the Andean core region.

In the vast forested lowlands of the Amazon and Orinoco basins east of the Andes indigenous agriculture focused on the cultivation of root crops, principally manioc and sweet potato, but also malanga, tannia or yautia (*Xanthosoma sagittifolium*), a yam (*Dioscorea trifida*), arrowroot, and leren. Chili peppers (*Capsicum* spp.) and a variety of tree fruits were also extensively cultivated. Very little archaeobotanical evidence on the antiquity of agriculture in the lowlands is available, but remains of manioc that postdate AD 400 have been found at Parmana near the mouth of the Amazon, and the excavator infers that manioc cultivation there dates back to the beginning of the sequence about 4000 ^{14}C years ago (Roosevelt 1980, 195, 235). However, the earlier existence of village settlements along the Amazon and Orinoco river systems implies a greater antiquity for root-crop cultivation in the region (Feldman and Moseley 1983; Lathrap 1970; Meggers and Evans 1983), and phytoliths derived from leren, recovered at the site of Pena Roja in southern Colombia and directly dated, suggest that this root crop was being cultivated on the western fringes of the Amazon basin by 8000 ^{14}C years ago (Piperno and Pearsall 1998, 203–6).

The tubers and tree fruits provided abundant carbohydrate but very little protein or oil and the lowland cultivators depended on fishing and hunting for their main dietary input of protein and fat—a subsistence system that was only slightly modified after maize was introduced to the northern lowlands—according to Roosevelt (1980, 195, 235), at Parmana about 2800 ^{14}C years ago. The lowland root-and-tree-crop/fishing/hunting system had no inherent tendency to expand into different environments, although it did become established throughout Amazonia.

Conclusion

This brief review of what is presently known about the origins of agriculture worldwide demonstrates that humankind's transition from hunting and gathering to agriculture was a very gradual process that started about 12,000 years ago and extended over many millennia. The assemblages of plants that came to be cultivated and domesticated varied greatly, but in most regions where agriculture arose the crops included grains, principally cereals and pulses, that provided staple supplies of carbohydrates, proteins, and oils. Root and tree crops were incorporated into many early agricultural systems, particularly in the tropics, and in some regions, for example Amazonia and highland

New Guinea, food production focused on them rather than on grain crops. In these regions people remained dependent on hunting and fishing for most of their dietary protein and fat, whereas in regions of early grain cultivation agriculture provided a more balanced diet, which freed cultivators from continuing dependence on wild foods. Thus it was the grain-based agricultural systems that tended to expand into new environments rather than those based on root and tree crops.

The uniquely productive and nutritionally well-balanced Southwest Asian agro-pastoral system provides the most dramatic example of such expansion, spreading latitudinally some 5000 km (3100 miles) from its core Levantine region westward to the Atlantic coasts of northwest Europe and eastward into central and south Asia. The grain-based systems of East Asia and North America also spread extensively in prehistoric times, but in neither region were domestic herd animals integrated with grain cultivation. The staples (respectively rice and millets and maize and beans) spread mainly longitudinally, and individually at different rates, rather than as agricultural "packages" as in the Southwest Asian system. Nevertheless, despite these contrasts in the manner and timing of early expansions, agriculture continued to spread, in later prehistoric and early historic times, at the expense of the hunting and gathering way of life; and after AD 1500 European trade and colonization accelerated the process, which led eventually to the present dependence on agriculture of almost the entire human population.

References and Further Reading

Amblard, S. 1996. Agricultural evidence and its interpretation on the Dhars Tichitt and Oualata, southeastern Mali. In *Aspects of African Archaeology*, edited by G. Pwiti and R. Soper. Harare: University of Zimbabwe Publications, 421–428.

Bar-Yosef, O., and Meadow, R.H. 1995. The origins of agriculture in the Near East. in *Last Hunters—First Farmers: New Perspectives on the Prehistoric Transition to Agriculture*, edited by T.D. Price and A.B. Gebauer. Santa Fe, NM: School of American Research Press, 39–94.

Bayliss-Smith, T. 1996. People-plant interactions in the New Guinea highlands: Agricultural hearthland or horticultural backwater? In *The Origins and Spread of Agriculture and Pastoralism in Eurasia*, edited by D.R. Harris. London: UCL Press, 499–523.

Bellwood, P. 1989. The colonization of the Pacific: Some current hypotheses. In *The Colonization of the Pacific: A Genetic Trail*, edited by A.V.S. Hill and W. Serjeantson. Oxford: Clarendon Press and New York: Oxford University Press, 1–59.

Butler, A. 1992. The Vicieae: Problems in identification. In *New Light on Early Farming: Recent Developments in Palaeoethnobotany*, edited by J.M. Renfrew. Edinburgh: Edinburgh University Press, 61–73.

Chang, K.C. 1986. *The Archaeology of Ancient China* (4th ed.). New Haven, CT: Yale University Press.

Chen, B., and Jiang, Q. 1997. Antiquity of the earliest cultivated rice in central China and its implications. *Economic Botany* 51, 307–310.

Choe, C.-P. 1982. The diffusion route and chronology of Korean plant domestication. *Journal of Asian Studies* 41, 19–29.

Cohen, D.J. 1998. The origins of domesticated cereals and the Pleistocene-Holocene transition in East Asia. *The Review of Archaeology* 19, 22–29.

Colledge, S.M. 2001. *Plant Exploitation on Epipalaeolithic and Early Neolithic Sites in the Levant*. Oxford: British Archaeological Reports International Series 986.

Crawford, G.W. 1992. Prehistoric plant domestication in East Asia. In *The Origins of Agriculture: An International Perspective*, edited by C.W. Cowan and P.J. Watson. Washington, DC: Smithsonian Institution Press, 7–38.

Crawford, G.W., and Shen, C. 1998. The origins of rice agriculture: recent progress in East Asia. *Antiquity* 72, 858–866.

D'Andrea, A.C., Klee, M., and Casey, J. 2001. Archaeobotanical evidence for pearl millet (*Pennisetum glaucum*) in sub-Saharan West Africa. *Antiquity* 75, 341–348.

Denham, T.P., Haberle, S.G., Lentfer, C., Fullagar, R., Field, J., Therin, M., Porch, N., and Winsborough, B. 2003. Origins of agriculture at Kuk swamp in the highlands of New Guinea. *Science* 301, 189–193.

Feldman, R.A., and Moseley, M.E. 1983. The northern Andes. In *Ancient South Americans*, edited by J.D. Jennings. San Francisco: Freeman, 138–177.

Fuller, D.Q. 2002. Fifty years of archaeobotanical studies in India: Laying a solid foundation. In *Indian Archaeology in Retrospect*, vol. III, *Archaeology and Interactive Disciplines*, edited by S. Settar and R. Kortisettar. New Delhi: Manohar, 247–364.

Fuller, D.Q., Korisettar, R., and Venkatasubbiah, P.C. 2001. Southern Neolithic cultivation systems: A reconstruction based on archaeobotanical evidence. *South Asian Studies* 17, 171–187.

Gade, D.W. 1969. The llama, alpaca and vicuña: Fact vs. fiction. *Journal of Geography* 68, 339–343.

Garrard, A. 1999. Charting the emergence of cereal and pulse domestication in south-west Asia. *Environmental Archaeology* 4, 67–86.

Glover, I.C., and Higham, C.F.W. 1996. New evidence for early rice cultivation in South, Southeast and East Asia, in *The Origins and Spread of Agriculture and Pastoralism in Eurasia*, edited by D.R. Harris. London: UCL Press, 413–441.

Golson, J. 1977. No room at the top: Agricultural intensification in the New Guinea highlands. In *Sunda and Sahul: Prehistoric Studies in Southeast Asia, Melanesia and Australia*, edited by J. Allen, J. Golson, and R. Jones. London: Academic Press, 601–638.

Golson, J. 1989. The origins and development of New Guinea agriculture. In *Foraging and Farming: The Evolution of Plant Exploitation*, edited by D.R. Harris and G.C. Hillman. London: Unwin Hyman, 678–687.

Hallam, S.J. 1989. Plant usage and management in Southwest Australian Aboriginal societies. In *Foraging and Farming: The Evolution of Plant Exploitation*, edited by D.R. Harris and G.C. Hillman. London: Unwin Hyman, 136–151.

Harris, D.R. 1976. Traditional systems of plant food production and the origins of agriculture in West Africa. in *Origins of African Plant Domestication*, edited by J.R. Harlan, J.M.J. de Wet, and A.B.L. Stemler. The Hague: Mouton, 311–356.

Harris, D.R. 1981. The prehistory of human subsistence: A speculative outline. In *Food, Nutrition and Evolution*, edited by D.N. Walcher and N. Kretchmer. New York: Masson, 15–35.

Harris, D.R. 1984. Ethnohistorical evidence for the exploitation of wild grasses and forbs: Its scope and archaeological implications. In *Plant and Ancient Man: Studies in Palaeoethnobotany*, edited by W. van Zeist and W.A. Casparie. Rotterdam: Balkema, 1984, 63–69.

Harris, D.R. 1989. An evolutionary continuum of people-plant interaction. In *Foraging and Farming: The Evolution of Plant Exploitation*, edited by D.R. Harris and G.C. Hillman. London: Unwin Hyman, 11–26.

Harris, D.R. 1995. Early agriculture in New Guinea and the Torres Strait divide. *Antiquity 69*, 848–854.

Harris, D.R. 1996a. Domesticatory relationships of people, plants and animals. In *Redefining Nature: Ecology, Culture and Domestication*, edited by R. Ellen and K. Fukui. Oxford: Berg, 437–463.

Harris, D.R. 1996b. The origins and spread of agriculture and pastoralism in Eurasia: an overview. In *The Origins and Spread of Agriculture and Pastoralism in Eurasia*, edited by D.R. Harris. London: UCL Press, 552–573.

Harris, D.R. 1998a. Beginnings of agriculture in tropical Africa: Retrospect and prospect. In *Africa: The Challenge of Archaeology*, edited by B.W. Andah et al. Ibadan: Heinemann, 101–114.

Harris, D.R. 1998b. The origins of agriculture in southwest Asia. *The Review of Archaeology 19*, 5–11.

Harris, D.R. 1998c. The spread of Neolithic agriculture from the Levant to western Central Asia. In *The Origins of Agriculture and Crop Domestication: The Harlan Symposium*, edited by A.B. Damania et al. Aleppo, Syria: ICARDA, 65–82.

Harris, D.R. 2002. Development of the agro-pastoral economy in the Fertile Crescent during the Pre-Pottery Neolithic period. In *The Dawn of Farming in the Near East*, edited by R.T.J. Cappers and S. Bottema. Berlin: ex oriente.

Harris, D.R. 2003. Climatic change and the beginnings of agriculture: the case of the Younger Dryas. In *Evolution on Planet Earth*, edited by A. Lister and L. Rothschild. New York: Academic Press.

Hather, J.G. 1991. The identification of charred archaeological remains of vegetative parenchymous tissue. *Journal of Archaeological Science 18*, 661–675.

Hather, J.G. 1994. The identification of charred root and tuber crops from archaeological sites in the Pacific. In *Tropical Archaeobotany: Applications and New Developments*, edited by J.G. Hather. London: Routledge, 51–64.

Hather, J.G. 2000. *Archaeological Parenchyma*. London: Archetype Publications.

Hillman, G.C. 1996. Late Pleistocene changes in wild plant-foods available to hunter-gatherers of the northern Fertile Crescent: Possible preludes to cereal cultivation. In *The Origins and Spread of Agriculture and Pastoralism in Eurasia*, edited by D.R. Harris. London: UCL Press and Washington, DC: Smithsonian Institution Press.

Hillman, G.C. 2000. The plant food economy of Abu Hureyra 1: the Epipalaeolithic. In *Village on the Euphrates: From Foraging to Farming at Abu Hureyra*, edited by A.M.T. Moore, G.C. Hillman, and A.J. Legge. New York: Oxford University Press, 327–399.

Hillman, G.C., Hedges, R., Moore, A., Colledge S., and Pettitt, P. 2001. New evidence for late glacial cereal cultivation at Abu Hureyra on the Euphrates. *The Holocene 11*, 383–393.

Holl, A. 1985. Subsistence patterns of the Dear Tichitt Neolithic Mauritania. African Archaeological Review 3:151–162.

Hope, G., and Golson, J. 1995. Late Quaternary change in the mountains of New Guinea. *Antiquity 69*, 818–830.

Imamura, K. 1996. Jomon and Yayoi: The transition to agriculture in Japanese prehistory. In *The Origins and Spread of Agriculture and Pastoralism in Eurasia*, edited by D.R. Harris. London: UCL Press, 442–464.

Jones, M., and Brown, T. 2000. Agricultural origins: The evidence of modern and ancient DNA. *The Holocene 10*, 769–776.

Kaplan, L. 1980. Variation in the cultivated beans. In *Guitarrero Cave: Early Man in the Andes*, edited by T.F. Lynch. New York: Academic Press, 145–148.

Kaplan, L., and T. Lynch. 1999. *Phaseolus* (Fabaceae) in archaeology: AMS radiocarbon dates and their significance for pre-Columbian agriculture. *Economic Botany 53*, 261–272.

Kenyon, L. 1960. Jericho and the origins of agriculture. *The Advancement of Science 66*, 118–120.

Lathrap, D.W. 1970. *The Upper Amazon*. London: Thames and Hudson.

Lebot, V. 1999. Biomolecular evidence for plant domestication in Sahul. *Genetic Resources and Crop Evolution 46*, 619–628.

Long, A., Benz, B.F., Donahue, D.J., Jull, A.J.T., and Toolin, L.J. 1989. First direct AMS dates on early maize from Tehuacán, Mexico. *Radiocarbon 31*, 1035–1040.

Lu, T.L.-D. 1999. The transition from foraging to farming in China. *Bulletin of the Indo-Pacific Prehistory Association 18*, 77–80.

MacNeish, R.S. 1958. Preliminary archaeological investigations in the Sierra de Tamaulipas, *Transactions of the American Philosophical Society 48*, Part 6. Philadelphia: American Philosophical Society.

Meadow, R.H. 1998. Pre- and proto-historic agricultural and pastoral transformations in northwestern South Asia. *The Review of Archaeology 19*, 12–21.

Meggers, B.J., and Evans, C. 1983. Lowland South America and the Antilles. In *Ancient South Americans*, edited by J.D. Jennings. San Francisco: Freeman, 286–335.

Moore, A.M.T., and Hillman, G.C. 1992. The Pleistocene to Holocene transition and human economy in Southwest Asia: The impact of the Younger Dryas. *American Antiquity 57*, 482–494.

Munson, P.J. 1976. Archaeological data on the origins of cultivation in the southwestern Sahara and their implications for West Africa. In *Origins of African Plant Domestication*, edited by J.R. Harlan, J.M.J. de Wet, and A.B.L. Stemler. The Hague: Mouton, 187–209.

Murdock, G.P. 1959. *Africa: Its Peoples and Their Culture History*. New York: McGraw Hill.

Murra, J.V. 1965. Herds and herders of the Inca state. In *Man, Culture and Animals*, edited by A. Leeds and A.P. Vayda. Washington, DC: American Association for the Advancement of Science, 185–215.

Neumann, K. 1999. Early plant food production in the West African Sahel: New evidence. In *The Exploitation of Plant Resources in Ancient Africa*, edited by M. van der Veen. New York: Kluwer, 73–81.

Neumann, K., Ballouche, A., and Klee, M. 1996. The emergence of plant food production in the West African Sahel: New evidence from northeast Nigeria and northern Burkina Faso. In *Aspects of African Archaeology*, edited by G. Pwiti and R. Soper. Harare: University of Zimbabwe Publications, 441–448.

Pearsall, D.M. 1989. Adaptation of hunter-gatherers to the high Andes: The changing role of plant resources. In *Foraging and Farming: The Evolution of Plant Exploitation*, edited by D.R. Harris and G.C. Hillman. London: Unwin Hyman, 318–332.

Pearsall, D.M. 1992. The origins of plant cultivation in South America. In *The Origins of Agriculture: An International Perspective*, edited by C.W. Cowan and P.J. Watson. Washington, DC: Smithsonian Institution Press, 173–205.

Piperno, D.R., and Flannery, K.V. 2001. The earliest archaeological maize (*Zea mays* L.) from highland Mexico: New accelerator mass spectrometry dates and their implications. *Proceedings of the National Academy of Sciences 98*, 2101–2103.

Piperno, D.R., and Holst, I. The presence of starch grains on prehistoric stone tools from the humid Neotropics: Indications of early tuber use and agriculture in Panama. *Journal of Archaeological Science 25*, 765–776.

Piperno, D.R., and Pearsall, D.M. 1998. *The Origins of Agriculture in the Lowland Neotropics*. San Diego, CA: Academic Press.

Piperno, D.R., Ranere, A.J., Holst, I., and Hansell, P. 2000. Starch grains reveal early root crop horticulture in the Panamanian tropical forest. *Nature 407*, 894–897.

Portères, R. 1950. Vielles agricultures de l'Afrique intertropicale. *L'Agronomie Tropicale 5*, 489–507.

Roosevelt, A.C. 1980. *Parmana: Prehistoric Maize and Manioc Subsistence along the Amazon and Orinoco*. New York: Academic Press.

Smith, B.D. (Ed.) 1992. *Rivers of Change: Essays on Early Agriculture in Eastern North America*. Washington, DC: Smithsonian Institution Press.

Smith, B.D. 1997. The initial domestication of *Cucurbita pepo* in the Americas 10,000 years ago. *Science 276*, 932–934.

Smith, B.D. 1998. *The Emergence of Agriculture*. New York: Scientific American Library.

Smith, B.D. 2001. Documenting plant domestication: The consilience of biological and archaeological approaches. *Proceedings of the National Academy of Sciences 98*, 1324–1326.

Smith, C.E., Jr. 1967. Plant remains. In *The Prehistory of the Tehuacán Valley*, vol. I, *Environment and Subsistence*, edited by D. Byers. Austin: University of Texas Press, 220–255.

Smith, C.E., Jr. 1980. Plant remains from Guitarrero Cave. In *Guitarrero Cave: Early Man in the Andes*, edited by T.F. Lynch. New York: Academic Press, 87–119.

Smith, C.E., Jr. 1986. Preceramic plant remains from Guilá Naquitz. In *Guilá Naquitz. Archaic Foraging and Early Agriculture in Oaxaca, Mexico*, edited by K.V. Flannery. Orlando, FL: Academic Press, 265–274.

Spriggs, M.1996. Early agriculture and what went before in Island Melanesia: Continuity or intrusion? In *The Origins and Spread of Agriculture and Pastoralism in Eurasia*, edited by D.R. Harris. London: UCL Press, 524–537.

Stahl, A.B. 1993. Intensification in the West African Late Stone Age. In *The Archaeology of Africa: Food, Metals and Towns*, edited by T. Shaw, P. Sinclair, B. Andah, and A. Okpoko. London: Routledge, 261–273.

Ugent, D., Pozorski, S., and Pozorski, T. 1982. Archaeological potato tuber remains from the Casma Valley of Peru. *Economic Botany 36*, 417–432.

Underhill, A.P. 1997. Current issues in Chinese Neolithic archaeology. *Journal of World Prehistory 11*, 103–160.

van Zeist, W., and Bakker-Heeres, J.A.H. 1982. Archaeobotanical studies in the Levant. 1. Neolithic sites in the Damascus Basin: aswad, Ghoraifé, Ramad. *Palaeohistoria 24*, 165–256.

Watson, P.J. 1989. Early plant cultivation in the eastern woodlands of North America. In *Foraging and Farming: The Evolution of Plant Exploitation*, edited by D.R. Harris and G.C. Hillman. London: Unwin Hyman, 555–571.

Wetterstrom, W. 1993. Foraging and farming in Egypt: The transition from hunting and gathering to horticulture in the Nile valley. In *The Archaeology of Africa: Food, Metals, and Towns*, edited by T. Shaw, P. Sinclair, B. Andah, and A. Okpoko. London: Routledge.

Wing, E.S. 1977. Animal domestication in the Andes. In *Origins of Agriculture*, edited by C.A. Reed. The Hague: Mouton, 837–859.

Woodman, P. 2000. Getting back to basics: Transitions to farming in Ireland and Britain. In *Europe's First Farmers*, edited by T.D. Price. Cambridge and New York: Cambridge University Press, 219–259.

Yan, W. 1993. Origins of agriculture and animal husbandry in China. In *Pacific Northeast Asia in Prehistory*, edited by C.M. Aikens and N.R. Song. Pullman: Washington State University Press, 113–123.

Yen, D.E. 1991. Domestication: The lessons from New Guinea. In *Man and a Half: Essays in Honour of Ralph Bulmer*, edited by A. Pawley. Auckland: Polynesian Society, 558–569.

Yen, D.E. 1995. The development of Sahul agriculture with Australia as bystander. *Antiquity 69*, 831–847.

Zhao, Z. 1998. The Middle Yangtze region in China is one place where rice was domesticated: Phytolith evidence from the Diaotonghuan Cave, northern Jiangxi. *Antiquity 72*, 885–897.

Zohary, D. 1996. The mode of domestication of the founder crops of Southwest Asian agriculture. In *The Origins and Spread of Agriculture and Pastoralism in Eurasia*, edited by D.R. Harris. London: UCL Press, 142–158.

Part II
The Migration of Plants

GHILLEAN T. PRANCE

Once agriculture was well underway and people had settled in towns and cities, the need for a variety of plants increased rather than decreased. The plants that had been used in the wild were gradually brought into cultivation as vegetables, fruits, medicines, ornamentals, and so on. People also began to move plants from one area to another, often far from the plant's natural range. This is nothing new. One thing that has particularly impressed me in my ethnobotanical work with Amazon Indians is the extent to which they carry plant germplasm around. If you meet Yanomami Indians walking a trail on a visit to another village they will invariably be carrying plant germplasm of some sort as a gift for their hosts. It has been estimated that the Kayapó Indians moved plants around over an area the size of western Europe. Once people became agriculturalists in any society, the movement of plants was no longer just by natural dispersal, but also through human agents.

As agriculture developed, both trading and warfare increased. Chapters in this part of the book focus on some of the major uses of plants. It is interesting and perhaps sad to see how in so many cases territorial and commercial interest dominated. Spices were a major cause of warfare between European nations as they sought to break the Asian monopoly. This desire for spices started at a very early date; for example, in 300 AD, Alaric the Goth ransomed Rome for 3,000 pounds of pepper. Spices were important in the days before refrigeration because they made decaying food palatable. More recently the antibacterial properties of many spices have been demonstrated; our ancestors were both killing germs and disguising unpleasant tastes with the spices they used, and so it is not surprising that these spices were popular and expensive commodities.

The natural dispersal of plants from one place to another by wind, animals, or sea currents has been occurring as long as plants have existed. However, once agriculture had been invented both the products of plants and the germplasm to grow plants outside their natural range began to be transported from one place to another by humans. More recently, humans have caused a much greater migration of plants, often with extremely deleterious effects. The history of Hawaii is a good illustration of this. This archipelago began to rise out of the ocean some 5 million years ago. Gradually plant seeds began to arrive, either carried by birds or wind or washed there by ocean currents. It is estimated that between 270 and 280 original dispersals to Hawaii occurred. Gradually these evolved into the 956 species that are now known to be native to Hawaii. The process of adaptive radiation worked well, and plants were able to occupy the many niches offered by these new islands. Today, since human colonization, the flora of Hawaii lists 1,678 species native and naturalized. Some 861 species have been added to the flora. The first human-caused introductions were made by the original Polynesian settlers, who brought with them in their canoes their crops, including taro and breadfruit, and many other useful plants. However, the biggest influx of new species began after Captain Cook's discovery of the islands. From then on many other alien species were introduced, some deliberately and many accidentally as weeds or as accidental passengers on ships, clothing, animals, and so on. Today one of the greatest threats to the native flora is the invasive aliens that are discussed in more detail in a chapter in the third part of this volume.

As people have traveled, traded, and migrated they have always moved their crops with them; often people have seen crops or products on their travels that they have coveted and tried to move elsewhere. All this has led to much intrigue and strife, as can be seen in the history of some of the crops described here. The Dutch and the British both went to great lengths to smuggle quinine seeds out of Peru and Bolivia to establish their own supplies in their eastern colonies. In many cases plants prospered better when moved away from their native areas, away from the pests and diseases with which they had evolved. For example, rubber cannot be grown in plantations in its native Amazon region because of several native diseases such as the leaf rust fungus *Dothidella ulei*. When transported to tropical Asia, rubber prospered well in plantations and the Amazon monopoly on the product was broken. Similarly, coffee was taken from Africa to South America, and it did far better away from its homeland diseases. Many crops produce much better when far removed from their native habitats, and that is why migration is so much a part of the cultural history of our crops.

These chapters deal both with major crops that have been transported around the world and with the considerable amount of use that still continues of wild species in their original habitats such as for local medicines and foods. They also show the huge variety of uses that humans have found for plants. We are a truly plant-dependent race whether we live in cities or in the jungle. At the Eden Project in Cornwall, England, where I work, there is an exhibit called "plant take away." It is a scene of a family in their kitchen. Gradually anything made from plant material begins to disappear. First the obvious, such as flowers and fruit on the table, disappear, but soon the table and chairs, the window frames, and the contents of the refrigerator go. Eventually the clothes disappear and the family is left nude; finally the dog expires from lack of oxygen. This is a powerful way of demonstrating the vital importance of plants to people. These chapters will also demonstrate this fundamental fact, since they deal with the plants that are the food we eat, the wood with which we build our houses, the medicines that cure our ills, the fibers that we use for clothing, the spices that add flavor to our foods, and even the perfumes and fragrances that make life more enjoyable.

3
Gathering Food from the Wild

ANDREA PIERONI

Introduction

Gathering food from the wild represents one of the most complex aspects of the use of wild plants, and was closely intertwined with the history of the first human communities. Although past hunter-gatherers are often thought of primarily as dependent on the hunting of wild animals, archaeological and ethnographic evidence shows that plant foods always formed the bulk of their diet. The only exception is in areas such as the Arctic, where it is too cold for most wild food plants to grow. Even in agricultural communities today, the gathering of wild plants frequently remains important for nutrition and food diversity.

In recent years it has become obvious that food and medicine are closely linked; a food plant may be used for medicine, and vice versa. Moreover, eating food from the wild is not simply an essential response in times of famine or food shortages, or an easy way to obtain primary nutrients, but more often a complex evolutionary process, involving different aspects of the relationship between humans and their natural environment. Non-cultivated gathered food plants are often weedy and grow in environments disturbed and managed by man. In addition, eating these plants provides many micronutrients and phytochemicals that are now known to play a central role as antioxidants in the prevention of various illnesses, especially age-related diseases.

The use of such plants reflects local tastes and customs, and is often a strong force for identity and social cohesion, particularly among women. In many cultures women organize the gathering of wild plants and the management of home-gardens.

It is impossible to list and discuss here the huge number of wild and weedy plants traditionally collected and consumed. This chapter covers some of the important species throughout the world, with a special emphasis on edible greens. These are mostly collected in the spring, when the leaves, stems, and buds of wild plants are softer and less bitter. There is little archaeological evidence relating to edible greens compared to nuts and seeds, which are more likely to survive. However, evidence from the diet of primates suggests that consumption of young leaves has always been a feature of the diet of modern humans and our hominid ancestors. Nuts, berries, and grains, also gathered from the wild both before and after domestication, are discussed in separate chapters.

See: Nuts, Seeds, and Pulses, pp. 133–52; Fruits, pp. 77–96; Grains, pp. 45–60

Pomo woman using seed beater to gather seeds into a burden basket, California, ca.1924. Library of Congress, Prints & Photographs Division, Edward S. Curtis Collection.

Africa

Aba *Ceropegia* spp.
Asclepiadaceae

Various species of *Ceropegia* provide an important food source for many populations in southwest Africa. The tuber is gathered and eaten raw throughout the year. It has its highest water content during the rainy season, becoming drier and sweeter in taste during the dry season. The leaves are also eaten raw.

African locust *Parkia biglobosa*
Fabaceae

African locust is a large tree native to Sudan, where local populations have used its seeds for many centuries. The seeds are roasted, bruised, fermented in water, and then pounded into powder and made into cakes. A beverage is also made from the pulp of the fresh pod. The leaves have many medicinal uses.

African spider flower, Bastard mustard *Cleome gynandra*
Brassicaceae

The leaves of this herbaceous species are often gathered and eaten as vegetables in the savanna regions of southern Africa, and are commonly dried and stored. The bitterness of the leaves is tempered by cooking in milk or butter. The leaves have also been used in the treatment of rheumatism, and the juice is claimed to be a treatment for earache.

See: Herbs and Vegetables, p. 128

Baobab *Adansonia digitata*
Bombacaceae

The baobab is one of the most versatile trees of tropical Africa, and its preeminent role in tribal mythology protects it from being cut. The trunk can reach 19 feet (6 m) in diameter, and some trees are over 1000 years old. The hollow trunks of living trees are often used as water tanks. The young leaves of the baobab are commonly gathered and eaten as vegetables in many African regions. The fruit, with its aromatic and sour flavor, is also edible and frequently used in western Africa, either raw or in beverages; the pulp is often mixed with water to prepare a juice that can be sweetened with sugar, if available. Seeds of the baobab have been ground and made into meal in times of famine in Angola.

Beggar's ticks, Spanish needle *Bidens pilosa*
Asteraceae

Native to temperate and tropical America, *Bidens pilosa* has spread to the Pacific, Asia, and Africa. The prickly seed vessel has hooks and clings to clothing. The leaves have a strong, resinous flavor and are eaten raw in salads, or steamed and added to soups and stews. They can also be dried for later use. It is one of the most important wild greens (*michicha*) in eastern Africa. In Australia and Hawaii the young shoot tips are used to make a tea. A juice made from the leaves is traditionally used all over the world to dress wounds and ulcers.

Bitter leaves *Vernonia amygdalina* and *V. cinerea*
Asteraceae

In central Africa the leaves are often used as a vegetable, although they must be washed prior to eating to get rid of their very bitter taste. They are claimed to stimulate the digestive system and to reduce fever. The leaves are also used as a topical medicine against bilharzia-transmitting leeches, and are also used instead of hops to make beer in Nigeria. Chimpanzees chew on the pith from young shoots if they have been attacked by parasites.

Meat dishes prepared with the bitter leaves are popular in many African restaurants worldwide and the dried herb is often available in major cities where there is a local African community.

Cape myrtle *Myrsine africana*
Myrsinaceae

Aerial parts of this evergreen shrub are collected and used as additives in meat and milk-based soups by the Batemi and Masai of east Africa. Saponin-like compounds contained in Cape myrtle, which forms a significant part of the Masai diet, are believed to inhibit absorption of dietary cholesterol, thus helping the indigenous people, who consume large amounts of meat, to remain healthy. The flowers of this species are also eaten, whereas the fruit is said to be used as a treatment for intestinal worms.

Gallant soldier, Guascas *Galinsoga parviflora*
Asteraceae

Native to South America, this annual weed has been introduced and naturalized to North America, Europe, Africa, and Asia. In eastern Africa, especially Tanzania, where the species is most commonly gathered as a wild green, the leaves, stem, and flowering shoots are collected and eaten. The plant is often dried and ground into a powder for use as a flavoring in soups and stews.

See: Herbs and Vegetables, p. 104

Ice plant *Mesembryanthemum crystallinum*
Aizoaceae

Originating in the Cape of Good Hope area, this succulent plant was introduced to Europe in 1727; by 1881 it was already being promoted (ultimately unsuccessfully) in the United States as a beneficial vegetable, to be boiled like spinach. The aerial parts have an acid flavor, being thick and very succulent with a slightly salty tang. The leaves and stems are still gathered from the wild in southern Africa, to be pickled like cucumbers or used as a garnish.

See: Herbs and Vegetables, p. 120

Jew's mallow, Jute *Corchorus olitorius*
Tiliaceae

Best known as a fiber plant, jute is also an important leafy green. Pliny recorded that the aerial parts of this species were frequently gathered and eaten by the ancient Egyptians. Possibly originating in tropical Asia, and grown by the Jews in the Near East (hence the name), the plant grows in many tropical areas. Gathered from the wild in eastern Africa and India, the species has been domesticated in Mauritius, Jamaica, and even in France, where its tender leaves are used in cooking.

See: Natural Fibers and Dyes, pp. 295–296; Herbs and Vegetables, p. 122

Umdoni tree *Syzygium cordatum*
Myrtaceae

Native throughout Africa, this tree produces pinkish-purple fruits, about twice the size of a peanut, which have a tart flavor and apple-like texture with a large pit. These fruits are often gathered from the wild in many parts of Africa, especially Zambia and Swaziland, where they are called *umncozi* and are the most commonly gathered wild fruit by adults and children alike. Fruits of the brush cherry (*S. paniculutum*) are gathered from the wild and eaten—raw or cooked—in Australia (see later).

Vangueria *Vangueria* spp.
Rubiaceae

Various species of *Vangueria* (*V. infausta* or wild medlar in Namibia, *V. madagascariensis* or Spanish tamarind in Madagascar, *V. cyanescens* in Swaziland) are gathered from the wild by indigenous people in Africa. The raw, soft-flesh fruit is eaten and tastes similar to a wild apple. When the fruits start drying out, from April onwards, they are soaked in water then boiled and mashed slightly and eaten as a kind of porridge. The fresh fruits cannot be stored for more than a week, but they can be dried in the sun and then stored for almost a year.

Americas

Agave *Agave americana* and related species
Agavaceae

These species play an extremely important role in the culinary traditions of Mexicans and Amerindians of the Southwest deserts. Just before the flower stalks appear, the plant is dug up and all the leaves are chopped at their base, leaving a cylindrical white and pulpy trunk, which is cooked in a fire pit for several days. The flesh is then eaten from around the fibers that grow in it. The same cooked crown is often sold as a candy in Mexican markets, or a sweet juice can be obtained from the cooked crown, which is used for making a syrup. Today the cooked crown is mashed, mixed with water, left to ferment, and then distilled, to produce *mezcal*. The flower stalks and buds of *Agave* were also thought of as a vegetable delicacy, and the seeds were at one time ground into powder. *Pulque* is a milky fermented drink produced mainly from *A. atrovirens*. Other *Agave* species have also been used as food in Central America.

See: Natural Fibers and Dyes, pp. 301–302; Caffeine, Alcohol, and Sweeteners, p. 181

Algaroba, Mesquite *Prosopis* spp.
Fabaceae

Indians of Peru, Chile, and California eat the sweet pulp contained in the pods of *Prosopis juliflora* (honey mesquite). The pods are sometimes dried and ground to make bread; in the past the pods were chewed to quench thirst during journeys. *P. dulcis* fruits are gathered from the wild in tropical

Screwbean mesquite (*Prosopis pubescens*). M. Kat Anderson @ USDA-NRCS PLANTS Database.

South America, *P. pubescens* (screwbean mesquite) pods and seeds were used as fodder and food by Mexican Indians, and the sweetish substance which surrounds the seeds of *P. spicigera* is considered a food in Iran and northwestern Pakistan.

Amaranth, Inca wheat *Amaranthus* spp.
Amaranthaceae
Cultivated from time immemorial for food purposes, *A. caudatus* seeds were a staple food in the diet of the Aztecs, who also consumed the aerial parts as greens. In North America, leaves of a number of species were consumed, and today the leaves of some of these species, naturalized in many other tropical and subtropical regions, are still an important wild food in eastern Africa. *A. retroflexus* is important in parts of southern Italy and North Africa.

See: Grains, p. 58; Herbs and Vegetables, p. 113

Cow tree *Mimusops elata*
Sapotaceae
Native to the Brazilian Amazon, the fruits of this species are similar to small apples and full of creamy milk (hence the common name) with an unusual taste. In the state of Para the fruit is very popular and sold in the streets. Natives of Amazonia also collect and drink the milk that exudes from the bark, but this coagulates very quickly, forming a "glue." Fruits of other species are gathered from the wild and eaten in southeastern Asia.

Izote *Yucca guatemalensis*
Agavaceae
The flowers of this species are widely gathered in Central America. After removing the bitter anthers and ovaries, the flowers are dipped in egg batter and fried or lightly boiled. The boiled flowers are eaten with lemon juice in Guatemala. The tender stem tips stripped of their leaves (*cogollo de izote*) are very popular in El Salvador. Flower stalks and buds of many other *Yucca* species are gathered and consumed; a few are eaten in southern Europe, where the species grows in many arid soils.

Pacaya palm *Chamaedorea tepejilote*
Arecaceae
The young flowers of this species are gathered from the wild and sold in many markets in Central America. They are used raw in salads, or boiled, or fried in egg batter to form a fritter called *recado de pacaya* in Central America. Usually cooked in several changes of water first to remove their

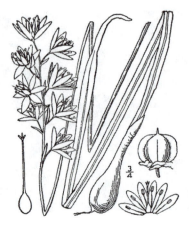

Camassia scilloides. USDA-NRCS PLANTS Database/Britton, N.L., and A. Brown. 1913. *Illustrated flora of the northern states and Canada.* Vol. 1, p. 509.

bitterness, they are commonly used as a garnish for *fiambra*, a Guatemalan cold salad served on All Saints' Day.

Quamash, Camas, Wild hyacinth *Camassia quamash* and related species
Hyacinthaceae
Bulbs of this species and of *C. leichtinii* have been an important food for Native Americans, who moved to quamash fields in the early autumn. The bulbs were placed in a fire pit and left to cook slowly for two days. The raw bulbs have a sweetish, mild, starchy flavor, but a gummy texture; when cooked, however, they develop a delicious sweet taste reminiscent of sweet chestnuts, and they are a highly nutritious food. Quamash is also dried and made into a powder, which is used as a thickener in stews or mixed with cereal flours when making bread.

Yampa *Perideridia* spp.
Apiaceae
The thick rootstock of many species, such as *P. oregana* (squaw potato), has been gathered and cooked in a fire pit, eaten raw, or dried and ground into flour by many North American indigenous peoples, especially Californian Indians. The Nez Percé Indians collected and boiled the tuberous roots, which have a cream-like flavor. The roots are also said to have various medicinal properties.

Asia

A'kub *Gundelia tournefortii*
Asteraceae
The perennial thistle A'kub is gathered in early spring from the wild by several indigenous groups in Palestine, Israel, and the surrounding arid areas. Its immature inflorescence heads are cooked in the same way as artichokes, covered with mincemeat, fried briefly in olive oil, than simmered in a lemon juice–based sauce. In the recent past mature fruits have also been used as a source of oil. Charred fruits at Neolithic sites in Iraq and Turkey are evidence that oil extraction dates back at least 10,000 years.

Bistort, Snakeweed *Polygonum bistorta* and related species
Polygonaceae

Leaves of snakeweed were consumed by northern populations in Europe and Asia; in the north of England, for example, young shoots of *P. bistorta* were used as an ingredient of a savory herb pudding. In northern Russia the roots have been gathered for many centuries, and eaten roasted, and the roots of Alpine bistorta (*P. viviparum*) are still used by Samoyed peoples. *P. japonicum* (*amatokoro*) and *P. multiflorum* are frequently used in Japanese and Chinese cooking; occasionally use of these plants has spread to Europe, where they are gathered and collected for sweets in the European nouvelle cuisine.

Leaves of *P. cognatum* are frequently collected in Anatolia. Often *mercimelek* (as the species is known in Turkey) is sun-dried in the spring and stored for winter. Bottles of the preserved herb are widely available in Turkish communities within Germany.

Bracken fern *Pteridium aquilinum*
Dennstaedtiaceae

In the past in many parts of the world, the rhizome was ground and added to flour to bake bread. In the Canary Isles (La Palma and La Gomera) up to the 1930s the rhizomes were ground and mixed with barley meal to prepare a kind of porridge called *gofio*. It is the young shoots of the plant that are important in Japanese and Korean cooking; the shoots are soaked for a day in water and ashes (an archaic detoxification method), then steamed or boiled and eaten as a vegetable or in soups. Sometimes the shoots are preserved in salt, in lees of *sakè* or in *miso*. Bracken fern shoots have also been used in Siberia to produce a kind of beer, and by native peoples in North America. Leaves are commonly used by shepherds in the Mediterranean to filter sheep's milk and to store freshly made ricotta cheese.

Caltrop, Devil's thorn *Tribulus terrestris*
Zygophyllaceae

The leaves and young shoots of caltrop are gathered and cooked in eastern Asia. The leafy stems have been used to thicken buttermilk—it is said that buttermilk sellers often diluted their merchandise with water and then thickened the mixture with this plant. The seeds are said to have various medicinal properties, and have been used for the removal of intestinal worms, to reduce flatulence, and as an aphrodisiac, astringent, and diuretic.

Gogd *Allium ramosum*
Alliaceae

Similar to cultivated Chinese chives (*A. tuberosum*), this species is a staple ingredient of the traditional diet of northern Chinese and nomadic Mongol peoples. Large quantities—up to 9 to 11 pounds (4 to 5 kg) fresh weight—of the aerial parts of *gogd* (the Mongolian name for this species) are gathered from May until July by each nomadic family then preserved with salt, ready to be used during the winter months. In this way *gogd* leaves are added to pots of boiled mutton, or used to make dumplings, which are eaten raw, steamed, or boiled. Sometimes the plant blossoms (*soriz*) are collected in late July and August and preserved in salt. *Gogd* is also used as a tonic for stomach ailments.

Hackberry, Nettle tree *Celtis* spp.
Ulmaceae

Fruits of several species of *Celtis* have been eaten by man for many centuries. A thin, sweet flesh surrounds the large stone. Stones of *C. tournefortii* have been found in large quantities in many Neolithic archaeological excavations in the Near East and probably formed a significant part of the prehistoric diet. They are still gathered from the wild and consumed as a snack in central Anatolia.

Celtis australis is found in the south of Europe, whereas *Celtis occidentalis* is native to the United States. Native Americans used *Celtis occidentalis* either as a fresh fruit, to flavor meat, or by pounding the berries and mixing them with fat and parched corn.

See: Origins and Spread of Agriculture p. 20

Oleaster, Russian olive *Elaeagnus angustifolia*
Elaeagnaceae

Fruits of oleaster are gathered and sold in the Near East (particularly in the local markets of Istanbul) and Iran, where a dessert made from the bittersweet flesh of the fruit is known as *zinzeyd*. In Nepal, the fruits are also consumed fresh or dried.

See: Ornamentals, p. 280

Salep *Orchis* spp.
Orchidaceae

For many eastern Asiatic populations and especially in Turkey, the dried roots of *Orchis* species and other genera are the source of *salep* (*sahlab*). This yellowish powder has been an important food in Istanbul as a hot beverage (*salep* powder is added to milk) and for *salep* ice cream; these both had great social and cultural significance. Today the gathering, commercialization, and export of many threatened *Orchis* species is forbidden, and true *salep* powder is often substituted by manioc flour or other artificial carbohydrate sources. Nevertheless, its use is still common.

Shepherd's purse *Capsella bursa-pastoris*
Brassicaceae

This species—one of the most common weeds worldwide—is of European origin. It accompanied Europeans during their explorations and is today ubiquitous in Europe, Asia, and America. It has frequently been used as a wild food (cooked), especially in China, Japan, and Korea, where the young leaves are gathered in the spring and sold in local markets. Whole plants (*naeng-i*) are used in Korea as cooked vegetables (*namul*) and the species has occasionally been introduced as a food crop. In Korean markets in California it is common to find the plant sold frozen.

Australia and Oceania

Corkwood *Hakea eyreana* and *H. suberea*
Proteaceae

Flowers of these trees growing in arid areas of Australia contain considerable amounts of sweet nectar that can be sipped with a straw or mixed with water to produce a beverage. Aborigines also ate the seeds of the fork-leaved corkwood.

Desert cynanchum *Cynanchum floribundum*
Asclepidaceae

Unripe pods of this shrub, growing in desert zones in Australia, are eaten raw by Aborigines. Older pods and leaves are steamed and eaten.

Kurrajong *Brachychiton* spp.
Sterculiaceae

Aborigines eat the seeds of several of these species raw or after having roasted them to remove the yellow hair surrounding the seeds, which is an irritant. In addition, the young tuberous roots of some species have been a popular food item with indigenous peoples of Australia.

Shepherd's purse (*Capsella bursa-pastoris*). USDA-NRCS PLANTS Database/Britton, N.L., and A. Brown. 1913. *Illustrated flora of the northern states and Canada.* Vol. 2, p. 158.

Lilly pilly *Syzygium australe* & spp.
Myrtaceae
Fruits of *Syzygium australe, S. luehmanii, S. oleosum,* and *S. paniculatum* are traditionally eaten fresh in Australia, or used in modern times for jellies, syrups, tarts, and puddings. *Syzygium paniculatum* (brush cherry) fruits would have been the first plant eaten by Captain Cook at Botany Bay. *S. luehmanii* (riberry) has a distinctive clove-like flavor and is eaten as an accompaniment for emu, kangaroo, and wallaby meat.

Macrozamia *Macrozamia* spp.
Zamiaceae
Several species of *Macrozamia*, all endemic in Australia, are an important food source for Australian Aborigines, once the plant has been processed to remove toxins. The crushed seeds can be soaked in water, where they break down and the poison is dissolved. Alternatively the seeds can be dried out and then leached in running water for 3 to 5 days. Aging the seeds is also sufficient to remove the toxicity so the seeds can then be eaten raw or cooked. More recently *M. spiralis* has been used for the production of alcohol and adhesive pastes and the manufacture of laundry starch.

Nicobar Islands breadfruit *Pandanus leram*
Pandanaceae
The fruits of this species, native to Southeast Asia and Polynesia, are gathered by the indigenous people, baked in hot sand or ashes, and the pulp is eaten. Occasionally the pulp is eaten raw, or is beaten from the fruits and soaked for a few days to make a mild alcoholic drink.

Noni, Indian mulberry *Morinda citrifolia*
Rubiaceae
Noni, the fruit of *M. citrifolia*, is traditionally eaten in its native area of Indonesia and Polynesia and today also worldwide. Noni fruits were exported by Polynesians at first to Tahiti, then to Hawaii, and from there noni juice reached the continental United States, where it is now a popular ingredient of many health food supplements. A large number of medicinal properties (to treat arthritis, diabetes, high blood pressure, muscle aches and pains, menstrual difficulties, mild and severe headaches, heart disease, AIDS, cancers, gastric ulcers, sprains, mental depression, senility, poor digestion,

arteriosclerosis, blood vessel problems, drug addiction, and more) have been claimed, but the medicinal properties of the plant have still to be researched.

Water lily *Nymphaea* spp.
Nymphaeaceae

In Australia a few species of *Nymphaea* (especially *N. gigantea* and *N. violacea*) are gathered from the wild by Aboriginal women. They collect the tubers, which are eaten roasted (they need to be leached in water several times before being eaten), and also the buds and flower stalks, which are commonly eaten raw. The unripe pods are used in traditional foods: they are first roasted, then the tiny seeds are extracted and eaten, or ground into flour.

See: Psychoactive Plants, p. 203; The Hunter-Gatherers, p. 9

Yam *Dioscorea* spp.
Dioscoreaceae

The large, fleshy, tuberous roots of several species of *Dioscorea* are cultivated today in many tropical countries. The majority of the species originated in Oceania and southeast Asia, and many of them still grow wild and are gathered by local peoples with digging sticks. The air potato or bitter yam *D. bulbifera* is the focal point of a well-known ceremony—known as *kulama*—of the Tiwi of Australia. The roots of this species (as of many other yam species) are poisonous and have to be prepared carefully to remove the poisons. In the *kulama* ceremony, while the yams soak in fresh water, the earth oven is prepared by pushing sand and grass outward from the center of the ceremonial ground and digging a large hole. Dry sticks about 3 feet (1 m) long are pushed upright into the ground around the oven and a fire is built up of sticks, grasses, and crumbled termite mounds; when the fire has burnt down to a bed of coals the yams are placed in the coals and covered with paper bark and sand. On the third day the yams are eaten. During this feast many new songs and dances are performed; it was traditionally one of the duties of new initiates to create new dances and songs.

See: The Hunter-Gatherers, p. 9; Roots and Tubers, p. 66; Origins and Spread of Agriculture, pp. 19, 20, 22, and 23

Europe

Borage *Borago officinalis*
Boraginaceae

In the culinary traditions of some Mediterranean areas, aerial parts of borage are the main ingredient of boiled mixtures of greens, generally used in soups (as in the *Prebuggiun* or wild greens of Genoa, in northwest Italy), or sometimes fried in olive oil and garlic (*erbucci*). The cultural use of borage in consumption of wild greens seems to mirror the spread of olive tree cultivation along many coastal areas, as has been modeled in areas bordering northwestern Tuscany and Liguria (central northern Italy). In central Europe, borage is often cultivated in gardens and the young leaves are used in mixed salads to add their distinctive cucumber taste. It is also quite common for the blue flowers to be used as a decoration for salads and desserts.

See: Herbs and Vegetables, pp. 99–100

Cow parsnip, Hogweed, Eltrot *Heracleum sphondylium* (synonym *Heracleum lanatum*)
Apiaceae

The aerial parts of wild cow parsnip have been used for a long time in central and eastern Europe and were the original ingredients of the famous Russian and Polish sour soup *borscht* (or *barszcz*). This soup was originally made by heating up the liquid that resulted from the natural lactic fermentation of the aerial parts of *H. sphondylium* (similar to the German tradition of fermenting a few varieties of cultivated *Brassica oleracea* and producing sauerkraut). In eastern Europe the name of this soup and of the plant, *H. spondylium,* are in fact the same—*barszcz.* In the past, particularly during times of famine, the succulent stems of cow parsnip have been gathered from the wild, eaten as green vegetables, or even transformed into a low-alcohol fermented drink, *raka.* The young stems were also used as a vegetable by western North American natives and occasionally gathered and eaten in the outer Hebrides. Today consumption seems to be restricted to a few areas in Siberia.

Lesser calamint *Calamintha nepeta*
Lamiaceae

Lesser calamint grows south of the Alps and is sometimes referred to as having "magic" aromatizing properties. Lesser calamint is the most important aromatic wild herb in central Italian cookery, and is used for cooking wild mushrooms (especially *Boletus edulis*) and cultivated zucchini. In Basilicata (southern Italy) lesser calamint is added to rennet during the making of a goat's cheese called *casieddu,* characterized by its unique wild mint taste derived from the essential oils of *Calamintha nepeta.*

Perennial wall rocket, Wild arugula *Diplotaxis tenuifolia*
Brassicaceae

This variety of arugula (rocket) is gathered and eaten raw in southeastern Italy (Apulia) and France (Languedoc). In Apulia it is often sold in local markets during the spring, and is the most common wild vegetable in southern Italy, used in salads or added to homemade pasta (*orecchiette*). Occasionally, the aerial parts of *D. erucoides* are collected from the wild and consumed in the Mediterranean area.

See: Herbs and Vegetables, p. 113

Spanish oyster *Scolymus hispanicus*
Asteraceae

Open-air markets selling *S. hispanicus* still survive today in some Mediterranean areas. This wild herb, which has a mild artichoke flavor, has been used for many centuries in cooking throughout the Mediterranean region. The young leaves are removed by hand and only the tender leaf stalks are cooked. In a few southern Italian communities it is traditionally gathered only during Holy Week, and cooked in a pie with lamb meat, cheese, ricotta, and eggs, to be eaten on Easter Day.

Tassel hyacinth *Muscari comosum*
Hyacinthaceae

Gathering wild bulbs of tassel hyacinth is still a common practice in Greece and in Apulia and Basilicata in Italy. In southern Italy the bulbs are traditionally eaten fried in olive oil having been soaked in cold water overnight to remove their bitterness, or pickled in olive oil. The eating of these bulbs spread to northern Italy with the labor emigration during the 1960s. Nowadays it is possible to buy the bulbs (mainly from North Africa) in small open-air markets of Florence, Milan, and parts of Germany and Switzerland if there is a sizeable community of southern Italians. Pliny refers to them being eaten with vinegar, oil, and *garum* (the characteristic sauce of the ancient Romans, made from fermented fish).

Wild asparagus *Asparagus acutifolius* and related species
Asparagaceae

Collecting wild asparagus (mainly *Asparagus acutifolius,* but also *A. albus, A. aphyllus, A. stipularis,* and *A. verticillatus*) during spring is a very common pastime for the rural population in central Spain, southern France, and central and southern Italy. Young shoots of wild asparagus are often sold in local markets and—being relatively expensive—provide an additional source of income for rural populations. Young shoots of other *Asparagus* species are also collected from the wild and consumed in Asia (*A. acerosus*) and in southern Africa (*A. laricinus*).

See: Herbs and Vegetables, p. 113

Wild chicory, Blue sailor *Cichorium intybus*
Asteraceae

Wild chicory has been used from time immemorial as a vegetable in the Mediterranean. However it was not until the 17th century that chicory was first described as cultivated. Cultivated varieties of chicory are now well known as vegetables. The roots have often been dried and ground for use as a coffee substitute. Young whorls of wild chicory are still gathered today and eaten cooked, in many regions of north Africa, southern Europe, and the Near East. The bitter taste of wild chicory is often claimed in folk cultures to be "healthy" and a "cleansing agent" for the blood, especially if the plant is consumed during the spring. Sometimes the water in which chicory has been boiled is drunk and is believed to be a medicine.

See: Herbs and Vegetables, p. 117

Wild fennel *Foeniculum vulgare* subsp. *piperitum*
Apiaceae

Although today the cultivated edible form of fennel characterized by its broad white, sweet leaf stalks and bulb is widely grown, collecting wild fennel to eat is an important activity in many Mediterranean areas. Young shoots of wild fennel are the main ingredient of the well-known Sicilian dish *pasta con le sarde* (noodles with fresh sardines), and fennel seeds are collected during the fall and used to flavor homemade sausages.

See: Herbs and Vegetables, p. 103; Plants as Medicines, p. 214

References and Further Reading

General

Couplan, F. 1998. *The Encyclopedia of Edible Plants of North America*. New Canaan, CT: Keats Publishing.

Etkin, N.L. (Ed.). 1994. *Eating on the Wild Side: The Pharmacologic, Ecologic, and Social Implications of Using Non-Cultigens*. Tucson: University of Arizona Press.

Etkin, N.L. 1996. Medicinal cuisines: diet and ethnopharmacology. *International Journal of Pharmacognosy 34*, 313–326.

Facciola, S. 1998. *Cornucopia II—A Source Book of Edible Plants*. Vista, CA: Kampong Publications.

Gibbons, E. 1987. *Stalking the Wild Asparagus*. Chambersburg, PA: Alan C. Hood.

Howard, P.L. (Ed.). 2003. *Women & Plants. Gender Relations in Biodiversity Management & Conservation*. London and New York: Zed Books.

Johns, T. 1990. *With Bitter Herbs They Shall Eat It: Chemical Ecology and the Origins of Human Diet and Medicine*. Tucson: University of Arizona Press.

Johns, T. 1999. Plant constituents and the nutrition and health of indigenous peoples. In *Ethnoecology—Situated Knowledge, Located Lives*, edited by V.D. Nazarea. Tucson: University of Arizona Press.

Phillips, R. 1983. *Wild Food*. London: Macmillan.

Scoones, I., Melnyk, M., and Pretty, J.N. 1992. *The Hidden Harvest: Wild Foods and Agricultural Systems, A Literature Review and Annotated Bibliography*. London: International Institute for Environment and Development.

Sturtevant, E.L. 1972. *Sturtevant's Edible Plants of the World*, edited by U.P. Hedrick. New York: Dover.

Tanaka, T. 1976. *Tanaka's Cyclopedia of Edible Plants of the World*. Tokyo: Yugaku-sha.

Africa

Asfaw, Z., and Tadesse, M. 2001. Prospects for sustainable use and development of wild food plants in Ethiopia. *Economic Botany 55*, 47–62.

Etkin, N.L., and Ross, P.J. 1994. Pharmacological implications of "wild" plants in Hausa diet. In *Eating on the Wild Side*, edited by N.L. Etkin. Tucson: University of Arizona Press.

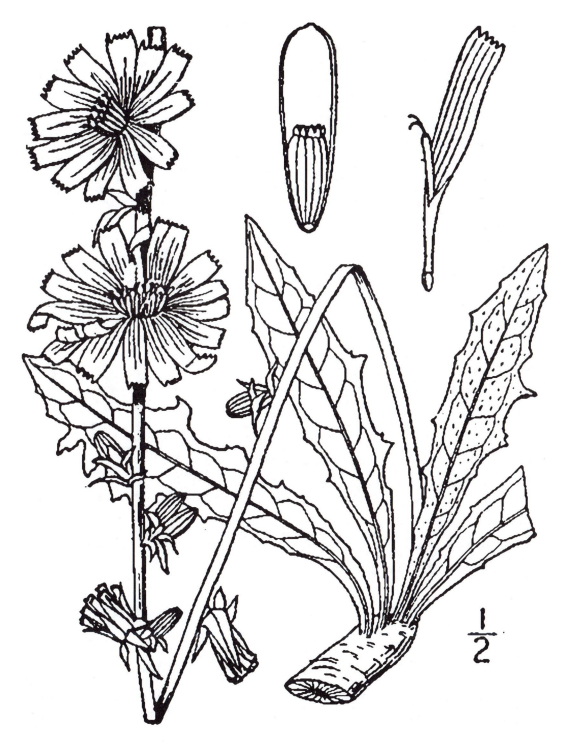

Wild chicory (*Cichorium intybus*). USDA-NRCS PLANTS Database/Britton, N.L., and A. Brown. 1913. *Illustrated flora of the northern states and Canada*. Vol. 3, p. 305.

Fleuret, A. 1979. The role of wild foliage plants in the diet: a case study from Lushoto, Tanzania. *Ecology of Food and Nutrition 8,* 87–93.

Fox, F.W., and Young, M.E.N. 1982. *Food from the Veld: Edible Wild Plants of Southern Africa Botanically Identified and Described.* Johannesburg: Delta Books.

Grivetti, L.E., and Ogle, B.M. 2000. Value of traditional foods in meeting macro- and micronutrients needs: the wild plant connection. *Nutrition Research Review 13,* 31–46.

Humphrey, C.M., et al. 1993. Food diversity and drought survival. The Hausa example. *International Journal of Food Science and Nutrition 44,* 1–16.

Johns, T., and Kokwaro, J.O. 1991. Food plants of the Luo of Siaya District, Kenya. *Economic Botany 45,* 103–113.

Johns, T., Mhoro, E.B., and Sanaya, P. 1996. Food plants and masticants of the Batemi in Ngorongoro District, Tanzania. *Economic Botany 50.* 115–121.

Marshall, F. 2001. Agriculture and use of wild and weedy greens by the *Piik ap Oom* Okiek of Kenya. *Economic Botany 55,* 32–46.

Maundu, P.M., Ngugi, G.W., and Kabuye, C.H.S. 1999. *Traditional Food Plants of Kenya.* Nairobi: Kenya Resource Centre for Indigenous Knowledge, National Museums of Kenya.

Ogle, B.M., and Grivetti, L.E. 1985. Legacy of the chameleon: edible wild plants in the Kingdom of Swaziland, southern Africa. A cultural ecological, nutritional study. Part I—Introduction, objectives, methods, Swazi culture, landscape and diet. Part II—Demographics, species, availability and dietary use, analysis by ecological zone. Part III—Cultural and ecological analysis. *Ecology of Food and Nutrition 16,* 193–208; *17,* 1–30; *17,* 31–40.

Ogoye-Ndegwa, C., and Aagaard-Hansen, J. 2003. Traditional gathering of wild vegetables among the Luo of Western Kenya—A nutritional anthropology project. *Ecology of Food and Nutrition 42,* 69–89.

Shackleton, S.E., et al. 1998. Use and trading of wild edible herbs in the Central Lowveld savanna region, South Africa. *Economic Botany 52* (3), 251–259.

Vainio-Mattila, K. 2000. Wild vegetables used by the Sambaa in the Usambarë Mountains, NE Tanzania. *Annales Botanici Fennici 37,* 57–67.

van Wyk, B.-E., and Gericke, N. 1999. *People's Plants: A Guide to Useful Plants of Southern Africa.* Arcadia: Briza.

von Maydell, H.-J. 1990. *Trees and Shrubs of the Sahel: Their Characteristics and Uses.* Weikersheim: Josef Margraf.

Americas

Bye, R.A. 1981. Quelites—ethnoecology of edible greens—past, present, and future. *Journal of Ethnobiology 1,* 109–123.

Cheatham, S., and Johnston, M.C. 1995. *The Useful Wild plants of Texas, the Southeastern and Southwestern United States, the Southern Plains and Northern Mexico.* Austin, TX: Useful Wild Plants.

Couplan, F. 1998. *The Encyclopedia of Edible Plants of North America.* New Canaan, CT: Keats Publishing.

Dunmire, W.W., and Tierney, G.D. 1997. *Wild Plants and Native Peoples of the Four Corners.* Santa Fe, NM: Museum of New Mexico.

Elias, T.S., and Dykeman, P.A. 1990. *Edible Wild Plants: A North American Field Guide.* New York: Sterling Publications.

Hodgson, W.C. 2001. *Food Plants of the Sonoran Desert.* Tucson: University of Arizona Press.

Kuhnlein, H.V., and Turner, N.J. 1991. *Traditional Plant Foods of Canadian Indigenous Peoples: Nutrition, Botany and Use.* Philadelphia: Gordon and Breach Science Publishers.

Ladio, A.H. 2001. The maintenance of wild edible plant gathering in a Mapuche community of Patagonia. *Economic Botany 55,* 243–254.

Moerman, D.E. 1998. *Native American Ethnobotany.* Portland, OR: Timber Press, 1998.

Turner, N. 1997. *Food Plants of Interior First Peoples.* Vancouver: University of British Columbia Press.

Turner, N. 2003. *Food Plants of Coastal First Peoples.* Vancouver: University of British Columbia Press.

Vierya-Odilon, L., and Vibrans, H. 2001. Weeds as crops: the value of maize field weeds in the Valley of Toluca, Mexico. *Economic Botany 55,* 426–443.

Asia

Aitchison, J.E.T. 1890. Notes on the products of western Afghanistan and of north-eastern Persia. *Transactions of the Botanical Society of Edinburgh 18,* 1–228.

Ertug, F. 2000. An ethnobotanical study in Central Anatolia (Turkey). *Economic Botany 54,* 155–182.

Johnson, N., and Grivetti, L.E. 2002. Environmental change in Northern Thailand: impact on wild edible plant availability. *Ecology of Food and Nutrition 41,* 373–399.

Johnson, N., and Grivetti, L.E. 2002. Gathering practices of Karen women: questionable contribution to beta-carotene intake. *International Journal of Food Sciences and Nutrition 53.* 489–502.

Khasbagan, Huai, H.-Y., and Pei, S.-J. 2000. Wild plants in the diet of Arhorchin Mongol herdsmen in inner Mongolia. *Economic Botany 54,* 528–536.

Leimer-Price, L. 1997. Wild plant food in agricultural environments: a study of occurrence, management, and gathering rights in Northeast Thailand. *Human Organization 56,* 209–221.

Lev-Yafun, S, and Abbo, S. 1999. Traditional use of A'kub (*Gundelia tournefortii,* Asteraceae), in Israel and the Palestinian Authority area. *Economic Botany 53,* 217–223.

Moreno-Black, G., et al. 1996. Non-domesticated food resources in the marketplace and marketing system of Northeastern Thailand. *Journal of Ethnobiology 16,* 99–117.

Ogle, M.B., et al. 2003. Food, feed or medicines: the multiple functions of edible wild plants in Vietnam. *Economic Botany 57,* 103–117.

Pemberton, R.W., and Lee, N.S. 1999. Wild food plants in South Korea; market presence, new crops, and exports to the United States. *Economic Botany 50* (1), 57–70.

Tukan, S.K., et al. 1998. The use of wild edible plants in the Jordanian diet. *International Journal of Food and Nutrition 49,* 225–235.

Australia and Oceania

Bindon, P. 1997. *Useful Bush Plants.* Perth: Western Australian Museum.

Brooker, S.G., Cambie, R.C., and Cooper, R.C. 1988. *Economic Native Plants of New Zealand.* Christchurch: Botany Division, D.S.I.R.

Cherikoff, V. 1997. *The Bush Food Handbook: How to Gather, Grow, Process and Cook Australian Wild Foods.* Boronia Park, New South Wales: Bush Tucker Supply Australia.

Dixon, A.R., McMillen, H., and Etkin, N.L. 1999. Ferment this: the transformation of noni, a traditional Polynesian medicine (*Morinda citrifolia,* Rubiaceae). *Economic Botany 53,* 51–68.

Isaacs, J. 1997. *Bush Food: Aboriginal Food and Herbal Medicine.* The Rocks, New South Wales: Lansdowne.

Parham, B.E.V. 1972. *Plants of Samoa: A Guide to their Local and Scientific Names with Authorities; with Notes on their Uses, Domestic, Traditional and Economic.* Wellington: Department of Scientific and Industrial Research.

Walter, A. 1999. *Fruits d'Océanie.* Paris: Institute de recherche pour le développement.

Europe

Bonet, M.A., and Vallès, J. 2002. Use of non-crop food vascular plants in Montseny biosphere reserve (Catalonia, Iberian Peninsula). *International Journal of Food Sciences and Nutrition 53,* 225–248.

Casoria, P., Menale, B., and Muoio, R. 1999. *Muscari comosum,* Liliaceae, in the food habits of south Italy. *Economic Botany 53,* 113–117.

Couplan, F. 1989. Le Régal Végétal, Plantes Sauvages Comestibles—Encyclopédie des Plantes Comestibles de l'Europe—Volume 1. Flers, France: Éditions Équilibres.

Forbes, M.H.C. 1976. Gathering in the Argolid: a subsistence subsystem in a Greek agricultural community. *Annals New York Academy of Sciences 268,* 251–164.

Guarrera, P.M. 2003. Food medicine and minor nourishment in the folk traditions of Central Italy. *Fitoterapia 74,* 515–544.

Koschtschejew, A.K. 1990. *Wildwachsende Pflanzen in unserer Ernährung.* Leipzig: VEB Fachbuchverlag.

Maurizio, A. 1927. *Die Geschichte unserer Pflanzennahrung.* Berlin: Verlagsbuchhandlung Paul Parey.

Pieroni, A. 1999. Gathered wild food plants in the upper valley of the Serchio river (Garfagnana), central Italy. *Economic Botany 53,* 327–341.

Pieroni, A. et al. 2002. Ethnopharmacology of liakra, traditional weedy vegetables of the Arbëreshë of the Vulture area in southern Italy. *Journal of Ethnopharmacology 81,* 165–185.

Riviera Núñez, D., and C. Obón de Castro. 1991. La Guía de incafo de las plantas útiles y venenosas de la Península Ibérica y Baleares. Madrid: INCAFO.

4
Grains

MARK NESBITT

Grains are the edible, starchy seeds (strictly, *caryopses*) of grasses. They account for over half of the world's food energy, or even more counting grain consumed indirectly as animal food. Cereals are the dominant food for a number of reasons. Diverse members of the grass family (Poaceae) will grow in virtually all the world's human habitats. Cereals are not only easily grown by farmers, but are also very nutritious. The cereal grain consists of the starchy endosperm; the oil-, vitamin-, and mineral-rich aleurone layer surrounding it; and the outermost, fibrous layers of bran. The embryo (germ) is also rich in oil. The grain has excellent keeping qualities prior to milling, enhanced in many species by the closely-fitting or attached husk (in some species the glumes, in others the lemma and palea). Grains generally lack the toxins so prevalent in the pulses, and nutritional deficiencies of species such as corn (maize) and sorghum can be mitigated through fermentation. Malting—the deliberate sprouting then killing of grains—results in the conversion of starch to sugar and is another powerful tool for enhancing the nutritional value of grains.

The domesticated cereals are often classified as temperate or tropical. Many of the temperate cereals, which fail to thrive at high temperatures, can be grown in the tropics, but only at high altitudes. Various of the tropical cereals, including corn, sorghum, and the millets, have C4 metabolism, based on a distinctive type of leaf anatomy and a modified photosynthetic pathway, which is well adapted to the high light levels of the tropics. Some tropical cereals, including sorghum and corn, are also viable crops in warmer temperate areas. Almost all cereals are annuals, sown in the autumn or spring and harvested in the summer. The leaves and stems of cereals are often as important as the grain, for animal feed, fuel, and construction materials such as thatch.

Rice, corn, and wheat are by far the most important of any crops, each accounting for about 600 million metric tons of grain each year (total world cereal production is 2000 million metric tons). These are the crops that underpinned the "green revolution" of the 1960s. New cultivars of cereals that responded well to increased inputs of water and fertilizer were developed. As a result, global agricultural production outpaced the doubling of the world's population that occurred between 1950 and 1990. The less important species of cereals are now attracting increased interest. Similar efforts applied to arid-land crops such as sorghum may benefit areas bypassed by the green revolution. Minor crops are important for other reasons; they are often well-adapted to cold or dry areas unsuitable for the major crops, offering a means of subsistence for farmers who would otherwise

Rice planting, Philippine Islands, ca.1890. Library of Congress, Prints & Photographs Division LC-USZ62-113571.

migrate to cities. Minor cereals frequently have important local culinary or religious significance, representing cultural diversity that, like biological diversity, should be preserved.

Domesticated cereals differ from their wild ancestors in a number of features (see Table 2 in Conservation of Crop Genetic Resources). In particular, grains are usually larger in domesticated forms, and the ears retain the grain until harvest, rather than shattering and releasing the grain, as in wild cereals. The different parts of the cereal ear usually survive charring or desiccation and are thus often well-preserved in archaeological deposits. By studying characteristics relating to grain size and dissemination, the domestication status of cereals can usually be established. In the Near East intensive archaeological and genetic studies have resulted in a good understanding of the domestication of wheat, barley, and rye (see Origins and Spread of Agriculture).

In other areas, such as Mesoamerica, south and east Asia, and Africa, both the archaeological records and our knowledge of the wild ancestors of cereals are poor. The problem is greatest in Africa, explaining why many cereals of African origin have their earliest archaeological records outside Africa, for example in the better-explored Indian subcontinent. A similar problem exists in Mexico, where no plant remains have been collected from early farming sites in the mountainous area where teosinte, the wild ancestor of corn, grows. The earliest corn yet found, from valleys outside this area, must substantially postdate the time of domestication.

All the species covered in this chapter belong to the grass family (Poaceae), except for three genera classified as pseudocereals: amaranths, buckwheat, and quinoa. These have small starchy seeds and hence are often considered as grains in culinary and agricultural terms.

Grass based cereals (Poaceae)

Archaeological evidence shows that in some areas, such as the Near East, the collection of wild grains for food ceased soon after the beginning of farming. Grains could be more easily obtained by cultivating domesticated cereals than by gathering from the wild. In other parts of the world, wild grains have continued to be important foods to the current day. This is mainly the case in areas too wet or too dry for farming, such as sea- and lakeshores and desert margins. In these areas wild grasses are still important resources, sometimes for hunter-gatherers, but more often as a supplementary resource for farmers. A wide range of grasses is used, often harvested from mixed stands of different species. Examples are given here of grasses harvested in the past, particularly in Europe (where their use has long since ceased) but also in North America, Australia, and sub-Saharan Africa, where some are still major food plants.

Crabgrass *Digitaria sanguinalis*

Crabgrass has a long history of collection as a wild grain in Europe, and came late to cultivation. It was domesticated in eastern Europe in the Middle Ages, and grown by Slavic peoples in a belt stretching from eastern Germany to the Ukraine for use in porridge and soups. However, its cultivation had ceased by the 20th century. Wild *Digitaria* species are today more important as forage grasses or for erosion control.

Floating sweetgrass *Glyceria fluitans*

This is a wild perennial grass of wet or marshy habitats. Known as manna grass or sugar grass, on account of its slightly sweet grains, it was extensively harvested in central Europe and Sweden up until the mid-19th century. The small grains were appreciated for their sweet taste and traded as far away as England. For consumption, the grains were pounded in a wooden mortar and eaten as gruel.

Lyme-grass *Leymus arenarius*

Lyme-grass was an important wild grain in Iceland from at least the 12th century, its use continuing until the mid-20th century. This perennial grass grows on the sandy shores of Iceland's southern coast, and was harvested with small sickles and threshed by lashing the sheaves against a wall.

After parching in a kiln, the resulting brittle florets were dehusked by treading in a cask. The grains were then ground, mixed with milk or whey, and consumed as an unbaked dough. Such wild resources were important until very recently because Iceland is too cold for most crops.

Maygrass *Phalaris caroliniana*

An annual grass of the southeast United States, maygrass has small seeds that are often found on prehistoric sites in the region, from 4500 ^{14}C years ago. Although there is only one ethnographic report of maygrass use by native Americans, its use as human food in the past is shown by its occurrence in many prehistoric coprolites (desiccated human feces) from cave sites in Kentucky. Prehistoric finds of maygrass occur well north of its current distribution, for example in Kentucky and Tennessee, suggesting that although never domesticated, maygrass might have been cultivated.

Panic grasses *Panicum* spp.

Although best known for the domesticated broomcorn millet (*P. miliaceum*, see later separate entry), the 500 species of the genus *Panicum* include several important wild cereals in dry areas of the southwest United States, the interior of Australia, and sub-Saharan Africa.

In the Sonoran desert of the American Southwest, seed was collected from wild populations of *Panicum sonorum* and *P. hirticaule* growing on the banks of the Colorado River. A large-seeded domesticated form of *P. sonorum*, which was sown in early summer on moist soils as the river subsided, also existed, although it was nearly extinct by the late 20th century. The grains were parched to remove the husk, then the seeds were ground to flour and used as sun-dried cakes. In the arid Great Basin, east of the Sierra Nevada mountains of California, a wide range of wild grasses was harvested, including species of *Panicum*, *Eragrostis* (see Teff later) and *Oryzopsis*. Many of these were sown, and the Owens Valley Paiute of the Great Basin used irrigation to enhance the yields of wild stands. Seeds were harvested by beating the grain into baskets and then grinding the seeds to flour.

In the dry interior of Australia, a similar range of wild grasses, including species of *Panicum*, *Setaria* (see Foxtail millet later), and *Eragrostis*, were recorded as being harvested by hunter-gatherers in the 19th century by hand-stripping of seeds or uprooting of whole plants. Vast ricks of uprooted grasses were observed along the Darling River. The seeds were wet-ground on a millstone, then formed into a cake for cooking.

Harvesting of wild grains is increasingly uncommon in North America (save for wild rice) and Australia, but in the Sahara desert and sub-Saharan Africa, large-scale use of wild grasses by farming communities is still an important food resource. *Panicum turgidum* (merkba, afezu) is used by the Tuareg peoples of southern Algeria. The grass is protected from grazing until after the grain is harvested; the seeds are used for porridge. *Stipagrostis pungens* (drinn, tessiya, tullult) and *Cenchrus biflorus* (kram-kram) are also used in the desert zone. South of the Sahara, different grasses are harvested from seasonally flooded lakes and swamps of west Africa. These include wild rice, *Oryza barthii*, and *O. longistaminata*, *Paspalum scrobiculatum* var. *commersonii*, and *Echinochloa stagnina*. In the savanna zone south of Lake Chad, wild grass grains are known as kreb, and derive from a mixture of grasses dominated by *Panicum* and *Eragrostis*.

See: Origins and Spread of Agriculture, p. 17

Wild rice *Zizania palustris*

Wild rice is an aquatic grass native to the Great Lakes region of eastern North America, entirely unrelated to the Asian and African rices (*Oryza*, see Rice later). It has been harvested for millennia by local native American peoples, including the Ojibway, Cree, Menomini, and Huron. It is traditionally harvested by flail-beating the grains into canoes. The harvested grain was dried, often over a fire, then dehusked by trampling. The grains were cooked in soups or boiled with fish, corn, or meat.

Cultivation of wild rice in diked fields began about 1950. In the 1960s wild rice was domesticated by plant breeders who selected shatter-resistant types that were less likely to shed grain before harvest. Domesticated wild rice is harvested by combine harvesters after the field is drained. Wild rice is an annual grass, but minimal resowing is required as even shatter-resistant cultivars shed some grain. Today about one-tenth of the harvest is from natural stands; the remainder is from fields, mainly in Minnesota and northern California. Of the three other species in the genus *Zizania*, *Z. aquatica* and *Z. texana* are small-seeded forms not harvested for food, and the Chinese species *Z. latifolia* is gathered as a wild grain in Manchuria.

See: The Hunter-Gatherers, pp. 10–12

Temperate cereals

Barley *Hordeum vulgare*

Although mainly grown today for animal feed and for brewing, barley was an important cereal food in the past. Even today, naked barley (a hull-less variety that is easy to process) is still a staple cereal in the Himalayas. The importance of barley in such high-altitude areas reflects barley's hardiness; like rye, it is a more dependable crop than wheat in areas that are particularly dry, wet, or cold, and in areas with soils affected by salinity.

The wild ancestor of barley is *H. spontaneum*, a large-seeded barley that grows alongside wild wheat in the Fertile Crescent of the Near East (see Figure 1 in Origins and Spread of Agriculture). Both wild barley and the first domesticated barley were similar in appearance, with two rows of grain on the ear, and both had hulled grains, in which the husk (lemma and palea) adheres to the grain. This husk is rich in silica and must be removed by pounding before consumption for food. In naked forms, the lemma and palea do not adhere to the grain and are easily removed by threshing.

Domesticated barley is abundant at Near Eastern archaeological sites from 9500 ^{14}C years ago, and the crop spread to Europe, Egypt, and south Asia from about 8000 ^{14}C years ago, with the other Neolithic crops of the Fertile Crescent. Six-rowed forms of barley, with three fertile spikelets on each node of the ear, appear at about 8000 ^{14}C years ago. Today barley is grown in most areas for animal feed, but the abundance of barley in archaeological deposits, often in kitchens, shows that it was an important human food, probably even outranking wheat, and was used for both porridge and bread. It is perhaps only since medieval times that barley's role as a staple food has disappeared in most regions. Food use of barley appears to have declined first in warmer regions such as classical Rome, and to have survived longest in northern, colder areas such as Scandinavia. In the Orkney and Shetland Isles, bere barley—distinctive six-row hulled barley—is still grown for the production of bannocks, a kind of soft flat bread.

Although naked forms of barley make sporadic appearances in the Fertile Crescent and Europe, it is only in East Asia that they became important foods. Although naked barley does not need dehusking prior to consumption, hulled barley has higher yields, and this may explain why it has stayed the dominant form.

See: Origins and Spread of Agriculture, pp. 14–17; Caffeine, Alcohol, and Sweeteners, p. 182

Canary grass *Phalaris canariensis*

Widely cultivated for bird feed, canary grass is a rare crop in Italy and, in the Canary Isles, where it is one of a number of cereals used to make a local cereal dish known as *gofio*. *Phalaris caroliniana* (maygrass, as described earlier) was probably cultivated as a cereal in prehistoric times in eastern North America.

Common millet, broomcorn millet *Panicum miliaceum*

Both the wild ancestor and the location of domestication of broomcorn millet are unknown, but it first appears as a crop in both Transcaucasia and China about 6000 ^{14}C years ago, suggesting that it may have been domesticated independently in each area. Like foxtail millet, it is a summer crop, sown in late spring and harvested in late summer. The grains are eaten whole after boiling, or ground into flour for porridge or bread. Broomcorn millet is an important bird feed in many countries. Sama (*P. sumatrense*) was domesticated in the Indian subcontinent and is present at Harappan sites from 3000 BC. It is cultivated in south Asia, particularly in the eastern Ghats of India, and eaten as whole grains or as bread.

See: Origins and Spread of Agriculture, p. 17

Foxtail millet *Setaria italica*

This millet of temperate regions was domesticated from its wild ancestor, *S. viridis*, in eastern Asia about 7000 ^{14}C years ago. Foxtail millet was a staple food in the Neolithic period in northern China, in contrast to the dominance of rice in the south. Genetic evidence suggests that foxtail millet may have been independently domesticated in central Asia and China. The major areas of production today are China and India. Foxtail millet is eaten as whole, boiled grains, or ground into flour. Like most other millets, it is a summer crop, sown in late spring and harvested after a growing season of as little as forty days. Foxtail millet is increasingly being displaced by broomcorn millet and other cereals. Yellow foxtail millet (*S. pumila*) is an Indian domesticate; it has been cultivated as a minor cereal in India since 2000 BC.

See: Origins and Spread of Agriculture, p. 17

Mango *Bromus mango*

Not to be confused with the tropical fruit, this cereal was cultivated in Chile and Argentina, but was displaced by corn, wheat, and barley after the Spanish Conquest. Cultivation ceased by about 1860. It was a biennial crop so only produced seed in its second year of growth. The Araucani Indians used it as a dual-purpose crop, as forage for grazing animals the first year and for grain the second year. After roasting, grains were ground into flour for bread, or for a fermented drink known as *chicha*.

Oats *Avena sativa*

Oats are descended from *A. sterilis*, a wild oat that spread as a weed of wheat and barley from the Fertile Crescent to Europe. In the wetter, colder conditions of Europe, in which oats thrive, it was domesticated about 3000 years ago, and soon became an important cereal in its own right on the cooler fringes of Europe. In medieval Britain oats were widely grown for bread, biscuits, and malting, but they now hold their importance only in the wetter parts of northern Europe. Oats are still an important food in Scotland, where uses include porridge, oatcakes, and the filling for haggis. Oats have also had an important role since the Roman period as feed for horses. British emigrants introduced oat cultivation to North America in the 17th century, but they have always been a minor cereal outside Europe. Oat bran is rich in a type of dietary fiber that has been shown to reduce cholesterol levels, and this has led to increased interest in its consumption.

Red oat or Turkey oat, *Avena byzantina*, is also descended from *A. sterilis*. Although sometimes considered to be the same species as *A. sativa*, red oats are genetically distinct and have a different distribution including, as the name suggests, Turkey. However, red oats have been largely replaced by *Avena sativa* in recent years.

Four minor cultivated species derived from wild forms of *Avena strigosa* grow in the western Mediterranean. Bristle oat, *A. strigosa*, is a fodder plant in central and northern Europe, still grown in the Shetland Islands but almost extinct. *A. brevis* and *A. hispanica* are now very rare crops of southwest Europe. *A. nuda* is a naked form of oat that threshes free of the tough husk. It has low yields and is not widely cultivated.

Avena abyssinica is only found in Ethiopia. It grows as a tolerated weed of other cereals, mainly barley, and is now in the course of domestication.

Rye *Secale cereale*

Rye was first domesticated in the Fertile Crescent of the Near East, with the earliest records dating to 8400 ^{14}C years ago, but did not become an established crop until the Bronze Age in Europe, about 3500 years ago. Like oats, rye only became important in the cold, wet conditions of central and northern Europe, in which it grows better than other cereals. It was used for bread and for thatching, and was (and in Turkey, still is) often planted with bread wheat as a mixture or *maslin*. In poor years the increased yield of the hardier rye compensates for the more vulnerable wheat. In western Europe rye was displaced by wheat as a staple cereal in the 18th century, and is now primarily grown for animal feed. Rye breads such as pumpernickel are still important in central Europe.

See: Origins and Spread of Agriculture, p. 16

Triticale × *Triticosecale*

Triticale results from the crossing of macaroni wheat or bread wheat with rye. It was originally a laboratory curiosity, discovered in the 19th century. Intensive breeding programs were carried out in the 1950s, in the hope of developing a new cereal species that would combine the grain quality of wheat and the hardiness of rye. However, problems with grain shriveling and with ergot (a fungus that produces harmful alkaloids) have slowed uptake as a crop. Triticale is grown today for animal feed, mainly in France, Poland, Russia, and Australia.

Wheat *Triticum* spp.

Wheat is by far the most important food grain of temperate regions. Its role in human subsistence is matched by the deep significance of wheat in religion and daily life. Wheat, in the form of bread, is central to Jewish and Christian rites. Although barley was domesticated at the same time, has higher yields, and was the most important cereal in antiquity, wheat has always been more highly valued, probably because of its better taste and more versatile culinary properties.

Papago winnowing wheat, ca.1907. Library of Congress, Prints & Photographs Division, Edward S. Curtis Collection.

The wheats divide into two groups, hulled and free-threshing. On threshing, the ear (spike) of hulled wheats breaks up into individual spikelets, in each of which one to three grains are tightly enclosed by tough husks or hulls (glumes). Before the grains can be consumed, they must be dehusked, traditionally by pounding in a mortar. In contrast, the glumes of free-threshing wheats are light and break away during threshing, releasing the naked grains immediately. Most wheats cultivated today are free-threshing; all wild wheats and many species cultivated in the past were hulled.

The earliest cultivated wheats were hulled forms. Einkorn wheat (*T. monococcum*) and emmer (*T. dicoccum*) were domesticated from wild einkorn (*T. boeoticum*) and wild emmer (*T. dicoccoides*) respectively, in the Fertile Crescent of the Near East. The earliest securely dated finds of domesticated einkorn and emmer are at Neolithic (Pre-Pottery Neolithic B) sites in Syria, Jordan, and southeast Turkey, dating from 9500 to 9200 ^{14}C years ago. Genetic evidence suggests that present-day einkorn and emmer derive from one or two domestications, probably in southeast Turkey. Emmer was the main wheat species grown by early farmers in Europe and the Near East and the only wheat grown in ancient Egypt, where emmer bread and beer were staple foods. Einkorn has always been less important. Because of their ability to thrive on poor soils and to resist fungal diseases, emmer and einkorn are still grown in wet mountainous areas stretching from the Pontic mountains of Turkey to the Carpathian mountains of eastern Europe, and in Italy and Spain. In Italy emmer is known as *farro* and cooked whole with beans or tomatoes to make soup. Einkorn is used as animal feed and for thatching, and in Turkey is favored as a food grain for cracked wheat (bulgur). Unlike emmer and spelt, einkorn has not become a popular health food, and it is in danger of becoming extinct as a crop.

Macaroni wheat (*T. durum*) is the free-threshing form of emmer wheat, with hard, flinty grains. It first appeared in the Near East about 9000 ^{14}C years ago. Macaroni wheat has always been important in Mediterranean areas. Its best-known use is for pasta, a food of uncertain origin, perhaps from the Arab world and not, as often claimed, brought back from China by Marco Polo. In the 19th century the Ukraine became the leading exporter of macaroni wheat for pasta making, but it lost this position to the United States during World War I. Macaroni wheat also makes a delicious bread that is a staple food in Sicily. Rivet wheat (*T. turgidum*) is a closely related species that is well adapted to the cooler conditions of northern Europe. It was popular during the medieval period, but bread wheats proved better adapted to the threshing machines introduced in the 19th century, and were better suited to industrialized baking. Rivet wheat has notably soft, floury grains, and produced a good bread-making flour that was mixed with flour from rye and bread wheat for daily baking. Miracle wheat (*T. turgidum* var. *pseudocervinum*) is a form of rivet wheat with branched ears, first recorded by the Roman author Pliny some 2000 years ago. Extravagant claims are often made for its yield but, sadly, miracle wheat has yields of grain well below those of ordinary wheats. It is often claimed that miracle wheat or mummy wheat derives from grains found in ancient tombs and subsequently germinated. However, except when stored at very low temperatures, cereal grains lose their ability to germinate within a few years or, at best, decades.

Spelt (*T. spelta*) is a hulled wheat that became widely cultivated much later than most other wheats. Genetic evidence shows that it is the result of the hybridization of emmer wheat and a wild goatgrass, *Aegilops tauschii*, some 8000 ^{14}C years ago near the Caspian Sea. However, spelt never became part of the Near Eastern crop complex that spread to Europe through the Balkans. Instead, it suddenly appears as an important crop in central Europe about 4000 years ago. Spelt displaced emmer as the major wheat of antiquity in Europe, before being displaced in turn by free-threshing wheat. Spelt has continued in local cultivation in Germany and Switzerland, where it is much appreciated for use in baking bread known as *Dinkelbrot*. Spelt grains are widely sold in health food shops in Europe and North America, and cultivation in western Europe continues to expand.

Spelt is often and mistakenly thought to have been grown in Pharaonic Egypt or the ancient Near East; this error results from confusion with emmer wheat, which was widely grown in both areas in antiquity.

Bread wheat (*T. aestivum*) is free-threshing and closely related to spelt. As with spelt, genes contributed from goatgrass (*Aegilops*) give bread wheat greater cold hardiness than most wheats, and it is cultivated throughout the world's temperate regions. Bread wheat is by far the most important wheat species today. Wheat first reached North America with Spanish missions in the 16th century, but North America's role as a major exporter of grain dates from the colonization of the prairies in the 1870s. As grain exports from Russia ceased in the First World War, grain production in Kansas doubled. Worldwide, bread wheat has proved well adapted to modern industrial baking, and has displaced many of the other wheat, barley, and rye species that were once commonly used for bread making, particularly in Europe. Compact wheats (*T. compactum*, in India *T. sphaerococcum*) are closely related, but have a much more compact ear, with spikelets packed closer together.

Modern wheat varieties have short stems, the result of *RHt* dwarfing genes that reduce the plant's sensitivity to gibberellic acid, a plant hormone that lengthens cells. *RHt* genes were introduced to modern wheat varieties in the 1960s from Norin cultivars of wheat grown in Japan. Short stems are important because the application of high levels of chemical fertilizers would otherwise cause the stems to grow too high, resulting in lodging (collapse of the stems). Stem heights are also even, important for modern harvesting techniques.

See: Origins and Spread of Agriculture, pp. 16 and 17

Tropical cereals

Adlay *Coix lacryma-jobi*

This species exists in two forms, Job's tears and Adlay. Job's tears have shiny bracts, and are often used as beads in botanical jewellery. Adlay (var. *ma-yuen*) has papery bracts and starchy grains, and is grown on a small scale as a food grain in east and southeast Asia. The grains are eaten whole in soup, or ground into flour and eaten as porridge or cakes. In Nagaland, northeast India, and in some parts of southeast Asia, adlay is used for brewing local beers. The fruits have a variety of traditional medicinal uses, including reputed anticancer properties. Cultivation is decreasing as adlay is replaced by rice and corn. The date and region of domestication are unknown, but adlay is found at archaeological sites in northeast India from about 1000 BC. The wild ancestor (var. *lacryma-jobi*), native to tropical Asia, is thick-shelled; the thin-shelled edible form evolved under domestication. Unlike most other cereals, adlay plants are cultivated on a small scale, often in home gardens. Each plant bears a large number of ears, which mature at different times; these are hand-picked as they mature.

See: Materials, pp. 343–344

Browntop millet *Brachiaria ramosa*

This rare millet is now only cultivated in a few dry areas of South India, but it is a good example of crop evolution in action. Three forms illustrate the transition from wild plant to crop: the wild form retains fully wild characteristics; a weedy form is grown as fodder and is harvested for grain in drought years; and the fully domesticated type, taller and with larger and non-shattering ears, is cultivated in pure stands. The grain is consumed as boiled whole grains, porridge, or unleavened bread. Today the browntop millet is threatened with extinction as farmers adopt new commercial crops, and food habits change.

See: Origins and Spread of Agriculture, p. 19

Corn, maize *Zea mays*

No other cereal has spread so widely or been used in such diverse manners as corn. In its homeland of Mesoamerica corn is the staple food, usually eaten as griddle-cooked *tortilla* bread. In North America, traditional corn landraces are still widely grown by native American peoples of the Southwest, but corn's greatest importance today is as an agro-industrial crop, mostly used for industrial starch, alcohol production, corn syrup, and animal feed. Corn was taken by the Portuguese in the 16th century to Africa and rapidly became established as a staple food over much of the continent. Today corn ranks second in world production behind rice. With its high yields, corn is likely to increase in importance in warm parts of the world.

The domestication of corn has long been controversial. The flowers on the cob of modern-day corn are arranged in a distinctive fashion, very different from that of its wild ancestor, teosinte (subsp. *parviglumis*). The female flowers are arranged in rows, which are pollinated through the long, thin styles (the "silk") that can be seen emerging from the cob. The female flowers mature into the corn grains that we eat. The male flowers grow together on the tassel at the tip of the corn stem, well placed to shower pollen onto the female cobs below. In contrast, teosinte has numerous much smaller cobs, with just two rows of grain, and a male inflorescence at the end of each cob. These differences led botanists to increasingly complex explanations of corn domestication, involving crossing of different species, or an extinct wild ancestor. However, recent studies of DNA have shown that teosinte is the only ancestor of corn. Teosinte grows today as a wild plant in river valleys of southern and western Mexico, and this area is believed to be where corn was first domesticated.

The earliest domesticated corn cobs have been found at archaeological sites in the Tehuacán and Oaxaca valleys of southern Mexico, dating to 6000 to 5000 ^{14}C years ago. Similar cobs are found on a primitive form of popcorn cultivated today on a small scale in Argentina. With its small cobs and branched stems, this variety is intermediate in appearance between teosinte and modern corn. By about 3000 years ago corn had spread from Mexico with early farmers, south to the Andes and north to eastern North America. Corn arrived in Europe and Asia after Columbus reached the Americas in 1492, and spread rapidly. Although many claims have been made for pre-Columbian dispersal of corn in the Old World, these are firmly contradicted by the complete absence of corn in the Old World archaeological record before 1492.

An astonishing range of landraces of corn has evolved in response to selection for culinary attributes, particularly in the southwest United States. Flour corn has large, soft, starchy grains, with blue, pink, or white flour used for cornmeal. Sweet corn is used for flour (*pinole*) or eaten on the cob. A genetic defect in metabolism prevents the sugars in the kernel from being fully converted to starch, hence the sweet taste. Popcorn has flinty, hard kernels that explode on toasting. The chemistry of popcorn is still poorly understood, but it appears that the high protein content of popcorn grains binds together the starch particles. When heated, the grain thus resists expansion, until the pressure of heated water vapor inside the grain forces it to explode.

In Mesoamerica the process of nixtamalization developed about 1500 BC, and is still widely practiced both in villages and in industrial food production. Corn grains are soaked, then cooked with lime or wood ash. This process both enables the husk (pericarp) of the grain to be easily removed, and improves the availability of a B vitamin, niacin, in the grain. Niacin deficiency leads to the disease pellagra, with symptoms including skin lesions and mental confusion. Pellagra was widespread in areas such as the southern United States, in which the diet of poor cotton shareholders depended heavily on corn, and the disease still occurs in parts of south Africa, Egypt, and India. Corn is also particularly deficient in two essential amino acids, lysine, and tryptophan. In the Americas *Phaseolus* beans complement corn by meeting this deficiency; elsewhere other pulses take the same role.

See: Age of Industrialization and Agro-industry, pp. 357–359; Origins and Spread of Agriculture, p. 21

Fonio *Digitaria* spp.

White fonio (*Digitaria exilis*) is the most important of a diverse group of wild and domesticated *Digitaria* species that are harvested in the savannas of west Africa. Fonio has continued to be important locally because it is both nutritious and one of the world's fastest growing cereals, reaching maturity in as little as six to eight weeks. Fonio is a crop that can be relied on in semi-arid areas with poor soils, where rains are brief and unreliable. The grains are used in porridge and couscous, for bread, and for beer. Black fonio (*Digitaria iburua*) is a similar crop grown in Nigeria, Togo, and Benin. Raishan (*D. compacta*) is a minor cereal, only grown in the Khasi hills of northeast India, with glutinous flour used to make bread or porridge.

Finger millet *Eleusine coracana*

Finger millet is an important tropical cereal in eastern and southern Africa and in India. It was domesticated in east Africa from *E. africana*, which grows in Ethiopia and Uganda. Claims of identifications of prehistoric seeds of finger millet from India and Africa are poorly documented, but it does appear to be one of the cereals, along with sorghum and pearl millet, that reached India in the second millennium BC. So far, the earliest certain record from Africa is from Axum, Ethiopia, dating to 600 AD.

Finger millet is not as drought tolerant as pearl millet and competes with corn for the best agricultural land in Africa. The grain is ground and made into porridge and a local bread, and it is also malted for beer. As the grain can be stored for long periods, it is an important crop during times of famine. Rich in calcium and iron, it is often recommended as a healthy food for pregnant women, children, and sick people. In recent years cultivation of finger millet has declined in its African heartland, causing concern that a valuable crop may be lost through lack of investment in plant-breeding programs. In India finger millet (known as ragi) has increased in importance as a dryland crop.

Guinea millet *Urochloa deflexa*

This millet is cultivated locally on the Fouta Djalon plateau of Guinea, west Africa. Liverseed grass (*U. panicoides*) is cultivated on a small scale in Gujurat, India. The grain stores well and is often kept as a buffer against famine years.

Kodo millet *Paspalum scrobiculatum*

Kodo millet was domesticated in India by 2000 BC, and is still cultivated in central India. It is generally regarded as an indigestible, low quality food, but is a useful crop on poor soils. The grains are consumed whole, like rice, or are roasted and ground into flour.

Pearl millet *Pennisetum glaucum*

The wild ancestor of cultivated pearl millet, *P. violaceum*, is harvested as a wild cereal during times of scarcity. Archaeological evidence suggests that it was harvested as a wild cereal before the advent of agriculture in tropical west Africa some 3000 to 4000 years ago. Genetic evidence points to west Africa as the most likely region of domestication; the earliest archaeobotanical finds are in Mauritania and Nigeria, dating to about 1000 BC. Sporadic records of pearl millet occur at Indian archaeological sites from about 2000 BC.

Pearl millet is an important grain crop in Africa and India. It tolerates drought and heat and grows mainly under rainfed conditions, and thus has a vital role on land too dry for sorghum or corn. It is mostly consumed as a porridge or gruel in Africa and as flat unleavened bread in India. The stalks are used for thatching and building. In other parts of the world (the United States, Canada, and Australia) it is grown as a green fodder crop and as feed grain for animals. As with finger millet, pearl millet is in danger of being displaced by crops, such as corn, that have been the subject of successful crop improvement programs.

See: Origins and Spread of Agriculture, p. 19

Rice *Oryza sativa*

Rice is the staple food of about half the world's population, mainly in Asia, and is now cultivated in most areas that have abundant water and hot summers. The history of rice domestication is still unclear. Its wild ancestor, *O. rufipogon*, grows throughout south and southeast Asia, and finds of rice at early sites in the region may derive from harvesting of wild grain. Dating of sites is often unclear. The earliest records of domesticated rice are probably those from the Yangtze river valley of southern China, dating to about 8000 to 6000 ^{14}C years ago. There are two main groups of rice varieties: the *japonica* group, which has short grains, and the *indica* group, which has long grains. Genetic evidence suggests that each group may have been independently domesticated. However suggestions that *indica* rice was domesticated in northeast India are not supported by archaeological evidence, which dates the first rice cultivation in the Indian subcontinent to about 2500 BC. It was cultivated in the Near East in the Hellenistic period (from 300 BC) and was traded throughout the Roman empire. However its cultivation in southern Europe did not begin until the medieval period. Rice was also a late arrival in Japan, coming from Korea at the beginning of the Yayoi culture in 400 BC. Archaeological and historical evidence suggests that rice spread slower than most crops, in part owing to its specialized need for abundant water. For example, rice did not become an important crop in North America until the late 17th century.

In dry or upland cultivation, rice is grown on hillsides as a rainfed crop similar to other cereals. Dry cultivation is mainly important in South America and Africa. In wet or lowland systems, most important in Asia, rice is grown on irrigated or flooded paddies. The seeds are often sown in a nursery and then transplanted into paddies. Deep-water rice is grown in water 30 inches (50 cm) or more deep, and is important in Bangladesh and other areas with deeply flooded river valleys. The rice plant has fast-growing stems that grow in pace with the rising water, up to a maximum of 14 feet (4 m). Much of the nutrition for deep-water rice is provided by the silt deposits borne by the flood water.

After harvest, rice must be dehusked to remove the inedible hull (lemma and palea). This is often carried out using a wooden mortar and pestle. The resulting grains are usually eaten as white, polished rice from which the bran has been removed. Whole "brown" rice is mainly popular as a "health food" in Western countries. Rice is most often consumed as whole grains boiled or steamed in water. Rice flour lacks gluten and so is usually consumed as noodles; this absence of gluten results in poor quality bread. Rice cultivars with starch that is low in amylose are waxy or glutinous, and are used industrially as a thickening agent for sauces and puddings, and in East Asia for snack foods such as rice crackers and cakes. In India rice is often parboiled prior to dehusking. This partially cooks the starch, which eases dehusking and milling.

Red rice (*O. glaberrima*) was an important locally domesticated cereal in west and central tropical Africa, but is increasingly being replaced by the Asian rice (*O. sativa*), which reached Africa in the 16th century. The wild ancestor of red rice, *O. barthii*, was probably domesticated in the valley of the Niger river 2000 to 3000 years ago.

Sawa millet *Echinochloa frumentacea*

This millet is widely grown as a cereal in India, Pakistan, and Nepal. Its wild ancestor is the tropical grass *E. colona*, but the exact date or region of domestication is uncertain. It is cultivated on marginal lands where rice and other crops will not grow well. The grains are cooked in water, like rice, or boiled with milk and sugar. Sometimes it is fermented to make beer. The closely related Japanese barnyard millet (*Echinochloa esculenta*) is cultivated on a small scale in Japan, China, and Korea, both as a food and for animal fodder. It is grown in areas where the land is unsuitable or the climate too cool for paddy rice cultivation. However, the development of rice varieties that can withstand cold has led to a sharp decline in the cultivation of sawa millet, in favor of that rice. Sawa millet's wild ancestor is *Echinochloa crus-galli*, a widespread temperate grass; the earliest records of the domesticated form date to 2000 BC from the Jomon period of Japan.

Sorghum *Sorghum bicolor*

Sorghum is an ancient staple food plant in Africa and Asia, and a relatively recent introduction to the Americas, where it is used as animal feed and for industrial purposes. Sorghum's wild ancestor is *S. verticilliflorum*, a wild grass that grows throughout tropical Africa. The earliest archaeobotanical records of domesticated sorghum in Africa are from Sudan and Cameroon, dating to 100 to 500 AD. However, finds from India that show sorghum had reached south Asia by 2000 BC indicate that domestication must have occurred earlier in Africa. Botanical evidence points to Ethiopia or Chad as the most likely area of origin.

Domesticated sorghums are highly diverse in appearance and food properties, but can be grouped into five broad races or groups within the species *S. bicolor*. *Durra* is the most important race in India and is widely grown in arid sub-Saharan Africa, as is *Caudatum*. *Guinea* sorghum is grown in humid west Africa, *Kafir* sorghums in southern Africa, and the low-yielding, rather primitive, *Bicolor* sorghums on a small scale in Africa and Asia. Sorghum is the most important cereal in sub-Saharan Africa, where it has proved highly adaptable to extremes of temperature and drought. However, like corn (*Zea mays*), the grain needs processing to improve its nutritional value. About sixty percent of its protein is in the form of prolamine, of low nutritive value. Additionally, dark-seeded forms contain tannins that protect them from bird predation but taste bitter and interfere with digestion. To counter these, sorghum is often consumed in fermented form, as beer or sourdough bread, as fermentation breaks down tannins and prolamine to more digestible forms. Sorghum is also eaten as porridge, dumplings, breads, and whole grains.

Sorghum was introduced to the Americas in the 18th century, and is an important crop in the Great Plains of North America, in Mexico, and in Argentina. In these countries intensive breeding and production have led to high yields of grain, used primarily for animal feed. Forms of sorghum in the *Bicolor* group with sugar-filled stems (sometimes known as sorgho or *S. saccharatum*) were introduced to the southern United States in the mid-19th century. For a hundred years sorghum molasses was a popular sweetener in the region; it remains popular in China. Another *Bicolor* sorghum is broomcorn. As its name suggests, this has very long bristles on the flower head, which are used in the manufacture of brooms. It is first recorded in 16th-century Europe, and is still important in southern Europe and the United States.

Sorghum has not yet attracted the same amount of research as the other major cereals, wheat, rice, and corn. However, with rapidly increasing interest in its food and industrial properties, cultivation of this crop, ranking fourth amongst cereals in world production, is likely to expand. In addition to their importance as foods, sorghum stems and leaves are important sources of animal fodder and building materials.

See: Origins and Spread of Agriculture, p. 19; Caffeine, Alcohol, and Sweeteners, p. 181

Teff *Eragrostis tef*

Teff is native to and (until recently) only grown in Ethiopia, where it is the most important and ancient cereal, grown at altitudes from 4300 to 9200 feet (1300 to 2800 m). Some 2000 varieties are known. The grain is ground to a flour used for making a bread, *injera*, the staple food of Ethiopia.

The grains of teff are unusually small for a cultivated cereal, but are exceptionally rich in minerals and other nutrients. Increased interest in novel foods, particularly those that are gluten-free, has led to the cultivation of teff in North America.

Pseudocereals (non grasses)

Amaranth *Amaranthus* spp.
Amaranthaceae

Inca wheat, *Amaranthus caudatus*, was domesticated in the high Andes and is still cultivated by the Quechua Indians sparsely, but over a wide region. It is mainly intercropped with corn or quinoa. The seeds are usually roasted or popped, and are often consumed as balls mixed with molasses.

A. cruentus and *A. hypochondriacus* were domesticated in Mesoamerica. Amaranth, known as *huautli*, was as important as beans and corn to the Aztec civilization of Mexico. The early Spanish conquerors noted that it was used in tortillas made from popped grains, and in a range of drinks. Popped amaranth tamale breads were offered to the fire god Xiuhtecutli in Aztec rituals. A mixture of popped amaranth and the syrup of maguey cactus (*Agave cantula*) played a key role in a form of communion that honored the gods. Cultivation rapidly declined after the Spanish conquest, perhaps in part because of Christian disapproval of these practices. The domestication and early history of the New World amaranth species is not well documented, but it is likely that cultivation of all three species dates back at least 4000 years. All three species are increasingly cultivated in south Asia.

Many amaranths are gathered or cultivated elsewhere, but as leafy vegetables.

See: Gathering Food from the Wild, p. 33; Herbs and Vegetables, p. 113

Buckwheat *Fagopyrum esculentum*
Polygonaceae

Buckwheat is thought to have been domesticated in China about 1000 BC and reached Japan by 722 AD. Its heartland of cultivation is the mountainous area stretching from northern India through Nepal and China to Korea and Japan. Buckwheat was cultivated in the steppes north of the Black Sea by the Iron Age (500 BC), arriving during the medieval period in Europe, where it was eventually widely grown.

The plant is an erect annual herb, bearing its fruits in clusters. The fruits are three-sided, dark-brown nutlets. Buckwheat is particularly well adapted to cold climates and poor, light soils. Its cultivation in Europe began to decline with the introduction of fertilizers in the early 20th century, because other cereals respond better to increased soil fertility. However increasing interest in health foods has led to a revival of interest in buckwheat in the West. Cultivation remains important in Brittany because of the use of buckwheat flour in local savory pancakes known as *galettes*. The best-known buckwheat dish is *kasha*, an eastern European and Russian porridge made from cooked grains. In Japan buckwheat noodles are known as *soba*. Tartary buckwheat (*F. tataricum*) is an uncommon grain crop of Siberia.

Quinoa *Chenopodium quinoa*
Chenopodiaceae

Quinoa bears very small lens-shaped, black seeds in dense terminal clusters. It was taken into domestication in the high Andes of South America, probably from wild *C. hircinum* plants in southern Peru or Bolivia. Unlike another crop of the high Andes, the potato, quinoa has not spread far from its area of origin. Domesticated quinoa has been found at sites in the region dating to 4000 ^{14}C years ago, and is still an important staple food. It grows at altitudes—up to 13,000 feet (4000 m)—that are too high for corn. After removal of the toxic saponin-rich seed coat, the seeds are traditionally washed in an alkaline solution to remove any remaining toxicity. Saponin content varies markedly between different landraces.

Quinoa seeds have a wide range of uses, including as whole grains and as flour for bread, tortillas, and soups. With a high protein content, averaging sixteen percent, and a good range of essential amino acids, quinoa is becoming popular as a health food and it is now widely available in specialist shops in Europe and North America. Increased trade may help to ensure continued cultivation of this ancient species. Canihua (*C. pallidicaule*) also originated in the high Andes and has similar uses. Huauzontle (*C. nuttalliae*) was domesticated in Mexico. The seeds are used, and the immature seed heads are also eaten, fried in batter. In eastern North America *C. berlandieri* was domesticated about 3500 ^{14}C years ago; however, despite being recorded by travelers in the 1720s, the domesticated plant was apparently extinct by the end of the 18th century.

Fat hen (*C. album*) is the Old World counterpart of quinoa. Domesticated forms are cultivated on a small scale in Nepal and northern India for bread, gruel, and fermented beverages.

See: Origins and Spread of Agriculture, p. 22; Herbs and Vegetables, p. 119; Plants as Medicines, p. 234

References and Further Reading

General

Coe, S. 1994. *America's First Cuisines.* Austin: University of Texas Press.
Foster, N., and Cordell, L.S. (Eds.). 1992. *Chilies to Chocolate: Food the Americas Gave the World.* Tucson: University of Arizona Press.
Grubben, G.J.H., and Partohardjono, S. (Eds.). 1996. *Plant Resources of South-east Asia. 10. Cereals.* Leiden: Backhuys.
National Research Council (Ed.). 1989. *Lost Crops of the Incas: Little-known Plants of the Andes with Promise for Worldwide Cultivation.* Washington, DC: National Academy Press.
National Research Council (Ed.). 1996. *Lost Crops of Africa. 1. Grains.* Washington, DC: National Academy Press.
Walker, H. (Ed.). 1990. *Staple Foods.* Proceedings of the Oxford symposium on food and cookery. London: Prospect Books.

Wild grains

Cane, S. 1989. Australian Aboriginal seed grinding and its archaeological record: A case study from the Western Desert. In *Foraging and Farming. The Evolution of Plant Exploitation*, edited by D.R. Harris and G.C. Hillman. London: Unwin Hyman, 99–119.
Cowan, C.W. 1978. The prehistoric use and distribution of Maygrass in eastern North America: cultural and phytogeographical implications. In *The Nature and Status of Ethnobotany*, edited by R.I. Ford. Ann Arbor: Museum of Anthropology, University of Michigan, 263–288.
Donkin, R.A. 1980. *Manna: An Historical Geography.* The Hague: W. Junk.
Guõmundsson, G. 1996. Gathering and processing of lyme-grass (*Elymus arenarius* L.) in Iceland: An ethnohistorical account. *Vegetation History and Archaeobotany 5*, 13–23.
Harlan, J.R. 1989. Wild-grass seed harvesting in the Sahara and sub-Sahara of Africa. In *Foraging and Farming. The Evolution of Plant Exploitation*, edited by D.R. Harris and G.C. Hillman. London: Unwin Hyman, 79–98.
Harris, D.R. 1984. Ethnohistorical evidence for the exploitation of wild grasses and forbs: Its scope and archaeological implications. In *Plants and Ancient Man. Studies in Palaeoethnobotany*, edited by W. van Zeist and W.A. Casparie. Rotterdam: A.A. Balkema, 63–69.
Nabhan, G., and de Wet, J.M.J. 1984. *Panicum sonorum* in Sonoran Desert agriculture. *Economic Botany 38*, 65–82.

Temperate cereals

Bonjean, A.P., and Angus, W.J. (Eds.). 2001. *The World Wheat Book: A History of Wheat Breeding.* Andover: Intercept.
Briggs, D.E. 1978. *Barley.* London: Chapman & Hall.
Caligari, P.D.S., and Brandham, P.E. (Eds.). 2001. *Wheat Taxonomy: The Legacy of John Percival.* Linnean Special Issue 3. London: Linnean Society.
Gupta, P.K., and Priyadarshan, P.M. 1982. Triticale: Present Status and Future Prospects. *Advances in Genetics 21*, 255–345.
Hillman, G.C. 1978. On the origins of domestic rye—*Secale cereale*: The finds from Aceramic Can Hasan III in Turkey. *Anatolian Studies 28*, 157–174.
Lockhart, G.W. 1997. *The Scots and Their oats.* Edinburgh: Birlinn.

Nesbitt, M., and Summers, G.D. 1988. Some recent discoveries of millet (*Panicum miliaceum* L. and *Setaria italica* (L.) P. Beauv.) at excavations in Turkey and Iran. *Anatolian Studies 38*, 85–97.

Padulosi, S., Hammer, K., and Heller, J. (Eds.). 1996. *Hulled Wheats. Proceedings of the First International Workshop on Hulled Wheats.* Rome: International Plant Genetic Resources Institute.

Pomeranz, Y. (Ed.). 1988. *Wheat: Chemistry and Technology.* [Two volumes.] St. Paul: American Association of Cereal Chemists.

Welch, R.W. 1995. *Oat Crop: Production and Utilization.* London: Chapman & Hall.

Tropical cereals

de Wet, J.M.J., et al. 1983. Domestication of sawa millet (*Echinochloa colona*). *Economic Botany 37*, 283–291.

Dirar, H.A. 1993. Indigenous Fermented Foods of the Sudan: A Study in African Food and Nutrition. Wallingford: CAB International.

Doggett, H. 1988. Sorghum. Harlow: Longman.

Fussell, B. 1992. The Story of Corn. New York: Alfred A. Knopf.

Grist, D.H. 1986. Rice (6th ed.). London: Longman.

Hilu, K.W., et al. 1997. Fonio millets: Ethnobotany, genetic diversity and evolution. South African Journal of Botany 63, 185–190.

Johannessen, S., and Hastorf, C. (Eds.). 1994. Corn and Culture in the Prehistoric New World. Westview, Boulder, CO.

Kimata, M., Ashok, E.G., and Seetharam, A. 2000. Domestication, cultivation and utilization of two small millets, Brachiaria ramosa and Setaria glauca (Poaceae), in South India. Economic Botany 54, 217–227.

Luh, B.S. (Ed.). 1991. Rice (2nd ed.). New York: AVI.

Owen, S. 1993. The Rice Book: The Definitive Book on the Magic of Rice Cookery. London: Doubleday.

Portères, R. 1955. Les céréales mineures du genre Digitaria en Afrique et en Europe. *Journal d'Agriculture Traditionnelle et de Botanique Appliquée 2,* 349–386, 477–510, 646–675.

Seetharam, A., Riley, K.W., and Harinarayana, G. (Eds.). 1990. *Small millets in global agriculture.* London: Aspect Publishing.

Williams, J.T. (Ed.). 1995. *Cereals and Pseudocereals.* London: Chapman & Hall.

Yukino, O. 2002. Domestication and cultivation of edible job's tears (*Coix lacryma-jobi* subsp. *ma-yuen*) under the influence of vegeculture. In *Vegeculture in Eastern Asia and Oceania,* edited by S. Yoshida and P.J. Matthews. Osaka: Japan Center for Area Studies, National Museum of Ethnology, 59–75.

Pseudocereals

Campbell, C.G. 1997. *Buckwheat.* Rome: International Plant Genetic Resources Institute.

National Research Council (Ed.). 1984. *Amaranth: Modern Prospects for an Ancient Crop.* Washington, DC: National Academy Press.

5
Roots and Tubers

HELEN SANDERSON

Root crops are an important but poorly understood group of food plants. There are three main groups: the yams (Dioscoreaceae), the edible aroids (Araceae), and potatoes (Solanaceae). Most root crops originated either in high, arid areas such as the Andes, where the tubers allow the plant to survive through difficult winter conditions, or in the tropical forest, where tuber plants are adapted to long dry seasons and quickly mature at the onset of the rainy season. Root crops are particularly important in areas that are too high or too wet for seed crops.

The term *root crop* is misleading as the "root" may be a swollen stem (taro) or a swollen root (sweet potato, carrot, beet). Other forms of modified stem include rhizomes (ginger), corms, tubers (potato), and bulbs (onion). This chapter covers the starchy root crops; others, such as ginger and onion, are covered elsewhere (See Spices; Herbs and Vegetables).

Because of their bulk and perishability, most root crops are consumed locally; their low profile in international trade is one reason why root crops are less studied than many other crops. Root crops have, however, dispersed in cultivation far outside their area of origin. A further factor in our ignorance of the early history of roots is that relatively little archaeological research has been carried out at early sites in the high Andes or the wet tropics. Thus, much of our archaeological evidence for potatoes in South America comes from the arid coastal zone of Peru, where many excavations have been done. Virtually no material has been collected in the Andes, where we know that the potato must have been domesticated earlier than in coastal Peru. Preservation of plant remains is also poor in tropical areas, forcing archaeologists to use evidence from phytoliths and starch residues that are difficult to identify or date. Where root remains are found, recent advances allow much more secure identification of even small fragments. This type of evidence has, for example, confirmed that sweet potato grew in the Pacific region in pre-Columbian times.

In the absence of firm evidence, there has been a great deal of speculation regarding the antiquity of root cultivation, with frequent suggestions that it predates the domestication of grain crops such as rice or corn. It is certainly true that the tubers can be easily propagated and there is good evidence that current-day hunter-gatherers, such as Australian aborigines, actively manage wild yam patches. However, there is a big difference in food output between a gathered wild food, even if managed, and an intensively cultivated crop. Evidence for intensive agriculture, in the form of fields

Woman with oca, near Cuenca, Ecuador. Copyright Edward Parker, used with permission.

or irrigation canals, is strikingly absent from most areas of the wet tropics until perhaps 5000 or 6000 years ago. Although the great antiquity of the use of wild tubers is beyond doubt, the date of their domestication and intensive cultivation is unlikely to precede that of cereals.

Potatoes and Andean Tuber Crops

Bitter potatoes, Rucki *Solanum* × *juzepczukii* and *S.* × *curtilobum*
Solanaceae

Bitter potatoes are ancient foods of the high Andean plateau spanning Peru and Bolivia. They are highly frost-resistant natural hybrids, *Solanum* × *juzepczukii* between *S. acaule* and *S. stenotomum*, and *S.* × *curtilobum* between *S.* × *juzepczukii*, and *S. tuberosum* subsp. *andigena*. They are often grown as security crops and are used mainly to produce chuño, a food that is processed by freeze-drying in order to remove the bitter glycoalkaloids. Geophagy (the consumption of clay) is also used to neutralize the bitter taste of these species. Both black and white versions of chuño are made; the former is a dehydrated product that stores indefinitely and is used widely in soups and stews, and the latter is more expensive and is eaten on festive occasions.

Mashua *Tropaeolum tuberosum*
Tropaeolaceae

Mashua grows wild and is cultivated in the high Andes of South America, from Venezuela to Argentina. Its history in unknown, but it is an ancient Andean tuber crop and has been cultivated since early times. It is probably the fourth most important root crop of the Andes after potato, oca, and ulluco. Cultivated types, with larger tubers than wild mashua, exist in numerous local landraces. Some types are bitter due to glucosinolates, which diminish after the tuber is frozen and pounded to a powder (in a similar way to bitter potatoes). The tubers are boiled, baked, or fried for savory and sweet dishes.

Oca *Oxalis tuberosa*
Oxalidaceae

This species is an important tuber crop of the Andes, second in production to the potato. Although the closest wild relatives exist in Peru and Bolivia, the wild ancestor is unknown. This and other *Oxalis* species were introduced to Europe in the 1830s as a rival to the potato, but never became popular. Oca has been successfully introduced to Mexico and New Zealand where the tubers are marketed as "New Zealand yam." Many different cultivated types are recognized which are highly diverse in size and color. Some types have a strongly acidic taste when first harvested. The acidic taste disappears if the plant is exposed to sunlight for several days or is traditionally freeze-dried. Oca can be boiled, baked, or fried, and traditionally was added to stews and soups.

Potato *Solanum tuberosum*
Solanaceae

Potato is one of the most important food crops of the world, following wheat, corn, and rice, and is cultivated in over 150 countries, mainly in the northern hemisphere. Major producers include Russia, China, Poland, Germany, and India. The most widely grown cultivated species is the "Irish" or "European" potato (*S. tuberosum*); although several other species (including bitter potatoes, see above) are grown in highland South America. International trade (except in seed tubers) is negligible as the crop is usually consumed locally.

Potatoes, and their wild relatives, are in the genus *Solanum*, which includes some 1700 species. The genus is considered native to the Andes and highland plains of Central and South America. The cultivated potato probably originated in the Peru-Bolivia region. They are perennial herbs, with fibrous roots and many rhizomes (underground stems) that become swollen at the tip to form the edible tubers. There are hundreds of potato cultivars, divided into early (new) and main crops. The cultivars show variation in tuber shape and skin colour, ranging from deep purple through to pale brown.

Remains of wild potatoes dating back to 12,500 ^{14}C years ago have been found at the site of Monte Verde in Chile. Remains of cultivated *S. tuberosum*, identified on the basis of their distinctive starch grains, have been found at coastal sites in Peru dating from 2000 BC. Domestication of potatoes must have occurred earlier, but no reliably dated material has yet been excavated in upland areas. Genetic evidence is still unclear, but suggests that the high altitude *S. tuberosum* subsp. *andigena* evolved through a complex domestication and hybridization process in the Andes of Peru and Bolivia. The lowland subspecies, *S. tuberosum* subsp. *tuberosum*, subsequently evolved from subspecies *andigena* in Chile and Argentina.

Potatoes were first brought to Europe in about 1570, but precise details are obscure. The stories attributing this introduction to Sir Francis Drake and Sir Walter Raleigh are apocrypal. It is probable that the first potatoes were brought to Spain from the port of Cartagena in Colombia. The earliest known records of potato cultivation in Europe date from between 1573 and 1576 in Seville, which produced a late crop during the months of December and January. Records from the Canary Islands indicate that potatoes arrived there at the slightly earlier date of about 1562.

The first potatoes to be introduced were of the *S. tuberosum* subsp. *andigena* complex which were ill-adapted to Europe—they are only able to produce tubers under a twelve-hour day length or less—and so were initially cultivated only in mild areas of Spain, southern Italy, and south-western France. It was not until almost two hundred years later, in the mid 18th to early 19th centuries, that the species was selected and developed to produce cultivars with higher yields and an earlier cropping season, so that it could be grown throughout the cooler parts of Europe. Potatoes were first introduced to North America in the 17th century. They were brought from Bermuda in 1621, having been exported from England in 1613. The crop spread to India, China, and Japan in the late 17th century. In 1770, Captain Cook introduced the potato to Australasia, where it became common by the 1850s.

Potato breeding developed rapidly in Europe and North America during the 19th century. However, in 1845 and 1846, the potato crop in Ireland was largely wiped out by the potato blight, leading to the Irish Potato Famine, during which over a million people died and a further 1.5 million emigrated to England and the United States. Apart from the tremendous social and historical impact of the disease, the blight imposed a major new selective factor on the potatoes of the time. Although some resistance to the disease had been built up by the 1870s, the episode is likely to have significantly narrowed the genetic base. From the mid-19th century onwards, repeated introductions of the Chilean subspecies *andigena* played a vital role in introducing greater variability in plant characteristics to plant breeding.

Potatoes are particularly nutritious, and are so versatile they have become almost ubiquitous. They are featured in a wide range of cuisines around the world. They are eaten boiled, fried, and roasted as a vegetable, or used as an ingredient in a variety of dishes. They can be processed into crisps or potato flour, or dehydrated and eaten or stored as dried potato. Storage methods include canning and freezing. Individual potato products include starch, alcohol (the basis of vodka and schnapps), glucose, and dextrin. The tubers are also used as an animal feed. All green parts of the plant, including potato tubers that have been exposed to light, contain poisonous solanines, so the eating of tubers with green patches should be avoided.

Aspects of the early history of the potato, of previous perceptions of its nature, and of its wide distribution, are reflected by the many names that exist for this plant. In Europe, names of the "earth apple" type are widespread. The modern French name *pomme de terre*, the Dutch *aardappel*, and the obsolete German *erdapfel* are among many variants. In Persia, the name *seb-i-zaminee* also means earth apple. Wild types continue to be eaten in South America, where a great diversity has been maintained. They are known as *papas criollas, papa* being the generic term for potato in South America.

See: Origins and Spread of Agriculture p. 22

Ulluco *Ullucus tuberosus*
Basellaceae

Ulluco is an ancient Andean crop native to central Peru and northern Argentina. Today, the tubers are grown commercially in an area from Venezuela to Chile and northwestern Argentina. Numerous cultivated types exist. The tubers, colored white, yellow, green, and purple, have a wide range of culinary uses and can be boiled like potatoes in stews, freeze-dried and ground into flour, or pickled.

Sweet Potatoes and Yams

African yam bean *Sphenostylis stenocarpa*
Fabaceae

The species is native to tropical Africa. The plants have edible tubers, leaves, pods, and seeds. It is both collected from the wild and cultivated, particularly for the seeds and tubers. It is grown as a garden crop or in fields, particularly in Central Africa (Zaire, Congo). The cultivated crop may be transplanted from the wild, or highly domesticated local races. However, cultivation of this plant is declining, as is the case with many other African indigenous crops. The tubers are twice as rich as the seeds in protein, and are eaten boiled, fried, or roasted. It is unclear whether its common name has arisen from its association with yam in cultivation, or from the fact that it produces yam-like tubers.

Sweet potato, Yam *Ipomoea batatas*
Convolvulaceae

Sweet potatoes are the world's seventh-largest food crop in terms of production, and are widely cultivated in the tropics and subtropics. In the tropics, they are the most important root crop in terms of volume consumed. The number of wild *Ipomoea* species is estimated at about 500. A direct wild ancestor of the domesticate, *I. batatas*, has not yet been positively identified, but it may have originated in tropical northwestern South America, probably from species related to the wild *I. trifida*. Due to domestication, artificial selection, and the occurrence of natural hybridization and mutations, there are a large number of cultivars. There is more diversity in the sweet potato than in, for example, cassava, yam, or cocoyam.

Archaeobotanical evidence shows that the cultivated crop was widespread in coastal Peru about 4000 ^{14}C years ago. The lack of earlier records, particularly from the presumed area of origin, reflects poor recovery of plant remains in the region. Sweet potato cultivation had spread over much of the Americas and across the Caribbean by the time the first Europeans arrived The means of dispersal of sweet potato from South America to Polynesia has been the subject of much speculation as there is no archaeological evidence for regular contact across the Pacific at this time. It is known that the sweet potato spread in two waves of dispersal; the earlier line ("kumara" line) was possibly introduced to Polynesia about 2000 years ago. The earliest archaeobotanical evidence is of a charred tuber from the Cook Islands dating to 1000 AD. The second wave ("batatas" line) occurred when Columbus introduced the starchy carrot-like type of sweet potato to Spain after his first voyage. Subsequent Spanish voyages brought back a less bitter type of sweet potato preferred by Europeans. These were introduced into Africa by the Portugese during the 16th century, and to the United States by 1648. Except for Polynesia (where the plant was already extensively grown), the Spaniards and the Portuguese spread the plant throughout the remainder of Asia and Malaysia.

Two main types dominate, the moist, sweet type popular in the United States, and the yellow or white, dry flesh type preferred in New Zealand, Japan, and other countries. The swollen roots are the main edible parts of the plant. They are usually eaten boiled, baked, or fried, and may be candied with syrup or used as a puree. They can be processed by canning and dehydrating, are suitable for flour manufacture, and are a source of starch, glucose, syrup, and alcohol. They are used as livestock feed and are an important industrial crop for the production of ethanol.

Yam *Dioscorea* spp.
Dioscoreaceae

The genus *Dioscorea*, to which yams belong, includes many tubers, bulbils, and rhizomes that are of economic importance. About sixty species are gathered or cultivated for their edible tubers, but only four of the most commercially significant are discussed in detail here. They are an important staple food grown in the humid and sub-humid tropics, with 95 percent of the global production being in Africa. The most commercially important species have been isolated into three main continental groups, Asiatic, African, and American, each representing independent centers of diversity and domestication. There are four domesticated yams that are important commercially: white yam (*D. rotundata*) and yellow yam (*D. cayenensis*) from West Africa, and greater or water yam (*D. alata*) and lesser or Chinese yam (*D. esculenta*) from Southeast Asia. A fifth domesticated yam, cushcush yam (*D. trifida*), originated in the Americas.

In Africa, yams have served played an important part in traditional religious and ritual ceremonies, which are likely to have been consequential in their domestication. There are uncertainties concerning the origin of the two African species, *D. rotundata* and *D. cayenensis*. It is thought that domestication occurred along the valley of the Niger, although an earlier theory suggests that *D. rotundata* is a hybrid of two ancestral species (*D. cayenensis* with *D. praehensilis* and/or *D. abyssinica*) that evolved within the 'yam zone' in the eastern part of West Africa. This zone is the only area in Africa today where the cultivation of yams is significant, and corresponds with the area inhabited by people speaking languages of the Eastern Kwa linguistic group. There is no direct evidence of yam cultivation in prehistoric Africa, but if cultivation began at the same time as that of the oil palm, cultivation dates back at least 5000 ^{14}C years. The West African yams were not dispersed from west Africa until they were taken to the tropical New World, in association with the slave trade.

The two main species domesticated in Southeast Asia have been major constituents of root crop agriculture in the region and throughout Oceania from before European contact. Indeed, linguistic evidence suggests that domestication of these yams may date back at least 6000 years. *D. alata*, now widely grown throughout the world, is unknown in the wild state but was possibly developed from *D. hamiltonii* or *D. persimilis* in Southeast Asia. *D. esculenta* is a native of Indochina where it occurs in both wild and domesticated forms. Little is known of the New World yam domestication, but *D. trifida* is thought to have developed on the borders of Brazil and Guyana. From there it spread through the Caribbean, where it is now widely grown.

Most of the world's yam production is from the *D. cayenensis* complex in Africa, although the most widely cultivated and important species in Southeast Asia is *D. alata*. *D. esculenta* is also an important cultivated species in Asia; it has the most easily digestible starch making this species suitable for specialist diets. Other species cultivated to an appreciable extent include *D. bulbifera*, *D. nummularia*, and *D. pentaphylla*. The one disadvantage of yams is that they grow deep into the ground, and harvesting can be difficult.

According to the species, plants may possess one or several tubers. Tubers from all species have to be cooked before consumption, to destroy the bitter, toxic dioscorine, which they contain in the raw state. The tubers are eaten fried, roasted, or boiled, and can be used for the production of chips, flakes or flour, and "fufu," a starchy paste popular in West Africa (also prepared from plantains, cassava, and cocoyams). Some species are used medicinally.

See: The Hunter-Gatherers, p. 9; Origins and Spread of Agriculture, pp. 19, 20, 22, and 23; Gathering Food from the Wild, p. 38

Yam-bean *Pachyrhizus erosus*
Fabaceae
The common name is applied to several leguminous plants with edible tubers, but the most commercially important is *P. erosus*. The species originated in Mexico and Central America, and it is found at Peruvian coastal sites dating from 3000 ^{14}C years BC. The Spanish first introduced it to the Far East by the end of the 17th century, and it has since spread to most tropical and sub-tropical regions. It is cultivated as a tuber crop in Mexico, the West Indies, south and Southeast Asia, and the Pacific Islands, mostly as a garden plant but also commercially. The tuber may be thinly sliced and eaten raw, or cooked or pickled. In the United States and China, it has been used as an alternative to water chestnut. In Indonesia, it is used to make a sharp, spicy salad called *rujak*. The young pods are also eaten, but the mature seeds are poisonous.

Aroids and Starchy Rooted Plants Used for Making Arrowroot and Tapioca

Achira, Queensland arrowroot *Canna edulis*
Cannaceae
Achira is a crop native to tropical South America with fleshy rhizomes that produce a yellow starch. It has long been a foodstuff consumed in the Andean region. In Peru abundant remains of the plant dating back to 2500 ^{14}C BC have been excavated. The rhizomes are used as a starchy vegetable and as a spice, and for starch production, animal fodder, and dye. Achira is now cultivated pan-tropically and in other warmer regions of the world. In Vietnam, during the late 20th century, large areas of this plant were cultivated for the production of the highly prized transparent noodles that are an indispensable part of that country's cuisine. Other uses for this species include use of the seeds as beads and the leaves as a food wrapping.

Arrowroot, Bermuda arrowroot *Maranta arundinacea*
Marantaceae
The exact origin of arrowroot is unknown, but it is indigenous to Central America and northern South America. The distinctive starch grains have been found at sites in Panama dating to 5000 ^{14}C years ago. It is now widely cultivated throughout the tropics, but is important mainly in the West Indies. In Southeast Asia it is mainly cultivated in home gardens. Its young rhizomes can be eaten boiled or roasted as a vegetable, and is the primary source of "arrowroot starch," which is used as a thickener for savory and sweet dishes. The starch is also used industrially in the manufacture of paper, board, powders, glues, and soaps. The rhizomes are used as fodder and fertilizer, and plants are also grown as ornamentals for their striking brown-purple leaves. Mashed rhizomes are used medicinally to treat wounds and, in French Guyana and Dominica, were used to draw poison from wounds inflicted by poisoned arrows, hence the name.

Babai, swamp taro *Cyrtosperma merkusii*
Araceae
The possible origin of babai is western Malesia or the Solomon Islands, but the exact location is not known. It is now distributed wild and cultivated from peninsular Malaysia, throughout most of Malesia to Oceania and is now a staple food on several Micronesian and Polynesian coral atolls. It is the dominant aroid crop in Micronesia, whereas in Polynesia and Melanesia, *Colocasia esculenta* is the major aroid crop. The species is cultivated on swampland, mainly by smallholders, for the starchy tubers, which are eaten after peeling and thorough boiling, steaming, or baking. Various other parts of the plant are also utilized and the species is grown as an ornamental.

Manioc being toasted in large iron pan. Flooded forest near Manaus, Brazil. Copyright Edward Parker, used with permission.

Cassava, Manioc, Yucca *Manihot esculenta*
Euphorbiaceae

Cassava (also known as yucca in Latin American countries, though not related to the genus *Yucca*) is a tropical root crop native to Central or South America. It is the only member of the Euphorbiaceae family that provides food. It also has the unique distinction of being the only highly poisonous major staple crop. As a tropical root crop, it is outranked in volume consumed only by the sweet potato. Cassava has been as significant to the historical evolution of tropical countries, such as Brazil and Zaire, as the potato has to countries such as Ireland and the United States.

The tubers are cigar-shaped, with a brown, often pinkish, rind that is usually hairy, with ivory-white flesh. They vary considerably in size, but are typically 10 inches (25 cm) long and 2 inches (5 cm) thick and are borne in clusters of up to ten. Selection of plants for cultivation would have been directed at reducing cyanogenic glucosides, which liberate toxic hydrogen cyanide (HCN) by enzymic breakdown. The reaction speeds up after the tubers have been cut, but toxicity can be removed by cooking or, more effectively, by grating and drying. They can be divided into two main groups based on the glucoside content of the central part of the storage roots: "bitter" and "sweet". However, this distinction is unjustified because all kinds of intermediates occur. Cultivation and domestication since ancient times has resulted in an enormous number of cultivars. A particular advantage of cassava is that its roots can be left in the ground for up to three years without deteriorating, making it a useful emergency food.

Although wild relatives occur in both Central and South America, genetic evidence suggests that the most closely related is *M. flabellifolia*, which grows on the southern border of the Amazon basin. Remains of the plant, dating from 2000 ^{14}C BC, have been found in coastal Peru, but domestication must have occurred earlier in Brazil. The Portuguese encountered cassava after 1500 AD on the coast of Brazil, and employed Indian and African slaves in its cultivation. It became the food staple for slaves in Brazil, and also became important in the slave diet in the Caribbean, after the Dutch set up a system similar to that of the Portuguese.

From Brazil, the Portuguese transported cassava to West Africa and by the 1660s it was an important food in northern Angola. Extensive cultivation did not occur in West Africa, however, until the early 19th century when former slaves returned from Brazil. During the mid-1700s, the French were likely to have been the first to introduce cassava to East Africa and islands near Madagascar, initially with disastrous consequences, as Africans there were unfamiliar with processing techniques and some died of poisoning. After this, the French and Africans learned how to process cassava, and introduced it to Réunion and Madagascar, where it has become an important staple.

In 1740, cassava raised in Java was introduced to Mauritius, shortly after the French brought the Brazilian plants there. Thus, the plant had traveled full circle, as Brazilian plants met those developed in Asia.

Cassava is now important in the economies of many tropical countries and major producers include Nigeria, Zaire, Brazil, and Indonesia. Worldwide, 65 percent of cassava is used directly for human consumption and the remaining 35% for animal feed, starch, and industrial uses. For human consumption, the storage roots are first peeled and chopped and then boiled, steamed, fried, or roasted. The peeled and chopped roots may also be dried or fermented for later use. It is mainly marketed in the form of cassava flour, but can be eaten as a vegetable (as can the young leaves) or processed into a range of other products, including chips, porridge, beverages, and a dry meal. The only cassava product familiar to most people outside the tropics is tapioca, a refined starch.

See: *Origins and Spread of Agriculture, p. 22*

Curcuma spp.
Zingiberaceae

Species of this genus are mainly found in the Indo-Malesian region from India throughout Southeast Asia to the southern Pacific. The rhizomes and tubers contain starch, which is extracted from some wild and cultivated species, often in times of food shortage. The starch present in the tuberous roots of *C. angustifolia*, *C. aromatica*, and *C. pierreana* is used in the same way as true arrowroot.

See: *Spices, p. 170*

East Indian arrowroot *Tacca leontopetaloides*
Taccaceae

Also known as Polynesian arrowroot, the species is a pan-Pacific cultigen, believed to have originated in Malesia. It is now widely distributed wild and cultivated in Africa, the Indian subcontinent, and islands of the Indian Ocean and Australia. The starchy tubers traditionally provided an important staple food. The bitter compounds are removed by repeated soaking in fresh water, and the starch is extracted and used to make cakes, as a thickener in foods, and for starching clothes. At present, its cultivation is declining. It has lost its role as a staple food (cassava has largely replaced East Indian arrowroot) and is mainly consumed as an emergency food.

Giant taro *Alocasia macrorrhizos*
Araceae

Giant taro probably originated in peninsular Malaysia, but has been introduced and naturalized in the Malesian region and Oceania. The species is grown in south and Southeast Asia and particularly Oceania for its large, thickened underground stem. It is a source of easily digested starch and is capable of storing starch in the stems for a longer period than most aroids. The stems are peeled and cut into pieces to be baked or boiled. It must be cooked thoroughly as the outer layers contain calcium oxalate crystals, which would otherwise cause irritation and swelling of the mouth and throat. Concentrations increase when the crop is left in the ground for too long, so some societies consider the plant an emergency resource only. In Tonga, Wallis, and Papua New Guinea, varieties with very low oxalate content have been selectively grown. It has medicinal uses and is an important ornamental.

Tannia, New cocoyam *Xanthosoma sagittifolium*
Araceae

Tannia is native to tropical South America and the Caribbean. It was cultivated for its edible corms in the Greater Antilles before Columbus arrived. During the slave-trading era, it was taken to Africa where it became the "new" cocoyam. In the 19th and early 20th centuries it spread throughout Oceania and into Asia. It is widely cultivated throughout the tropics for its edible corms, which are dried, peeled, and ground into flour, or are boiled, baked, steamed, or fried for a variety of dishes. Tannia is now considered to be a complex polymorphic species, comprising many cultivated forms of *Xanthosoma*, which is more of an agricultural unit than a taxonomic entity.

Taro *Colocasia esculenta*
Araceae

Although it is the prime staple in many Pacific islands, the edible corm of this species is also an important food crop in the tropics and sub-tropics, where it can be grown under submerged or flooded conditions. It probably originated in mainland Southeast Asia, where related wild species grow. Linguistic evidence suggests taro is likely to have been cultivated and exchanged for over 5000 years in Southeast Asia, including Thailand, Malaysia, Indonesia, and the Philippines. From there it may have been transported to Papua New Guinea and into Micronesia and Polynesia, where it is highly valued as a food. This easterly spread and cultivation of taro is closely associated with the development of Lapita culture. In its westward spread, taro reached Madagascar, where it became widely established between the 1st and 11th centuries AD. It was carried farther westward in medieval times, along the Mediterranean and across to sub-Saharan Africa. It has become an important plant in West Africa.

Although its use is widespread globally, production of taro is now of marginal importance, taro having been largely replaced by rice. There are many different cultivated types; for example, 70 local names were recorded in Hawaii. Each form is distinguished by morphological characteristics and time taken to mature. The corm must be peeled and cooked thoroughly to remove the calcium oxalate crystals. It is fried, baked, steamed, and roasted. It can be made into flakes or chips or mashed and fermented to make "poi."

See: Origins and Spread of Agriculture, p. 19

Umbelliferous Root Crops

Arracacha, apio, Peruvian parsnip *Arracacia xanthorrhiza*
Apiaceae

Arracacha is possibly one of the oldest cultivated Andean plants. The species is likely to have originated in the Andean highlands from Venezuela to Bolivia, where wild relatives occur. It spread into the Caribbean, probably after the Spanish conquest, and was introduced to southeastern Brazil towards the end of the 19th century. It is also grown in some parts of Central America, Africa, India, and Sri Lanka, though attempts to grow it elsewhere have been unsuccessful. It is now popular in the northern part of South America and parts of the Caribbean, where it is grown for local use in home gardens, instead of potato. Its secondary tubers, which resemble those of a small parsnip, are starchy and are eaten boiled or fried. They are also used as an ingredient in a number of popular dishes.

Carrot *Daucus carota* subsp. *sativus*
Apiaceae

Different forms of wild carrot, usually recognized as subspecies of *D. carota*, grow over much of western Asia and Europe and have tiny, acrid-tasting roots. *D. carota* subsp. *sativus*, the domesticated form, has enlarged roots and is referred to in classical texts 2000 years ago. The original domesticated carrot evolved in the Near East, in the region between Afghanistan and Turkey. It had dark purple roots that were often branched. Orange, carotene-colored forms evolved later and are illustrated in a Byzantine herbal dating to 512 AD. The orange types displaced the purple forms in Europe and the Mediterranean by the 17th century. Carrots can be used raw or cooked as a vegetable, in savory or sweet dishes, and, in Asia, are often preserved in jams and syrups.

Celeriac *Apium graveolens* var. *rapaceum*
Apiaceae

Apium graveolens is cultivated in two different forms. One is the so-called true celery (*A. graveolens* var. *dulce*), grown for its stems; the other is celeriac (*A. graveolens* var. *rapaceum*), cultivated for the swollen, underground, base of the stem. In its wild form, the species is native to the Mediterranean region and the Middle East. Celeriac was probably domesticated much later than celery. It is first mentioned in European and Arab sources in the 16th century. Celeriac, also known as knob celery and turnip-rooted celery, has a mild celery- and parsnip-like flavor. It is eaten cooked or grated raw in salads and has been widely used as an ingredient of French salads and in soups, stuffings, and purées in Russia and northern Europe.

See: Herbs and Vegetables, p. 116

Parsnip *Pastinaca sativa*
Apiaceae

Parsnip grows wild in southern and central Europe and western Asia and has been domesticated to produce an edible root. It was introduced into northern Europe where it was domesticated in Roman times.The fleshy forms were not developed until the Middle Ages, so the parsnip may have initially been used as a flavoring or sweetener. In medieval Europe, the sweet, starchy parsnip was popular and became a staple food (potatoes had not yet arrived in Europe). It was introduced to Virginia in 1609 and was widely grown by native Americans. It is now eaten mainly in northern Europe; consumption in the United States is small. It is eaten as a vegetable, roasted, fried, baked, or boiled. It can also be used to make wine and is used as animal fodder.

Turnip-rooted chervil *Chaerophyllum bulbosum*
Apiaceae

This species is native to central and eastern Europe and western Asia. It has no major economic importance, although there has been a recent revival in interest. It is mainly eaten in eastern Europe, Germany, and Holland. The tuberous edible root is similar to a carrot and is round with gray skin and yellowish white flesh. It has the texture of a floury potato and is usually boiled as a vegetable.

Brassicaceous and Asteraceous Roots

Black salsify, Spanish salsify *Scorzonera hispanica*
Asteraceae

Black salsify is native to central and southern Europe. It was first used as a medicinal plant. It was probably first cultivated in Spain and Italy in the 16th century, appearing later in England. It is now cultivated in temperate and sub-tropical areas and at higher altitudes in the tropics, and is particularly popular in Europe. The roots are very distinctive being carrot-shaped with black skin and white flesh. They are cooked as a vegetable and also used, after drying and roasting, as a coffee substitute. The roots contain inulin, composed of fructose units, which can be eaten by diabetics in place of glucose. The leaves are also used in salads and to feed silkworms.

See: Herbs and Vegetables, pp. 114–5

Jerusalem artichoke *Helianthus tuberosus*
Asteraceae

This is an ancient domesticate of eastern North America, closely related to the sunflower *H. annuus*. Its tubers were eaten by indigenous people, especially in what is now Canada. Archaeological evidence shows that it was domesticated by 1100 AD. It was probably taken to Europe, initially France, in the early 1600s. It is not a major crop, but is cultivated around the world in many temperate countries, and in some parts of the tropics. The irregular and knobbly tubers contain inulin, and so are suitable for diabetics. The tubers are eaten baked, steamed, mashed, in soups, or raw in salads. The species used to be grown for alcohol production in parts of France and Germany. The tubers are also used as cattle fodder in China, France, and Italy.

Salsify *Tragopogon porrifolius*
Asteraceae

Salsify is native to the eastern Mediterranean region and the roots were probably eaten in classical times. Cultivation of the roots began in the 16th century, following a similar history to black salsify (*Scorzonera hispanica*). Today, it is grown commercially on a small scale in the United States Its white roots are boiled and eaten with melted butter or cream and cheese.

See: Herbs and Vegetables, p. 127

Swede, Rutabaga *Brassica napus* subsp. *rapifera*
Brassicaceae

This is the root form of oilseed rape. The species is only known in cultivation, and is of European origin. It evolved as a hybrid between turnip (*Brassica rapa* subsp. *rapa*) and cabbage (*B. oleracea* var. *capitata*). Historical records indicate that it was grown in Finland in the 17th century and spread from Sweden to central Europe and had reached England by 1800. It is cultivated most in northern European countries and around the Baltic Sea and Russia. It was grown in the United States at the beginning of the 19th century. Swedes are grown as a food for humans, used mainly in stews or mashed as a vegetable, and have also been an important fodder crop for ruminants during winter. They can be distinguished from turnips in having ridged leaf base scars forming concentric rings at the top of the hypocotyl.

Turnip *Brassica rapa* subsp. *rapa*
Brassicaceae

The turnip is thought to originate in the Mediterranean and east Afghanistan. It is an ancient crop, but the time and place of domestication are unclear. For oilseed forms (subsp. *oleifera*), there is evidence to suggest multiple domestication from the Mediterranean to India from about 2000 BC. The cultivation of turnips as an edible crop predates recorded history, but is likely to have originated in northern Europe, between the Baltic and the Caucasus. Turnips were an important food for the Romans, and by the classical period many varieties were available. The earliest archaeological record of turnip root is from 13th century Byzantine Sparta, Greece. The turnip was introduced through Asia to northern China, where it had become common well before the Middle Ages. Accounts also suggest it was taken to Japan by 700 AD. In France, particular care is taken in growing turnips. There they are picked when small, and consumed braised, fried, and glazed. In the Middle East, turnips are pickled and dyed pink using beetroot. They are important forages for livestock in northern Europe and New Zealand. Other forms have been developed for their leaves (e.g., bok choi) and seed oil.

Salad Roots

Beetroot, Mangold, Mangel-wurzel *Beta vulgaris* subsp. *vulgaris*
Chenopodiaceae

All of today's commercially cultivated beets are descended from *Beta maritima*, a seashore plant from the Mediterranean and Atlantic coasts of Europe and North Africa. Subspecies *vulgaris* includes beetroot, mangold, mangel-wurzel, and sugar beet. No definite indications of a form of garden beetroot with a fleshy swollen hypocotyl are found until 13th century Italy. The modern prototype of beetroot is regarded to be "Beta Roman," dating back to 1587. Beetroot is eaten boiled as a vegetable, in dishes or pickled, and is an essential ingredient of borscht. The scarlet color is due to the combination of a purple pigment, betacyanin, and a yellow one, betaxanthin, which are much more stable than most red plant colors and are sometimes extracted for food colorings. Forage beets (including mangolds and mangel-wurzels) possess very large swollen roots that were developed as cattle feed, probably in the 18th century. The closely related foliage beets (leaf beet, spinach beet, Swiss chard) are classified as subspecies cicla.

See: Herbs and Vegetables, p. 116

Radish *Raphanus sativus*
Brassicaceae

The wild ancestor of radish may be *R. raphanistrum*, possibly crossed with other wild relatives, from western Asia. The wild taxa occur throughout Europe and Asia, and are introduced weeds in America. Radish is widely referred to in Greek and Roman texts, and the latter suggest it was important in Roman Egypt as an oilseed. Forms with the swollen hypocotyl (root) may have originated in medieval times. Spanish and Portuguese colonists introduced it to the New World in the late 16th century. It is now grown virtually worldwide. There are two basic types: small-rooted European types, grown throughout temperate regions; and large-rooted Asian types. They vary widely in shape, color, and strength of flavor. In Europe, they are eaten raw in salads, and in Asia are ubiquitous in oriental dishes and garnish. Other types have been developed for their leaves, sprouted seeds, and seed 'pods'.

Other Tubers

Chinese artichoke *Stachys sieboldii*
Lamiaceae

Chinese artichoke is native to northern China and has been cultivated for its tubers since ancient times in both China and Japan. It has been cultivated in France on a small scale since the end of the 19th century and in Malaysia since the 1980s. The crop is occasionally cultivated in other countries. After harvesting, the tuber must be eaten quickly as it rapidly discolors and deteriorates. Because it can remain in the ground for a long period, it provides a useful winter vegetable. For consumption, it is cleaned well (rather than peeled) and eaten raw and grated in salads. Alternatively, it may be plain boiled, fried, or pickled.

Chinese water chestnut *Eleocharis dulcis*
Cyperaceae

This is a widespread and variable species of the Old World tropics, distributed from tropical West Africa and eastwards to Southeast Asia, Japan, Pacific islands, and northern Australia. The place of domestication is uncertain, but is possibly southern China. Its cultivation in shallow marshes, lakes, and flooded fields is widespread, particularly in China, Taiwan, and Thailand. Corms are used as a vegetable, either raw or cooked; they have a crisp texture and are used in many local dishes.

Hausa potato, Country potato *Solenostemon rotundifolius*
Lamiaceae
Hausa potato is cultivated as a food crop in tropical Africa—where it probably originated—and India. The Arabs may have taken it to India, from where it spread on to the East Indies. The starchy, slightly aromatic tubers are eaten raw or boiled or steamed. The leaves are also eaten as a vegetable and parts of the plant are used medicinally.

Lotus, Hasu *Nelumbo nucifera*
Nelumbonaceae
This species is a symbol of purity to Buddhists and has been held sacred in the Far East for over 5000 years. It originated in continental Asia, but is now distributed worldwide. It has been cultivated in China for over 3000 years and is exported globally. It is a member of the water lily family and is prized for its aquatic rhizomes and seeds. The rhizomes are eaten raw or roasted, sliced and fried as chips, pickled, or candied. In China, the root is also ground into a starchy paste for use as a thickener. The leaves are used as a vegetable and as a wrapping. Seeds from ancient lake beds have been successfully germinated. The oldest is radiocarbon dated to 1350 ^{14}C years old and is by far the most ancient seed of any plant species to have germinated.

See: Nuts, Seeds, and Pulses, pp. 138–9

Ti plant *Cordyline fruticosa*
Agavaceae
This species is found in India, Indonesia, Malaysia, northeast Australia, New Guinea, New Zealand, and the Pacific Islands, though it probably originates from New Guinea. It is especially prized in Hawaii for its large, sweet, fleshy rhizome. These are eaten cooked or baked and are used as a natural sweetener or for the production of alcoholic beverages.

Tiger nut, Chufa *Cyperus esculentus*
Cyperaceae
This species of perennial sedge is widespread in tropical and temperate zones throughout the world. The cultivated form is thought to have originated in the Mediterranean, perhaps from the wild *C. esculentus* var. *aureus*. Tubers of a wild form of *C. esculentus* are abundant at the site of Wadi Kubbaniya in upper Egypt, dating to 18,000 ^{14}C years ago. Tubers, perhaps from the domesticated form, are abundant at Pharaonic sites from about 3000 BC. Tiger nuts are cultivated in tropical Africa and southern Europe for their small, sugar and oil-rich tubers, but have naturalized globally throughout tropical, sub-tropical, and temperate areas, becoming a troublesome weed. The tubers are eaten raw or cooked by boiling or roasting, and are dried and made into flour. In some countries, tubers are used in sweetmeats or to produce drinks such as the Spanish 'horchata'. The tuber oil has potential as an important source of bio-diesel fuel.

Wild mung bean *Vigna vexillata*
Fabaceae
Vigna vexillata is native to the Old World tropics, but it is now found across the tropics. The edible tubers are collected from wild plants in Africa and India. However some cultivation for the tubers may take place, particularly in Ethiopia and the Sudan. The tubers are soft, easy to peel, and have a creamy, white, flesh. They are eaten boiled or raw.

References and Further Reading

General

Advisory Committee on Technology Innovation Board on Science and Technology for International Development National Research Council. 1989. *Lost Crops of the Incas: Little Known Plants of the Andes with Promise for Worldwide Cultivation*, Washington, DC: National Academy Press.

Duke, J.A., and DuCellier, J.L. 1993. *CRC Handbook of Alternative Cash Crops*. Florida: CRC Press.

Flach, M., and Rumawas, F. (Eds.). 1996. Plant Resources of South-East Asia No. 9. Plants Yielding Non-seed Carbohydrates. Leiden: Backhuys.

Food and Agricultural Organisation of the United Nations. 1990. *Roots, Tubers, Plantains and Bananas in Human Nutrition*, Rome: FAO.

Hernándo Bermejo, J.E., and J. León (Eds.). 1994. *Neglected Crops: 1492 from a Different Perspective*. Plant Production and Protection Series 26. Rome: FAO.

Onwueme, I.C. 1978. The Tropical Tuber Crops: Yams, Cassava, Sweet Potato and Cocoyams, Chichester: Wiley.

Onwueme, I.C., and W.B Charles. 1994. *Tropical roots and tuber crops: production, perspectives and future prospects*. FAO Plant Production Paper 126. Rome: FAO.

Scott, G.J., Rosegrant, M.W., and Ringler, C. 2000. *Roots and Tubers for the 21st Century: Trends, Projections and Policy Options.* Washington, DC: International Food Policy Research Institute.

Yoshida, S., and P.J. Matthews (Eds.). 2002. *Vegeculture in Eastern Asia and Oceania*. Japan Center for Area Studies, Symposium Series 16. Osaka: National Museum of Ethnology.

Potatoes and Andean Tuber Crops

Burton, W.G. 1989. *The Potato* (3rd ed.). Harlow, Essex: Longman.

Dodge, B.S. 1970. *Potatoes and People: The Story of a Plant*, Boston: Little, Brown.

Grun, P. 1990. The evolution of cultivated potatoes. *Economic Botany 44*(Supplement), 39–55.

Harris, P.M. (Ed.). 1992. *The Potato Crop: The Scientific Basis for Improvement*. London: Chapman Hall.

Hawkes, J.G. 1998. The introduction of New World crops into Europe after 1492. In *Plants for Food and Medicine*, edited by H.D.V. Prendergast, N.L. Etkin, D.R. Harris, and P.J. Houghton. Kew: Royal Botanic Gardens, 147–159.

Wilson, A. 1993. *The Story of the Potato through Illustrated Varieties*. Wisbech, Cambridgeshire: Balding and Mansell.

Zuckerman, L. 1998. The Potato: From the Andes in the Sixteenth Century to Fish and Chips, the Story of How a Vegetable Changed History. London: Macmillan.

Sweet Potatoes and Yams

Ayensu, E.S. 1972. Guinea yams: The botany, ethnobotany, use and possible future of yams in West Africa. *Economic Botany 26*(4), 301–318.

Coursey, D.G. 1967. Yams: An Account of the Nature, Origins, Cultivation and Utilisation of the Useful Members of the Dioscoreaceae. London: Longman.

Degras, L. 1993. *The Yam: A Tropical Root Crop*. London: Macmillan.

Potter, D. 1992. Economic botany of *Sphenostylis* (Leguminosae). *Economic Botany 46*, 262–275.

Woolfe, J.A. 1992. *Sweet Potato: An Untapped Food Resource*. Cambridge and New York: Cambridge University Press.

Aroids and Starchy Rooted Plants Used for Making Arrowroot and Tapioca

Cock, J.H. 1985. *Cassava: New Potential for a Neglected Crop*. Boulder, CO: Westview Press.

Gragson, T.L. 1997. The use of underground plant organs and its relation to habitat selection among the Pumé Indians of Venezuela. *Economic Botany 51*, 377–384.

Jones, W.O. 1959. *Manioc in Africa*. Stanford, CA: Stanford University Press.

Matthews, P.J. 1995. Aroids and the Austronesians. *Tropics 4*, 105–126.

Milliken, W., and Albert, B. 1997. The use of medicinal plants by the Yanomami Indians of Brazil, part II. *Economic Botany 51*, 264–278.

O'Hair, S.K., and Asokan, M.P. 1986. Edible aroids: botany and horticulture. *Horticultural Reviews*, Volume 8, edited by J. Janick. Connecticut: AVI Publishing Company, 43–99.

Pollock, N.J. 1990. Arrowroot as a Pacific foodstuff. *Ethnobotany 2*, 1–10.

Spennemann, D.H.R. 1994. Traditional arrowroot production and utilisation in the Marshall Islands. *Journal of Ethnobiology 14*, 211–234.

Umbelliferous Root Crops

Langley, J. 1991. Histories of the carrot. *Henry Doubleday Research Association Newsletter 126*, 34–35.

Brassicaceous and Composite Roots

Ahokas, H. 2002. Cultivation of *Brassica* species and *Cannabis* by ancient Finnic peoples, traced by linguistic, historical and ethnological data; revision of *Brassica napus* as *B. radice-rapi*. *Acta Botanica Fennica 172*, 1–32.

Hather, J.G., Peña-Chocarro, L., and Sidell, E.J. 1992. Turnip remains from Byzantine Sparta. *Economic Botany 46*, 395–400.

Hawkes, J.G. 1998. The introduction of New World crops into Europe after 1492. In *Plants for Food and Medicine*, edited by H.D.V. Prendergast, N.L. Etkin, D.R. Harris, and P.J. Houghton. Kew, UK: Royal Botanic Gardens. 147–159.

Other Tubers

De Vries, F.T. 1991. Chufa (*Cyperus esculentus*, Cyperaceae): A weedy cultivar or a cultivated weed? *Economic Botany 45*, 27–37.

Ehrlich, C. 1989. Special problems in an ethnobotanical literature search: *Cordyline terminalis* (L.) Kunth, the "Hawaiian ti plant." *Journal of Ethnobiology 9*, 51–63.

Merlin, M. 1989. The traditional geographical range and ethnobotanical diversity of *Cordyline fruticosa* (L.) Chevalier. *Ethnobotany 1*, 25–39.

Pascual, B., Maroto, J.V., López-Galarza, S, Sanbautista, A., and Alagarda, J. 2000. Chufa (*Cyperus esculentus* L. var. *sativus* Boeck.): An unconventional crop. Studies related to applications and cultivation. *Economic Botany 54*, 439–448.

Zhang, H.Y., Hanna, M.A., Ali, Y., and Nan, L. 1996. Yellow nut-sedge (*Cyperus esculentus* L.) tuber oil as a fuel. *Industrial Crops and Products 5*, 177–181.

6
Fruits

CHARLES R. CLEMENT

The biblical tale of the Garden of Eden describes how fruit trees provided most of Adam and Eve's sustenance. Upon their expulsion from the Garden, they and their descendants had to learn agriculture and live "by the sweat of their brows." Modern archaeological research about the transition to agriculture shows that pre-agricultural subsistence included numerous fruit trees in most parts of the world, including Southwest Asia. The more affluent hunter-gatherers, who generally became the first farmers (Hayden 1995, 273–99), used a wide range of fruits and nuts in their subsistence strategies. Fruits were not abandoned during the transition to agriculture, with wild fruits an important resource in most farming communities until recently. However, in temperate areas fruits were generally domesticated several millennia later than the annual cereals and pulses. The timing might in some cases may reflect the difficulty of vegetative propagation of certain trees.

Fruits come in an enormous variety of forms and chemical compositions. In general terms, they can be divided into juicy, starchy, and oily. The vast majority of fruits grown commercially are juicy, the major exceptions being banana, date, and avocado. But in a worldwide review of subsistence fruits, a large number are starchy and oily, and offer significant energy to the diet. Some of the most important oil crops are fruits, such as olive (*Olea europaea*), coconut (*Cocos nucifera*), and African oil palm (*Elaeis guineensis*). In addition, there are dozens of minor oil crops are found throughout the tropics and subtropics.

As might be expected, fruits are more important where they are naturally more abundant, especially in wooded and forested areas of the humid and semi-humid tropics, subtropics, and temperate zones. This is evident from the information that we have about the origins of the fruits about which we know the most. Southeast Asia is an especially important region of origin, both on the mainland and on the adjacent islands. The better known temperate fruits generally come from Asia, although some come from its European extension. Curiously, the African humid tropics did not yield any fruit domesticates, perhaps because fruits were so easy to obtain from the forest, but the savannas yielded two important fruits (melons and watermelons). Both the Mesoamerican and South American humid tropics contributed numerous species, many of which are widely distributed today. The species discussed here will be organized by regions of origin.

Starfruit or carambola (*Averrhoa carambola*). Photo by Scott Bauer. ARS/USDA.

Today, fruit species are extremely important in both world trade and subsistence. Of the thirty crops that feed the world (the "short list;" Harlan 1995), seven are fruits. This short list was widely criticized for underrecognition of crop diversity, so Prescott-Allen and Prescott-Allen (1990) generated a longer list, (the "long list") with 103 crops, of which 23 are fruits. Although the numbers are different, the proportion is the same: 22 to 23 percent. Both certainly underestimate the true importance of fruit crops, especially in the less developed countries and in the tropics, because fruit crops are widely used for subsistence and traded in local markets, from where data on their production and consumption never make it to the national and international statistics. Brazil, for example, is both the world's largest producer and consumer of bananas, but less than half of national production is included in national statistics!

The fact that Brazil is the world's largest producer of bananas even though they originated in Southeast Asia, raises an interesting issue. Fruit crops, and crops in general, tend to be most important outside of their centers of origin, especially when produced commercially. This is because they temporarily escape from the pests and diseases that have evolved with them, thus enabling the crops to be more productive in other lands. While this is generally true commercially, subsistence practices often retain the pre-modern pattern of locally originated fruits being locally important. This is possible because of the high species diversity of traditional arboricultural systems (as seen in the forest fruit orchards of Borneo, for example) in which individual plants escape from their pests and diseases by 'hiding' behind other plants or by being rare in the system (Altieri 1995, 433). Pests also more readily multiply in monocultural stands, which often provide a very large food supply for pest species.

Vegetative propagation is the norm in modern commercial fruit culture and has been suggested as a prerequisite for domestication (Spiegel-Roy 1986). However, this idea is questionable because, in most fruit species, vegetative propagation is either a labor intensive or a technically demanding operation, or both. Hence, it is likely to be used only when the horticulturist identifies a particularly high-quality fruit, not just a slightly larger or slightly sweeter fruit, but something truly worth working to maintain. During the early millennia of fruit domestication, it was more likely that all propagation was sexual and mass selection was practiced in natural populations while hybrid swarms were created by planting trees from different populations in the same locality. After generations of mass selection in these conditions, particularly high quality fruit are more likely to appear. In fact, throughout the humid tropics, sexual propagation by small farmers is still the norm, which makes them the curators of the majority of the world's fruit genetic resources.

While fruits represent only 20 percent of the crops that feed the world, they are much more important than these statistics suggest. They supply a good deal of the vitamins and minerals that are essential to healthy diets, as well as a wide diversity of colors, shapes, and flavors that make local and world cuisines attractive. More importantly, as an Australian horticulturalist once commented to this author, "Agriculture makes modern civilization possible, but horticulture makes it bearable!"

The Fruits of Southeast Asia

This region is the most important center of origin for fruit species grown commercially, both large-scale and small-scale. It was initially identified as such by Alphonse de Candolle (1908) and confirmed by the Russian studies lead by Nicolai I. Vavilov (1992). Subsequently, Harlan (1995) proposed that this region is not really a center of origin and diversity, but a mosaic of origins and concentrations of diversity, which seems a better definition for such a large and variable region. In this chapter, Southeast Asia is defined as extending from eastern India to southern China and includes the larger archipelagos of eastern Oceania, from the Philippines to Indonesia.

Banana *Musa acuminata, M. acuminata × balbisiana*
Musaceae

The bananas are far and away the most important fruits in the world (Simmonds 1995). They also have a complex genetic history. *Musa acuminata* appears to have originated in Southeast Asia and the immediately adjacent larger islands. Wild *M. acuminata* has very seedy fruits which were selected very early for parthenocarpy (development of fruit without seeds) and seed sterility, resulting in the seedless fruits typical of modern bananas. At this early stage in its domestication, *M. acuminata* was diploid but soon triploids (with three sets of chromosomes rather than two) appeared. The triploids were immediately recognized and conserved because of their greater vigor and larger fruit; they are the table bananas of world commerce today. As these *M. acuminata* diploids and triploids were distributed and cultivated in various parts of Southeast Asia, they came into contact with *M. balbisiana* and new hybrid polyploids appeared, were recognized, and conserved. All of this initial development occurred in Southeast Asia and the larger islands of western Oceania during the millennia before the Christian era.

During the first millennium AD, bananas were distributed both west, to Madagascar and then Africa, and east, to the smaller islands of Oceania, eventually reaching Hawaii. Bananas were among the first crops taken to the New World during the European conquest, quickly being adopted by Amerindians; so quickly, in fact, that bananas occasionally arrived before the European explorers. Throughout this distribution they quickly became major components of subsistence. The Yanomamo tribe of northwestern Amazonia, for example, adopted bananas early and transformed them into their staple subsistence crop. In the humid tropics all types of bananas do well and are important.

Today bananas are major subsistence resources between 40° N and S, often being the major source of starch, hence energy, for rural and poor urban populations. The export trade is based on modern high technology commercial plantations in tropical America, from where they are exported to North America, Europe, and East Asia. In 2002 the FAO measured world banana production at over 68 million tons, but this is surely an underestimate because subsistence use is seldom included in national statistics. Africa probably produces 50% of the world total but is notorious for poor statistics.

See: Natural Fibers and Dyes, p. 302

Citrus
Rutaceae

The citrus species are the most important group of fruit species in the world. They also originated in Southeast Asia (Roose et al. 1995). Their taxonomy is an endless source of discussion: Swingle and Reece (1967) proposed sixteen species in the genus, while Tanaka (1977) proposed 162! Recently, Scora (1988) proposed that the basic biological species are *C. medica* (citron), *C. reticulata* (mandarin), and *C. maxima* (pummelo), and that the other familiar 'species' are of hybrid origin (Roose 1988), most of which also originated in Southeast Asia, although grapefruit (*C. paradisi*) originated in the Caribbean quite recently (18th century). Unfortunately, little is known about the specific origin, domestication, and early history of these important species, at least until they started appearing in Southwest Asia and Europe several millennia after they entered the domestication process in Southeast Asia.

The citron appeared in Persian historical records in 300 BC, but it took more than a millennium for the other species to appear in Southwest Asia and Europe. The sour orange, *Citrus aurantium*, appeared in the 11th century, the pummelo (also called pomelo) and the lemon *C. limon* in the 12th, the lime *C. aurantifolia* in the 13th, the sweet orange *C. sinensis* in the 15th and finally the mandarin in 1805 (Roose et al. 1995). These dates correspond with the expansion of European trade with Asia. Lemon and sweet orange arrived in the Americas with Columbus' second voyage, and the others later.

Sweet orange is the citrus produced in the largest quantities today, with world production estimated at 62 million tons (FAO 2002); mandarin, lime, lemon, pummelo, grapefruit, and sour orange are also produced commercially but on a slightly smaller scale (Prescott-Allen and Prescott-Allen 1990). Southeast Asia is responsible for only 15% of world production, consistent with the tendency for commercial production to be remote from the center of origin of the crop, although Asian subsistence production is probably greater than commercial production there. Southern Europe, the southern United States, and Brazil (most of this used in manufacturing orange juice) are major commercial producers today.

Mango *Mangifera indica*
Anacardiaceae

The genus *Mangifera* contains approximately forty species, of which more than half have fruit eaten by people in various parts of Southeast Asia (Gruèzo 1992). The best known is mango, which has been cultivated for several thousand years in semi-humid to semi-arid parts of India (Gruèzo 1992, 203–6). The mango is thought to have originated in what is now eastern India (where it has been cultivated for at least 4000 years as mentioned in mythology) and Burma, from where it spread throughout Southeast Asia. Unlike the other important Southeast Asian fruit species, mango is diploid and its domestication was straightforward. Today hundreds of cultivars are recognized in India alone, as well as dozens in other parts of the world. Nonetheless, the majority of mango trees in the world are seedlings, so that fruit diversity is impressive in many regions far from the origin of the crop.

About 1000 AD, mango was taken to east Africa, from where it spread into the semi-humid to semi-arid parts of the continent. Because it is a more tropical crop than *Citrus*, for example, it was never successfully introduced into Europe, despite attempts by many botanical gardens. Mango was introduced into the Americas well after the initial European conquest, arriving in Bahia, Brazil, in the early 1700s. From there it was taken to the Caribbean in about 1782 and to Mexico and Florida in the early 19th century (Corrêa 1974).

Based on FAO data, mango is the sixth most important world fruit crop, with 26 million tons being produced each year (FAO 2002). Again, this is certainly a major underestimate of its importance given its common occurrence in domestic gardens throughout the tropics and subtropics.

See: The Hunter-Gatherers, p. 3

Breadfruit, Breadnut *Artocarpus altilis*
Moraceae

The breadfruit originated in what is now Papua New Guinea and became a starchy staple of the peoples of Oceania (Ragone 1997). In Papua New Guinea the diversity of breadfruit extends from fully seeded types (the primitive type, breadnut, whose seed was more important than the pulp in human subsistence) to fully seedless types (the full domesticate that depends upon vegetative propagation). From here the breadfruit was distributed throughout central and eastern Oceania reaching Hawaii with the Polynesians. Curiously, the breadfruit only reached western Oceania, specifically the larger archipelagos of the Philippines and Indonesia, after European contact, perhaps because the Oceanian peoples did not invade these more populous regions after domestication of the breadfruit. The Europeans were fascinated by breadfruit, since a baked fruit tasted of bread (hence the name) and a boiled fruit was reminiscent of potatoes (*Solanum tuberosum*). The Europeans, in the person of Captain Bligh and the crew of the Providence, took breadfruit to the Americas in 1793, where it became popular (though banana was preferred) in the Caribbean.

Durian *Durio zibethinus*
Bombacaceae

The durian, with its very distinctive musky aroma, is the most controversial of tropical fruits. All who come across it either love it or hate it because of that aroma. This aroma has been described as a mixture of rotten onions and cheese, with turpentine (León 1987). In contrast, the delightful flavor of the aril around the seed has been described as a mixture of caramel and strawberries with cream (idem). Hence, those who hate it have been unable to get a morsel past their nose, while those who love it use the aroma to find the fruit.

The genus *Durio* is restricted to Southeast Asia, where at least seven species are fruit crops, the most important of which is *D. zibethinus* (Subhadrabandhu et al. 1992). In the humid tropics below 800 m (2600 feet), the durian is often found on small farms, either planted in the home garden and along the edges of fields, or in arboriculture systems (Foresta and Michon 1993, 709–724). Regional production was estimated at one million tons in the late 1980s, even though only a fraction of this came from commercial systems, showing its enormous popularity in local and regional markets. The majority of the commercial crop is grown in Thailand, Malaysia, and Indonesia.

Carambola, Star fruit *Averrhoa carambola*
Oxalidaceae

The origin of the carambola is Southeast Asia, but precisely where and when is not yet clear. The common name originated in Malabar, western India, where it was an early introduction since it also has a Sanskrit name: karmara (Donadio et al. 2001). The tree is naturally prolific, and produces fruits that are five-pointed or star-shaped in cross section, with yellow to orange rind and mildly perfumed pulp. These characteristics have transformed it into one of the Southeast Asian fruits that is most widely planted today, both commercially and for subsistence. It spread beyond Southeast Asia rather late, arriving in Brazil in the 19th century.

Mangosteen *Garcinia mangostana*
Clusiaceae

Garcinia is a large Old World genus, with a few cultivated species in Southeast Asia (e.g., Sunda Islands, Malay peninsula), of which only the mangosteen is economically important. The species is thought to be an allotetraploid (a hybrid between species made viable by chromosome doubling) between two Malaysian species (Verheij 1992, 177–181), which may explain why it is only found in cultivation. Mangosteen was restricted to the lowland humid tropics of Southeast Asia until very recently; it has spread as far as Sri Lanka, south India, northern Australia, and the New World tropics only in the last two centuries (idem).

The mangosteen has been historically considered the queen of tropical fruits, with a delicious, sweetish-tart flavor, and mild perfume. In the late 1980s, Thailand was the major producer, followed by Malaysia and the Philippines. Given the strong demand for the fruit, a major planting program took place in the 1990s around the tropics, even though seedlings and even grafted plants take 7 to 12 years to bear fruit.

Lychee *Litchi chinensis*
Sapindaceae

The lychee originated in southern China and northern Vietnam (Menzel 1992, 191–5), although subspecies exist in Indonesia and the Philippines. The earliest Chinese horticultural writings mention dozens of lychee varieties, attesting to its popularity and very early cultivation. It is reputed to have been cultivated by Malayan peoples as early as 1500 BC. The lychee is the most environmentally sensitive of all tropical fruit crops and fails to yield well, or at all, unless its requirements are met. Considering its popularity in Southeast Asia, its spread to India, Africa, and the New World occurred very late, starting in Jamaica in 1775, and in California in the late 1800s. Nonetheless, it is attracting considerable horticultural and commercial attention, given its attractiveness to modern consumers worldwide.

Rambutan *Nephelium lappaceum*
Sapindaceae

The rambutan originated somewhere in Southeast Asia (probably in the Malay Archipelago), and was soon distributed and cultivated throughout the region (van Welzen and Verheij 1992). The tasty fruit is quite similar to lychee, but rather than being knobby, it has filiform protuberances (red or yellow spinterns) on the rind that are 0.5 to 1.0 inches (1.0 to 2.5 cm) long. Rambutan is less demanding and generally more productive than lychee, which has encouraged its introduction into the American, Australian, and African humid tropics during the 20th century.

The Fruits of Southwest Asia

This region is the oldest center of origin of agriculture, where wheat, barley and peas were domesticated. The grape (*Vitis vinifera* L.), the second most important fruit crop on the short list, originated here and is treated elsewhere in this volume. A few commercially important and minor fruit species originated here, only two of which are included in the long list (date palm and fig). In this chapter, Southwest Asia extends from western India to the Mediterranean, including the so-called Fertile Crescent and Arabia.

Date palm *Phoenix dactylifera*
Arecaceae

The date palm was probably the first fruit tree to be domesticated in the Old World. The distribution of truly wild date palms is uncertain, as date cultivation now occupies most of the habitat of wild dates. It is likely that they were domesticated in Arabia. Early finds of date stones in the United Arab Emirates and Kuwait, dating to 5000 BC, may mark the beginnings of cultivation or trade (Beech and Shepherd 2001). By 3000 BC the date palm was a well established crop in Mesopotamia. Archaeological evidence of *madbasa* (date presses) for the extraction of *dibs* (date syrup) dates back to 1750 BC in the Arabian Gulf. Today it is distributed throughout Old World sub-tropics and tropics, where "it must have its feet in the running water and its head in the fire of the sky" (an Arab saying cited by Wrigley 1995). Throughout this immense region there are probably 100 million date palms, most of which were vegetatively propagated, even though the process is very inefficient. Vegetative propagation

Date palm (*Phoenix dactylifera*). Copyright Dave Webb, used with permission.

and hand pollination (mentioned in cuneiform texts of Ur from around 2300 BC) have been standard practice for thousands of years, especially in the Fertile Crescent. Each palm provides 40 to 60 kg (90 to 130 pounds), and occasionally up to 200 kg (440 pounds), per year of an energy-rich fruit that is often a staple food. The date was sacred to the inhabitants of ancient Mesopotamia and continues to be extremely important to Moslems, who appreciate it year round but especially to break fast during Ramadan and Hadj. Although taken to the New World and Australia in the last few hundred years, the date never became important outside its original distribution.

See Caffeine, Alcohol, and Sweeteners, p. 185

Fig *Ficus carica*
Moraceae

The fig was domesticated in the Fertile Crescent and the drier areas south of Mesopotamia very early during the development of agriculture in the region. The earliest finds, of wild figs, date to 7800–6600 BC, and there is evidence of cultivation in Mesopotamia and Egypt by 3000 BC (Zohary and Hopf 2000). It continues to be more important in this region than elsewhere, although it was spread early throughout the Mediterranean basin and eastwards to India and finally China. The majority of production occurs in the area where it was first distributed. It was spread to the New World in the 16th century. In some locations it is even considered an invasive species.

Zohary and Hopf (2000) consider the fig to be a prime example of a fruit crop whose domestication and importance depended on the development of vegetative propagation. The species is dioecious, hence cross-pollinated (by a single species of co-evolved wasp, *Blastophaga psenes*), so only the female plant is easily selected for important fruit traits. Open-pollinated figs yield widely segregating progeny, half of which are non-fruiting males; hence the demand for vegetative propagation, which is easily achieved by cuttings. Consequently, there are several hundreds of named clones in two groups: Smyrna figs (which require pollination for fruit set and consequently only fruit where the wasp is present) and common figs (which are parthenocarpic). Sycomore fig (*F. sycomorus*) originated in ancient Egypt. It is pollinated by a wasp, but produces seedless fruits.

Jackfruit, Jakfruit *Artocarpus heterophyllus*
Moraceae

The largest of the cultivated fruits, the jackfruit originated in the rain forests of the Western Ghats, India, from whence it was distributed in cultivation throughout India and into Southeast Asia (Soepadmo 1992, 86–91). Like the breadnut (*A. altilis*), the jackfruit offers both seeds and edible pulp,

Jackfruit (*Artocarpus heterophyllus*). Copyright Dave Webb, used with permission.

the latter generally strongly flavored and scented in the soft juicy types, and less strongly flavored and scented in the crispy varieties. The crispy varieties are often more easily accepted by Westerners. As both an energy and protein source, jackfruit is an important compliment in subsistence diets throughout Southeast Asia and often becomes a staple during food shortages. During the 1st millennium AD it was taken to Africa. During the 16th century, the Portuguese took it to Brazil. Subsequently it spread into the Caribbean, reportedly being introduced to Jamaica in 1782. In both of these later distributions it became important in the subsistence economy but not in commerce.

The Fruits of Temperate Eurasia

Temperate Eurasia extends from Europe to China across the Old World and is home to dozens of temperate fruit species, including the apple. The majority of these originated in temperate Asia, some in the western-central part, some in the eastern part, especially China. Many of them originated early, but some originated quite late and continue to appear. Several berries were domesticated recently in Europe and the most recent berry crop, the kiwifruit, was domesticated in the second half of the 20th century.

Apple *Malus domestica*
Rosaceae

Genetic studies have shown that the wild *Malus sieversii*, native to the Tien Shan mountains of Kyrgyzstan and northwest China (Harris et al. 2002), is the wild ancestor of both *M.* × *domestica* (which is the major apple west of the center of origin and is now spread around the world) and *M. asiatica* (the major apple of China, although *M.* × *domestica* is now being widely planted in Asia). Hybridisation with other wild apple species may have been rarer than previously thought (cf. Watkins 1995), although some taxonomists still classify apple as *Malus* × *domestica*, in other words as a hybrid species. Archaeological evidence for the spread of apple from central Asia is scanty, but the apple was doubtless spread by travellers on the great trade routes running from central Asia to the Danube.

Although the modern apple is always grafted today, at least commercially, early propagation was by seed. Selection was practiced on populations of wild apples that resembled today's 'crab apples,' and slow progress was achieved from continued mass selection. In fact, 'crab apples' are a frequent find in early Neolithic and later archaeological sites (Zohary and Hopf 2000). These crab apples were of European and near Eastern species such as *Malus sylvestris*, collected from the wild and often dried and threaded on strings for winter consumption. Only later, when the Greek civilization flourished, did apples start looking distinctly modern, since the Greeks were familiar with the art of grafting (White 1970).

In the last several hundred years apples started to be selected and grafted throughout Europe, and were taken by colonists to North America. The story of Johnny Appleseed (based on the life of the arborist and evangelist from New England, John Chapman) shows that good European stock started to be seed propagated again, giving rise to new hybrid swarms and allowing the selection of new apple varieties unlike those in Europe. Today, the European and North American varieties are the basis for the worldwide apple business, with dessert apples, cooking apples, cider apples, and even crab apples. Apple production is estimated at 56 million tons (FAO 2002), making it the third most important fruit crop worldwide.

See: Caffeine, Alcohol, and Sweeteners, p. 181; Ornamentals, p. 266

Apricot *Prunus armeniaca*
Rosaceae

The apricot originated in China (Watkins 1995, 423–9) and was mentioned in an early Chinese text of Shan-hai-king (2205–2198 BC). It was introduced into the Fertile Crescent in historic times, about 100 BC (Hopf 1973), from Iran or Armenia (hence the species epithet). It was spread throughout the eastern Mediterranean, even though it has rather exacting ecological requirements. It is considered by many to be one of the most delectable of *Rosaceae* tree fruits.

Kiwifruit, Chinese gooseberry *Actinidia chinensis* (syn. *A. deliciosa*)
Actinidiaceae

Kiwifruit is perhaps the most recently domesticated of all internationally trade fruit crops, having been introduced to New Zealand in 1904 and fully developed as a crop by farmers in New Zealand by the 1930s. It is derived from *A. deliciosa* (Ferguson 1995), although both *A. deliciosa* and *A. chinensis* are still collected from the wild in China. The common name was coined in 1959, to allow marketing of the Chinese gooseberry in the United States. The kiwifruit has become the classic example of "new" crop creation, highlighting the importance of varietal selection, agronomic practices, post-harvest handling and packaging, and, perhaps most important, clever marketing (Beutel 1990, 309–16).

Pear *Pyrus communis*
Rosaceae

Like the apple, the pear originated in central Asia and was distributed both east and west (Watkins 1995). It has been cultivated in China for over 4000 years. Several western pears may have been culti-vated early, before hybridizing with *P. communis*, especially *P. nivalis* and *P. serotina*. Several Chinese pears were also cultivated (*P. pyrifolia* Nakai, the Chinese sand pear; *P. ussuriensis*, the Ussurian pear; *P. breschneideri*, the Chinese white pear; Zohary and Hopf 2000). The pear hybridized with local populations and related species en route, resulting in abundant variation for local selection. By Greek times, several grafted pear varieties were described and open-pollination was known to result in inferior types (Theophrastus, ca. 300 BC, cited by Zohary and Hopf 2000).

Again like the apple, Europeans started selecting and propagating pears in the last several hun-dred years and spread them to North America. Although there was no Johnny Pearseed, pears were initially propagated in North America by seed, resulting in an explosion of variability for later selection and propagation by grafting. In 2002, 17 million tons of pears were produced worldwide, most derived from European and North American varieties, but also a good portion (at least 20%) from Chinese varieties.

Peach *Prunus persica*
Rosaceae

The peach originated in China and Tibet and was distributed westward to India, the Fertile Cres-cent, and Europe, arriving in Greece in about 300 BC (Zohary and Hopf 2000). During the west-ward dispersion, the nectarine (the smooth-skinned mutation of the peach) may have originated in Turkestan, just north of Persia (Hesse 1975). The Romans spread the peach around the Mediterra-nean starting in the 1st century AD, generally as seed, even though grafting was known to them. This allowed local selections to be identified that were ecologically better adapted to local condi-tions. The Iberians were fond of peaches and took them to the Americas early. In the Americas they were enthusiastically accepted by Indian populations, who spread them to their current ecological limits (Hesse 1975). Today, the United States is a major world producer, with major areas on both coasts. Production is highest in China, Italy, the United States and Spain.

See: Ornamentals, p. 269

Plum *Prunus domestica*
Rosaceae

The European plum (*P. domestica* subsp. *domestica*) originated in eastern Europe and western Asia. The Damson plum (*P. domestica* subsp. *insititia*) originated in western to central Asia (Zohary and Hopf 2000). Both were collected from the wild before being domesticated in these two areas, and started to be grafted in Roman times. Since their origin, they have also hybridized with other Eurasian *Prunus* spp., as well as, after their introduction to the New World, with the American plum (*P. americana*). The Japanese plum (*P. salicina*) originated in China, spread to Japan. Itwas introduced to the U.S. in 1870 and hybridized with the American plum.

Persimmon, Date plum, Kaki *Diospyros kaki*
Ebenaceae

The kaki or persimmon is native to East Asia (northern China to Indo-China) and appears to have been cultivated and first domesticated in China before the Christian era (IBPGR 1986). It was introduced into Japan during the 8th century AD and is often called the Japanese persimmon. There are other Asian species (especially *D. discolor* and *D. lotus*) and a North American species (*D. virginiana*) that have been cultivated. The kaki is the major species of commerce, although still a minor fruit. Wild types or primitive kakis may have astringent fruit, due to excessive tannins. The best cultivars are non-astringent and sweet.

Red currant *Ribes rubrum* and Black currant *R. nigrum*
Grossulariaceae

The red currant was first described as a crop about 500 years ago in Germany. The black currant was first recorded in 17th century British herbals (Keep 1995). Hence, both are the first truly European contributions to world fruit culture. The red currant hybridizes easily with other European *Ribes* spp., especially *R. sativum* and *R. petraeum*, both of which contributed considerable diversity and quality to modern red currant varieties. Some North American species have recently been included in improvement programs, but North American production is still based primarily on European stock.

Raspberry *Rubus idaeus* and Blackberry *R. ulmifolius*
Rosaceae

These berries are often called brambles because of their growth habit (Ourecky 1975). Their taxonomy is extremely complex and the two names used here are simply convenient; some bramble students use the subgeneric names *Idaeoatus* (raspberries) and *Eubatus* (blackberries) instead of specific names, given the ease with which species in each group hybridize. Both subgenera are temperate and circumpolar.

The raspberries were first cultivated in Europe in the 1500s, but the earliest named cultivars only appeared in the 1800s in both Europe and North America (Ourecky 1975; Jennings 1995). By this time both European and North American 'species' were being cultivated, hybridized, and selected. The blackberries appear to have been cultivated in Europe only a few years before they were cultivated in North America; a European cultivar was introduced into North America in 1850. Soon thereafter, it 'escaped' and hybridized with American blackberries, resulting in new variation that was selected locally. Although the brambles are not among the most important world fruit crops, they are locally important in many parts of the northern hemisphere and a Latin American species (*R. glaucus* Benth.) has recently started to be cultivated and selected.

Sweet cherry *Prunus avium* and Sour cherry *P. cerasus,*
Rosaceae

The sweet cherry originated in western to central temperate Asia and appears to have been introduced rather late into the Mediterranean basin (about 300 BC) where it may have been used principally for its wood (Fogle 1995, 348–66). Soon after, the Romans reported the introduction of a single sweet cherry variety to Rome (White 1970). The sour cherry is an allotetraploid between the sweet cherry and the ground cherry (*P. fruticosa* Pallas), which is native to extreme western Asia and eastern Europe. Hence, the sour cherry originated somewhere in this region. Although numerous cherry species are indigenous to North America, only the sweet and sour cherries are commercially planted there, after their introduction in the colonial era.

The Fruits of Africa

Although Africa contributed several important crops to world agriculture, it only contributed two important fruit crops. Both of these are savanna natives, as expected by Harlan's (1995) proposal that agricultural crops originated in a mosaic of environments across a wide area of transition between the forest and the savanna.

Melon *Cucumis melo*
Cucurbitaceae

The melon was probably first domesticated from the wild *C. metuliferus*, one of great variety of annual melons that are grown throughout the drier parts of subtropical and tropical Africa and Asia (Zohary and Hopf 2000). Melon seeds have been reported from Egypt during the 2nd millennium BC (Zohary and Hopf 2000), at about the same time as watermelon, and soon thereafter in Southwest Asia, indirectly supporting an African domestication. By the 1st millennium AD, the major groups of melons were evident. The two most important melon groups are the Cantalupensis group (cantaloupe, muskmelon, and Persian melon, with orange aromatic flesh) and the Inodorus group (winter, honeydew, crenshaw, casaba, with white or green, only mildly aromatic flesh) (Bates and Robinson 1995). By the 15th century, melon varieties are mentioned in European sources and must have been introduced into the New World soon after contact. These two groups account for the majority of world melon production, although the other five groups (not listed here) contain melons that are locally important.

Watermelon *Citrullus lanatus*
Cucurbitaceae

The watermelon may have originated in tropical and southern Africa, where wild fruits are still harvested in the Kalahari desert, including both bitter and sweet types used for water and the edible seeds, and where several locally domesticated cultivars are grown (Bates and Robinson 1995, 89–96). The closely related *C. colocynthis* appeared earlier than the watermelon in Egypt (at 3800 BC), followed by the watermelon at 2000 BC (Zohary and Hopf 2000). The watermelon spread to Southwest Asia, India, China by the 11th century AD and to the New World in the centuries since contact. Throughout this distribution, locally selected cultivars are numerous, varying in size, shape, color and patterning of the rind, color and density of the flesh, and color of the seeds (Bates and Robinson 1995). China, the United States, and Turkey are major producers, although most subtropical and tropical countries produce enough for self-sufficiency.

See: Nuts, Seeds, and Pulses, p. 152

The Fruits of North America and Mesoamerica

The North American continent can usefully be divided into two parts when dealing with cultivated plants. Central Mexico to Panama is known as Mesoamerica, usually including the Caribbean. Central Mexico to the Arctic is North America. Mesoamerica was home to several well known Amerindian civilizations, the most recent being the Mayas and the Aztecs; but the region has a much longer history, especially in terms of crop domestication. Curiously, the crops of the Caribbean have a strong South American influence because the islands were colonized by Arawak Indians in the 1st and early 2nd millennium. De Candolle (1908),

Vavilov (1992), and Harlan (1995) all agreed that Mesoamerica was an important center of crop origin. In contrast, North America contributed relatively little to the world crop repertoire, the most important being the sunflower (*Helianthus annuus* L., Compositae). In the last two centuries, North American berries have become important crops that are now grown in Eurasia and South America.

Avocado *Persea americana*
Lauraceae

The avocado is one of the oldest Mesoamerican domesticates, with good archaeological evidence for use by 6000 BC, cultivation by 5000 BC, and changes due to selection by 750 BC in southern Mexico (Smith 1969). The attractiveness of the avocado as a subsistence and modern food is due to good vitamin and mineral contents, high energy content derived principally from mono-unsaturated fats, and good supply of soluble and insoluble fibers, all of which make it one of the most nutritious fruits available (Bergh 1995, 240–5). The crop was domesticated in southern Mexico and adjacent Guatemala, soon differentiating into three geographical ecotypes: Mexican (var. *drymifolia*); Guatemalan (var. *guatemalensis*); West Indian or Lowland (var. *americana*). These ecotypes are adapted to different altitudes and, consequently, to different latitudinal extremes, temperatures, and rainfall requirements in their modern distribution. The Lowland ecotype was being introduced into Amazonia at the time of European conquest (1540) and was renamed "palta" in the low elevation Andes (avocado is a corruption of the Aztec ahuacatl, variants of which occur throughout its Mesoamerican distribution). Avocado is produced primarily in the Americas, although Israel, Ethiopia, Spain and Indonesia are also producers.

Papaya, Pawpaw *Carica papaya*
Caricaceae

The papaya originated in the drier lowlands of southern Mexico and Guatemala. There plants with golf-ball size fruits with little flesh and packed with seeds grow in disturbed and transitional ecosystems, especially in and adjacent to Mayan ruins (Morshidi 1996, 284). Earlier hypotheses suggested its center of origin extended from southern Mexico to northwestern South America (León 1987). Although associated with Mayan ruins today, the papaya was certainly domesticated much earlier, but perhaps not quite as early as the avocado. By the mid-16th century, the papaya was present in an enormous variety of sizes (up to 5 kg), shapes (round to long and thin), flesh colors (yellow to red), and sugar contents (insipid to very sweet). By that time it had also spread throughout the humid and semi-humid lowland American tropics. Subsequently, the papaya has spread around the world. Three other *Carica* spp. are very minor fruit crops of the central and northern Andes in South America.

Acerola, Barbados cherry *Malpighia emarginata*
Malpighiaceae

The acerola grows wild throughout the Caribbean and is reputed to be truly wild in the Yucatan Peninsula of southern Mexico (IBPGR 1986). Its domestication is certainly pre-Colombian (1490s) but it is not clear that significant changes in fruit size and quality resulted from this selection. Reports state it was widely used by the native population during the conquest, when the Spanish compared it to European cherries. Unlike the true cherry, the acerola is extremely rich in vitamin C, with 1 to 4 g per 100 g useable pulp. This is ten times more than guava and forty times more than lemon. Although its flavor is unexceptional, the vitamin C attracted commercial interest, and plantations were established in Puerto Rico in 1945 and later throughout tropical and subtropical lowland America.

Acerola (*Malpighia emargina*). Copyright Dave Webb, used with permission.

Cranberry *Vaccinium macrocarpon* and Highbush blueberry *V. corymbosum*
Ericaceae

The genus *Vaccinium* is large (possibly 400 species) and widespread; many species have long been harvested from the wild (Hancock 1995, 121–3). The cranberry was first cultivated in Massachusetts in the early 19th century, initially by management of water in cranberry swamps. Cranberry breeding started in the mid-20th century, although most selections are little removed from the wild type. The highbush blueberry and its close relative the rabbiteye blueberry (*V. ashei*) started to be domesticated at the end of the 19th century and breeding started in the early 20th century. The lowbush blueberry (*V. angustifolium*) is an even later addition to the *Vaccinium* crop list, with breeding and cultivation starting in the late 20th century. The majority of world *Vaccinium* production is still based in North America, primarily in the USA, although Europe (including Russia), Australia, and New Zealand are becoming important producers.

Cranberry (*Vaccinium macrocarpon*). USDA-NRCS PLANTS Database/Britton, N.L., and A. Brown. 1913. *Illustrated flora of the northern states and Canada*. Vol. 2, p. 705.

Prickly pear, Barbary fig *Opuntia ficus-indica* (syn. *Opuntia maxima*)
Cactaceae

The prickly pear—also known as barbary fig or Indian fig—was cultivated and perhaps semi-domesticated in Mexico well before the conquest (León 1987). Both its tasty red fruit and its fleshy "leaves" (in reality photosynthesizing stems with widely flattened cross-sections) are used, the latter either cooked, baked, or roasted as vegetables. Today, prickly pear is common in semi-arid southern Europe and North Africa as a fruit, and is used as cattle forage in semi-arid northeastern Brazil.

See: Invasives, pp. 382–3

The Fruits of South America

The South American continent can usefully be divided into two major sections in terms of crop plants, with numerous minor regions. The first section is the Andes Mountains, which extend along the western edge and are home to numerous important food crops, although to only a few minor fruit crops. The second section is the Amazon, which occupies the center and is home to cassava (*Manihot esculenta*) and numerous fruit crops. Clement (1999b) reevaluated De Candolle (1908), Vavilov (1992), and Harlan (1995) and proposed a mosaic of crop origins and regions of diversity that occupy large sections of the various biomes of South America. In Amazonia, sixty-three of the 138 crops that were present at the time of the European conquest are fruits (Clement 1999a), a much higher proportion than on the world lists. The semi-arid northeast of Brazil also contributes fruit crops to the South American list, as does the humid Atlantic Forest, which extends from the northeast of Brazil southward to Uruguay. Nonetheless, South America has contributed fewer fruits to the world lists than Asia, although it has the potential to contribute more in the future.

Pineapple *Ananas comosus*
Bromeliaceae

The pineapple was probably domesticated in the Parana/Paraguay River basin, in the frontier between Brazil and Paraguay (Leal 1995), although the Orinoco River basin cannot be ignored as a possible center of origin. Well before European conquest, the pineapple was distributed throughout the lowland American tropics. This wide distribution was due to its rusticity (good growth and acceptable yield on sands to clay soils, with 500 to 5000 mm of rainfall), fruitfulness, excellent flavor, and ease of vegetative propagation. During this distribution, the Amerindians developed numerous cultivars. These have been grouped into five horticultural varieties: Spanish, Cayenne, Queen, Pernambuco, and Maipure. Today the vast majority of the international trade is based on one cultivar and its sports: Smooth Cayenne. This cultivar appears to have been found in the upper Orinoco River basin, Venezuela, and was taken to Cayenne, the capital of French Guyana, from where it was sent to France in 1820. Shortly thereafter it was distributed around the world. In the 20th century, the Smooth Cayene cultivar became the most important pineapple for canning, because of its uniformly cylindrical shape. The other varieties often have better flavor and color, however, and are becoming more important as the international fresh pineapple market expands.

Strawberry *Fragaria* × *ananassa*
Rosaceae

The modern strawberry has an unusual history. The genus *Fragaria* contains forty-six species with edible fruit found on all continents except Australia (Jones 1995). Many have been collected from the wild for millennia and some have even been cultivated locally for centuries. One of these species, *F. chiloensis*, grows wild from Alaska to Chile along the Pacific coast, and was cultivated and certainly semi-domesticated before the European conquest. Both naturally and as a result of human selection, *F. chiloensis* has larger fruit than other species of the genus. Another species, *F. virginiana*, grows wild throughout central and eastern North America, and has typically smaller fruit, which were popular with indigenous peoples before the conquest. Both are octoploids. Although natural hybrids between these species have since been reported in western North America, the modern strawberry (*F. × ananassa*) is a hybrid that was identified, selected, and cultivated in European gardens and first described in 1759, although it may have originated in 1750 in France (Jones 1995). Hence, the species originated within the modern period, a most unusual occurrence for a fruit crop, most of which have long histories. Also unusually, the fruit crop is of European origin, but derived completely from native American species. Finally, within 150 years, its importance expanded from a single garden to a world production of just over 3 million tons (FAO 2002).

Cherimoya *Annona cherimola,* Soursop *A. muricata,* and Sweetsop *A. squamosa* Annonaceae

These three annonas are the best of the numerous species that occur in the Neotropics and Africa (Pinto et al. 2004). The cherimoya was domesticated in the mid-altitude Andes of Ecuador, Peru, Bolivia, and Chile, the soursop in the lowland humid tropics of northern South America, and the sweetsop in the lowland sub-humid tropics of Mesoamerica. The sweetsop is immensely popular in India and has escaped from cultivation. Because the annonas were domesticated without vegetative propagation, they all come sufficiently true from seed to satisfy the home market. As the most temperate annona, the cherimoya was taken to Europe soon after contact, where it is currently cultivated in Spain and Italy, as well as in temperate Chile, Mexico, and the United States. As the most tropical annona, the soursop was spread throughout the lowland humid tropic quite early, the same occurring with the sweetsop in the lowland semi-humid tropics. As a consequence of this distribution, the cherimoya has become reasonably well known in Europe and the United States, while the soursop and sweetsop are still the domain of the tropical countries where they are produced. Annona breeding programs have existed in many countries since the beginning of the 20th century, when the 'atemoya' (*A. squamosa* × *A. cherimola*) was created nearly simultaneously in Florida and the Philippines (IBPGR 1986). Several variants of this hybrid are now cultivated in Florida, South Africa, India, Egypt, and other minor producing regions.

Sweetsop (*Annona squamosa*). Copyright Dave Webb, used with permission.

Guava *Psidium guajava*
Myrtaceae

The Myrtaceae is one of the most typical tropical plant families and has offered dozens of minor fruits to human societies throughout the lowland humid, sub-humid, and semi-arid tropics; only the guava is moderately important. The origin of guava is controversial. De Candolle (1908) identified the region between Mexico and Peru. But it only occurs in cultivation in Pacific Peru, where it appeared very early (before 2500 BC) in the archaeological record, and it is not mentioned as one of the earliest crops in Mexico. Hence, Clement (1999) identified the area of origin as northern South America, principally because of the numerous other *Psidium* with similar fruits in this large region. Because of the ease with which guava escapes from cultivation, however, it will be a difficult task to identify its true origin. Escapes are now found throughout the lowland humid and semi-humid world tropics. Guava is a locally important fruit crop in India, Pakistan, Mexico, and Brazil. In all of these areas, numerous cultivars have been selected with red, pink, or white flesh and varying degrees of vitamin C (although always at least five times more than *Citrus* fruits).

See: Plants as Medicines, p. 233; Invasives, pp. 383–4

Passionfruit *Passiflora edulis*
Passifloraceae

There are 370 species of passionfruit, 350 in the Americas and 20 in Asia and Australia (IBPGR 1986). The most important is *P. edulis*, although other species are locally important in tropical America. All are vines that produce berries, ranging in size from 1 to 8 inches (2 to 20 cm) in diameter. The passionfruit considered here come in two forms, the purple (f. *edulis*) and the yellow (f. *flavicarpa*). Both forms have aromatic, tart, delightfully flavored pulp around the seeds. The purple *edulis* is slightly sweeter and less tropical, having originated in southern Brazil, and has been spread more widely around the world, with Hawaii, Australia, and South Africa mentioned as important production areas (IBPGR 1986). The yellow *flavicarpa* is less sweet. It is also larger than the purple, more tropical, and currently more important in Brazil for juice and flavorings, which are now being exported to Europe, North America, and Japan.

See: Plants as Medicines, p. 233

Peach-palm, Pejibay *Bactris gasipaes*
Arecaceae

The peach palm is the only domesticated palm in the Americas. It appears to have been initially cultivated for its wood. Cultivation brought different populations into contact, allowing an increase in variability for native peoples to select. Once they started selection for fruit, this was transformed from a small (1 to 2 g), oily (25 to 30 percent of fresh weight) fruit with hundreds of fruits per bunch into a large (20 to 200 g) starchy (25 to 40 percent) fruit with only one hundred fruits per bunch (Clement 1995, 383–388). The initial selections probably occurred in lowland southwestern Amazonia. From there, the species was distributed both northeastward into central and eastern Amazonia, and then northwards into Mesoamerica as far as Honduras. At the time of European conquest, the peach palm was a starchy staple in many areas of the lowland humid tropics. At that time, the peach palm was of such extreme importance to the indigenous population of southern Mesoamerica that a group of Spanish adventurers cut down 20,000 plants in the Talamanca River valley of Costa Rica in order to subjugate the native population (Patiño 1963). The cooked starchy fruits were fermented to make a thick juice with varying alcoholic content or to preserve the pulp from one harvest season to the next, or ground and dried into flour for elaboration of breads and cakes. So critical was this fruit to indigenous diets that a majority of births occurred nine months after the two to three month peach palm harvest. This was in part because of the inebriating effect of the juice and in part because of the better nutritional status of the population after three months of eating these energy and vitamin A rich fruits. Since the conquest, however, the peach palm has become a marginalized fruit crop throughout tropical America, even though Brazilian, Colombian, and Costa Rican research institutions have developed considerable information about modern uses and cultivation (Mora Urpí et al. 1997, 83). Commercially, the peach palm is now a major source of hearts-of-palm because of the palm's clustering growth habit. This helps conserve wild populations of other palm species, which have been harvested unsustainably.

References and Further Reading

Altieri, M.A. 1995. *Agroecology: The Science of Sustainable Agriculture* (2nd ed.). Boulder, CO: Westview Press.

Bailey, C.H., and Hough, L.F. 1975. Apricots. In *Advances in Fruit Breeding*, edited by J. Janick and J.N. Moore. West Lafayette, IN: Purdue University Press, 367–383.

Bates, D.M., and Robinson, R.W. 1995. Cucumbers, melons and water-melons, *Cucumis* and *Citrullus* (Cucurbitaceae). In *Evolution of Crop Plants*, edited by J. Smartt and N.W. Simmonds. London: Longman, 89–96.

Beech, M. and E. Shepherd. 2001. Archaeobotanical evidence for early date consumption on Dalma Island, United Arab Emirates. *Antiquity* 75: 83-9.

Bergh, B.O. 1995. Avocado, *Persea americana* (Lauraceae). In *Evolution of Crop Plants*, edited by J. Smartt and N.W. Simmonds. London: Longman, 148–151.

Beutel, J.A. 1990. Kiwifruit. In *Advances in New Crops*, edited by J. Janick and J.E. Simon. Portland, OR: Timber Press, 309–316.

Clement, C.R. 1995. Pejibaye *Bactris gasipaes* (Palmae). In *Evolution of Crop Plants*, edited by J. Smartt and N.W. Simmonds. London: Longman, 383–388.

Clement, C.R. 1999a. 1492 and the loss of Amazonian crop genetic resources. I. The relation between domestication and human population decline. *Economic Botany* 53(2), 188–202.

Clement, C.R. 1999b. 1492 and the loss of Amazonian crop genetic resources. II. Crop biogeography at contact. *Economic Botany* 53(2), 203–216.

Corrêa, M.P. 1974. *Dicionário das plantas úteis do Brasil e das exóticas cultivadas*. Rio de Janeiro: Ministério de Agricultura.

Cowan, C.W., and Watson, P.J. (Eds.). 1992. *The Origins of Agriculture: An International Perspective*. Washington, DC: Smithsonian Institution Press.

de Candolle, A. 1908. *Origin of Cultivated Plants*. New York: Appleton. (Translation of the 1882 edition.)

Donadio, L.C., et al. 2001. *Carambola (Averrhoa carambola L.)*. Série Frutas Potenciais. Jaboticabal, São Paulo, Brazil: Sociedade Brasileira de Fruticultura.

Ferguson, A.R. 1995. Kiwifruit, *Actinidia* (Actinidiaceae). In *Evolution of Crop Plants*, edited by J. Smartt and N.W. Simmonds. London: Longman, 1–4.

Fogle, H.W. 1995. Cherries. In *Advances in Fruit Breeding*, edited by J. Janick and J.N. Moore. West Lafayette, IN: Purdue University Press, 213–255.

Foresta, H., and G. Michon. 1993. Creation and management of rural agroforests in Indonesia: potential applications in Africa. In *Tropical Forests, People and Food: Biocultural Interactions and Applications to Development. 13. Man & the Biosphere*, edited by C.M. Hladik et al. Paris: UNESCO and London: The Parthenon Publishing Group, 709–724.

Gruèzo, W.S. 1992. *Mangifera* L. In *Plant Resources of South-East Asia. 2. Edible Fruits and Nuts*, edited by E.W.M. Verheij and R.E. Coronel. Bogor, Indonesia: Plant Resources of South-East Asia (PROSEA), 203–206.

Hancock, J.F. 1995. Blueberry, cranberry, etc., *Vaccinium* (Ericaceae). in *Evolution of Crop Plants*, edited by J. Smartt and N.W. Simmonds. London: Longman, 121–123.

Harlan, J.R. 1995. *The Living Fields: Our Agricultural Heritage*. Cambridge, U.K.: Cambridge University Press.

Harris, S.A., Robinson, J.P. and Juniper, B.E. 2002. Genetic clues to the origin of the apple. *Trends in Genetics* 18(18): 426-430.

Hayden, B. 1995. A new overview of domestication. In *Last Hunters, First Farmers: New Perspectives on the Prehistoric Transition to Agriculture*, edited by T.D. Price and A.B. Gebauer. Santa Fe, NM: School of American Research Press, 273–299.

Hesse, C.O. 1975. Peaches. In *Advances in fruit breeding*, edited by J. Janick and J.N. Moore. West Lafayette, IN: Purdue University Press, 285–335

Hopf, M. 1973. Apfel (*Malus communis* L.); aprikose (*Prunus armeniaca* L.). In *Reallexikon der Germanischen Altertumskunde, Vol 1*, edited by H. Beck et al. Berlin: Walter de Gruyter, 368–372.

International Board for Plant Genetic Resources (IBPGR). 1986. *Genetic Resources of Tropical and Sub-tropical Fruits and Nuts (Excluding Musa)*. Rome: IBPGR.

Jennings, D.L. 1995. Raspberries and blackberries, *Rubus* (Rosaceae). In *Evolution of Crop Plants*, edited by J. Smartt and N.W. Simmonds. London: Longman, 429–434.

Jones, J.K. 1995. Strawberry, *Fragaria ananassa* (Rosaceae). In *Evolution of Crop Plants*, edited by J. Smartt and N.W. Simmonds. London: Longman, 412–418.

Keep, E. 1995. Currants, *Ribes* spp. (Grossulariaceae). In *Evolution of Crop Plants*, edited by J. Smartt and N.W. Simmonds. London: Longman, 235–239.

Leal, F. 1995. Pineapple, *Ananas comosus* (Bromeliaceae). In *Evolution of Crop Plants*, edited by J. Smartt and N.W. Simmonds. London: Longman, 19–22.

León, J. 1987. *Botanica de los cultivos tropicales.* San José: IICA.

Menzel, C.M. 1992."*Litchi chinensis* Sonn. In *Plant Resources of South-East Asia. 2. Edible Fruits and Nuts,* edited by E.W.M. Verheij and R.E. Coronel. Bogor, Indonesia: Plant Resources of South-East Asia (PROSEA), 191–195.

Mora Urpí, J., Weber, J.C., and Clement, C.R. 1997. *Peach palm.* Bactris gasipaes *Kunth. Promoting the conservation and use of underutilized and neglected crops. 20.* Gatersleben/Rome: Institute of Plant Genetics and Crop Plant Research/International Plant Genetic Resources Institute.

Morshidi, M. 1996. *Genetic Variability in* Carica papaya *and Related Species.* Thesis (Ph.D.), Dept. Horticulture, University of Hawaii at Manoa, Honolulu, HI.

Ourecky, D.K. 1975. Brambles. In *Advances in Fruit Breeding.* Edited by J. Janick and J.N. Moore. West Lafayette, IN: Purdue University Press, 98–129.

Patiño, V.M. 1963. *Plantas cultivadas y animales domésticos en América Equinoccial.* Tomo I. Frutales. Cali, Columbia: Imprenta Departamental.

Prescott-Allen, R., and Prescott-Allen, C. 1990. How many plants feed the world? *Conservation Biology* 4(4): 365–374.

de Queiroz Pinto, A.C., Rocha Cordeiro, M.C., de Andrade, S.R.M., Ferreira, F.R., da Cunha Filgueiras, H.A., and Elesbão Alves, R. 2004 *Five Important Species of Annona.* Southampton, U.K.: International Centre for Underutilised Crops.

Ragone, D. 1997. *Breadfruit,* Artocarpus altilis *(*Parkinson*) Fosberg. Promoting the conservation and use of underutilized and neglected crops. 20.* Rome: International Plant Genetic Resources Institute (IPGRI).

Roose, M.L., Soost, R.K., and Cameron, J.W. 1995. Citrus, *Citrus* (Rutaceae). In *Evolution of Crop Plants,* edited by J. Smartt and N.W. Simmonds. London: Longman, 399–403.

Roose, M.L. 1988. Isozymes and DNA RFLPs in citrus breeding and systematics. In *Proceedings of the Sixth International Citrus Congress,* edited by R. Goren and K. Mendel, Philadelphia: Balaban, 57–67.

Scora, R.W. 1988. Biochemistry, taxonomy and evolution of modern cultivated citrus. In *Proceedings of the Sixth International Citrus Congress,* edited by R. Goren and K. Mendel. Philadelphia: Balaban, 277–289.

Simmonds, N.W. 1995. Bananas, *Musa* (Musaceae). In *Evolution of Crop Plants,* edited by J. Smartt and N.W. Simmonds. London: Longman, 370–375.

Singh, S., Krishnamurthi, S., and Katyal, S.L. 1967. *Fruit culture in India.* New Delhi: ICAR.

Smith, C.E. 1969. Additional notes on pre-Conquest avocados in Mexico. *Economic Botany 23,* 135–140.

Soepadmo, E. 1992. *Artocarpus heterophyllus* Lamk. In *Plant resources of South-East Asia. 2. Edible fruits and nuts,* edited by E.W.M. Verheij and R.E. Coronel. Bogor, Indonesia: Plant Resources of South-East Asia (PROSEA), 86–91.

Spiegel-Roy, P.1986. Domestication of fruit trees. In *The Origin and Domestication of Cultivated Plants,* edited by C. Barigozzi. Amsterdam: Elsevier, 201–211.

Subhadrabandhu, S., Schneemann, J.M.P., and Verheij, E.W.M. 1992. *Durio zibethinus* Murray. In *Plant resources of South-East Asia. 2. Edible fruits and nuts,* edited by E.W.M. Verheij and R.E. Coronel. Bogor, Indonesia: Plant Resources of South-East Asia (PROSEA), 157–161.

Swingle, W.T. and Reece, P.C. 1967. The botany of Citrus and its wild relatives. In *The Citrus Industry,* revised edition, edited by W. Reuther, H.J. Webber and L.D. Batchelor. Vol. I. Berkeley, CA: University of California, Division of Agricultural Sciences, 190–430.

Tanaka, T. 1977. Fundamental discussion of *Citrus* classification. *Studia Citrologia 4,* 1–6.

van Welzen, P.C.and Verheij, E.W.M. *Nephelium lappaceum* L. In *Plant resources of South-East Asia. 2. Edible fruits and nuts,* edited by E.W.M. Verheij and R.E. Coronel. Bogor, Indonesia: Plant Resources of South-East Asia (PROSEA), 235–240.

Vavilov, N.I. 1992. The phyto-geographical basis for plant breeding. In *Origin and geography of cultivated plants,* edited by V.F. Dorofeyev. Cambridge: Cambridge University Press, 316–366.

Verheij, E.W.M. 1992. *Mangifera indica* L. In *Plant resources of South-East Asia. 2. Edible fruits and nuts,* edited by E.W.M. Verheij and R.E. Coronel. Bogor, Indonesia: Plant Resources of South-East Asia (PROSEA), 211–216.

Verheij, E.W.M. 1992. *Garcinia mangostana* L. In *Plant resources of South-East Asia. 2. Edible fruits and nuts,* edited by E.W.M. Verheij and R.E. Coronel. Bogor, Indonesia: Plant Resources of South-East Asia (PROSEA), 175–177.

Watkins, R. 1995. Cherry, plum, peach, apricot and almond, *Prunus* spp. (Rosaceae). In *Evolution of Crop Plants,* edited by J. Smartt and N.W. Simmonds. London: Longman, 423–429.

Watkins, R. 1995. Apple and pear, *Malus* and *Pyrus* spp. (Rosaceae). In *Evolution of Crop Plants,* edited by J. Smartt and N.W. Simmonds. London: Longman, 418–422.

White, K.D. 1970. *Roman Farming.* Ithaca, NY: Cornell University Press.

Wrigley, G. 1995. Date palm, *Phoenix dactylifera* (Palmae). In *Evolution of Crop Plants,* edited by J. Smartt and N.W. Simmonds. London: Longman, 399–403.

Zohary, D., and Hopf, M. 2000. *Domestication of Plants in the Old World: the Origin and Spread of Cultivated Plants in West Asia, Europe and the Nile Valley.* Oxford, U.K.: Oxford University Press.

Zohary, D. 1995. Fig, *Ficus carica* (Moraceae). In *Evolution of Crop Plants,* edited by J. Smartt and N.W. Simmonds. London: Longman, 366–370.

7

Herbs and Vegetables

JANE M. RENFREW AND HELEN SANDERSON

Herbs and vegetables play an important part in the human diet by adding flavors and vitamins. This chapter is concerned with the origins and uses of culinary herbs and of vegetables (other than the pulses, roots, and tubers which are considered separately). For convenience this chapter will be divided into two sections.

Culinary Herbs

Probably the best answer for the question "What is an herb?" is the quote attributed to the emperor Charlemagne, "The friend of the physician and the praise of cooks." It is the culinary uses of herbs that will be examined here (see also Plants as Medicines chapter). On the whole herbs are aromatic plants, and very many of them are native to the flora of the Mediterranean. A large number of herbs belong to the onion family, Liliaceae (Alliaceae); the mint family, Labiatae or Lamiaceae; the parsley family, Umbelliferae or Apiaceae; and the tarragon family Asteraceae (formerly Compositae). Quite a few were known and used by the ancient Greeks and especially by the Romans and were spread by the latter throughout their empire in Europe. Many herbs reached the New World from Europe after Columbus's arrival, especially with the Pilgrim Fathers.

Aniseed *Pimpinella anisum*
Apiaceae

Aniseed is an annual herb native to the eastern Mediterranean and was cultivated by the ancient Egyptians as well as the Greeks and Romans. The ripe dried fruit, which contains a volatile oil, is the part of the plant used. It occurs wild but has been grown in herb gardens in Britain since the 15th century, and seeds were carried to North America by the first European settlers. The seeds are strongly licorice-flavored and are used in cooking fish, poultry, and creamy soups. It is used principally to flavor drinks such as ouzo, liqueurs such as Pernod and Ricard, and also in confectionary. As an herbal tea, aniseed has been used medicinally for bloating and colic.

See: Spices, p. 157

Arugula/rocket (*Eruca vesicaria* subsp. sativa). USDA-NRCS PLANTS Database/Britton, N.L., and A. Brown. 1913. *Illustrated flora of the northern states and Canada.* Vol. 2, p. 192.

Basil *Ocimum basilicum*
Lamiaceae

Basil is native to tropical Asia and Africa and has been cultivated in Europe for centuries. The leaves are used fresh in cooking being added to soups, salads, especially tomato salad, and meat and fish dishes. Fresh basil is an essential ingredient in pesto sauce, made by pounding fresh leaves in a mortar with salt, olive oil, and garlic. Parmesan cheese, pine nuts, or walnuts may also be added, and the resulting sauce is stirred into minestrone soup or hot pasta. Thai basil (*O.* × *citriodorum* "Thai") is an annual native to Thailand and Burma. It has a darker leaf than common basil and a slight anise flavor. In Thailand it is used in salads and as a garnish.

Bay *Laurus nobilis*
Lauraceae

The bay tree is native to Asia Minor and the Mediterranean. It is the aromatic leaves which are valued for their flavoring properties. The Greeks and Romans made laurel leaf crowns as symbols of wisdom and glory for their athletes and emperors. In cooking the dried leaves are used in meat pâtés and stews (drying reduces bitterness of the leaves) and are a key ingredient in *bouquet garni*. They are also used for packing licorice and dried figs to discourage weevils.

Bearberry *Arctostaphylos uva-ursi*
Ericaceae

The red berries of this small shrub were cooked with meat as a seasoning for the broth by North American Indians. It grows in Canada and as far south in the United States as New Jersey and Wisconsin, as well as in the northern latitudes of Europe and Asia.

Bergamot, Beebalm *Monarda didyma*
Lamiaceae

The bergamots are native to North America. The whole plant is pleasantly fragrant, and it is popular with bees on account of the quantity of nectar to be found in its blossom. It was used by the North American Indians and the early settlers as Oswego tea, which tastes like a scented China tea and is most refreshing when taken cold. The plant was brought to Europe and is cultivated as an ornamental garden plant. *M. fistulosa*, or wild bergamot, is used in herbal teas. Neither should be confused with the bergamot tree, *Citrus bergamia*, which yields the oils used in Earl Grey tea.

Borage *Borago officinalis*
Boraginaceae

Borage is indigenous to the Mediterranean area but has been cultivated in Britain and North America, with a long history of use as a medicinal herb and used by beekeepers to attract bees and flavor honey. Young fresh borage leaves are rich in vitamin C and can be used in salads. The pretty blue

Borage (*Borago officinalis*). USDA-NRCS PLANTS Database/Britton, N.L., and A. Brown. 1913. *Illustrated flora of the northern states and Canada*. Vol. 3, p. 93.

Burnet (*Sanguisorba minor*) Joe F. Duft @ USDA-NRCS PLANTS Database/USDA NRCS. 1992. *Western wetland flora: Field office guide to plant species.* West Region, Sacramento, CA.

flowers can be candied and used for decoration. The flowers and leaves also make an attractive addition to a wine and fruit cup when freshly picked.

See: Gathering Food from the Wild, p. 38

Burnet *Sanguisorba minor, S. officinalis*
Rosaceae

This is the name for two common European herbs that bear dark crimson or deep brown flowers, hence the name, which is derived from the French word brunette. Salad burnet, *Sanguisorba minor*, is the species usually cultivated in gardens. It is native to the Mediterranean countries, Asia Minor, Iraq, Iran, Afghanistan, and Middle Asia. Today it is cultivated sporadically in Europe, including Britain, Germany, and France, and in North America and Asia. The leaves have a pleasant cucumber flavor and have been eaten since classical Greek times and were commonly used in salads from the 15th to 19th centuries. In recent times they have been used as ensalada italiana in Spain, especially in Catalonia. The leaves are also added to cold drinks in the same way as the better-known borage. The upper parts of the plant and its roots have been used in folk medicine for digestive disorders. In arid areas it is occasionally cultivated as fodder for sheep.

Great burnet, *Sanguisorba officinalis*, is a larger plant of similar uses. It originates from the temperate zones of Eurasia and North America and is cultivated mainly in Asia and Japan. The young leaves are used for salads and are also served with other vegetables or as a spice for soups. In folk medicine, extracts of the roots are used in Russia, China, and Japan as an antiseptic.

Capers *Capparis spinosa*
Capparaceae

Capers are the unopened flower buds of a spiny trailing shrub, which occurs in the Mediterranean region and in North Africa and Asia Minor. It is cultivated mainly in Mediterranean areas, but also in Tibet, India, and Southeast Asia. The biggest producers are Italy and Spain. The plant is sprawling and is covered in thorns, though highly domesticated races without thorns are also cultivated. As the buds develop rapidly, they must be harvested daily which increases the costs of the product. The buds are pickled in vinegar or preserved in salt and have been used since at least the time of the ancient Greeks as a condiment to add a salty-sour flavor to sauces (e.g., tartare sauce), cheeses, salad dressings, stews, pasta sauces, and various other meat and fish dishes. The fresh buds are not eaten, because the characteristic slightly bitter flavor is due to the presence of capric acid which is only developed by pickling. Pickled mature fruits are also eaten in southern Europe, known as cornichons de câprier. The bark of the root was used medicinally in classical times, and it is still considered a medicinal plant in India.

See: Spices, pp. 157–8

Chamomile, Roman camomile *Chamaemilum nobile* (syn. *Anthemis nobilis*)
Asteraceae
Chamomile is native to southern Europe and northern Africa and grows wild as a perennial in parts of Britain. It yields an essential oil used to flavor ice cream. The flower heads are used to make Chamomile tea. German chamomile, *Matricaria recutita,* also of the Asteraceae family, has a wider native distribution in Europe and is naturalized in North America. It is used in similar ways. Both species are grown commercially in Europe and the former USSR.

See: Plants as Medicines, p. 213

Chervil *Anthriscus cerefolium*
Apiaceae
A native of Asia Minor, the Caucasus, and southern Russia, chervil was introduced and spread throughout Europe by the Romans. Chervil leaves have a slightly sweet, aniseed taste, and are used fresh or dried in France in omelettes, soups, and salads.

Chives *Allium schoenoprasum*
Alliaceae
Chives grow wild in most of Europe, in Russia as far east as Siberia, in Canada, and the northern part of the United States. Although used in antiquity, the herb was probably not cultivated until the Middle Ages. The chopped young leaves have a fresh onion flavor which is delicious in salads, in omelettes, and as a garnish for boiled new potatoes. Its purple flowers are sometimes added to salads.

Comfrey *Symphytum officinale*
Boraginaceae
This plant is native to Europe, western Siberia, Caucasus, and North America. Since ancient times it has been used as a medicinal plant, but it has also been used sporadically as a flavoring herb, particularly as a potherb, for homemade wine, and for butter in Bavaria. Both the leaves and stems have been eaten and are usually boiled, and it can be eaten like spinach as a vegetable. In Bavaria the leaves are dipped in batter and fried. The stems can be blanched by earthing them up, which makes the astringent flavor milder. Although comfrey has been advocated as a health food for humans, cultivation has mainly been for animal fodder.

Costmary, Alecost, Bible leaf *Tanacetum balsamita* subsp. *balsamita* (syn. *Chrysanthemum majus*)
Asteraceae
Costmary is native to the Near East and was used by the ancient Egyptians, Greeks, and Romans. By the 16th century it was common in gardens in Britain though it was probably first introduced by the Romans. The colonists took it to North America where it is common in the eastern and midwest states. It gets its American name, bibleleaf, from the long leaves being used as page markers in the bibles of the early colonists. It is used for flavoring ale (beer) and can be used for flavoring soups, game, poultry, veal, and stuffing. It has a soft balsam flavor and should be used sparingly in cooking. The related *C. coronarium* (crown daisy, Shingiku, or chop suey green) has piquant leaves which may be used in salads or cooked in Oriental cuisine.

Curry leaf *Murraya koenigii*
Rutaceae

This species grows wild in much of South and Southeast Asia, and it is cultivated in India, Sri Lanka and Southeast Asia, Australia, the Pacific islands, and Africa. The small, shiny, pungent leaves are used for flavoring in cookery, in curry dishes, and in many kinds of chutney in south India, Sri Lanka, and in some parts of Southeast Asia. The leaves are used in a similar way to the bay leaf in western countries and are employed in fresh, dried, and powdered forms. When they are used fresh, they are fried first to make them brown and crisp before adding to the dish. When dried, the leaves retain their aroma and can be bought either whole or powdered.

See: Spices, p. 162

Dandelion *Taraxacum officinale*
Asteraceae

The dandelion is native to Europe and Asia and was introduced by European colonists to North America. The young leaves of this familiar wild plant can be eaten raw in salads or tossed in melted butter as a vegetable or cooked as a potherb. The flowers are used to make dandelion wine and the leaves to flavor herb beers. The dried root can be ground and roasted as a substitute for coffee.

See: Caffeine, Alcohol, and Sweeteners p. 177

Daun salam, Indonesian bay leaf *Syzygium polyanthum*
Myrtaceae

This tree is native to Malaysia, Indonesia (Java and Sumatra), and Thailand, and has edible flowers and fruits. The leaves are used as a flavoring and play an important part in Indonesian cuisine, where a single leaf placed in the cooking pan gives a subtle aromatic flavor to dishes. Its leaves can be used dried or powdered. It is used, for example, in nasi goring (fried rice), and its role is generally compared to the curry leaf in Indian cuisine. Young leaves are also frequently cooked like greens, often with meat.

Dill *Anethum graveolens*
Apiaceae

Dill is native to the Mediterranean countries and southern Russia and is now cultivated widely in Europe, India, and North America. Dill has been used since classical times in herbalism to aid digestion. The leaves and seeds have a similar flavor, reminiscent of caraway, though the seed

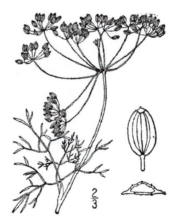

Dill (*Anethum graveolens*).USDA-NRCS PLANTS Database/Britton, N.L., and A. Brown. 1913. *Illustrated flora of the northern states and Canada.* Vol. 2, p. 634.

has a more pronounced flavor. Its principal culinary use is in pickling cucumbers (dill pickles) when the whole plant together with green seeds are used and are also used in sauerkraut. In parts of northern Europe dill sauce (made with leaves or "dill weed") is served with fish. Dill goes well in celery or zucchini soups. Dill seed is used as a condiment and can be used to flavor root vegetables, cakes, and sweets. The feathery leaves have a more delicate flavor; they should be used raw or added at the end of cooking time to keep their flavor.

See: Spices, p. 163

Elecampane *Inula helenium*
Asteraceae

Elecampane is native to Asia and parts of Europe. In the past the roots of elecampane were used in Roman cooking for sauces or salted were served as an hors d'oeuvre or at the end of a meal to aid digestion. From Medieval times onwards they were candied and used in confectionary. The aromatic leaves have also been used as a potherb.

Fennel *Foeniculum vulgare*
Apiaceae

Fennel is a native perennial of southern Europe and the Caucasus and is widely naturalized in Europe and North America. The leaves were very popular as a herb with the Romans. The leaves and seeds both have a strong anise flavor. Fennel leaves have been used in Britain for centuries with fresh or salted fish, as fennel sauce, or chopped in mayonnaise or in stuffing. In Italy it is used to flavor pork. Seeds are used to flavor bread, pastries, confectionary, liqueurs, and fish dishes. They contain an essential oil whose main constituent is anethole. The cultivated form, Florence fennel (see following, subsp. *vulgare* var. *azoricum*), has an enlarged, sweet tasting leaf base or bulb that is cooked as a vegetable or sliced raw in salads.

See: Gathering Food from the Wild, p. 40; Spices, p. 163; Plants as Medicines, p. 214

Fitweed, Shado beni *Eryngium foetidum*
Apiaceae

The origin of this species is not known, but it is native to Central and South America, from southern Mexico to Panama, Colombia, Bolivia and Brazil, and from Cuba to Trinidad. It has been introduced into Florida and the Old World tropics where it has naturalized in many places. It was introduced to Southeast Asia by the Chinese as a substitute for coriander, both as a garnish and in cooking. Although it is collected from the wild or gardens in most places, it is also cultivated in South America and occasionally elsewhere in Thailand, Cambodia, Vietnam, the Philippines and Japan. The leaves are aromatic and smell like coriander. Its fresh leaves are used as a flavoring in soups, curries, stews, rice, and fish dishes. The tender, young leaves are eaten raw or cooked as a vegetable. It also has many medicinal uses throughout its distribution and has been used as an aphrodisiac.

Garden angelica *Angelica archangelica*
Apiaceae

Angelica grows wild over much of Asia and Europe and is cultivated in Belgium, Hungary, and Germany. The young green stems are commonly candied and used to decorate cakes and puddings. In some countries it is eaten as a green vegetable, and sometimes the young shoots are blanched and used in salads. It has been added to rhubarb and to marmalade as a flavoring. It has been used to flavor drinks such as vermouth, chartreuse, and especially gin.

Garlic *Allium sativum*
Alliaceae

Garlic was cultivated in Mesopotamia and Egypt in antiquity (remains of garlic bulbs have been found at Deir el Medina and in the tomb of Tutankhamun for example), and it was well known to the Greeks and Romans. The Romans used it to strengthen laborers and to make their soldiers more courageous; but Horace, in his third Epode, describes how infuriated he was to be given some by Maecenas at a feast:

> "If ever man with impious hand
> Strangled an aged parent,
> May he eat garlic, deadlier than hemlock!
> (Ah, what hardy stomachs reapers have!)
> What is this venom savaging my frame?
> Has viper's blood, unknown to me,
> Been brewed into these herbs. . . ."

The common name is Anglo-Saxon in origin. Today garlic is cultivated widely around the world, the best-flavored bulbs coming from warm countries. The bulbs are broken up into individual cloves for culinary use. The strong, onion-like, pungent flavor is not to everyone's taste, and it should be used in very small quantities if it is not to overpower other flavors in cooking. It can also be used raw in garlic butter as a dressing for cooked meat.

See: Plants as Medicines, p. 219

Gallant soldier, Guascas *Galinsoga parviflora*
Asteraceae

This species is native to South America, particularly the Andean region and Colombia, and has naturalized into parts of North America, Europe, Africa, and Asia to become a cosmopolitan weed. In its native countries it is considered of culinary importance. The steamed young tops of the plant can be eaten as a vegetable. In Colombia it can be bought dried and ground into a green powder which adds a delicious flavor to soups and stews. It also makes good fodder for animals.

See: Gathering Food from the Wild, p. 31

Horseradish *Armoracia rusticana*
Brassicaceae

Horseradish is probably indigenous to the east and southeastern Europe, where it has been cultivated since antiquity, and spread to other parts of Europe in medieval times. Today, it is cultivated in many places within the temperate zones of the Old and New World, mainly in Europe and North America. Although it is cultivated on a larger scale in Europe, North America, and South Africa, it is mostly grown by smallholders and in home gardens for both culinary and medicinal uses.

The roots, when grated, yield a volatile oil which is very pungent. Its pungent taste is due to the volatile essential oils similar to those of mustard, which are lost in cooking. The peeled and grated roots are used fresh to make a hot, spicy sauce by mixing with salt, vinegar, and oil and are eaten with meat and fish dishes. In Britain it is most famously served with roast beef. The leaves are used medicinally. The hot, flavored young leaves are one of the bitter herbs used for the Jewish Passover.

Hyssop *Hyssopus officinalis*
Lamiaceae

Hyssop is native to the Mediterranean and Central Asia and has naturalized in North America. It is an attractive shrub with blue flowers that is chiefly cultivated for its medicinal properties. The aromatic, slightly bitter flavor of hyssop counteracts fatty, oily meat and fish. A few chopped leaves go well with stuffings or sausages, or they can be added to salads, stewed fruit, or fruit pies. The tops are used to flavor liqueurs such as Chartreuse.

Lemon balm *Melissa officinalis*
Lamiaceae

Lemon balm is native to southern Europe and has been in cultivation for the past two thousand years. It was brought to Britain by the Romans. The leaves, which have a pleasant lemon flavor, are used in salads and added to cool drinks and wine cups. Chopped leaves can be added to omelettes. In Belgium and Holland they are used when pickling herrings and eels. It is the basic ingredient of Melissa cordial, *eau-de-melisse des carmes*, and Benedictine liqueur. It is much loved by honeybees.

Lemon verbena, Lemon beebrush *Aloysia citrodora*
Verbenaceae

A native of South and Central America, this deciduous shrub was introduced to Europe in the 18th century by Spanish explorers. Its sweetly lemon-scented leaves are used chopped in seasonings or to flavor fish, poultry, jams, jellies, and puddings. The leaves also make a refreshing tea.

See: Fragrant Plants, p. 219

Lovage *Levisticum officinale*
Apiaceae

Lovage is native to southern Europe (a related plant, *Ligusticum scoticum*, grows wild in northern Britain and on the Atlantic coasts of North America). It was used by the Greeks and Romans to aid digestion. The seeds, leaves, and leafstems have a strong, earthy, celery flavor and are chiefly used to flavor soups and stews. It is particularly useful in vegetarian dishes, with rice, vegetable seasonings, and nut roasts. Its stems can be used in salads or candied like angelica. The seeds are used on bread or cheese biscuits.

Makrut lime, Kaffir lime *Citrus hystrix*
Rutaceae

This species of *Citrus* is native to tropical Southeast Asia, but is now grown throughout Southeast Asia, Central America, and the Mascarene and Hawaiian Islands for its leaves and fruit juice, which are used as a flavoring. The leaves are a more common ingredient in Southeast Asian dishes and are particularly ubiquitous in Thai cuisine. It is usual to tear the lime leaves before adding them to the cooking pot and to remove them once the dish is cooked. The leaves are used both fresh and dried in soups, curries, sauces, and gravies. Usually sold fresh in the native countries, dried whole and powdered leaves of makrut lime can also be found in Asian and western supermarkets. The dried or candied peels and the juice of the fruits also serve as a flavoring, and essential oils obtained from the leaves and fruit peel are utilized in the cosmetics industry.

Marsh mallow *Althaea officinalis*
Malvaceae

The native distribution of this species extends from the Middle Asian steppes through the Ukraine, the Balkans, and the Mediterranean region. From the Atlantic coast of Europe it reached southern England, the Netherlands, and central Europe. Although its leaves are edible, the main use of the plant is its roots, which yield a mucilaginous substance which is the traditional basis for the sweet confection known as marshmallow, but it has now been almost completely replaced by gum arabic. It was formerly cultivated in France, Italy, Germany, Balkan, Hungary, and southern parts of Russia; however, in recent years cultivation has decreased. The roots of this species have been used medicinally, and in India the leaves are eaten as a green vegetable.

Milfoil, Yarrow *Achillea millefolium*
Asteraceae

This plant is native to Europe but is now widely naturalized in North America, New Zealand, and Australia. It has pungent, strong-flavored leaves used in salads and soups, and a tisane is also made from the plant, and it has been used to flavor beef. Most commonly, this species has been used as a medicinal plant. It has also been used as a substitute for tobacco, nutmeg, cinnamon, and hops.

See: Plants as Medicines, p. 218

Mint *Mentha* spp.
Lamiaceae

There are a number of different species of mint found, but they hybridize easily and may be difficult to identify botanically. Mint was used by the Greeks and Romans for its scent; for the Athenians in particular it indicated strength as they used to rub it on their arms to make them stronger. The Romans are said to have brought mint to Britain and to have introduced mint sauce to British cuisine. The following are among the most common wild mints that have contributed to cultivated forms.

Water mint (*Mentha aquatica*) is a strongly-flavored mint found in very damp conditions. It was crossed with spearmint (*M. spicata*) to give rise to the hybrid peppermint (*M.* × *piperita*). It is believed to have been used and grown since at least Roman times. In the Middle Ages it was a popular strewing herb known as *menastrum*.

Cultivated varieties of peppermint, known as black peppermint with dark stems and white peppermint with green stems, are grown chiefly for the production of peppermint oil which is mainly used in the manufacture of sweets and candies. Peppermint oil contains menthol which has anesthetic properties and produces a cool sensation in the mouth. It is also used in the production of peppermint-flavored liqueurs such as *crème de menthe*. Pliny says that the Greeks and Romans crowned themselves with peppermint at their feasts, and that their cooks flavored both sauces and wines with it.

Pennyroyal (*M. pulegium*) is a low-growing species of mint that grows wild in damp, shady parts of temperate Europe and Asia. It has a pungent peppermint smell and a sharp taste. It was used by the Romans to drive away fleas and is often mentioned in Anglo-Saxon herbals as a cure for whooping cough, asthma, and indigestion. It is an essential flavoring herb in black pudding, the English north country delicacy. It was also used to make pennyroyal tea, an old-fashioned remedy for colds.

Spearmint (*Mentha spicata*).USDA-NRCS PLANTS Database/Britton, N.L., and A. Brown. 1913. *Illustrated flora of the northern states and Canada.* Vol. 3, p. 149.

The well-known cultivars of round-leaved mint *M. rotundifolia*, apple mint, Bowles' mint, and pineapple mint are amongst the best-flavored of all the culinary mints. Spearmint (*M. spicata*) is native to the Mediterranean and was probably introduced to Britain by the Romans. It has been grown in America for well over two centuries. It is the commonest form of mint grown in gardens, and there are a number of different varieties based on the shape and color of the leaves, the color of the stem, and the hairiness of the plant. As with other forms of mint it is inclined to spread vigorously in cultivation. Gerard says of it, "The smell of Mint does stir up the minde and the taste to a greedy desire of meate." It was also thought to prevent milk from curdling. This form of mint is used in British cooking to flavor new potatoes and peas and to make mint sauce, traditionally served with lamb. It is also much used in the Levant, Middle East, and in India as a flavoring.

The flowers and buds of mountain mint, *Koellia virginiana*, were used by North American Indians to flavor meat and soups.

See: Plants as Medicines, pp. 215–6; Fragrant Plants, pp. 249–50

Mitsuba, Japanese parsley *Cryptotaenia japonica*
Apiaceae

This species is native to far eastern Russia, Japan, Korea, and China. It is cultivated in Japan, Korea, China, and occasionally Southeast Asia and North America as a culinary herb. The main parts used are the leaves, the green stems, and the highly-prized blanched white stems, which look similar to coriander but have a milder flavor. Mitsuba comes from two main varieties, kansai (green) and kanto (whiter). In Japan, great care is taken in its cultivation and to techniques such as winter and summer blanching to make the stems more tender. It has a similar role in Japan to parsley in western countries and coriander in most other Asian countries, with a wide range of uses. These are eaten, fresh or blanched, to season clear and fish soups and a wide range of Japanese dishes. They should never be cooked for more than a couple of minutes or the flavor is lost. The seeds are also used as a seasoning. The plant has been widely employed for its medicinal properties.

Mustards: Black mustard *Brassica nigra,* Brown mustard *B. juncea,*
White mustard *Sinapis alba*
Brassicaceae

The condiment mustard is based on the seeds of these three plants of the cabbage family. Black mustard has been in cultivation for more than two thousand years and is native to Europe and Asia and has become naturalized in North America. It was the main ingredient of mustard until World War II, but because it sheds its seeds readily when ripe it is unsuited to mechanical harvesting and has now been largely replaced by brown mustard whose seed is not so strongly flavored. Black mustard is still grown in areas where hand harvesting persists. White mustard (known as yellow mustard in the United States) is the mustard of "mustard and cress." It is used in American mixed mustards and to some extent in English mustard but is forbidden in Dijon mustard.

The young leaves of all mustard varieties can be eaten. In Europe and Asia black mustard is cultivated for its young leaves, eaten raw as salad or cooked as a potherb. Brown mustard, known as mustard greens, is also eaten in young tender form as a salad green, or cooked in soups and stews. Mizuna or Japanese mustard (*Brassica juncea* var. *japonica*) has a mild, sweet earthy flavour and is a common component of mesclun salad.

See: Spices, p. 165

Nasturtium *Tropaeolum majus*
Tropaeolaceae
Nasturtium seeds were first brought to Europe from Peru by the Spanish in the 16th century. The succulent leaves, buds, and flowers have a sharp, peppery taste and can be added to salads. The unripe fruits can be pickled like capers but should only be eaten in small quantities.

Parsley *Petroselinum crispum*
Apiaceae
Parsley is one of the best-known and most widely grown herbs. In England and America the curly-leaved varieties are most popular; elsewhere the plain, flat-leaved forms are favored, and they do have a better flavor. The Greeks used it to make victory wreaths to crown the athletes at the Isthmian Games and for offerings at the tombs of the dead. The Romans were the first to use it for flavoring food. It is native to the Mediterranean region and was introduced into Britain in the 16th century and later to the United States. The Hamburg variety is grown for its enlarged fleshy taproot and was introduced to Britain from Holland in 1727 by Miller. Fresh or dried leaves are used, finely chopped, as flavorings for sauces, soups, seasonings, rissoles, and mince, and sprinkled over vegetables and salads.

Poke root *Phytolacca decandra*
Phytolaccaceae
This is regarded as one of the most important of indigenous North American plants, with the plant being used as a dye and the dried root used in many traditional herbal remedies. It is now also common in the Mediterranean region. The young shoots are used as a good substitute for asparagus, and poultry are very happy to eat the berries (hence the common name pigeon berry), although eating too many can give their flesh an unpleasant flavor. In Portugal the juice of the berries was used to color port wines, but this was discontinued as it affected the flavor. As the plant matures the whole plant becomes poisonous.

Rosemary *Rosemarinus officinalis*
Lamiaceae
Rosemary is native to the rocky limestone hillsides around the Mediterranean and was introduced to Britain by the Romans. It is fairly tender and needs winter protection in the northern parts of Britain and the United States. Sprigs of leaves are used in cooking lamb, stews, and strong game and have a camphor-like flavor. Leaves may be infused in milk for sweet puddings and custards. Rosemary was one of the most popular herbs for flavoring ale.

Rue *Ruta graveolens*
Rutaceae
Rue is native to the Mediterranean region, although it has naturalized elsewhere now. It was cultivated by the ancient Greeks and Romans as a spice and medicinal plant and was also commonly used during the Middle Ages and the Renaissance. In Britain, it grew wild in parts of Lancashire and Yorkshire in 1597, possibly as a remnant of Roman cultivation. The leaves are blue-green and fleshy and have a strong aroma and bitter flavor. In the kitchen it was used as an occasional flavoring to soups and stews and was often pickled to use as a relish with meat. Today, the plant is cultivated in several countries of Europe, Asia, Africa, and America. The bitter pungent leaves are used, fresh or dried, in small quantities for salads and as a flavoring for bread, meat, vinegars, and various dishes. It is also used to impart bitterness to wines. Rue oil, obtained by distillation of the leaves, is mainly produced in Spain and Portugal. It is used as a spice and perfume.

Sage *Salvia officinalis*
Lamiaceae

Garden sage is native to the north Mediterranean and Dalmatian coasts and has long been cultivated for culinary and medicinal purposes. It is the leaves that are valued, and there are many varieties with different flavors. The most commonly used in cooking is the narrow-leaved form with blue flowers. The flavor is strong and camphor-like. It is used in stuffing for duck, goose, and pork, in sausages and sometimes with eel. It is not much used in French cooking but is often used in Italy with veal and game; in the Middle East it is used with onions in making up kebabs. There are a number of cheeses which are flavored with sage: Sage Derby and the Vermont sage cheeses for example.

See: Plants as Medicines, p. 221; Ornamentals, p. 271

Screwpine *Pandanus amaryllifolius*
Pandanaceae

The name screwpine refers to the spirally-arranged leaves at the top of the stem characteristic of members of the Pandanaceae. These plants are native to the Asian tropics, from India through Southeast Asia to northern Australia and Oceania. Several species are used including *P. tectorius* whose flowers are the source of keora (or kewda) essence, used in parts of India and Sri Lanka as a flavoring. The most important species used for flavoring, though, is *P. amaryllifolius.* It is an ancient cultigen that has never been found in the wild, though it may originate in the Moluccas. It is cultivated in western Indonesia, Malaysia, Thailand, New Guinea, Sri Lanka, the Philippines, and, more recently, in Hawaii. Female flowering plants of this species are not known, therefore it is always propagated vegetatively. The leaves are used in cooking to impart flavor and color (chlorophyll) to rice, sweets, jellies, and many other food products; the leaves are removed before consumption. It is widely used to flavor ordinary rice, as a substitute for expensive aromatic rice cultivars. Containers for desserts are also made from the leaves. For example, fried chicken wrapped in pandan leaves is a delicacy. Juice is pressed from the leaves for flavoring and coloring cakes.

Shiso, Beefsteak plant *Perilla frutescens*
Lamiaceae

This plant is native to the Himalayas, northern India, China, and Japan. It has naturalized in many parts of this region as well as in other parts of its cultivation area including the Ukraine and the United States. It was probably first introduced into cultivation in China in ancient times, but is now

Shiso (*Perilla frutescens*). Douglas Ladd @ USDA-NRCS PLANTS Database/USDA SCS. 1989. *Midwest wetland flora: Field office illustrated guide to plant species.* Midwest National Technical Center, Lincoln, NE.

grown in eastern Asia, Southeast Asia, India, Iran, Caucasus, southeastern Europe, South Africa, and the United States. The several cultivars of this herb include both green-leaved and red- or purple-leaved forms and are grown as an oil crop, potherb, and for its medicinal properties. For culinary purposes, the leaves and sometimes fruits, seedlings, and flowers are eaten as a condiment with seafood and raw fish and used in teas. It is an important plant in Japanese cooking as a garnish and food colorant, especially in the national dish shisho. Red-leaved varieties are used in the coloring and preservation of pickled fruits. The seeds are roasted and crushed to produce edible perilla oil, which was formerly used only locally, but has now attracted wider interest due to its high levels of polyunsaturates.

Smartweed, Tade *Polygonum hydropiper, P. odoratum*
Polygonaceae
This annual herb grows wild in temperate and subtropical areas of the Northern hemisphere. It is also known as water pepper on account of its spicy leaves. In Japan (where it is known as tade) it is consumed on a large scale as a garnish and accompaniment for many dishes. It was also formerly cultivated in Europe (where it was grown especially during war times as a substitute for pepper) and China, but only has medicinal uses in Europe now. In Japan, the plants are usually marketed as seedlings, just a few days after the seeds have germinated. The leaves, which may be purplish or green, broad or narrow, are very pungent. The pungency is reported to be different to that of wasabi and pepper, and fresh leaves are used as a garnish for Japanese favorites such as sashimi, tempura, and sushi. It is a popular herb in summer cooking.

    P. odoratum, also known as rau ram, is native to Indo China. It is a traditional crop in Vietnam and Cambodia and has also been introduced to the United States and Australia. The leaves have a taste reminiscent of lemon and coriander leaves and are used for salads, as a potherb and spice. The leaves are also used medicinally, and the essential oil is of great interest for the flavor and fragrance industry.

Stinging nettle *Urtica dioica*
Urticaceae
Stinging nettles are to be found throughout the temperate regions of the world from Japan to the Andes. Young nettle leaves can be cooked like spinach (they do not sting once boiled!) or can be added to spring soups and stews as a potherb. Nettle beer is a traditional home brew. The leaves are very nutritious, containing more iron than spinach and quantities of Vitamins A and C.

Summer savory *Satureja hortensis*
Lamiaceae
Summer savory is native to the Mediterranean region and was used by the Romans as a sauce mixed with vinegar. Traditionally it was eaten with broad beans. Today it is used to give a spicy, warming flavor to soups, poultry, meat and egg dishes, and with peas, lentils, and boiled vegetables. It is closely related to winter savory, *Satureja montana,* which is similar but inferior in flavor. It was one of the first herbs to be taken to North America by the early settlers.

Sweet cicely *Myrrhis odorata*
Apiaceae
A native of high ground in Europe and Russia, the leaves and fruits have a smell of anise or licorice. When cooked with sour fruit such as rhubarb or gooseberries they reduce the need for additional sugar. As a natural form of sweetening it is a useful addition to a diabetic diet. A North American species, *Osmorhiza longistylis*, looks similar but has broader leaves and can be used in similar ways.

Sweet marjoram *Origanum majorana*, Pot marjoram *O. onites*
Lamiaceae

Marjoram is thought to have originated in Cyprus and adjacent southern Turkey, although it has naturalized over much of the Mediterranean region. It was cultivated in Egypt and by the Greeks and Romans in antiquity. In classical times it was regarded as the herb of happiness. It was probably introduced to Britain during the Middle Ages and is today cultivated in Mediterranean areas and in several other countries in Europe, America, Africa, and Asia including Southeast Asia. It was only on return of the American GIs from Italy after World War II that sweet marjoram became well known in the United States. It was used to flavor ale before the popularity of hops, and it is still used as a beer preservative. Its aromatic leaves may be used either fresh or dried in much the same way as thyme, although it has a sweeter flavor. There was an old belief that sprays of marjoram and thyme laid beside milk in a dairy would prevent it from going sour in thundery weather. As its flavor is more delicate than that of thyme it is usually added to dishes just before the end of cooking. It may also be used raw in salads.

Pot marjoram, *M. onites*, originated in eastern Sicily, southern Greece, and southern Turkey. It is hardier than sweet marjoram and has been locally cultivated as a flavoring in the Mediterranean area and western and central Europe at least since the 16th century and is now cultivated in North America. In cooler regions it is grown in pots as an edible indoor plant.

Tansy *Tanacetum vulgare*
Asteraceae

Tansy grows wild throughout Europe, Asia, North America, and New Zealand. The leaves and shoots were formerly used in puddings and omelettes and tansy cakes customarily made at Easter. It is said to have been in the drink given to Ganymede by order of Jupiter to immortalize him as cupbearer to the gods.

Tarragon *Artemisia dracunculus*
Asteraceae

Tarragon grows wild through most of the temperate regions of the northern hemisphere from Europe to China and North America. It belongs to the same botanical family as wormwood, southernwood, and mugwort, all used as herbal remedies. Tarragon is much used in French cooking. Tarragon vinegar is made by putting fresh tarragon leaves in vinegar and allowing them to infuse for some time. It is essential for making Bearnaise sauce and for all dishes *a l'estragon*. It is quite a strong flavor and must be added carefully to delicate dishes. *A. brotanum* (southernwood or garderobe) is strongly aromatic and used to repel moths but is not used in food.

Thyme *Thymus vulgaris*
Lamiaceae

Thyme grows wild in mountainous areas bordering the Mediterranean. It is one of the great European culinary herbs and was used by the ancient Greeks. They regarded it as the herb of courage, elegance, and grace. It was used with burnt sacrifices, which gave rise to its name (*thymon*, Greek = to fumigate). The Romans grew it beside their beehives to flavor honey. They also used thyme to flavor cheeses and liqueurs. It may have been introduced to Britain by the Romans, but it was certainly commonly cultivated in England before the middle of the 16th century. There are many varieties which are used for culinary purposes such as lemon thyme (*T. × citriodorus*), orange thyme, Corsican thyme, Sardinian thyme, and the broad-leaved, narrow-leaved, and variegated forms. In cooking it is used to flavor stuffing and is a key ingredient in *bouquet garni*. It is used in soups, vegetables, fish, and chiefly with wine and onions to flavor meat dishes, especially those cooked slowly in the oven. It is often used in the pickle for olives. The preservative qualities of thyme are valuable in sausages and salamis, potted meats, and strong potted cheeses. Bees are very fond of it and it flavors their honey. Thyme is also one of the flavorings in the liqueur Benedictine.

Vervain *Verbena officinalis*
Verbenaceae

Vervain probably originated in the Mediterranean area but is now naturalized in Europe, North Africa, and West Africa. It was used by druids as a holy offering. In central, west, and southern Europe it used to be commonly cultivated as a medicinal plant, and recent cultivation has been reported in Germany. The plant has no perfume, but it does have a slightly bitter taste and is used to make herbal teas. Blue vervain, *V. hastate,* is indigenous to the United States and was used medicinally by Native Americans.

Watercress *Nasturtium officinale*
Brassicaceae

Watercress grows in fresh, running water in most temperate climates where there is a limestone or chalk subsoil. The leaves have a hot tangy taste and make an excellent soup or salad. It is native to southern Asia and the Middle East, and has been cultivated for a long time in Europe. It reached Britain in the 16th century and was then carried to North America by the early European settlers.

Wild ginger *Asarum canadense*
Aristolochiaceae

The roots of this herbaceous plant, not related to ginger (*Zingiber officinale*), were regarded by the Indians of North America as an appetizer and were added to any food that was cooking to give it additional flavor. It grows in rich soils in North America, especially in North Carolina and Kansas.

Kitchen Garden Vegetables

It is not easy to define a culinary vegetable: normally they are not sweet and are eaten with savory dishes, fish, meat, eggs, and cheese. In this section we are not considering root vegetables, (discussed in the chapter Roots and Tubers) or nuts and seeds, or pulses such as peas, beans, and lentils (all discussed in Nuts, Seeds, and Pulses chapter), but those that we are considering include leaf vegetables used either in salads or cooked, leaf stalks such as celery, fruits including tomatoes and avocadoes, squashes and cucumber, bulbs such as onions and leeks, shoots such as asparagus, flowers such as

Wild ginger (*Asarum canadense*). USDA-NRCS PLANTS Database/Britton, N.L., and A. Brown. 1913. *Illustrated flora of the northern states and Canada.* Vol. 1, p. 642.

cauliflower and sprouting broccoli, fleshy bracts and receptacles as in globe artichokes. They are valuable in a balanced diet for providing a source of vitamins especially B, C, and E and minerals including iron and potassium. With the exception of avocado they do not provide much fat. The pulses are valuable sources of vegetable protein and carbohydrate.

Alexanders *Smyrnium olusatrum*
Apiaceae
This species is native to the Mediterranean region but is able to thrive further north and throughout Western Europe since its introduction by the Romans. It is now almost completely forgotten as a food although it still grows wild in much of Europe, including the British Isles where it has naturalized. It resembles celery and was grown as a potherb for its leaves, young shoots, and roots probably since the Iron Age. It became very popular in the 4th century during the reign of Alexander the Great and was also widely used in the Middle Ages as a vegetable in the same way as the then bitter celery. It has been used medicinally since antiquity for its diuretic and antiscorbutic properties. In the 18th century the use of alexanders as a food was surpassed by improved types of celery.

Amaranthus spinach *Amaranthus* spp.
Amaranthaceae
Several members of this family are grown as important leaf vegetables in tropical Southeast Asia, Africa, and the Caribbean area. *Amaranthus tricolor*, Chinese or vine spinach, is one of the most widely used for its leaves. It seems to have been native to India. The leaves are often sliced and stir fried. Some species of *Amaranthus* are traditionally grown as pseudo-cereal crops for their seeds especially in Central and South America, for example *A. caudatus* Inca wheat. *A. cuentas* is a popular leaf vegetable in most of sub-Saharan Africa and is also grown as a grain.

See: Gathering Food from the Wild, p. 33; Grains, p. 58

Arugula, Rocket, Roquette *Eruca sativa* (syn. *Eruca vesicaria* subsp. *sativa*)
Brassicaceae
Arugula (rocket) is native to the Mediterranean region and western Asia and has spread to North America. Its pungent leaves have been used as a salad plant since classical times. It is also grown as an oilseed crop in India, Pakistan, and Iran. Wild rocket (*Sisymbrium tenuifolium* or *Diplotaxis tenuifolia*) has a stronger flavor.

Asparagus *Asparagus officinalis*
Alliaceae
Asparagus is native to marshy habitats in central and southern Europe, North Africa, and western and central Asia. It was grown in classical times and then lost its popularity until the 17th century. Nowadays it is cultivated all around the world with the United States being the largest producer. The young shoot or spear is eaten. Selective breeding has given rise to a greatly-thickened fleshy shoot, which is much prized as a delicacy. The young stem is prepared by lightly boiling or steaming and serving with melted butter, hollandaise sauce, or vinaigrette. Much of the crop is canned or deep frozen.

See: Gathering Food from the Wild, pp. 39–40

Avocado *Persea americana*
Lauraceae
A native of Central America, this species has recently become an important food crop in subtropical and tropical regions, with an export trade to temperate zones. It is cultivated chiefly in California, Florida, Brazil, South Africa, Israel, Australia, and Southeast Asia. The fruits vary in shape from

being pear-shaped to rounded. The green, yellow, or red fruit skin encloses light green flesh, which is very nutritious with some fat and protein, and a single stone. It is eaten raw, sometimes with a little lemon juice or vinegar, or in salads.

See: *The Hunter-Gatherers, p. 7; Origins and Spread of Agriculture, p. 21; Fruits, p. 89*

Balsam pear, Bitter melon *Momordica charantia*
Cucurbitaceae

India is considered to be the possible centre of origin for this Old World vine, though it is believed the Portuguese carried it from Africa to Brazil in the 16th or 17th century. The species is now widespread in the tropics from the Philippines to the Caribbean due to seed dispersal by birds. It is also widely cultivated, particularly in China and Southeast Asia, for its edible fruits. The cucumber-like fruits are preferred when immature and green and are prepared by halving and scooping out the seeds. The Chinese braise or steam them, and in India it is eaten in curries, picked, boiled or fried, or raw after being steeped in water to remove the bitter taste. The tender shoots and leaves of the plant can also be cooked as a kind of spinach. It is highly regarded throughout much of Asia where it is thought to have medicinal properties.

Bamboo shoots *Phyllostachys, Bambusa* spp.
Poaceae

Bamboo shoots are the newly emerged, edible shoots of several species of bamboo, in genera of the subfamily Bambusoideae, including *Phyllostachys*, *Bambusa*, and *Dendrocalamus*. Bamboo plants are native to tropical Asia, Africa, and America, but are cultivated throughout Asia for the thick, pointed shoots that emerge from the ground beneath the bamboo plant, which would, if left, develop into a new culm. The usual practice is to cover the bases of the plants with mud and manure in winter and to cut the new emerging shoots in spring. After removing the leaf sheaths, the stems are boiled for about half an hour to remove any bitterness (cyanogenic glycosides) but to retain their crisp texture. They are a particularly common food item in China, Japan, Korea, and Southeast Asia, and in many other countries due to the popularity of Chinese cuisine, and they are often sold canned. The species most commonly used in China and Japan are *Phyllostachys pubescens*, along with several other members of the genus, and *Sinocalamus* species. *Thyrsostachys siamensis* is an important species in Thailand for bottling and canning.

Basella, Malabar, Ceylon spinach *Basella alba*
Basellaceae

Basella is usually considered a native of southern Asia (India), but its exact origin is not known. In Southeast Asia and China it has been grown since ancient times and is now widely cultivated as a minor vegetable in most tropical and temperate regions around the world. It is commonly grown for its young shoots, which make a succulent, mucilaginous vegetable, used as a potherb in soups and stews, boiled, fried in oil, or sometimes as a green salad. In the Western Hemisphere, basella is grown in Mexico and parts of North America for its leaves that are used in the same way as spinach. An early use of the fruits in China seems to have been for dyeing purposes. The red fruit juice can be used as ink, cosmetic, and for coloring foods, and a number of medicinal applications have been reported. The red forms are commonly planted as ornamentals, even becoming popular in Europe as a pot plant.

Black salsify, Spanish salsify *Scorzonera hispanica*
Asteraceae

Black or Spanish salsify, *Scorzonera hispanica*, like salsify (see following) is grown predominantly for its roots but has edible leaves. It is native to central and southern Europe and was first used as a medicinal plant. Similar to salsify, it was probably first cultivated in Spain and Italy in the 16th century, appearing later in England. It is now cultivated in temperate and sub-tropical areas and at higher altitudes in the tropics. The leaves are consumed raw in salads or lightly steamed. They are also used to feed silkworms.

See: Roots and Tubers, p. 71

Broccoli, Calabrese *Brassica oleracea* var *italica*
Brassicaceae

There appears to be no evidence for this crop before the 18th century when it was reported growing wild on the cliffs near Naples. A number of types have evolved with different colored flower heads: by the 19th century it was possible to grow red-, cream-, green-, purple-, and brown-headed types. Broccoli reached the United States in the late 18th century but did not become popular until the 20th century. Ninety percent of the American broccoli crop is now produced in California. Today green and purple sprouting forms are the most popular. It is similar in structure to the cauliflower but produces a loose cluster of flower heads rather than a single compact head.

Bok choi, Pak-choi *Brassica rapa* subsp. *chinensis* (syn. *Brassica chinensis*)
Brassicaceae

This is one of the Chinese cabbages which evolved in China and has been in cultivation there since the 5th century AD. It is now widely grown in the Far East and Southeast Asia. It is an open rather than a hearted vegetable with fairly smooth oval, green leaves on thick stalks. Both the stalks and the leaves shrink greatly on cooking. They have a mild cabbage-like flavor. It is used in soups and stir-fry dishes but is seldom eaten raw. *Brassica pekinensis*, Pe-Tsai, is another form of Chinese cabbage also domesticated early in China and is now grown throughout the world. There are several forms: one is long and narrow like a Cos lettuce, another is less long and barrel shaped in profile. The form with a compact, elongated head is best known in western countries where it is sometimes known as "Chinese leaf." It is used as a salad plant, a vegetable, in soups, and in pickles.

Brussels sprouts *Brassica oleracea* var *gemmifera*
Brassicaceae

This crop appears to have originated near Brussels around 1750, and it was not grown in Britain until the early 19th century. Thomas Jefferson planted some in the United States in 1812. The dense, compact axillary buds are borne close together all along a tall single stem. They are best harvested in the early winter months. In Britain they are the traditional accompaniment to chicken or turkey at Christmas.

Cabbage *Brassica oleracea* var *capitata*
Brassicaceae

These head cabbages are very familiar with their short stems and greatly enlarged terminal buds. Green cabbages were eaten by the Greeks and Romans: they cut off the heads in the autumn, leaving the stalks in the ground to sprout in the spring as a spring green. Cabbage is usually eaten boiled; in parts of France it is stuffed with mincemeat; it may be served raw, very thinly sliced and mixed with grated carrot and mayonnaise as a coleslaw. Savoy cabbages, with their dense heads of crinkled leaves, had developed in Germany and by 1543 when three different types were described. They had reached Britain by 1597. The red cabbages are thought to have originated in Germany in the 12th century and to have reached Britain by the 14th century. Nowadays it is chiefly used for pickling. Cabbages reached North America in 1541 on the third voyage of Jacques Cartier. They came mainly from Germany and the Low Countries.

Cauliflower *Brassica oleracea* var *botrytis*
Brassicaceae

They appear to have been developed in the east Mediterranean, possibly in Cyprus. Cauliflowers were grown in Italy in the 15th century, but after that they do not seem to have been much cultivated until the end of the 17th century. Cauliflower seed was being regularly imported from Crete to Scotland as late as 1684. By the end of the 18th century it had become a much-valued vegetable. During the last couple of hundred years forms suitable for growing in hot, humid climates were developed in India. Cauliflowers develop a single large, swollen flower head which is edible. It consists of a tightly packed mass of white/cream-colored, underdeveloped flower buds. It is sometimes served raw, but more often boiled and served with a white or cheese sauce.

Collard greens *Brassica oleracea* var *viridis*
Brassicaceae

Collard greens are non-head forming and similar to kale in appearance. The name is a corruption of colewort, a generic term for cabbage plants. Collards were native to the eastern Mediterranean and have been cultivated for at least two thousand years, being introduced to America by early European settlers. They are a minor commercial crop in the United States, but a standard winter green in home gardens, and served as a boiled vegetable.

Celery *Apium graveolens*
Apiaceae

Wild celery grows in moist places, especially by the sea, in much of Europe and Asia. It did not come into widespread cultivation until the 16th and 17th centuries in France and Italy when the milder-tasting forms were selected for their leaf stalks. The best leaf stalk celery is obtained by earthing up the stalks as the plant grows, which has the effect of blanching the stalks and reducing their bitterness. Recent breeding has produced self-blanching forms which do not need earthing up. Celery can be eaten raw with cheese or in salads; it can be cooked by boiling or braising or can be made into soup. Celery leaves are an important vegetable in Southeast Asia where the Chinese developed cultivated varieties quite independently. These are thinner, juicier, and more strongly flavored than those in the west. They do not blanch the stems and always eat it cooked, often in a mixture of vegetables.

See: Roots and Tubers, p. 71; Spices, p. 159

Chard, spinach cress *Beta vulgaris* subsp. **vulgaris**
Chenopodiaceae

Beta is an Old World genus mainly confined to Europe and the Near East, and *B. vulgaris* probably originated in the Mediterranean region. The earliest forms of beets to be cultivated were the leaf beets and chards, which do not posses a swollen taproot, as many of their allied cultigens do. Their use as a leaf vegetable precedes Greek and Roman times, when, as early in the Orient, the leaves were used medicinally and as potherbs. Early descriptions of chards exist from the 2nd century. Today, chard is widely grown for its succulent leaves, which are used as a green vegetable like spinach. The whole leaf can be eaten, including the broad, white leaf stalk, up to several centimeters across, which is often eaten as a separate vegetable. Many types have been developed ranging from reddish-purple leaf stalks to bright yellow.

Chaya, Tree spinach *Cnidoscolus chayamansa*
Euphorbiaceae

This species is indigenous to Mesoamerica and is now cultivated throughout Central America and the West Indies as a hedge plant. The plants, which can attain the height of a human, have shoots and large leaves that are eaten as a vegetable. The leaves must be boiled to remove the toxic compounds and stinging hairs that are present in some types. In Mexico it is used to make Tzoto-bilchay, a regional specialty tamale from Yucatan made of maize dough, ground squash seeds, and chaya leaves.

Chayote, Christophine *Sechium edule*
Cucurbitaceae

This gourd probably originated in Central America and was cultivated by the Aztecs. It is now widely grown in the tropics and forms an important crop in Brazil, Mexico, Costa Rica, and the West Indies. The pear-shaped fruits are longitudinally furrowed and grow on vigorous vines. The whitish flesh encloses a single flat seed. The fruit can be boiled, baked, or fried as a vegetable, added to sauces, or used in salads.

Chickweed *Stellaria media*
Caryophyllaceae

Chickweed, so called because poultry like it, is a plant native to Europe although it now grows wild throughout the temperate regions of the world. It is principally regarded as a medicinal plant, but is likely to have also been used since ancient times as a potherb and vegetable on account of its ability to grow throughout winter months and is hence a good winter or famine vegetable. The classical Greeks and Romans sometimes cultivated chickweed, and it was also popular in ancient Japan. Today the leaves and stems are used mostly as a potherb or are eaten raw, the flavor being between that of spinach and cabbage. In North America it has a reputation for making an excellent salad, and in India it is cultivated for green manuring and is occasionally eaten as a vegetable. The seeds are also edible. The genus *Stellaria* was named after the Latin word for star on account of their star-shaped flowers.

Chicory *Cichorium intybus*
Asteraceae

The wild form is native to central Russia, western Asia, and southern and central Europe. Because there is some confusion between the identification of endive (see following) and chicory (especially in France where chicory is called *endive* and endive is known as *chicorée*) it is not known where it was first brought into cultivation. The root has been used as a substitute for coffee, the broad leaf forms are used in salads or cooked as a vegetable. *Chicons* are produced by digging up year-old roots, cutting off the root tips and all but an inch of foliage, and replanting them in a warm dark place where the blanched *chicons* are produced. They can be eaten raw or cooked.

See: Caffeine, Alcohol, and Sweeteners p. 177

Cucumber *Cucumis sativus*
Cucurbitaceae

Cucumbers have been cultivated in India for thousands of years and are thought to be derived from the wild *Cucumis hardwickii* which grows in the foothills of the Himalayas. It was grown in ancient Egypt, by the Greeks and Romans, and also had reached China by the 6th century AD. It reached America in 1494 when Columbus introduced it to Haiti. It is now grown throughout the world. The fruit is usually elongated in shape and may reach up to 90 cm (36 inches) in length, though there are some more globular forms. They are normally harvested in an immature state and are eaten raw, either peeled and sliced in salads or sandwiches, or added to cold yoghurt-based summer soups or sauces. The term gherkin applies to varieties of cucumber grown for pickling and includes not only the small fruits of dwarf varieties of *C. sativus* but also the fruits of *C. anguria*, the West Indian or burr cucumber. This species is grown from Brazil to the West Indies, Texas, and Florida.

Eggplant, Aubergine *Solanum melongena*
Solanaceae

Eggplants are native to tropical Asia and were first cultivated in India. The Arabs introduced it to Spain in the 8th century AD. It is now cultivated widely in the tropics and subtropics. The purple fruit is cooked as a vegetable boiled, fried, or stuffed, and it may be included in curries and moussaka.

Endive *Cichorium endivia*
Asteraceae

Endive may have originated in the eastern Mediterranean region and was known to the ancient Egyptians, Greeks, and Romans but does not appear to have reached central Europe until the 16th century. It was introduced to Britain by 1548, although it may have been grown here for some time before this. Barnaby Googe, writing in 1578, instructed that it be sown late and watered well, and that the leaves be tied up in bunches and blanched under crocks. It is now grown throughout the world, including the tropics. The leaves are usually used fresh in salads, although they may sometimes be blanched to reduce their bitterness. There are two forms: one with curled, divided leaves and the other with broad, almost entire, flat leaves, sometimes with red pigmentation. In France, *endive* refers to chicory (see preceding).

Epazote, Mexican tea *Chenopodium ambrosioides* and related species
Chenopodiaceae

Epazote is native to tropical and subtropical America, but in the Old World it has escaped cultivation and has frequently naturalized in tropical and subtropical areas. It is widely cultivated in warmer climates and is a traditional and necessary ingredient in Mexican and Central American cooking where it is used for the preparation of a tea and as a potherb. The aromatic (some would say unpleasant) leaves are found in markets in long green stalks as well as in chopped and dried forms. When added to dishes containing beans, epazote is believed to reduce gassiness, and it is also used to make Mexican tea, which is another name for the plant.

Other important related species include allgood (or good King Henry, wild spinach; *C. bonus-henricus*) and lamb's quarters (or fat hen; *C. album*). Allgood is a plant of eastern Mediterranean origin, which now grows in Europe and North America. It was formerly an important vegetable in

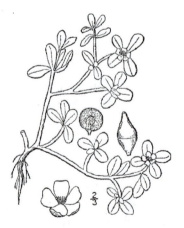

Mexican tea (*Chenopodium ambrosioides*). Robert H. Mohlenbrock @ USDA-NRCS PLANTS Database/USDA NRCS. 1992. *Western wetland flora: Field office guide to plant species*. West Region, Sacramento, CA.

Europe and had been grown since classical times for its tender young shoots. Its cultivation decreased after the introduction of spinach. Lamb's quarters is a cosmopolitan weed and ruderal plant found throughout the tropics frequently in mountainous areas. It is cultivated in India as a traditional leaf vegetable and also in the western Himalayas and Sichuan as a grain crop (the Old World counterpart of quinoa, to which it is related). It is likely to have been cultivated to a greater extent previously, as indicated by finds of seeds from the prehistoric lakeside and other pre-Roman settlements in Europe.

See: Plants as Medicines, p. 234

Florence fennel *Foeniculus vulgare* var *dulce*
Apiaceae

This is a native of the Azores, but it is best known from Italy where it is widely grown for its greatly swollen leaf bases which form a kind of false bulb with an aniseed flavor, much prized as a vegetable It may be eaten raw or cooked, often accompanying cheese dishes. Use of fennel leaves as a herb is described in the section preceding.

Garden cress *Lepidium sativum*
Brassicaceae

Garden cress is a native of western Asia, but its main use is in Europe and North America. Domestication presumably took place in the Middle East, and it has been cultivated since antiquity in Greece and Rome and from here is likely to have been introduced to other parts of Europe. Several varieties and sub-varieties are distinguished. Today it is mainly grown as a salad vegetable and is usually available in supermarkets growing in small plastic boxes.

Garden egg *Solanum aethiopicum* and other *Solanum* species
Solanaceae

This plant is very commonly cultivated throughout tropical Africa, and it originated from the hairy ancestral species *S. anguivi*, which grows wild throughout the area. The plants were introduced to Brazil during the slave trade and are also found rarely in Spain, Italy, Georgia, and India. Four main cultivar groups are recognized: the large orange fruits of the cultivar groups Gilo and Kumba are cooked in stews and even eaten raw, and the glabrous leaves of the cultivar groups Kumba and Shum are boiled as spinach. There are hundreds of names for this plant, used by the many tribes of Africa for the various groups.

Solanum is a very large genus containing both tuberiferous and non-tuberiferous species, mostly American in distribution. There are many locally important vegetables within this genus that are cultivated or wild-collected for their edible fruits and leaves.

See: Roots and Tubers, pp. 163–4

Globe artichoke *Cynara cardunculus* (syn. *C. scolymus*)
Asteraceae

The globe artichoke probably evolved in the Mediterranean region and was valued by the Greeks; Pliny the Elder said it commanded higher prices in Rome than any other vegetable, and supplies had to be imported from North Africa, and commoners were prohibited from eating them! They were probably spread throughout the Roman Empire but the earliest reference to their being grown in Britain is in the 16th century. It is the fleshy base of the large scales, or bracts, in a rosette surrounding the flower buds which are eaten together with the receptacle, before the flower has opened. They may be baked or boiled and served hot or cold with various sauces such as hollandaise sauce. Only the soft, inner basal part of each "leaf" is eaten. The large, fleshy heart that remains after each "leaf" has been removed from the stem is often bottled as "artichoke hearts." In Italy small flower heads are cooked and preserved in olive oil.

Ground elder *Aegopodium podagraria*
Apiaceae

A widespread perennial weed in Europe, northern Asia, and the United States that can be easily cultivated for its edible leaves. It is thought that the Romans introduced it to Britain as a vegetable eaten rather like spinach. It was widely cultivated in the Middle Ages, especially in the herb gardens of castles and monasteries. It can still be found growing in their ruins.

Gourd, Ngon melon *Cucumeropsis mannii*
Cucurbitaceae

This species is native to west and central tropical Africa from Guinea to Sudan, Uganda, and south to Angola. It is cultivated in its native area, especially in West Africa and rarely elsewhere, mainly for its seeds, rich in oil and protein, which are variously employed in local cooking. The fruits and leaves are also used as vegetables.

Hoja santa, Acuyo, Root beer plant *Piper sanctum*
Piperaceae

This species is native to Mexico and Guatemala and is cultivated in Mexico. The leaves are used in traditional medicine (the Spanish name hoja santa meaning holy leaf), herbal teas, as a wrapping for tamales and grilled fish, and as an essential ingredient of the Mexican sauce *mole verde*. The leaves, when crushed, have a strong sassafras or root beer odor but are not used in preparation of root beer.

Ice plant *Mesembryanthemum crystallinum*
Aizoaceae

This species is a native of southern Africa and was introduced to Europe and North America in the 18th century as a substitute for spinach, though it was never successful as a popular vegetable. It is cultivated in central Europe, the Mediterranean, and India as a potherb and a vegetable, cooked and eaten in a similar way to spinach. These plants have been noted to be so effective in deterring fire in South Africa that nurseries in South Africa have been promoting the use of ice plants in "firescaping" their properties. They are also grown in California as sand stabilizers. The plants were formerly grown on the Canary Islands and in some Mediterranean countries for the production of soda. Ice plant has wide, succulent leaves covered in crystal-clear, glittering surface cells capable of engorging water. The glistening leaves have an icy appearance—hence the name.

See: Gathering Food from the Wild, p. 31

Kale *Brassica oleracea* var *acephala*
Brassicaceae

The kales are non-heading Brassicas which are most closely related to the wild cabbage and are the most cold-tolerant members of the family. The frilled leaf types are the only ones grown for human consumption; other forms are grown for livestock feed and fodder. They were grown widely by the Greeks and may well have been brought to Britain by the Romans. It was the most commonly eaten green vegetable of country people in most parts of Europe until the end of the Middle Ages when headed cabbages evolved. In England kale was known as cole or colewort; kale or kail is the Scottish name, and it is still grown there in colder regions than cabbages. The young leaves and side shoots are picked in spring.

Lamb's lettuce *Valerianella locusta*
Valerianaceae

Lamb's lettuce is native to Europe, the Mediterranean region, and north Africa to India, but has naturalized in parts of North America and temperate eastern Asia. It was once widely used in Britain and Europe as the main winter and early spring salad vegetable, before winter lettuces were widely grown. It was particularly popular in the 17th century when it was served with cold boiled beetroot and celery, but began to fall out of favor in the 18th century. In Europe, several types with varying leaf characteristics are still in commerce.

Leek *Allium porrum*
Alliaceae

Leeks do not occur in the wild: they probably developed from the wild *Allium ampeloprasum* which is found in the Mediterranean, Azores, Canaries, Cape Verdes, and Madeira. Middle Eastern cultivated leeks differ from the European forms and are classified as *Allium kurrat,* kurrat being the Arabic name for leek. The Middle Eastern leek has narrower leaves than the European form and a distinct, often subdivided bulb. Leeks were much prized in ancient Egypt and even in Roman times Egyptian leeks were said to be the best. The Emperor Nero earned the nickname "porrophagus" because he ate raw leeks dressed with olive oil on several days a month to clear his voice. The Romans regarded leeks as far superior to onions and garlic, and Apicius gives four recipes for cooking them. They have been grown in Britain at least since Saxon times. It is claimed that Welsh warriors wore leeks in their helmets to show which side they were on in a victorious battle against the Saxons in the 7th century, and that is why the leek became the symbol of Wales. They are essentially a European crop. The leek has flat, sharply folded leaves that encircle each other forming an elongated, cylindrical bulb or pseudostem. The lower part is blanched by earthing up during growth. This along with the green leaf tops are the edible parts used not only as a vegetable but also in soups and stews. In parts of northeast England there are competitions to grow the largest leeks, the limit being about 30 cm tall and 7 cm in diameter. For culinary purposes smaller leeks are more flavorsome and tender.

Lettuce *Lactuca sativa*
Asteraceae

This is probably the most widely cultivated salad plant in the world. It seems to have originated in the Near East from the wild *L. serriola* and was very popular with the ancient Egyptians. The Greeks and Romans also grew lettuces, and the latter probably introduced it to Britain. Lettuces were taken to the New World in 1494, and all cultivated American lettuces have a European origin. Lettuces with firm hearts were developed in the 16th century. There are basically two forms of lettuce: the cabbage form with rounded heads and soft or crisp leaves (e.g. iceberg, Batavia) and the taller Cos type with longer crisp leaves forming upright heads. A special type of lettuce was developed in China with a fleshy stem, which is usually cooked rather than eaten raw.

Marrow, Zucchini, Summer squash *Cucurbita pepo*
Cucurbitaceae

Cucurbita pepo includes vegetable marrows, zucchini (courgettes), some squashes, and some pumpkins; all are eaten when young and do not store well. They were domesticated in Mexico before 5500 BC and formed part of the basic maize-bean-squash agriculture of pre-Columbian Central America. Cultivars have been developed that can tolerate cooler climates, and they are now widely distributed throughout the world. Zucchini are baby squashes or marrows, the fruit being cut when

12 to 25 cm in length. If allowed to grow to their full size they are vegetable marrows. These can be eaten as a boiled vegetable or stuffed with minced meat, onions, and tomatoes. *C. ficifolia* (fig leaf squash, chilacayote) is eaten boiled as a vegetable when unripe, and the ripe flesh is used to prepare sweets and drinks. In Mexico the seeds are blended with honey to make a desert, *palanquetas*. Zucchini flowers are also battered and fried.

Marsh samphire *Salicornia europaea*
Chenopodiaceae

Marsh samphire and its close relations (it is difficult to distinguish them apart) are coastal plants, found in the salt marshes and estuarine areas of Europe, Africa, temperate Asia, and northern America. In France and Britain, marsh samphire is gathered from the wild during the summer months, and the tips are boiled or steamed and eaten as a vegetable or are pickled. In both countries they are also gathered commercially for sale to restaurateurs and fishmongers. Some species of marsh samphire are cultivated offshore Saudi Arabia, Eritrea, and Mexico for the production of oil. These cultivated plants sometimes find their way to British delicatessens during the winter months when the natives are out of season. An alternative name for this plant is glasswort due to its former use in Britain when it was burned to provide alkali for the glass industry.

Not to be confused with rock samphire (see following, p. 126).

Melokhia *Corchorus olitorius*
Malvaceae

This is a species native of India that now grows in most of the world's tropical and subtropical regions. It is a relative of the jute plant and is grown worldwide as a source of fiber. Its young, tender, green leaves are edible raw and cooked and have long been used as a food. For culinary purposes the use of the leaves are widespread from West and North Africa and the Mediterranean islands through the Middle East to Malaysia, Australia, the Pacific islands, South America, and the Caribbean. In Egypt the leaves are used to make the national dish, a thick soup called molokhia to which they impart a mucilaginous quality. The leaves can be dried and stored and are frequently found in Middle Eastern food shops.

New Zealand spinach *Tetragonia tetragonoides* (syn. *T. expansa*)
Aizoaceae

This plant is native to New Zealand, Australia, Tasmania, the Pacific Islands, Japan, China, and Taiwan. It was brought back to Europe by Captain Cook in the 18th century. (He used it as a source of Vitamin C on his voyages). It is a fast-growing annual which tolerates dry soils and hot temperatures and can be used as a substitute for spinach.

Okra, Lady's fingers *Abelmoschus esculentus* (syn. *Hibiscus esculentus*)
Malvaceae

This member of the cotton family is probably native to tropical Africa; it is now cultivated in many tropical and subtropical areas from Thailand and India to Brazil. It seems to have reached the New World with the slave trade and is recorded in Brazil in 1658 and Dutch Guiana in 1686. It is the immature fruit pod which is eaten. It may be boiled or fried. The pods are highly mucilaginous and are used to thicken soups and stews. In America okra is characteristic of Creole and Cajun cookery. In India dried okra is cut into short sections which are fried like croutons and taste quite different from other forms. *Hibiscus* (*Abelmoschus*) *manihot* is cultivated in tropical Asia and Melanesia mainly for its leaves and flowers. The leaves are tender and sweet and have a very high protein content and may be eaten raw or steamed.

Onion *Allium cepa*
Alliaceae

Onions appear to have originated in Afghanistan and spread westwards with the Greeks and Romans. They were not widely grown in Europe until the Middle Ages and were introduced to America by Columbus on his second voyage to Haiti (1493–94). Nowadays they are an important crop with the United States, China, Russia, and India being the largest producers. They are valued principally for their flavor and have a wide range of culinary uses. They can be eaten raw, fried, boiled, or roasted, in soup, stews, curries, and in chutney. The edible bulb is composed of enlarged leaf bases, and it may be globular or spindle-shaped. The skin may be brown, yellow, white, red, or purple. Bermuda onions include a number of mild-flavored forms distinguished by their colors: Red Bermuda, White Bermuda, and Yellow Bermuda. Among the most popular American onions are two hybrid forms, both of the Yellow Granex type: Vidalia which is grown around the town of that name in Georgia, and Maui which grows in Hawaii and is very sweet. Another popular form is Walla-Walla which grows in Washington State and is said to have been introduced there from Corsica early in the 20th century. Shallots are considered to be a variety of onion in which the bulbs multiply freely. They are said to have been grown in 12th century France and were being cultivated in Britain from the beginning of the 18th century. They can be eaten raw or cooked, and their small size makes them useful for pickling. Their leaves have been used as a substitute for chives. The Canada onion *Allium canadense* was introduced to England in 1820, and its very small bulbs are cultivated for pickling and for cocktail onions. The characteristic flavor of onions is due to a complex of sulfur compounds. When an onion is cut the crushing of the cells and the exposure to the air allows an enzyme to work which develops the pungency and releases the volatile substance allicin which irritates the eyes. Cooking onions transforms the taste to a slightly sweet flavor; some of the starch in the onion is transformed by heat into sweet tasting dextrin and sugar.

Orache *Atriplex hortensis*
Chenopodiaceae

This is a native species of Europe as far east as Siberia and all around the Mediterranean coast, although it is now naturalized in North America. Early Mediterranean civilizations cultivated orache as a green vegetable, much like spinach. The plant is tall and spindly with small arrow-shaped leaves and grows well in poor, sandy soils. The leaves taste similar to spinach but are less succulent. There are various cultivars with red, white, and green leaves—the green-leafed types were used in Italy to color pasta. Another use was to mix orache and sorrel, thus alleviating the slightly acidic taste of orache. It was used in England in the 16th and 17th centuries and was particularly popular in soups and stuffings. Its popularity declined with the increasing use of spinach throughout Europe, and it is now seldom used. It is occasionally grown as an ornamental, and in Russia it is used as a dye plant.

Palm hearts *Cocos nucifera, Bactris gasipaes, Sabal palmetto*
Arecaceae

Palm hearts, also known as palm cabbage, are the edible young apical shoots of palms. There are more than 2,500 palm species broadly distributed over the tropical and warm temperate regions of the world, and many of these are exploited locally or are in international trade for their source of hearts. In many cases, particularly in single-stemmed species, the palm is destroyed once the palm heart is removed, as the heart is the apical growing bud. Other species are multi-stemmed, so the apical bud can be removed from individual stems within a cluster without destroying the entire plant.

Because palms are usually felled to obtain the heart, they are sometimes given the name of "millionaires salad" and are marketed as a luxury vegetable. Palms processed for canning in international trade include the coconut (*Cocos nucifera*) and peach palm (*Bactris gasipaes*), but other species are commercially exported from countries of Central and South America and from Thailand. In the United States, palm hearts have often come from the native palmetto palm (*Sabal palmetto*), but more recently, the hearts have often been derived from South American species. An important industry based on the production of palm hearts from the peach palm also exists in Costa Rica, where thousands of palms are planted to produce hearts for the export trade. Palm hearts are eaten raw, cooked, and canned in salads or as a snack or appetizer.

Pumpkin, Winter squash *Cucurbita moschata*
Cucurbitaceae

This was probably the first of the cucurbit family to be domesticated, remains have been found in Mexico dating to before 5000 BC, and it was grown in Peru by 3000 BC. It will tolerate hotter conditions than other cucurbits and is widely grown in the tropics today. The fruits can be used as a vegetable and can be stored in the winter because they have hard, protective shells. Popular varieties include acorn squash, butternut squash, hubbard squash, and spaghetti squash. Pumpkin blossoms were dried by the North American Indians and used to thicken soups. A similar species, *C. mixta*, was also domesticated early in Mexico.

Purslane, Little hogweed *Portulaca oleracea*
Portulacaceae

This plant originated in the Near East or Central Asia, and its stems and leaves have been eaten in those regions for more than two thousand years. It now has a worldwide distribution as a weed and by cultivation, and it is frequently naturalized in temperate and tropical countries. It was already in North America in pre-Colombian times. The original wild plant had small green leaves and a sprawling habit, but cultivated types are mostly upright with emerald green or golden leaves. The plant was cultivated and eaten in ancient Egypt and in classical Greece and Rome. In Britain, it was first mentioned in herbals in 1562 when different cultivated types were already being used as a salad, in pickle, and in medicine. Purslane is still recognized as a medicinal plant and was included in the World Health Organization's list of most useful medicinal plants.

Today, purslane is not frequently eaten in western countries, although it can be found in Mexican markets in the United States under the name verdolaga. It is more popular in the Middle East and the Indian subcontinents. The leaves are fleshy and mucilaginous and mild in flavor. They can be cooked like spinach or used raw as a salad vegetable, it also has thickening properties and is used for this purpose in cooking.

Other species of the genus *Portulaca* provide edible leaves in various parts of the world.

Rampion *Campanula rapunculus*
Campanulaceae

Rampion is distributed in Europe (excluding the northern countries) westwards to Siberia and in North Africa and southwest Asia. It used to be cultivated in Europe, and its leaves were a common ingredient of 16th-century soups and stews or as a salad, and its roots were boiled or steamed and eaten cold as a vegetable. By the early 19th century the use of rampions was dying out in Britain as the potato gained acceptance, though they were still being extensively grown in France and Switzerland. By the end of that century though, the use of rampions seem to have vanished. Today it is rarely cultivated for culinary purposes.

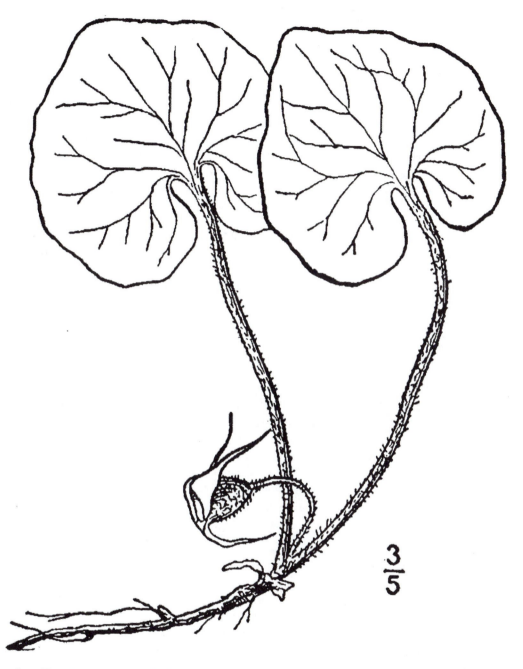

Purslane (*Portulaca oleracea*). USDA-NRCS PLANTS Database/Britton, N.L., and A. Brown. 1913. *Illustrated flora of the northern states and Canada.* Vol. 2, p. 40.

Rhubarb *Rheum × hybridum*
Polygonaceae

Rhubarb is the name given to the edible leaf stalks (petioles) of hybrids involving the species *R. rhabarbarum*. Wild rhubarbs are all native to Asia, though they are now cultivated in many countries with a temperate climate. They prefer a cool climate and particularly flourish in the general area of Mongolia and Siberia and in the vicinity of the Himalayas. The rhizomes from which the leaf-bearing stalks grow survive readily in ground that is frozen during the winter.

Rhubarb was known in classical Greece and Rome as an important dried root with medicinal properties. There is some debate about how far Chinese knowledge of rhubarb extends, but it does seem to be certain that it was known by 206 BC as an important medicinal plant that was in trade at that time. In England rhubarb became known, at first in a purely medicinal context, in the 16th century, though it was likely they were using the wrong species (the reported medicinal one was *R. palmatum*). It reached North America by the 17th century, shortly after the American Revolution. The idea of eating the stalk may not have occurred to people until later, and rhubarb didn't appear in English cookery books until the 1800s, with other recipes for sweet pies and tarts appearing soon after. New hybrids were being developed as early plants had mainly green stalks, and it was those with a red tinge that were selected to produce the modern types—numerous cultivars now exist.

Although, botanically speaking, rhubarb is a vegetable, in culinary terms it is considered a fruit. It is mainly used for pies (hence its name of pie plant in the United States), crumbles, fools, and jams. In some countries it is consumed as a vegetable. In Poland, rhubarb is cooked with potatoes and aromatics. It is used in an Iranian stew and in Afghanistan it is added to spinach. In Italy it is used to make rabarbaro, an apertif regarded as a health drink.

See: Plants as Medicines, p. 222

Rock samphire *Crithmum maritimum*
Apiaceae

Rock samphire, also known as sea or true samphire, is a plant native to sea cliffs, rocks, sand, and shingle of the Mediterranean, Black Sea, and Atlantic coasts from Scotland to the Canary Islands. It is collected from the wild as a leaf vegetable, and both the Greeks and Romans used it in salads and also lightly steamed to be eaten as a vegetable. It is still collected in some areas including Italy and Greece. Pickled rock samphire was once so common in England that people dangerously scaled cliffs to collect it, but its popularity declined in Britain by the end of the 19th century due to its scarcity. It is occasionally cultivated for its edible leaves in Italy, France, and the United States. The fleshy leaves are eaten raw in salads, cooked in butter, and may also be pickled.

Sago *Metroxylon sagu*
Arecaceae

Sago is originally a Javanese word that refers to the starch-containing palm stem and has become a generic term to refer to the stem starch from other palm species. The most important as a source of sago is *Metroxylon sagu*, a species whose center of diversity is likely to be New Guinea or the Malaccas. It has subsequently spread throughout Indonesia, Malaysia, and the Pacific Islands where it grows in low-lying, swampy areas. The trunk contains starch used as a staple food for humans in Southeast Asia, and, as with most other palms, nearly all the other parts of the plants are utilized in some way (see also palm hearts preceding). Although the use of sago for subsistence from wild or semi-cultivated stands is widespread in Indonesia, commercial production of the palms also occurs in Indonesia and Malaysia.

Traditional starch extraction involves felling the palm, chopping the pith, and repeatedly kneading and straining the pith in water. One sago palm may yield up to 400 kg of starch. The palm produces a light, almost pure starch that is used in various food products and industrially. When sago is consumed locally, it is made into a thick paste and consumed as it is or used in various dishes. When it is prepared for export, it is mixed to a paste with water, rubbed through a course sieve, and dried to produce "pearls." In western cooking its use has declined considerably since its popularity in the beginning of the 19th century, when it was considered a superior substance and was added to soups and dishes.

Salsify *Tragopogon porrifolius*
Asteraceae

Salsify is native to the eastern Mediterranean region, and the roots were probably eaten in classical times. Cultivation for the roots began in Italy and France in the 16th century, and today it is grown commercially on a small scale in Russia, France, and Italy for its edible roots. Sometimes the plant is grown for its edible young shoot and leaves alone. They are tender enough to eat raw or are steamed and served with butter and a dash of vinegar. The flavor is subtle and it has been served in this way as the first course to a meal, rather than as a vegetable to accompany meat.

See: Roots and Tubers, p. 72

Scallion, Welsh onion, Spring onion, Ciboule *Allium fistulosum*
Alliaceae

This member of the onion family appears to have evolved in China and Japan and to have reached Europe from central Asia at the end of the Middle Ages. The lower, fleshy part of the leaves barely swell into a bulb, but it forms many laterals with hollow cylindrical leaves. The green leaves are used in salads or chopped and used like chives; sometimes the whole plant is cooked. It has a delicate flavor.

Sea kale *Crambe maritima*
Brassicaceae

This species grows wild around the sandy coastlines in western Europe, the Baltic Sea, and the Black Sea. In the wild, when the young plant gets covered with sand, it is cut off from the light and develops a thick, unblanched leaf stalk with a small, undeveloped leaf at the tip. This is edible, mild, and tender and has been gathered from the wild as a vegetable that can be eaten like asparagus. Before cultivation, the wild population in Britain seems to have been enough to supply the markets there, but the method of forcing sea kale became widespread. On the beach, the young leaves were blanched by heaping sand or shingle over the crowns, and in the garden, ashes were used. Specially made seakale pots were also employed by covering the dormant plant with the pot and heaping manure up around it. Forced sea kale was considered a great delicacy until the 18th century. Today it is grown on a small scale in Britain and very rarely in other countries of Europe and North America.

An Asian relative of sea kale, *C. tataria*, grown largely for its edible roots from Siberia through towards Hungary, also has an edible stem.

Snake gourd *Trichosanthes cucumerina*
Cucurbitaceae

This is a tropical, climbing gourd growing wild in tropical east and Southeast Asia from Pakistan, India, and south China to tropical Australia. It has been grown since ancient times but is mostly cultivated in India and the Far East. The slender fruit is greenish white when immature and dark red when mature. The gourd can grow to a length of 2 m. Fruits of the wild plants are very bitter and inedible, but immature fruits of cultivated forms are eaten as a vegetable or in curries. Young shoots and leaves are also edible. In some forms, all young parts have an unpleasant smell and bitter taste, but these characteristics disappear after boiling. The fruits become inedible after ripening as they taste bitter and become fibrous like loofahs. In west Africa, the red fruit pulp is used as a kind of cheap tomato paste.

Sorrel *Rumex acetosa*, French sorrel *R. scutatus*, Herb patience *R. patientia*
Polygonaceae

Sorrel is the name given to several members of the genus *Rumex*, so called in reference to its sour tasting leaves and stems. The genus has a worldwide distribution, mainly in temperate regions, and several species are cultivated. Common sorrel, *R. acetosa*, French sorrel, *R. scutatus*, and herb patience, *R. patientia*, have been eaten as green vegetables since ancient times. At first, common sorrel was the most used, and it was often cultivated, but during the Middle Ages improved varieties of French sorrel were bred in Italy and France, and these became more popular. These new varieties arrived in England at the end of the 16th century. *R. acetosa* and *R. scutatus* are now the most widely grown species eaten in France and Italy. Many species of sorrel originated in Eurasia and were used by ancient Egyptians, Greeks, and Romans. These species have subsequently become naturalized in North America alongside native species that had been utilized by Native Americans.

The presence of oxalic acid gives sorrel its typical sour taste. When sorrel is eaten as a vegetable, it can be blanched first and the water discarded to reduce the acidity or steamed in a little butter. For serving, the sharpish taste may be softened by the addition of an egg yolk and cream. One old practice was to mix it with orache as it has a milder flavor. Sorrel is added to salads and used as an ingredient in soups, purées, and sauces, as an omelette filling, and as a stuffing for fish. An old English accompaniment to meat and fish was greensauce made from sorrel pounded to a paste with vinegar or lemon juice and sugar, and this name was also applied to the plant itself.

Spider plant, Bastard mustard *Cleome gynandra*
Brassicaceae

Spider plant is considered native to Asia but is widely distributed as a weed in the Old World tropics and has also been introduced into tropical America. It is cultivated in tropical Africa, Southeast Asia, and the Caribbean, but is only of local importance. It is a common market vegetable in Malaysia and Thailand where it is sold fresh or pickled in brine. The bitter leaves have a strong taste, described as being between radishes and asafoedita. They are eaten as a vegetable and cooking and fermentation both reduce the bitterness. They are also used as flavoring, and the leaves are salted and used as a pickle. In east and west Africa, it is used as a potherb and as flavoring for sauces. Both leaves and seeds are used medicinally, and the seeds are used as a substitute for mustard, hence the English name, and they also yield a good oil. The plant is commonly grown as an ornamental.

Spinach *Spinacia oleracea*
Chenopodiaceae

Spinach originated in Iran, spread to China about AD 600, and from there spread to Japan and Korea. The Arabs introduced it to Spain in the 11th century, but it was not widely cultivated in Europe before the 16th century. Nowadays it is cultivated worldwide. There are two forms: summer spinach with rounded "seeds" and winter spinach with "seeds" bearing two or three prickles. It can be eaten raw or boiled in a small amount of water, in soups, soufflés, green pastas, and *oeufs Florentine*.

Sugar cane, Pit pit *Saccharum edule*
Poaceae

Sugar cane, or pit pit, is a relative of commercial sugar cane *S. officinarum*. Its origin is unknown, and it only occurs in cultivation. It is probably derived from the wild *S. robustum*, which is also considered one of the possible ancestors of sugar cane. It is cultivated in Java, Kalimantan, Papua New Guinea, Melanesia, and the New Hebrides for the unopened buds of the inflorescences. They are unusual in that they do not normally emerge from their leaf sheaths, forming a compact mass about the size of a banana. In the market they are usually sold in a bundle tied together by the top of the sheaths. They may be eaten baked, steamed, or boiled, or are occasionally eaten raw, and in Papua New Guinea are an important seasonal food in some areas.

See: Caffeine, Alcohol, and Sweeteners, p. 186

Sweet pepper, Paprika *Capsicum annuum*
Solanaceae

Sweet peppers were widely grown in Mexico, the Caribbean, Central and South America in pre-Columbian times, and they were probably domesticated in Mexico around 7000 BC. Columbus brought them back to Europe in 1492, and they spread quickly to Africa and Asia. The fruit is a hollow berry which may be red, yellow, or green in color and contains many small white seeds. Sweet peppers may be eaten raw in salads or cooked, sometimes being stuffed with rice or mincemeat and tomatoes and baked in the oven.

Tomato *Lycopersicon esculentum*
Solanaceae

The cultivated tomato is thought to have originated from the cherry tomato, *Lycopersicon esculentum* var *cerasiforme,* which occurs wild in Peru, Ecuador, and elsewhere in tropical America. It was probably first domesticated in Mexico. Early in the 16th century it was brought to Europe by the Spaniards. Its name is derived from the Aztec name *xitomatl.* It was at first regarded as a poisonous plant in Europe and known as "Pomme d'amour," "love apple," and "golden apple," although it was clearly regarded as a food plant in Mexico and many varieties had been developed by the American Indians before the Spanish conquest. The culinary barrier seems to have been broken first in Italy, and gradually the taste for eating them spread around the Mediterranean and then northwards through Europe where they were long treated with suspicion. This may partly have been because of the color of their fruits and partly because they were related to poisonous plants of the nightshade family. Parkinson grew them as ornamentals even though he was aware that they were being eaten with delight in more southerly parts of Europe. Even in 19th century Britain caution persisted and their use seems to have been restricted to upper-class and Jewish households, and Cobbett writing in 1833 observed that they were being sold at high prices. Large scale production seems to have begun in Italy, in the region near Naples, in the early 19th century. It was not until the end of the 19th and beginning of the 20th centuries that tomato growing became big business in Europe and tomatoes were found in most salads and tomato sauce was the recognized accompaniment of all fish dishes.

Meanwhile in the 18th century they were introduced to the United States from Europe, where they soon gave rise to a canning industry and to the invention of tomato ketchup, America's national condiment. They also were gradually introduced into many other parts of the world making them now one of the most important vegetable and salad plants in the world. Part of its success is due to its ability to combine with other flavors such as basil, garlic, onion, thyme, oregano, peppers, cheese, egg, and meat flavors.

The fruit is a fleshy berry which may be red or yellow in color and can vary in size from the small cherry tomato (1.5 cm in diameter) to the "beef" type (10 cm diameter). They are usually more or less globular in shape, but Italian plum tomatoes are ovoid, and there are also pear-shaped forms. Its fruits are consumed raw or cooked and processed in a great variety of ways: as juice, soup, sauce, ketchup, paste, and they can also be dried. Green tomatoes are used to make chutney and pickles. Developments in plant breeding have led to greatly increased yields and disease resistance at the expense of its flavor; those lucky enough to be able to grow their own tomatoes at home know that there is a world of difference in the flavor of the home grown, freshly picked fruits.

Tomatoes are tolerant of a wide range of climates from tropical to temperate regions and there are many different cultivars. Large scale production is to be found in Russia, China, the United States, Egypt, and Italy.

Water spinach *Ipomoea aquatica*
Convolvulaceae

This species originated in tropical Africa (probably India) and can be found in south and Southeast Asia, tropical Africa, South and Central America and Oceania. Only in south and Southeast Asia is water spinach, or kangkong, an important leafy vegetable. It is intensively grown in water and is frequently eaten throughout Southeast Asia, Hong Kong, Taiwan, and in southern China. Several varieties have been developed, some with relatively thick stems and pale leaves and others with narrow stems and numerous leaves. The young stems and leaves are cooked lightly or fried in oil and eaten with various dishes. In Cantonese cuisine it is always stir fried with garlic and fermented bean curd or shrimp sauce. The vines are used as fodder for cattle and pigs, and the juice of the plant and crushed flowers are used medicinally.

See: Roots and Tubers, p. 65

Wild Cabbage *Brassica oleracea*
Brassicaceae

Wild cabbage is native to the seaside cliffs surrounding the Mediterranean and southwest Europe and can be found as far north as southern Britain. The wild form has evolved into a number of distinct varieties in which different parts of the plant (stems, leaves, buds, and flowers) have become palatable (see broccoli, Brussels sprouts, cabbage, cauliflower, collard greens preceding). Essentially wild cabbage is a temperate zone plant, but it is now cultivated throughout the world.

Winter cress *Barbarea verna*
Brassicaceae

This species is native to west and southwest Europe, but it has been introduced elsewhere and has naturalized in Africa, America, and Asia. It has been cultivated as a vegetable in west Europe, mainly in France, England, more seldom in central Europe or North America. Although its cultivation is not widespread, in former times the cultivation occurred mainly in the center of France, but today it is also popular in Belgium. It is also cultivated to a small extent in Malaysia. The leaves are used as a winter or spring vegetable for salads or as a garnish like cress.

References and Further Reading

Bown, D. 2002. The Royal Horicultural Society New Encyclopedia of Herbs and Their Uses. London: Dorling Kindersley.

Bunney, S. 1984. *The Illustrated Book of Herbs, Their Medical and Culinary Uses.* London: Octopus Books.

Darby, W.J., Ghalioungui, P., and Grivetti, L. 1977. *Food: The Gift of Osiris.* London: Academic Press.

Davidson, A. 1999. *The Oxford Companion to Food.* Oxford and New York: Oxford University Press.

De Guzman, C.C., and Siemonsma, J.S. (Eds.). 1999. *Plant Resources of South-East Asia No 13. Spices.* Leiden: Blackhuys.

Densmore, F. 1974. *How Indians Use Wild Plants for Food, Medicine & Crafts.* New York: Dover Publications.

Flach, M. 1997. *Sago palm. Metroxylon sagu Rottb. Promoting the conservation and use of underutilized and neglected crops.* 13. Gatersleben: Institute of Plant Genetics and Crop Plant Research and Rome: International Plant Genetic Resources Institute.

Garland, S. 1985. *The Herb and Spice Book.* London: Frances Lincoln.

Grieve, M. 1992. *A Modern Herbal.* London: Tiger Books International (reprint of 1931 edition).

Kiple, K.E., and Ornelas, K.C. (Eds.). 2000. *The Cambridge World History of Food.* Cambridge and New York: Cambridge University Press.

Lovelock, Y. 1972. *The Vegetable Book, an Unnatural History.* London: Allen & Unwin.

McVicar, J. 2002. *New Book of Herbs.* London: Dorling Kindersley.

Purseglove, J.W. 1991 (reprinted). *Tropical crops—Dicotyledons.* New York: Longman Scientific & Technical.

Renfrew, J.M. 1973. *Palaeoethnobotany: The Prehistoric Food Plants of the Near East and Europe.* New York: Columbia University Press.

Renfrew, J.M. 1985. *Food and Cooking in Roman Britain: History and Recipes*. English Heritage, Historic Buildings and Monuments Commission for England.

Schippers, R.R. 2000. African Indigenous Vegetables : An Overview of the Cultivated Species. Chatham: Natural Resources Institute.

Smartt, J., and Simmonds, N.W. (Eds.). 1995. *Evolution of Crop Plants* (2nd ed.). London: Longman.

Siemonsma, J.S., and K. Piluek (Eds.). 1993. *Plant Resources of South-East Asia No. 8. Vegetables*. Leiden: Backhuys.

Stobart, T. 1987. *Herbs, Spices and Flavourings*. Harmondsworth: Penguin Books.

Stuart, D. 1984. *The Kitchen Garden: An Historical Guide to Traditional Crops*. Gloucester: Alan Sutton.

Tindall, H.D. 1983. *Vegetables in the Tropics*. London: Macmillan Press.

Vaughan, J.G., and Geissler, C. 1999. *The New Oxford Book of Food Plants: A Guide to the Fruit, Vegetables, Herbs and Spices of the World*. Oxford and New York: Oxford University Press.

8
Nuts, Seeds, and Pulses

GEORGINA PEARMAN

It is not surprising that reproductive plant parts (disseminules), such as seeds, nuts, and fruits, are so important as food plants. They are often rich in nutrients, because the disseminule holds the food supply that allows the seedling to grow. Equally important, seeds and nuts are often adapted to survival in the soil before germination, through modifications such as hard seed coats. These adaptations allow storage of seeds and nuts by humans, sometimes over a period of years, making these foodstuffs an important resource for seasonal scarcities and famine years.

Terms such as *seeds*, *nuts*, and *fruits* have strict botanical definitions, but in practice are used more broadly. For example, true nuts—one-seeded dry indehiscent fruits (fruits that remain closed upon reaching maturity)–are only found in a few genera: *Quercus* (acorns), *Corylus* (hazelnut), and *Fagus* (beechnut). However, the term *nut* is often used more widely for any edible kernel within a hard shell, such as walnuts, Brazil nuts, almonds, and pine nuts. Some "nuts" are in fact tubers: the tiger nut (*Cyperus esculentus*) is described in the Roots and Tubers chapter. In this book such terms are used in the looser sense, with the term *fruit* limited to fleshy fruits.

This chapter covers nuts, pulses, and oilseeds. Fleshy fruits are covered in the Fruits chapter, grains in the Grains chapter, and some wild fruits and seeds in the chapter on Gathering Food from the Wild.

Edible Nuts

It is striking that so many nut-bearing plants have not been domesticated, or have only recently been taken into cultivation. Brazil nuts are still almost exclusively harvested from the wild, while the acorn, a staple food in many societies prior to the twentieth century, was likewise never domesticated. The long generation time of trees is a likely delaying factor to domestication; the intending breeder must wait years before selected seed returns a further generation. Most nuts are rich in fat (50 to 70 percent by weight). Only chestnut and acorn are starch-rich, and both have been widely used as flour supplements or replacements.

Acorn *Quercus* spp.
Fagaceae

Acorns have been a fat-rich (but protein-poor) staple foodstuff in temperate oak forests throughout the northern hemisphere. The best-known users were the hunter-gatherers of 19th-century California, for whom acorns, particularly from *Q. kelloggii*, were the most important food. Acorns were knocked from trees each autumn by shaking or beating with long poles, and then stored in pits or granaries. After removing the shell, acorns were pounded into flour in bedrock mortars. The bitter tannins were removed from the flour by leaching with water. Acorn flour was used to make acorn bread or, most often, eaten as acorn mush, a kind of soup. Acorns were also one of a number of nuts, including hickory nuts, collected by Native American farmers in the eastern United States. Here acorns were leached in water, or sometimes boiled in a lye made from wood ash to neutralize the tannins.

In the Mediterranean, particularly Spain and Sardinia, acorns from various species, including *Q. ilex*, were important foodstuffs until the mid-20th century. Acorns were most often consumed as bread after the tannins had been removed by boiling, or neutralized by use of lye or clay. In Spain and Portugal pigs are commonly fattened on acorns in oak forests. In the Near East sweet acorns from *Q. brantii* were a staple food in southeast Turkey until the 1940s, and are still widely eaten as a snack by shepherds. In Japan wild nuts were of great importance in mountainous areas until the 19th century. Acorns were leached, pounded into flour, and eaten as cakes. In Korea acorns are an important foodstuff, prepared as a jelly-like gruel known as "mook" and also used in noodles.

Archaeological evidence of nutshells and, in California, bedrock mortars, demonstrates that acorns have been important foods in these regions for thousands of years. They were never domesticated, probably because of their unpredictable tannin content, and because oaks do not produce acorns until the tree is twenty or thirty years old.

Almond *Amygdalus communis*
Rosaceae

One of the earliest nut trees cultivated in Old World agriculture, the almond was taken into cultivation from a wild form (subsp. *spontanea*) in the eastern part of the Mediterranean basin in the Early Bronze Age (ca. 3000 BC), at about the same time as the grape vine. Wild almonds have been eaten since the Neolithic period, but were probably roasted to destroy the high content of amygdalin, a cyanogenic glycoside that breaks down to release cyanide on consumption.

The almond is a small tree of up to 30 m (100 feet) high, of which there are many cultivars. Two botanical varieties occur: sweet almond, *P. dulcis* var. *dulcis,* and bitter almond, *P. dulcis* var. *amara.* Trees grown from sweet almond seeds can produce bitter almond seeds; traditionally orchardists have to rogue out these seedling trees. Almond cultivation in Europe has remained in the Mediterranean basin. Elsewhere, the Spanish developed orchards initially in Chile in the 16th century and then in California by 1800. California today produces two-thirds of the world's crop. It is also cultivated commercially in Australia and South Africa.

Beechnut *Fagus*
Fagaceae

The beechnut has been gathered and used for human food since prehistoric times. In Europe, the nuts from the predominant species *F. sylvatica*, which is found throughout Europe and temperate west Asia, have been used for several thousand years. The nuts from the main North American species, *F. grandifolia*, were once an important food for North American Indians. More often eaten by humans in times of famine or food scarcity, the beech has been neglected as a source of edible nuts. Its main use has been for feeding animals, especially pigs.

Betel nut, Pinang, Areca nut *Areca catechu*
Arecaceae

The seed of the areca palm *Areca catechu* contains a stimulant, arecaidine, and is widely used in the Indian subcontinent, Southeast Asia, and the Pacific islands of both Melanesia and Polynesia. The

Beech nuts (*Fagus grandifolia*). USDA-NRCS PLANTS Database/Britton, N.L., and A. Brown. 1913. *Illustrated flora of the northern states and Canada.* Vol. 1, p. 615.

nut is combined with lime and wrapped in a betel leaf (*Piper betle*), sometimes also with tobacco, and chewed or placed as a quid inside the mouth.

See: Psychoactive Plants, p. 193

Brazil nut, Para nut *Bertholletia excelsa*
Lecythidaceae

Brazil nuts are the seeds of one of the tallest trees (40 to 50 m or 130 to 160 feet) of the Amazon forest. The nuts are seed kernels. They fall from the tree packed twelve to twenty at a time in large fruits that weigh up to 3 kg (6.6 pounds)—a danger to anyone standing below. Considered amongst the finest of all nuts, they are commercially the most important of the many kinds that grow in South America. A small proportion is eaten locally but most are exported to the United States and Europe. The first shipments to the United States took place in the 19th century, with the Amazon River opening up for trade in 1866. When the price of Brazilian rubber collapsed in 1910 as a result of competition from Southeast Asia, Brazil nuts became a vital export crop in the Amazon region. Although there are a few plantations in Brazil, the bulk of the crop comes from wild trees, from which Brazil nuts are harvested after the large fruits have been allowed to fall to

Brazil nuts being shelled, Acre, Brazil. Copyright Edward Parker, used with permission.

the ground. Due to the unsophisticated methods of harvesting, the difficulties in cultivation, and the destruction of many of the wild trees in the rainforest as a result of deforestation, overall production is unlikely to increase.

Cashew nut *Anacardium occidentale*
Anacardiaceae

Native to the northern part of South America, the cashew's likely center of origin is semi-arid northeast Brazil. The nut is at the apex of the "fruit" or "apple," a fleshy and enlarged fruit stalk. The fruit is sweet-tasting and eaten raw or as juice or jam. Early Portuguese navigators found that the Indians valued both the nuts and "fruit" and soon after took the crop to India, the East Indies, and Africa. The Spanish probably took it to Central America and the Philippines. It remained a smallholder crop for the next 300 years in Asia and Africa. It is now a significant plantation crop in international trade, ranking third after almonds and hazelnuts. The main producer is Mozambique. The shell or pericarp of the nut contains an oil rich in phenols (similar to those found in poison ivy) that cause severe allergic reactions. The shells are therefore heated and removed from the nut before they enter trade.

Chestnut *Castanea sativa*
Fagaceae

The cultivated chestnut, a large tree up to 35 m (115 feet) high, is closely related to wild and feral forms occurring in the northern parts of the Mediterranean basin, north Turkey, and Caucasia. Initial domestication is likely to have originated in the latter two areas, with the cultivated and wild trees then spreading into southern Europe. The chestnut has been a staple food for thousands of years in traditional farming communities, and was also used for animal feed. Charred remains of this nut start to appear in several European countries in the first millennium BC. The Romans ground chestnuts into a flour that was used to extend wheat flour, a practice that survives in southern Europe.

From the 16th to 19th centuries the wild varieties continued to be a staple food for peasants in poorer regions, whereas the larger cultivated chestnut was regarded as a delicacy, although never traded in great quantities. Today these nuts are marketed to a limited degree but are mostly cultivated and used locally.

Outside Europe, chestnuts of several other *Castanea* species are used. *C. dentata*, the American chestnut, native to the United States, was cultivated widely until the last century when it was almost wiped out by fungal diseases. The Chinese chestnut, *C. mollissima*, has been cultivated in China for at least as long as its European counterpart and used in much the same way. It is now the main species grown in the United States.

Chinese water chestnuts *Eleocharis dulcis*
Cyperaceae

The Chinese water chestnut, *E. dulcis*, is the corm of an Asiatic sedge and is unrelated to the *Trapa* water chestnuts. Indigenous to the Old World tropics, its range stretches from tropical Africa to the Pacific islands. It is an important and popular food in China, eaten raw, or steamed or cooked quickly in a wok. It is the source of water chestnut flour, an important ingredient in China. The plant is cultivated in tanks or flooded fields and exported both fresh and canned to Europe and North America, where it is particularly popular in Chinese immigrant communities.

Coconut *Cocos nucifera*
Arecaceae

The coconut palm is an important crop throughout the humid tropics. Its origins are much debated, but it is likely to have originated in the Indo-Pacific region, the region of its greatest genetic diversity and where fossils have been found. During the Holocene, the range of wild coconut palms, dispersed by sea currents, probably spanned the tropical Indian and Pacific oceans from East Africa to Panama and Costa Rica. The most likely area for its domestication is Melanesia, on the coasts and islands between Southeast Asia and the western Pacific. Its subsequent spread is difficult to document but it has probably been due to a combination of human introduction and continued natural dispersal by the sea currents. The coconut became truly pan-tropical when Portuguese and Spanish explorers recognized its value in the 16th century, carrying it, respectively, across the Atlantic and Pacific oceans.

Traditionally selected for its water content, which is pure, sweet, and uncontaminated, the ripe nut (technically a drupe, or stone fruit) and all other parts of the palm have become useful under domestication. The oil extracted from the dried endosperm (copra) became the most important part of the plant. Large-scale planting of coconuts for commercial export of the oil and dried copra began suddenly in many regions during the mid-19th century. With the abolishment of slavery in 1835, many plantation owners planted coconuts as a low-labor crop. From the 1860s to the 1960s coconut oil was the leading vegetable oil in international trade until overtaken by soybean and palm oils. There is still a large international demand for its use in cooking, in high-quality soaps, and for various industrial applications. Grown throughout the tropics, major producers include the Philippines, Indonesia, India, Sri Lanka, the Pacific region, Malaysia, and Mexico. It is also of great local importance in subsistence economies.

See: Materials, p. 349

Cola nut *Cola acuminata* and *Cola nitida*
Sterculiaceae

Cola or kola is native to the rainforests of West Africa. The "nut" (not a true nut but the interior part of the fleshy seeds) is a masticatory, traditionally used as a stimulant in this region for hundreds of years. It contains caffeine and a smaller amount of theobromine, a heart stimulant. The origin of cola chewing is uncertain although it was thought to be sacred to Muslims in northern Nigeria at least 800 years ago.

Domestication took place in West Africa as its use increased. The cola trade flourished on ancient caravan routes from Nigeria to the Sahara. Of the two main species used, *C. nitida* is now cultivated over most of West Africa and *C. acuminata* is grown as far south as Angola. It was introduced to Jamaica and Brazil in the 17th century but has never reached the same popularity in those countries. Nigeria is the leading producer of cola and most of this crop is consumed within Nigeria itself and in neighboring West African countries. Ground cola nuts have been used in the past in the production of the popular soft drink Coca-Cola and concentrated extracts of the cola nut are still used for natural flavor in many soft drinks.

See: Caffeine, Alcohol, and Sweeteners, p. 179

Ginkgo nut *Ginkgo biloba*
Ginkgoaceae

The sole survivor of a group of primitive trees that were dominant in the northern hemisphere 125 million years ago, *Ginkgo biloba* is native to northern China. It has been valued in both China and Japan as a sacred tree, cultivated around Buddhist temples since about 1000 AD, and used for food, medicine, and ritual. After removing the putrid-smelling flesh, the cooked kernels of the female tree are enjoyed as a delicacy, often consumed with bird's-nest soup. The nut has also been used to make a cosmetic detergent, and the seed coat has been used as an insecticide. The tree was introduced to Europe as an ornamental in the 18th century.

See: Plants as Medicines, p. 223

Hazelnut *Corylus avellana*, Filbert *Corylus maxima*
Betulaceae

The common hazel of Europe and west Asia is *C. avellana*, a tree or bush growing up to 6 m (20 feet) high. Remains of the shells are commonly found in sites dating from Mesolithic to Medieval times in northern Europe, where hazelnuts have been an important part of the human diet for at least 9000 years. They were introduced to America in the 17th century and most cultivated nuts there are of European origin, although there has been some hybridization with the American wild hazel, *C. americana*. In the 19th century there was much hybridization with *C. maxima*, the filbert. This hazel is closely related to and interfertile with *C. avellana* and is native to the Balkans, north Turkey and Caucasia. Its larger nuts were and still are collected extensively from the wild as well as from cultivation. It also has higher yields.

Turkey is now the largest producer of hazelnuts. Other major producing countries include Spain and Italy. The nuts are generally eaten fresh, or used in cakes and confectionery.

Hickory nut *Carya ovata*, *Carya laciniosa*, and *Carya tomentosa*
Juglandaceae

Unlike the pecan (*C. illinoinensis*), other *Carya* species have no commercial importance, but there are several whose nuts are edible. The hickories are widespread throughout the temperate regions of the eastern and central United States and northern Mexico. The Algonquin Indians of Virginia traditionally pounded the kernels, adding water to make "hickory milk," which was used in savory dishes. The nuts were also popular with early settlers. *C. ovata*, *C. laciniosa*, and *C. tomentosa* (mockernut) are still eaten in parts of the United States but are not cultivated to any great extent. The hickories have remained among the least developed of the native North American nuts.

Lotus seeds *Nelumbo nucifera*
Nelumbonaceae

Nelumbo nucifera, the water lily, is native across Asia, from Iran through India, China, and Vietnam to Japan, Malaysia, and northeast Australia. It has been a sacred plant in both the Near and Far East for thousands of years. In Chinese cuisine it is traditionally the swollen rhizome that has been eaten, although the seeds also have a long history of use; both uses date back 3000 years. They are either eaten raw when immature, or roasted and boiled when mature. They can also be

Hickory nut (*Carya ovata*). USDA-NRCS PLANTS Database/Britton, N.L., and A. Brown. 1913. *Illustrated flora of the northern states and Canada*. Vol. 1, p. 583.

ground into flour, or dried for storage. The seeds remain popular in both sweet and savory oriental dishes.

See: Roots and Tubers, p. 74

Macadamia nut, Queensland nut *Macadamia ternifolia* and *M. tetraphylla*
Proteaceae

The only native Australian plant developed as a commercial food crop, Macadamias are prized for their unique delicate flavor and nutritious oil (the highest oil content of all nut crops). The smooth-shelled *M. ternifolia* and the hard-shelled *M. tetraphylla* are native to the rain forests of subtropical southeast Queensland and northern New South Wales and have been utilized there by Aborigines since ancient times. Cultivation of both species began in Australia in 1858, and by the 1880s it was an important crop in Hawaii. It has more recently been introduced to, and commercially grown in, southern California, Central America, Kenya, Malawi, and South Africa.

Oyster nuts *Telfairia pedata*
Cucurbitaceae

The oyster nut is the seed of the gourd of *T. pedata*. The fast-growing vine is native to the coastal forests of Tanzania and Mozambique, and is cultivated in east tropical Africa, where it is grown for its leaves and shoots as well as its nuts. The kernels are eaten raw, boiled, or roasted and are used in both savory and sweet dishes. The nuts are becoming increasingly important in international trade, both for their edible oil and for use in confectionery, and are sometimes used as a substitute for Brazil nuts. The West African species, *T. occidentalis* is also harvested for its seeds.

Paradise nuts, sapucaia nuts *Lecythis* spp.
Lecythidaceae

The paradise or sapucaia nuts of *Lecythis* (notably *L. pisonis* and *L. zabucajo*) are native to the Amazon rainforests of northeast Brazil and neighboring Guyana. The nuts are reputed to have a better flavor than the closely related Brazil nut and have been traditionally gathered from wild trees. Sapucaia are rarely found at the market because the fruit capsules open while still on the tree, providing a feast for wild birds and animals (especially monkeys) and leaving few on the forest floor for humans to gather. Additionally, they must be eaten soon after they fall from the tree; otherwise they turn rancid. As a result, large commercial paradise nut plantations have not yet been developed although trials exist in Brazil, the Guianas, the West Indies, and Malaysia.

Peanut, Groundnut *Arachis hypogaea*
Fabaceae

The peanut is an important crop grown for both its oil and nuts in many tropical and subtropical countries of the world. However, it is not a true tree nut. It is the seed of a tropical legume with flowers above ground but with pods that develop after fertilization below ground. Its wild ancestor is probably *A. monticola*, a species native to the eastern foothills of the Andes in northwestern Argentina and southern Bolivia. The crop did not enter archaeological record until it had been taken across the Andes to Peru. It was in cultivation there before 2000 bc and became common in the first millennium bc. However, the peanut did not reach Mexico until about the time of the Spanish conquest.

Spanish explorers took the peanut to Europe and the Philippines, and Portuguese explorers took it to East Africa. It is likely to have reached North America via the slave trade from Africa. Important producing countries now include India, China, the United States, Argentina, Brazil, Nigeria, Indonesia, Myanmar, Mexico, and Australia. In India and China it is grown primarily for its oil, whereas in the United States, it is grown for consumption of whole nut products.

Pecan nut *Carya illinoinensis*
Juglandaceae

The most important nut tree native to North America, this species has a huge range in the central southern region of the United States, extending south to eastern Mexico. The pecan was an important food for many Indian tribes and archaeological sites in this region are commonly associated with pecan groves. However, plantings outside its native habitat, in the "Pecan Belt" of the South (west of the Mississippi and east to Georgia and neighboring states) are all of post-Columbian times. By 1850, pecan trees were common in the gardens throughout this belt. Between 1905 and 1920, hundreds of cultivars were named but only twenty to thirty cultivars, planted in commercial orchards, have stood the test of time. Outside the Pecan Belt, pecan orchards are proving successful in California.

The trees can live for centuries and grow up to 50 m (164 feet) high. Most pecans now come from cultivated trees, although many old, wild trees continue to produce nuts, which are gathered and marketed. Although most of the world supply is still from North America, the pecan is now being grown in South Africa, Brazil, Australia, and Israel. It is mainly eaten raw and used in sweet dishes or confectionery.

Pili nut *Canarium ovatum*
Burseraceae

The pili nut is the most important of the nut-bearing species of *Canarium*. Native to the Bicol region of the Philippines, the nuts have been gathered from the wild since ancient times. Often grown as a home garden tree, the nuts remain an important food for Filipinos, who have a near-monopoly on the production of processed pili. Eaten raw or roasted, the kernels also yield good quality oil, suitable for culinary purposes but also used for lighting purposes. The pili nut has been introduced elsewhere in Southeast Asia and Central America.

Pine nuts *Pinus*
Pinaceae

The small edible seed kernels of the female cones of *Pinus* species are popular in many areas of the world. The seeds of the stone pine, *P. pinea*, have been an important ingredient in Mediterranean and Middle Eastern dishes for several thousand years. The tree grows throughout the northern Mediterranean from Portugal to Turkey and Lebanon. It is also cultivated to some extent in Spain, Portugal, and Italy.

The nuts are also imported to the United States, and are marketed in competition with the nuts gathered from native wild American species. These are known as pinyon nuts, and have been used by Native Americans for at least 6000 years, particularly in California and Nevada. The most important pinyon species of North America are *P. edulis* (two-needle pinyon), *P. monophylla* (single-leaf pinyon), and the Mexican *P. cembroides*. Chinese nuts, from *P. koraiensis,* are exported to Europe and the United States at a lower price. In the Himalayas Chilghoza pine (*Pinus gerardiana*) is an important food source. The immature cones are gathered and then heated to force the cones to open and release the seeds.

The *Araucaria* pines of the southern hemisphere also bear edible nuts. These include the Chile pine (or monkey puzzle tree), *A. araucana,* the Parana pine of Brazil, *A. angustifolia* and the bunya-bunya pine of Australia, *A. bidwilli.*

Pistachio *Pistacia vera*
Anacardiaceae

The pistachio is native to the Kopet Dagh mountains of central Asia and was introduced into Mediterranean Europe late, not before the classical period. It is one of the most drought-resistant fruit trees of west Asia, deciduous and up to 10 m (33 feet) in height. It is extensively grown for its oval, tasty nuts. Generally only producing a good crop in alternate years, it is now grown on a commercial scale for export mainly in Iran, Turkey, and California. It was introduced to California in the 1950s but only developed on a large scale in the 1970s. There is also cultivation in Syria, Afghanistan, Greece, India, and Italy.

Other *Pistacia* species grow wild in the eastern Mediterranean. Their fruits, usually referred to as terebinth nuts, are smaller and rounder than pistachio nuts. They have a soft shell and a distinct taste of turpentine, and are commonly roasted and eaten whole in the Near East. Terebinth nuts are also ground and used as a coffee substitute. The mastic tree *Pistacia lentiscus* produces resin used in chewing gum and picture varnish, and mastic gum was extensively burnt as incense in the New Kingdom period of ancient Egypt (1500 to 1100 BC).

See: Fragrant Plants, p. 249

Quandong nuts *Santalum acuminatum*
Santalaceae
Santalum acuminatum is a small parasitic tree or shrub native to the southwest and central desert regions of Australia. Aborigines have prized the aromatic kernels for many centuries. They are generally eaten roasted and have quite a pungent taste. They are fairly popular in Australia but virtually unknown elsewhere. Recently attempts have been made to develop quandong trees as a new crop in Australia. The hard shells have been made into jewelry.

Souari nuts *Caryocar nuciferum*
Caryocaraceae
Caryocar nuciferum is native to northern South America, where it has been a traditional food resource of forest-dwelling Indians. The soft, white kernels, which are larger than the Brazil nut, have a sweet almond-like flavor and can be eaten raw, roasted, or cooked in salt water. They also yield a good cooking oil. Like the Brazil and Paradise nuts, Souari nut production comes almost entirely from the large, wild trees. Cultivation has proven quite difficult but some trial plantings have been made in several tropical countries. *C. villosum* and *C. brasilliense* also has fruits that yield edible seeds and oil.

Walnut, Persian walnut *Juglans regia*
Juglandaceae
The wild Persian or English walnut is believed to be native to the area stretching from the Balkans, through central Asia to western China. The most plausible area of domestication is northeast Turkey, the Caucasus, and north Iran. Textual sources indicate that it was already widely grown in the time of the Roman Empire. Perhaps independent of this, cultivation in China began up to 2000 years ago. Significant planting in Northern Europe began in the 16th and 17th centuries where it was often also grown for its timber. Today it is cultivated commercially in many regions, California being the world's leading producer.

Water chestnuts *Trapa* spp.
Trapaceae
The kernel of the hard-shelled fruit of aquatic *Trapa* species has been collected as human food for thousands of years. *T. natans*, Jesuit's nut, is native to temperate and subtropical Eurasia and was probably domesticated in the Far East 5000 to 6000 years ago. It was traditionally cultivated throughout Europe, East Asia, and India and introduced to America only 100 years ago. Today it is only a relic crop in Mediterranean countries and China.

T. bicornis, the ling nut (not to be confused with *Eleocharis dulcis*, Chinese water chestnut, described earlier), is cultivated widely in China, Japan, and Korea, where it is eaten boiled, or preserved in honey and sugar. *T. natans* var. *bispinosa*, singhara nut, is an important food in Kashmir where it has been extensively cultivated. This species, native to tropical Asia, now grows over a very wide area from Africa to Japan. The nuts of all three species need to be cooked thoroughly before eating, to remove possible toxicity.

Pulses

Pulses (or legumes) are the dry, edible seeds of food plants in the Fabaceae (Leguminosae) family. Some Fabaceae fruits are eaten as whole pods, while still green; these are covered in the chapter on Herbs and Vegetables. Although the field yields of pulses are usually low compared to those of cereals, they are appreciated for their much higher protein content, 20 to 40 percent, compared to 10 to 15 percent in the cereals. Many pulse seeds contain toxic or anti-nutritional constituents, hence the requirement to soak and adequately cook them.

Adzuki bean *Vigna angularis*

Perhaps a more recent domesticate than many other pulses, the Adzuki bean was domesticated in Japan through selection from the wild form *V. angularis* var. *nipponensis*, which occurs in Japan, Korea, Taiwan, China, and Nepal. It is second to the soya bean in popularity in Japan and has been used in celebratory dishes both there and in China for many years. The dried pulse is either cooked whole or made into a meal used in soups, cakes, or confections. It has now been introduced in South America, the southern United States, Hawaii, New Zealand, Thailand, India, Kenya, Zaire, and Angola.

Broad bean, Faba or fava bean, Field bean *Vicia faba*

The wild ancestor of broad, or faba, bean has not been identified, but belongs to the *V. narbonensis* group of species, which grow in the Mediterranean and the Fertile Crescent. Broad bean may have been cultivated since the Neolithic, but is uncommon until the third millennium BC. By the Iron Age, it was well established in Europe. It spread to China much later, possibly not until the 13th century AD, brought there at the beginning of the spice trade. By the 16th century considerable cultivation had developed there, and today China produces 65 percent of the world's crop.

Predominantly a temperate crop, *V. faba* is an erect plant with pods containing large seeds that have a high nutritional content. The seeds vary considerably in size. In Western Europe, Central and South America, Russia, China, and Japan the large-seeded *major* type is more commonly grown; this evolved recently, perhaps about 500 AD. The more primitive small-seeded *minor* type, found in archaeological remains, is now found mainly in Afghanistan, Ethiopia, and India.

There are many myths and superstitions relating a "darker side" of the broad bean and their consumption has been a taboo in Indo-European cultures. The Greek philosopher Pythagoras (580–500 BC) considered beans impure and injurious to health. This taboo has sometimes been explained in nutritional terms. Eating broad beans causes a type of hemolytic anemia called favism in a small percentage of the population, mainly in the Mediterranean and Europe, who have a genetic deficiency of an enzyme that breaks down toxic components. However, there is no evidence that favism was recognized as linked to beans in antiquity, and similar taboos occur for other plants, such as black mungo bean, that are not associated with disease.

Cluster bean, Guar *Cyamopsis tetragonolobus*

The cluster bean is probably native to India where it has been cultivated for at least 1000 years. It was traditionally grown for its edible young pods and seeds and, as cultivation spread to Southeast Asia, the Pacific Islands, and East Africa, also for fodder and green manure. Nowadays the use of guar gum, obtained from the seed flower, has much greater economic importance. The gum has been increasingly used in the food, paper, textile, and cosmetics industries. Large-scale production occurs in India and the southwest United States, where it has been used for industrial purposes since the 1940s.

See: Materials, p. 337

Cowpea, Black-eyed bean, Yawa *Vigna unguiculata*

Cultivated cowpeas and their close relatives are classified as a single botanical species. The earliest archaeological records are in India and date to 4000 BC, but the wild progenitor (var. *dekindtiana*) grows in west Africa, and cowpea must have been domesticated there. The cowpea was introduced to the tropical New World from Africa during the slave trading of the 17th century, and is increasingly important in Brazil and Venezuela.

Cultivated cowpeas are annual herbs with growth habits ranging from climbing, prostrate, and intermediate to semi-erect and erect. They can be divided into three groups of cultivars. The *unguiculata* group, cowpea, is particularly important in Africa and is mainly grown as a pulse crop. The *biflora* group, catjang bean, is mainly cultivated in India, both for seed and as a green vegetable. The yard-long bean, *sesquipedalis* group, has long pods—to 90 cm (35 inches) long—which are eaten young, as a vegetable, in south and southeast Asia.

Garbanzo bean, Chickpea, Bengal gram *Cicer arietinum*

The garbanzo bean or chickpea is one of the world's three most important pulses. Its domestication can be dated to the start of food production in the Fertile Crescent around 7000 ^{14}C years BC. The wild progenitor, *C. reticulatum*, is restricted to southeast Turkey. The chickpea spread from there to the Mediterranean area by 6000 ^{14}C years BC, and to India by 3000 BC. It was carried to the New World, particularly South America, by the Spaniards and the Portuguese in the 16th century. More recently it has been introduced to Australia. The largest production today is in India, where it is grown as a winter crop. Two main types of cultivated chickpea have emerged: the large-seeded variety ("Kabuli") with relatively smooth, rounded, light-colored seed coats and pale cream flowers that predominates in the Mediterranean countries and Near East, and the variety with small wrinkled seeds ("Desi"), dark-colored seed coats, and usually purple flowers that prevails in India, Afghanistan, and Ethiopia. The Desi form closely resembles archaeological remains of chickpea, and must be the more primitive form.

Today, the chickpea is consumed mostly in the areas of production and only a small proportion is exported. It is the most important pulse in India, which is the world's largest producer of chickpeas; it is known as *channa* and makes a popular dhal. There is also considerable production in the Middle East, where it forms the basis of some of the most popular national dishes including hummus and falafel. The main exporting countries are Turkey, Mexico, and more recently, Australia.

Grass-pea *Lathyrus sativus*

This is a drought-resistant, high-protein-content pulse crop that originated in the Aegean region 6000 to 5000 ^{14}C years BC, after the first wave of pulse domesticates. Archaeological evidence shows that it was cultivated in southern Europe and the Near East, though in both areas it is now mainly used as animal feed. It is now mainly grown for food use in India, Pakistan, Bangladesh, and Ethiopia. However, grass pea contains an amino acid, ODAP, that can cause lathyrism. This is a neurodegenerative disease that results in consumers losing use of their legs. Under normal circumstances, boiling of grass pea in water breaks down the toxin. At times of drought, however, grass pea may be the only crop that survives, under circumstances in which water and fuelwood are in short supply. Inadequate cooking then results in outbreaks of lathyrism. Plant breeders have recently developed low-ODAP forms that may give a new lease on life to this crop.

Haricot bean, Kidney bean, Flageolet, Dwarf bean *Phaseolus vulgaris*

The haricot bean is the most important legume grown in Europe and North America. It is also grown in many other temperate and subtropical regions of the world. Wild populations (var. *aborigineus*) cover a wide latitudinal range from the Tropic of Cancer in Mexico to the Tropic of Capricorn in Argentina. Genetic evidence shows that the haricot bean was domesticated independently in North and South America. The oldest small-seeded forms have been found in Mexico at sites dating to 2300 ^{14}C years ago, and large-seeded forms found in Peru date to 4300 ^{14}C years ago. In both cases, domestication probably occurred much earlier. In Peru, cultivation did not spread from the highlands to the coast until about 500 BC, and similarly the Central American crop only spread into North America after 300 BC and east of the Mississippi after 1000 AD. In these areas, acceptance of the haricot bean as a staple food crop was even slower, but by Columbian times the cultivation of the beans rivaled that of maize over most of the New World.

In the 16th century the Spanish took mainly the southern, Andean, variety to Europe and from there it spread rapidly to the Middle East and Western Asia. From Europe, they were also distributed to the eastern United States and to Africa, although some direct introduction from South America could also have occurred. It is still the main pulse crop throughout tropical America and many parts of tropical Africa. Brazil is the largest producer. It is a minor crop in India and most of tropical Asia.

As well as being cultivated for the seeds, several types are grown for their young, edible pods, which have had the stringiness bred out of them. In temperate regions the plant is grown for its young pods. The haricot of French bean has a milder flavor and less showy flowers than the runner bean. French bean pods tend to snap easily, hence the common name snap bean.

Haricot bean plants vary from pole beans with a twining stem that needs support (mainly garden crops), to an erect low-growing bush (dwarf beans, mainly commercial crops). There are thousands of landraces, and both old and modern cultivars continue to be grown throughout the world.

Horse gram *Macrotyloma uniflorum*
Horse gram is a poorly known crop, common at Indian sites from 2500 BC onwards and perhaps domesticated in south India. Known in India as kulthi, this is regarded as a low-grade pulse, used in south Asia as a food, but grown more widely throughout the tropics as animal feed.

Jack bean *Canavalia ensiformis*, Sword bean *Canavalia gladiata*
Adaptable to a wide range of climates, jack bean is now cultivated on a small scale throughout the tropics, being especially popular in India, Africa, and Indonesia. It is used for fodder, as a green manure, or as a cover crop, and the young pods and immature and mature seeds can be eaten. The drought-resistant jack bean could be an increasingly important source of protein in areas where other pulses do not thrive. It was domesticated in Mexico or Central America. *C. plagiosperma* is a similar crop plant, originating in coastal Ecuador and Peru, and now close to extinction. Both species appear in the archaeological record from about 4000 ^{14}C years ago, in Mexico and Peru, respectively. The common name "jack bean" is sometimes used for *C. gladiata* (the sword bean), a species used for similar purposes in the Old World, with its origin in south or Southeast Asia.

Lablab bean, Hyacinth-bean *Lablab purpureus*
Widely cultivated in India since about 2500 BC, this legume is now known to be of east African origin. Due to its high protein content, high yield, ability to withstand droughts, and easy harvesting, it is an important crop now widely grown for both human food and animal fodder in tropical and subtropical regions. In Asia it is traditionally grown as a vegetable or pulse crop. Its use as a forage or pasture plant (in Africa and Brazil) or as a cover or green manure crop (e.g., in Sudan) is more recent.

Lentil *Lens culinaris*
Lentil ranks among the oldest and the most appreciated grain legumes of the Old World. Its cultivation is associated with the start of Neolithic agriculture some 9500 ^{14}C years ago in the Near East, along with wheat and barley. The wild ancestor (subsp. *orientalis*) grows throughout the Fertile Crescent.

Cultivation of the lentil spread into Mediterranean Europe and North Africa with the extension of agriculture in the sixth millennium ^{14}C BC, becoming an important and well-established crop. It was eventually introduced into most of the subtropical and warm temperate regions of the world. Today the Indian subcontinent is the largest grower, producing roughly 38 percent of the world's lentils, but lentils are grown in most subtropical and warm temperate areas, especially in Ethiopia, Syria, Turkey, and Spain. In India the crop is one of its most important dietary pulses; lentil seeds, entire or split, are largely used in soups and dhal. Lentil seeds are also a source of commercial starch for the textile and printing industries. After threshing, the residues can be fed to livestock; in the Near East this animal feed can sometimes fetch a higher price than the seed.

Lima bean, Butter bean, Madagascar bean *Phaseolus lunatus*

Lima or butter bean seeds vary in size from 1 to 3 cm (0.4 to 1.2 inches) in length. Archaeological evidence points to independent domestication in Central and South America. The larger-seeded bean is found at sites in Peru to dating to ^{14}C 5000 years old. The smaller-sized variety from Mexico dates back at least 1200 years. Cultivation spread rapidly over the Americas in post-Columbian times, via the Philippines to Asia and by the slave trade from Brazil to West and Central Africa. By far the largest producer today is the United States. Lima bean is also a staple food in many parts of tropical Africa and Asia, but is grown essentially as a subsistence crop. The dry seeds are used as a pulse and, in Asia, as sprouts.

See: Origins and Spread of Agriculture, p. 22 and 23

Mung bean, Green gram *Vigna radiata*

Like urd (black gram), the mung bean has been in use in the Indian subcontinent for at least 4000 years. The likely wild ancestor grows in the north and west of the Indian subcontinent. Despite its widespread distribution and local importance, it has never become a major commercial crop outside Asia. It is an important pulse crop in India, where most of the production is traded and consumed locally (after cooking the dried pulse). In China it is used to make bean sprouts and is a source of starch for mung bean vermicelli. Thailand leads the world trade supply.

Pea *Pisum sativum*

Like lentil, the pea is one of the oldest grain legumes. Its wild progenitor, *P. sativum* var. *pumilio*, is a twining winter annual, native to oak parklands from Anatolia through the Fertile Crescent. The pea was domesticated in the Near East around 8000 BC and spread with Neolithic agriculture to Europe, North Africa, Middle Asia, and India. It reached the Americas by the 17th century. Today the pea is the world's second most important pulse with Russia and China producing almost 80 percent of the world's production of dry peas and the United States and United Kingdom producing the largest number of green peas (which is only 10 percent of the production of dry peas).

P. sativum is an important source of protein to farming communities in the Near East, Middle East, Mediterranean countries, temperate Europe, Ethiopia, and northwest Europe. Here it is largely eaten as the mature dry seed. It is also used in animal feed. In many areas, particularly in developed countries, it is the immature seed that is harvested and eaten fresh, canned, or frozen.

Cultivated peas show a wide range of morphological variation. Varieties of the cultivated pea were first described in the 16th century, initially with a distinction between the field peas—with their colored flowers, small pods, and long vines—and the garden peas with white flowers and large seeds.

See: Origins and Spread of Agriculture, p. 17

Pigeon pea, Red gram, Congo pea *Cajanus cajan*

The world's fifth most important pulse crop, the pigeon pea has a long history of cultivation. Domestication took place in India, the center of diversity of the species, and the first archaeobotanical evidence dates to about 1500 BC. Although an African origin has been suggested, the wild ancestor *C. cajanifolius* only occurs in India. From India the crop was taken to Africa, thence to the

Caribbean with the slave trade. It is now grown throughout the tropical regions of the world, yet over 90 percent of the world crop is produced in India. It is the most commonly used legume in dahl after chickpea, and is also used as a fodder crop and for fuel.

Rice bean *Vigna umbellata*

The center of diversity of the rice bean and presumed center of domestication from the wild progenitor, *V. umbellata* var. *gracilis,* is Indo-China. It was extensively grown as a lowland crop following the harvest of traditional long-season rice varieties throughout Indo-China and extending to southern China, and into northeast India and Bangladesh. Changes to patterns of rice growing, with multiple cropping, have led to a decline in demand, but more recently there has been a renewed interest because of the rice bean's exceptionally high nutritional content. Its small seeds, when dried, are only slightly larger than rice and are eaten with or instead of rice, usually boiled. The whole plant is used as fodder, as a cover crop and as green manure.

Runner bean *Phaseolus coccineus*

Although the earliest archaeological record dates to 1100 ^{14}C years ago, it is highly likely that the runner bean has existed in cultivation in the central highlands of Mexico and Guatemala for at least 4000 years. American Indians still cultivate it as a perennial for its tender pods, green and dry seeds, and tuberous starch roots. The "scarlet runner" was introduced to Britain in the 16th century. At first it was grown for its flowers, which were often used in nosegays and garlands. By the 17th century various types with a diversity of characteristics were in cultivation in Britain. Today it is cultivated as an annual crop for its tender, green pods in the highland areas of Africa, South America, and many temperate areas, although production is largely for local use. Commercial production takes place in England, Argentina, and South Africa. It is usual to eat the pods, boiled, when still young and tender, although the dried seeds and tuberous roots are also sometimes eaten.

Soya bean, Soybean *Glycine max*

The soya bean is one of the most important staple foods of the world, particularly in the Far East and Southeast Asia, where it is the main source of protein in many areas. Its importance as a food plant in the Far East dates back thousands of years. It was cultivated in north China by 1000 BC, probably reaching central and south China by the 1st century AD, and spreading throughout neighboring countries to the southeast over the next fourteen or fifteen centuries. The soybean

Mature soybeans. Photo by Scott Bauer. ARS/USDA

was only introduced into the West on a productive scale at the beginning of the 20th century. The United States started production on a large scale during the 1940s and is now the largest producer, growing 75 percent of the world's soya bean crop. Despite its high protein content, attempts to introduce the crop to many poor regions of the world, such as India, Africa, and Central America, have historically met with little success, although southern South America is a notable exception.

G. max is a larger, more erect, plant than its vine-like wild progenitor, *G. soya*. Its larger seeds have a higher oil content and better flavor, and cook quicker and are more digestible. The importance of soya beans in Asian cuisine depends on diverse and sophisticated methods of preparation. Products common in the East including tofu (bean curd), soy sauce, kekap, Japanese miso (bean paste), and Indonesian tempeh (a fermented cake-like product) have become popular in the West. Soya milk is a popular alternative to milk from dairy animals. The U.S. soybean crop is grown mainly for its oil, which has surpassed other vegetable oils in popularity. The residual cake, high in protein, is widely used to feed livestock.

See: Age of Industrialization and Agro-industry, p. 364

Tamarind *Tamarindus indica*

Distributed throughout the tropical and subtropical regions of the world, tamarind is valued for its pods. The pulp surrounding the small beans has a sweet-and-sour flavor and provides an acidifying agent used widely in Indian and Southeast Asian cooking. In Jamaica and Latin American countries, it is made into syrup with the addition of sugar and diluted to make a soft drink. In Thailand it is eaten as a sweetmeat; there and in Indonesia, the roasted seeds are also used.

Tepary bean *Phaseolus acutifolius*

The tepary bean has been cultivated as a traditional pulse crop in central Mexico for 2300 ^{14}C years and in Arizona for 1000 ^{14}C years. Since the beginning of the 20th century the plant has been introduced as a drought-resistant crop and cultivated on a small scale in further parts of the United States, Central America, the West Indies, Chile, Australia, South Asia, and East and West Africa. It is a short-lived annual species and is a close relation to, and has largely been replaced by, the haricot bean. Its commercial value is minimal and it is largely grown for subsistence.

Urd, Black gram *Vigna mungo*

Archaeological evidence suggests that, along with its morphologically close relative the mung bean, urd has been cultivated in India for 4000 years. It was probably taken into domestication in northern India. World trade supply is dominated by Thailand, yet the largest producer is India, where it is grown predominantly for local use. Ground urd makes the best poppadoms and is also used in dumplings (idli), and pancakes (dosa) of South Indian cuisine. The seed is reputedly the origin of the very small Indian weight known as *masha*. The color black is an ominous color in Hindu symbolism, and black mungo beans play an important role in Hindu death rituals.

Winged bean *Psophocarpus tetragonolobus*

The origin of the cultivated winged bean is unclear, although it is possible that *P. grandiflorus*, of tropical Africa, is the progenitor. The main centers of diversity however are in tropical Asia, principally Papua New Guinea and Indonesia, where there is a long history of cultivation in subsistence systems and subsequently the highest varietal diversity. The large, indeterminate, twining climber is unusual as a crop plant in that all parts of the plant are edible and of high nutritional value. It is the immature pod that is most often eaten; utilization of the seed is limited by the hard, bitter outer covering (testa). Outside tropical Asia, the plant remains little known and underused, although more recently it has been introduced to the tropics and subtropics of West Africa, South America, and the

West Indies. It has been suggested that winged bean seed oil can be used as a substitute for soybean oil although the potential for this has yet to be realized.

Oilseeds

Oilseeds are perhaps some of the least familiar food plants: many consumers will only see the oil itself, which is used for cooking, as a major ingredient in many processed foodstuffs, and in many industrial applications (see Materials chapter). The major oilseeds of commerce not only have a high oil content, but also have an oil that is reasonably resistant to turning rancid through oxidation. The health impact of saturated fats, which comprise 90 percent of coconut oil and 50 percent of palm oil, and of the industrial process of hydrogenation, which hardens oils for food use in products such as margarine, remains uncertain.

Castor oil, Palma-Christi *Ricinus communis*
Euphorbiaceae

Castor is native to east Africa and probably originated in Ethiopia. The seeds have been found in Egyptian tombs dating from 4000 BC and its use has been recorded in North Africa from that time onwards, spreading through the Mediterranean, the Middle East, and India. An archaeological find in India dates to 1300 BC. The use of castor oil for medicinal purposes in China is recorded in the 6th century AD, although it is likely to have been introduced to China earlier. At present, the oilseed is largely used for industrial purposes, such as soap, varnish, crayons, and textile dyes. India dominates the world export market whereas in China, the other most important producer, production is mostly used locally. It is now cultivated as a minor crop in most drier areas of the tropics and subtropics. Because of its purgative effect (castor oil is a well-known laxative) the main uses today are industrial. The oil is rich in ricinoleic acid, a fatty acid used in paints, polymers, and lubricants.

Castor beans contain two toxins, ricinine and ricin, but these are not present in cold-pressed oil. Ricin is one of the most potent plant toxins, and castor beans are often responsible for accidental poisoning cases. The Bulgarian exile Georgi Markov was assassinated in London in 1978, by stabbing with a ricin-laden umbrella point.

Jojoba *Simmondsia chinensis*
Simmondsiaceae

A desert shrub, *S. chinensis* is native to an extensive area in the Sonoran Desert, including parts of Arizona, California, and northwest Mexico. *Simmondsia* is unique among plants in that its seeds contain an oil that is a liquid wax. The nuts have been traditionally eaten raw or roasted by native American Indians, who also used the oil as a hair conditioner and as a medicine. Following the ban in use of sperm whale oil in 1973, the cosmetic industry turned to jojoba oil for use in shampoos, moisturizers, and sunscreens. Plantations of the jojoba tree are still young and world production is low. Jojoba cannot yet be termed a domesticated plant, though with plant breeding in progress, domestication is likely to occur. The oil has further potential as an industrial lubricant.

See: Materials, p. 341

Olive *Olea europaea*
Oleaceae

Olives were collected from the wild long before their cultivation—as early as 17000 ^{14}C BC. Wild olives are distributed throughout coastal areas of the Mediterranean. The earliest evidence of cultivation comes from several Chalcolithic (4700 to 3200 ^{14}C BC) sites in Israel and Jordan. It was probably introduced to the west Mediterranean basin by Greek and Phoenician colonists. By Roman times, oleiculture seems to have reached its present-day geographical spread in the Old World.

Picking olives in Palestine, ca.1890. Library of Congress, Prints & Photographs Division, LC-USZ62-126412.

There are hundreds of distinct cultivated varieties of olives. These are traditionally separated into oil varieties and table olives. Unlike many other oils, the oil-rich part of the fruit is the fleshy part of the drupe rather than the hard, woody seed. Olives contain a bitter phenolic glycoside called oleuropein. This is not carried into olive oil during pressing, but must be removed from table olives before consumption. Removal is carried out by curing in water, brine, or oil, or by treatment with an alkali solution.

Oleiculture is still largely dependent on traditional cultivars. The bulk of the olive crop is grown in the Mediterranean basin. Successful introductions elsewhere have been essentially limited to regions with very similar climatic conditions. During the 16th century, the Spanish attempted to introduce olives to many areas in the New World. Most grew well but bore little or no fruit, with the exception of those in Peru. The most successful introduction was to Mexico in the 18th century. Plantings extended up to California and by the mid to late 19th century, olive planting boomed there.

Over the last 50 years the area of olive plantations in the Mediterranean has decreased. The crop has had to cope with increasing competition from modern oil-producing crops. Olives have remained popular amongst Mediterranean peoples. Recently, there has been a renewed interest in other parts of Europe, and in North America, due to the recognition of its nutritional value and the popularity of Mediterranean food. Olive oil is rich in monounsaturated fats.

Palm oil, Oil palm *Elaeis guineensis*
Arecaceae

The West African palm, *Elaeis guineensis,* is indigenous to the tropical rain-forest belt of West and Central Africa between Senegal and north Angola. Crude palm oil use has a long history in West Africa, with abundant archaeological finds in the last 5000 years, although the palms were not usually planted, but rather encouraged and managed. There was some movement of the palm from West Africa along slave routes and through colonial trading, but exploitation as a commercial export crop did not start until the late 19th century as the world market for vegetable oils grew. Palm oil and palm and kernel oils became widely used in cooking oils, margarine, and soap, and later in industry. Commercial plantations started to be established in Africa after 1920.

The oil palm has been expanding as a plantation crop over the past 30 years. It is now the second most important vegetable oil after soybean. The main producer is Malaysia, supplying 70 percent of world exports. Other important exporters include Indonesia and, to a lesser extent, Papua New Guinea, the Ivory Coast, Thailand, the Solomon Islands and the Philippines. Oil palm cultivation is also important in West Africa and South and Central America, although it is largely used in these regions for local consumption.

Pounding oil palm fruit and separating oil, Sierra Leone. Photo Capt. P. W. Clemens, 1928. Courtesy Royal Geographical Society, London.

The American oil palm, *E. oleifera*, had been used to some extent but the species has never become a plantation crop.

See: Materials, p. 341

Oilseed rape, Canola *Brassica napus* subsp. *oleifera*
Brassicaceae
First documented in herbals of the 17th century, like its tuber-bearing relative the swede, oilseed rape was used mainly for lighting and lubricants. It has become a major commercial oilseed crop in many European countries, owing to heavy subsidies from the European Union. China and India are also significant producing areas. A significant boost to production, particularly in Canada, was given by the breeding of low erucic acid cultivars in the late 1960s. These are suitable for food use, for example in margarine and cooking oils.

Oilseed turnip, Turnip rape *Brassica rapa* subsp. *oleifera*
Brassicaceae
The wild progenitor, *B. rapa* subsp. *campestris*, is native to temperate Eurasia. Primitive cultivars (usually considered as separate subspecies) were grown as oilseed crops in India since 2000 BC. It spread to China as a weed and an oilseed crop but its ancient history is not well known. Although the root crop form of this species, turnip, has been cultivated since classical times, oilseed turnip did not reach northern Europe as an oilseed crop until the Middle Ages. Its importance is declining, as the crop is replaced by oilseed rape.

Safflower *Carthamus tinctorius*
Asteraceae
The cultivated safflower originated in the Fertile Crescent (the arc around the Mesopotamian lowland from the southern Levant to the southern foothills of the Zagros Mountains). Possible archaeological finds of the domesticated form date to about 2500 BC. It was used in ancient Egypt as a yellow dye (from the flowers) and oilseed from about 1600 BC. Subsequently it became very widely used for these purposes throughout the subtropics. It was developed as an oil crop in Egypt and India at least 2000 years ago but nowhere else until the late 19th century. Production began on a commercial scale in the 1940s in the United States but has since declined worldwide due to

competition from other vegetable oils. Today, India is the largest producer. Other important sources include Mexico, the United States, Argentina, Australia, and Ethiopia.

See: Natural Fibers and Dyes, p. 311

Sesame seeds *Sesamum indicum*
Pedaliaceae

It is likely that the cultivated plant is of Indian subcontinent origin, as this is where its wild progenitor (*S. malabaricum*) grows. The oldest archaeological finds are from Abu Salabikh in Iraq, dating to 2500 BC, and roughly contemporary finds from Harappa in Pakistan. Sesame occurs fairly frequently at later sites in the Indian subcontinent, but is rather sporadic in the Near East, where flax was the major oilseed. Suggestions of an African domestication now appear unlikely. Sesame seeds were found in the tomb of Tutankhamen (ca. 1350 BC), but it is unclear whether it was an important crop.

An annual crop up to 2 m (6.5 feet) in height, sesame is today grown in China, India, Africa, and throughout the Americas for its highly nutritious oil. This is used primarily in the manufacture of margarine and cooking fats, but also in soaps and paints, and as a lubricant and illuminant. In India it is also used as the basis of scented oils used in perfumery. The residue, left after oil extraction, is used for animal feed. The seeds are also widely used in baking and confectionery.

Sunflower, Common sunflower *Helianthus annuus*
Asteraceae

The probable wild progenitor, *H. annuus* subsp. *jaegeri*, is found in the southern United States. It is possible that the sunflower became a camp-following weed and was not domesticated until it was introduced to the central part of that country. The earliest certain archaeological remains of the domesticated sunflower were, until recently, seeds from Arizona dating to 2800 ^{14}C years ago. Domesticated sunflower seeds have recently been found in Mexico dating to 4100 years ago, suggesting an independent domestication may have occurred there. By the 16th century, the sunflower was a minor garden crop in much of North America. The seeds (technically one-seeded fruits called achenes) were either ground or toasted although some Indian groups extracted the oil from them.

The plant was introduced into Europe as a curiosity and ornamental in the 16th century. It was developed as an oilseed in Russia in the late 18th century. By the late 19th century, Russia had increased the oil content to 50 percent and *H. annuus* became by far the most important oilseed crop in Europe. By 1950, the crop was planted extensively in China, and in temperate southern Africa and South America. In Argentina, the sunflower became an important crop when the country was cut off from its supply of olive oil during the Spanish Civil War. Argentina went on to become one of the world's largest producers along with Russia and the United States. The oil is high in polyunsaturates, and is used in cooking oil, margarine, and paints. Roasted seeds are a popular snack and are widely fed to birds and small animals.

Other Edible Nuts and Seeds

Pumpkin seeds *Cucurbita pepo*
Cucurbitaceae

Cucurbita pepo is one of the earliest New World Crops. Genetic and archaeological evidence points to two domestication events, one in Oaxaca, Mexico, with finds dating to 9000 ^{14}C years ago, and one in the eastern United States, with finds dating to 4400 ^{14}C years ago. Selection was initially for larger seeds and larger fruits rather than edible flesh. Over time, though, the species became an integral part of the pre-Columbian maize-bean-squash diet complex. By the 16th century, the pumpkin had reached Europe and soon spread to the rest of the world. The nutritious seeds have become important as a health and snack food in the United States and Europe.

Watermelon seeds *Citrullus lanatus*
Cucurbitaceae

Although the history of the watermelon fruit is well documented, it is not clear how the seeds developed to their present-day popularity. In Africa, China, and, to a lesser extent, India, the oily seeds are eaten either raw or roasted, following removal of the seed coat. They are popular in soups and stews or may be ground into meal to make bread. In some districts of western tropical Africa, watermelons with bitter flesh are grown solely for their seeds. In many parts of Africa, edible oil is extracted from the seeds.

See: Fruits, p. 88

References

Nuts

Duke, J.A. 1989. *CRC Handbook of Nuts.* Boca Raton, FL: CRC Press.
Howes, F.N. 1948. *Nuts: Their Production and Everyday Uses.* London: Faber and Faber.
Mason, S. 1995. Acornutopia? Determining the role of acorns in past subsistence. In *Food in Antiquity*, edited by J. Wilkins, D. Harvey, and M. Dobson. Exeter: University of Exeter Press, 12–24.
Rosengarten, F., Jr. 1984. *The Book of Edible Nuts.* New York: Walker.
Verheij, E.W.M., and Coronel, R.E. 1991. *Plant Resources of South-East Asia (PROSEA) 2. Edible Fruits and Nuts.* Wageningen, the Netherlands: Pudoc.
Wickens, G.E. 1995. *Non-wood Forest Products 5. Edible Nuts.* Rome: Food and Agriculture Organisation of the United Nations.
Woodroof, G. 1979. *Tree Nuts: Production, Processing, Products* (2nd ed.). Westport, CT: AVI.

Pulses

Duke, J.A. 1981. *Handbook of Legumes of World Economic Importance.* New York: Plenum Press.
van der Maesen, L.J.G., and Somaatmadja, S. (Eds.). 1989. *Plant Resources of South-East Asia (PROSEA) 1. Pulses.* Wageningen, the Netherlands: Pudoc.
Pickersgill, B., and Lock, J.M. (Eds.). 1996. *Advances in Legume Systematics. Part 8. Legumes of Economic Importance.* Kew: Royal Botanic Gardens.
Simoons, F.J. 1998. *Plants of Life, Plants of Death.* Madison: University of Wisconsin Press.
Westphal, E. 1974. *Pulses in Ethiopia: Their Taxonomy and Agricultural Significance.* Agricultural research report 815. Wageningen, the Netherlands: Pudoc.

Oilseeds

Salunkhe, D.K., et al. 1992. *World Oilseeds. Chemistry, Technology, and Utilization.* New York: Van Nostrand Reinhold.
Seegeler, C.J.P. 1983. *Oil Plants in Ethiopia: Their Taxonomy and Agricultural Significance.* Agricultural research report 921. Wageningen, the Netherlands: Pudoc.
Serpico, M., and White, R. 2000. Oil, fat and wax. In *Ancient Egyptian Materials and Technology*, edited by P.T. Nicholson and I. Shaw. Cambridge: Cambridge University Press, 390–429.
van der Vossen, H.A.M., and Umali, B.E. (Eds.). 2001. *Plant Resources of South-East Asia (PROSEA) 14. Vegetable Oils and Fats.* Leiden: Backhuys.
Vaughan, J.G. 1970. *The Structure and Utilization of Oil Seeds.* London: Chapman & Hall.
Weiss, E.A. 2000. *Oilseed Crops* (2nd ed.). Oxford: Blackwell Science.

9
Spices

BARBARA PICKERSGILL

There is no clear boundary between spices, herbs, and condiments. Is black pepper a spice or a condiment? Is coriander a spice, because the fruits are used as such, or a herb, because the leaves are used like parsley? In *The Book of Spices*, Rosengarten suggests that spices are primarily tropical and herbs primarily temperate, but caraway is a temperate spice, and tropical species of *Tagetes* and *Eryngium* are used as herbs in Latin America. The decision about which species to include in this chapter and which to exclude has thus inevitably been somewhat arbitrary.

Spices, like herbs, add interest to a diet containing mostly bland carbohydrates and mask unpleasant flavors in imperfectly preserved meat. They are characteristically aromatic, usually but not always because they contain volatile oils or resins. Virtually any part of the plant may be used: rhizomes (ginger, turmeric), bark (cinnamon), leaves (curry plant), flower buds (cloves, capers), stigmas (saffron), arils (mace), but most often fruits and/or seeds are used.

Spice-producing species are not distributed randomly, taxonomically or geographically. Some families provide many different spices, notably Apiaceae (aniseed, caraway, coriander, cumin, dill, and fennel) and Zingiberaceae (cardamom, ginger, and turmeric). In others, only one genus is involved, for example Solanaceae (*Capsicum*) and Orchidaceae (*Vanilla*). None of the major spices is native to Africa; the Americas provide only three (chili pepper, allspice, and vanilla); but the Far East is disproportionately rich and the Indian subcontinent contributes significantly.

Spices are usually dried before use. They can then be stored for long periods and are light and easy to transport. They have figured in long-distance trade since very early times and the great value placed on many of them means that spices have played a significant role in world history.

History of the Spice Trade

The beginnings of man's use of and trade in spices cannot be dated with certainty. Egyptians of the 3rd millennium BC used many spices (most native to the Mediterranean and Middle East) in their embalming techniques. Cardamom and turmeric, both probably of Indian origin, were known and grown in Assyria and Babylon by the 8th to 7th centuries BC. By the time of the Greek and Roman Empires, there were well-established overland trade routes across Asia, most notably the

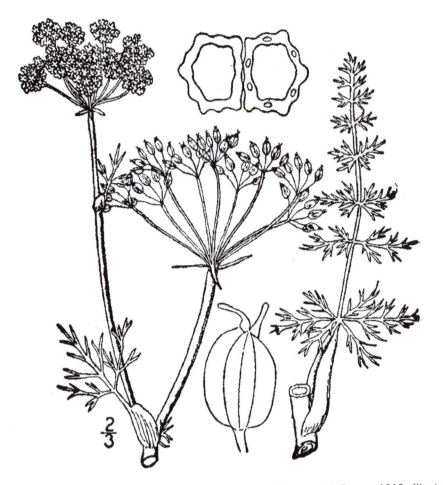

Caraway (*Carum carvi*). USDA-NRCS PLANTS Database/Britton, N.L., and A. Brown. 1913. *Illustrated flora of the northern states and Canada*. Vol. 2, p. 659.

Silk Road, which ran for over 4000 miles from China to the Persian Gulf. Ginger, cassia and other spices—as well as silk—were imported to the Mediterranean via the Silk Road and other routes. The Arabs controlled the western ends of these routes and the wealth of many Arabian cities was founded on the spice trade.

A sea trade also developed around the shores of the Indian Ocean, described by the Greek sea-captain who wrote, probably around 80 AD, *The Periplus of the Erythraean Sea*. At about the same time, Pliny described how "cinnamon" (probably cassia) was brought "over vast seas on rafts which have no rudders to steer them or oars to push or pull them or sails or other aids to navigation; but instead only the spirit of man and human courage . . . and they say that these merchant sailors take almost five years before they return and that many perish." In his book, *The Spice Trade of the Roman Empire*, Miller has reviewed at length evidence that these were Indonesian canoes, sailing from Indonesia to India, Madagascar, and the adjacent coast of Africa, and that this trade could date back sufficiently far to validate both ancient Egyptian and Old Testament references to Far Eastern cassia.

Around 50 AD, the Greek merchant Hippalus discovered the Indian Ocean monsoons (which, blowing in alternate directions over the same routes at half-yearly intervals, allowed open-sea navigation across the Arabian Sea, the Indian Ocean, and the Bay of Bengal). This enabled the Romans to make the round trip from their Red Sea ports to the Malabar coast of India in one year instead of at least two. This reduced the importance of the overland spice routes and, together with the collapse of the incense trade, contributed to the ruin of many Arabian cities on the old caravan trail. This breaking of the Arab monopoly enabled the Romans to become conspicuous consumers of spices. Spices were imported through Alexandria and Constantinople (present-day Istanbul) and spread to the Roman Empire north of the Alps. However, the fall of the Roman Empire and the occupation of Alexandria by the Arabs in 641 returned control of the spice trade to the Arabs.

During the Crusades, certain European cities, notably Venice and Genoa, were promised trading facilities in the Holy Land, once the landing points had been secured there. The wealth that under-pinned the Renaissance in Italy was made possible partly through the wealth created by the result-ant trade in spices and other eastern valuables. By the late Middle Ages, a pound of ginger was worth a sheep and a pound of nutmeg worth seven fat oxen. Curiosity about the source of these valuable commodities led the Venetian brothers Nicolo and Maffeo Polao, and Nicolo's son Marco, to explore the overland caravan routes to the Far East. The fall of Constantinople to the Turks in 1453, and the heavy duties imposed by the Arabs on the spice trade, added urgency to the efforts of various European nations to exploit the Polos' discoveries and find a sea route to the Indian and Far Eastern sources of spices.

Bartholomew Diaz and Vasco da Gama, financed by Prince Henry the Navigator of Portugal, rounded the Cape of Good Hope in 1486 and 1497–98. A sea route to India and trading relations with the Hindu king of Calicut were established. Pedro Alvarez Cabral of Portugal, charged with capitalizing on this king's offer to trade cinnamon, cloves, ginger, and pepper for gold, silver, corals, and scarlet cloth, tacked across the Atlantic Ocean, thereby discovering Brazil, and eventually estab-lished trading posts in India. Venice, which had been supplying Europe with about 1.5 million pounds of pepper annually, consequently lost her monopoly to Lisbon.

In the meantime, Christopher Columbus convinced Ferdinand and Isabella of Spain that he could reach China by sailing west. He reached the West Indies in 1492, the American mainland in subsequent voyages, and found chili peppers and allspice. Columbus' supposed discovery of a west-ern route to the Spice Islands led to Magellan's voyage around South America to the East Indies and the first circumnavigation of the globe. Only one ship survived this voyage, but the cloves, nutmeg, mace, and cinnamon brought back covered the costs of the expedition and led to the captain being awarded a coat of arms of two cinnamon sticks, three nutmegs, and twelve cloves.

By 1511, the Portuguese controlled the important spice producing and trading centers in Asia. The Dutch transshipped the imported spices from Lisbon to syndicates in Antwerp. England was also a naval power and, in 1600 and 1602, England and Holland founded their respective East India Companies to exploit this lucrative trade. The Dutch occupied the Spice Islands and subsequently Ceylon (now Sri Lanka), monopolizing the trade in many spices until the French established cinna-mon, cloves, and nutmeg in their Indian Ocean colonies from plants smuggled from Dutch-controlled areas. The British blockaded trade from the Dutch East Indian ports and eventually replaced the Dutch as the colonial power in much of this area.

Today, no country holds a monopoly of production of the most profitable spices. These are grown wherever suitable soils and climates exist. Ginger, originally from Asia, is now an important crop in Jamaica, while vanilla, originally from Mexico, is produced mainly in Madagascar. Spices have lost virtually all their medicinal importance. Their relegation to the kitchen makes it easy to forget their historical significance in leading to the discovery of the monsoons, funding the Renais-sance, provoking European voyages of exploration and the circumnavigation of the globe, as well as some of the more deplorable incidents of colonial history.

African pepper, Guinea pepper, Ethiopian pepper *Xylopia aethiopica*
Annonaceae

This tree is cultivated near villages or protected in the forest in the coastal regions of west Africa. The fruits were once exported to Europe as a spice, but are now traded only within Africa, where they are also used medicinally. In addition, they are pulverized with *Capsicum* and mixed with cola nut to prevent weevil attack. Pieces of the bark may be used to flavor palm wine and, when pulverized, are added to snuff to increase its pungency. The crushed seeds are mixed with other spices and used as a cosmetic and perfume for the body and for clothing.

Allspice *Pimenta dioica*
Myrtaceae

Allspice, so-called because it combines the flavors of cinnamon, nutmeg, and clove, is the dried unripe berry of *Pimenta dioica*. Allspice and clove both belong to the family Myrtaceae. Allspice is native to the Caribbean region, thus one of relatively few major spices to come from the Americas (chili pepper and vanilla being others).

The Mayas of Central America reputedly used allspice in embalming bodies of important people. Columbus may have encountered allspice on his first voyage, in 1492, undertaken to bring back spices for Ferdinand and Isabella of Spain. The Journal of that voyage records that he showed black pepper (*Piper nigrum*) to natives of Cuba, "and they recognised it . . . and by signs told him that in the neighborhood there was much of it." Dried allspice berries are about the same size as, and superficially resemble, black peppercorns. Possibly for this reason, the Spaniards called allspice "pimienta," which is their word for black pepper. This confusion was perpetuated in Linnaeus' generic name *Pimenta* and in the English common names of pimenta and Jamaica pepper. Today allspice is included as one of the five "peppers" in Five Pepper blends.

Allspice is thought to have reached England by 1601. Philip Miller, famed for his *Gardeners' Dictionary*, grew it in a stove house in the 1730s. Linnaeus gave the species the epithet *dioica*, which calls attention to the fact that, although the flowers appear to be bisexual, some trees are predominantly barren (functionally male), whereas others are "bearers" (functionally female, with flowers with fewer stamens, very little viable pollen, but good crops of fruit). This requirement for cross-pollination may partly explain why allspice has not been introduced successfully outside the American tropics, although from the 17th to 19th centuries the berries were often used in preservatives for meat on long voyages. Allspice is still used in Scandinavia in a preservative for fish.

Allspice is the only major spice whose commercial production is today confined to the Western Hemisphere. It has been produced continuously in Jamaica since the early 16th century and Jamaica continues to dominate world production.

Ajowan Ammi *Trachyspermum ammi*
Apiaceae

Ammi is cultivated in Mediterranean countries, Southwest Asia, the Indian subcontinent, and Northeast Africa. In Ethiopia, its aromatic fruits are sold in most markets. Local lore suggests that ammi reduces the pungency of chili peppers (also much used in Ethiopian cooking). In India, the fruit is a characteristic spice in curries and in pickles and beverages. Indians also use ammi as a household remedy for indigestion, often recommend it against cholera, and include it in many cough mixtures. The oil in the fruits, extracted by steam distillation, contains about fifty percent thymol, which is a powerful antiseptic and is used against fungal infections of the skin and in toothpastes and gargles. Ammi was a principal source of natural thymol, but this use has decreased with the advent of synthetic thymol.

Anise *Pimpinella anisum*
Apiaceae

A native of the Mediterranean region, anise was known to the Greeks and Romans. In mediaeval England, Edward I included it among commodities subject to toll, thereby raising funds to repair London Bridge. The Royal Wardrobe Accounts describe the use of little bags of orris root and anise to perfume the linen of Edward IV. Today anise is perhaps best known as flavoring for the Mediterranean *pastis*, the Turkish *raki*, and the Latin American *aguardiente*. Anise oil from the crushed fruits is more widely used than true licorice as a source of licorice flavoring.

See: Herbs and Vegetables, p. 97

Asafoetida, Hing *Ferula* spp.
Apiaceae

Asafoetida is a gum resin collected from cuts made low on the stem or at the top of the root of *Ferula assa-foetida* and other species of *Ferula* native to dry parts of Asia. Asafoetida is now used primarily in medicine, but is also used today in resinous and powdered form in cooking sweet and sour dishes, among others, and is available in Indian markets. The Persians called it "food of the gods," used it as a condiment, roasted the roots, and ate the leaves as greens. The fetid smell characteristic of the fresh plant disappears on cooking, taking on a garlicky-onion flavor. It is popular with Brahmins and Jains in India who are forbidden to eat onions and garlic. Asafoetida may have been imported to the Roman empire as a substitute for *laser*, a resin extracted from *silphium*, believed to be a species of *Ferula*. It was exported from ancient Cyrenaica (Libya) to ancient Greece and Rome, as a medicine, spice, and women's contraceptive. By the first century AD, the plant had become scarce, soon to be extinct.

Black cumin *Nigella sativa*
Ranunculaceae

Black cumin was probably domesticated in the Middle East, where it still occurs wild. It is mentioned in the Old Testament (Isaiah 28:27), was widely grown in Pharaonic Egypt (seeds were been recovered from Tutankhamun's tomb, dated about 1325 BC), and the Greeks and Romans used it as a condiment. The seeds are still used, whole or crushed, in or on bread, in eastern Mediterranean countries. Black cumin is also used to flavor vinegar and as a constituent of curry.

Candlenut *Aleurites moluccana*
Euphorbiaceae

Seeds of this tree, native to Malaysia, are best known as the source of a drying oil, originally used locally to produce candles. However, in parts of Southeast Asia, especially Indonesia, the seed kernels are crushed with garlic, shallots, chili peppers, shrimp paste, and other ingredients to produce a paste which is fried and used to flavor savory dishes.

See: Materials, p. 340

Capers *Capparis spinosa*
Capparidaceae

Capers are unopened flower buds of *Capparis spinosa*. The characteristic flavor develops only after pickling in vinegar. Capers are produced commercially in France, Spain, and Italy, where plants with large buds have been selected. Aristotle and Theophrastus refer to caper as uncultivated, but it was grown in gardens in France by 1536. In 1755, capers were introduced into North America and grown in Louisiana, where they were eventually killed by frost. The species was probably first used by man for its edible fruits, which are still collected and eaten in autumn in the Near East. Seeds occur, sometimes in quantity, in archaeological sites in this region from about 5800 BC onwards.

See: Herbs and Vegetables, pp. 100–1

Caraway *Carum carvi*
Apiaceae

Caraway is adapted to cooler climates than are such other umbelliferous spices as coriander, cumin, aniseed, or ammi. Wild caraway occurs from northern Europe and Siberia south to the Pyrenees, northern Italy, the Balkans, northern Iran, and the Himalayas.

It is claimed by some that caraway has been cultivated and consumed in Europe longer than any other condiment, but there is, at least as yet, no long archaeological record to support this. The ancient Greeks used caraway, along with poppy seed, in bread, foreshadowing one of the principal uses of caraway today. The first century Greek physician, Dioscorides, mentioned caraway as both a herb and a tonic. The Roman Apicius used caraway in some of his recipes. By the 12th century, caraway was known to the Arabs as *karauya* and listed among the plants cultivated in Morocco. In England, caraway appears in a record compiled in 1390 by the master cooks of Richard II, while, in 1629, Parkinson considered its fleshy taproot to be superior to that of parsnip. The leaves are also edible: the young leaves are said to make a good salad, while larger leaves may be cooked like spinach.

In the Orient, caraway is known by names such as Roman cumin or Persian caraway, suggesting that it was introduced from Europe or the Middle East, though the date of any such introduction is unknown. Caraway was cultivated in American gardens by 1806. Its fruits are much used in North America to flavor bread and cheese. Caraway is also used to make the liqueur Kümmel, to season sausages, and in pickling spice. Oil from the fruits is also used in flavoring and for medicinal purposes. Holland is the principal commercial source of caraway.

Cardamom *Elettaria cardamomum*
Zingiberaceae

Today, cardamom signifies the dried fruits and seeds of *Elettaria cardamomum*. Various species of the related genera *Amomum* and *Aframomum* are known as false cardamoms, for example *Amomum subulatum* (Nepal cardamom) and *Aframomum corrorima* (Ethiopian cardamom).

The confusion in names makes it difficult to trace the early spread of cardamom. True cardamom is native to southern India and Sri Lanka and is probably the cardamom of 4th century BC Ayurvedic medicine. In Greece, at about the same time, Theophrastus described the inferior *amomon*, the superior *kardamomon*, and a round reddish pepper with a capsule, which some have also identified with cardamom. Pliny likewise distinguished amomum, associated with northern India; cardamom, which had the same kind of fruit as amomum and came from Arabia; and *siliquastrum* or *piperitis*, "a kind of pepper" from southern India with seeds in "small pods . . . such as we see in beans." In the 15th centurym the Portuguese referred to *siliquastro* or "the pepper of Calicut," which may link Pliny's *siliquastrum* with true cardamom. *Aframomum melegueta* of west Africa is today known as melegueta pepper, so the identification of some classical "peppers" with cardamom is not as far-fetched as it might at first seem.

In China, amomum figured significantly in the economy of the Sung period (960–1279). Around 1150, Idrisi included cardamom among products brought to Aden from "Sindh, India and China," while in 1514 Magellan's brother-in-law, Barbosa, located cardamom production in Malabar.

Cardamom has remained extremely popular in Arab countries, where cardamom coffee is a symbol of hospitality. Cardamom is also much used in Scandinavian countries, in baking and to flavor meat dishes. Cardamom is the third most expensive spice, so sometimes the cheaper Ethiopian cardamom has been used in Scandinavia to replace true cardamom.

Cardamom was introduced to Guatemala in 1920 and Guatemala is now second to India as an exporter of the spice.

Celery *Apium graveolens*
Apiaceae

Wild celery occurs in Europe and Southwest Asia, and also in North America, where it has escaped from cultivation. Varieties grown for "seed" (actually the fruits) differ little from the wild plants, unlike the vegetable types grown for their petioles or swollen hypocotyls. Inflorescences of wild celery were included in the garlands in Tutankhamun's tomb. Fruits (possibly wild) have been recovered from 7th century BC Samos and Swiss lake dwellings of the Late Neolithic. Cultivation of celery, for seed, became widespread in the Roman Empire. Today celery is grown commercially for seed in France, India, and the United States.

See: Herbs and Vegetables, p. 116; Roots and Tubers, p. 71

Chili peppers *Capsicum* spp.
Solanaceae

Although now regarded as essential components of spicy stews and curries in Europe, Asia, and Africa, chili peppers were unknown in the Old World before 1493. The oldest archaeological specimens from Mexico and Peru resemble modern wild peppers, but before the Spanish Conquest, peppers varied greatly in fruit size, shape, pungency, and color. Friar Bernardino de Sahagun, who arrived in Mexico in 1529, wrote that Aztec nobility ate "frog with green chillis; newt with yellow chilli; tadpoles with small chillis; maguey grubs with a sauce of small chillis; lobster with red chilli . . ." These were almost certainly all *Capsicum annuum*, which is adapted to the cool dry highlands of Mexico and is today by far the most economically important species.

For Peru, Garcilaso de la Vega (son of an Inca princess and a Spanish nobleman) mentioned a thick pepper (*rocot uchu*), which is probably *C. pubescens*, still known as *rocoto* and widely used in the Andes. He wrote also of yellow and brown peppers, "though in Spain only the red kind has been seen." Dried orange fruits of one landrace of *C. baccatum* and brown fruits of one landrace of *C. chinense* are still sold in most Peruvian markets, along with red fruits of both species.

C. chinense and the closely-related *C. frutescens* are better adapted to the hot humid lowlands of Amazonia, the Caribbean, and Central America than is *C. annuum*. When Columbus wrote in his Journal of his first voyage that "there is much *axí*, which is their pepper, and it is stronger than pepper, and the people won't eat without it . . .," he probably referred to these species. *C. chinense* includes the Scotch Bonnet and Habanero peppers and is valued for aroma as well as pungency. *C. frutescens* includes the pepper used in Tabasco sauce, and also small, very pungent, peppers grown for export in various parts of the Old World tropics.

Several species probably reached Europe when the Spanish padres sent seeds of plants used by Amerindians for trial in monastery gardens. Chili consumption spread rapidly within the Iberian Peninsula, alarming Portuguese merchants who had a lucrative monopoly of the more expensive black pepper. Cultivation of *Capsicum* was therefore prohibited. Elsewhere, the herbalists Matthiolus (c. 1570) and Dodonaeus (1644) warned against *Capsicum* on health grounds. However, 17th century sailors customarily carried dried chilis on their transatlantic voyages. Chilis are an excellent source of vitamin C and may have helped protect sailors against scurvy, though the quantity needed to be consumed to prevent scurvy would have been significant.

Chili seeds remain viable for several years. Seeds from fruits not consumed on long voyages could thus be sown in new continents. The Portuguese probably spread chilis along their trade routes from Brazil to Africa and thence to India, where three types of *Capsicum* were recorded as early as 1542. The Arawak name *axí*, converted into *achi* in Portuguese, became *achar* in India. 'Chili' is derived from an Aztec word meaning red. From India, *Capsicum* probably spread via Arab and Hindu traders to Indonesia, New Guinea, and thence to Melanesia. Europeans sailing west across the Pacific erroneously concluded that *Capsicum* was native to the Old as well as the New World.

Chilis reached Eastern Europe via the Turks, probably with those other American introductions, maize and the turkey. Hungary is an important producer of paprika today, but *Capsicum* was not cultivated there until after the Turkish occupation, and then mainly for home consumption. During the Napoleonic Wars, the sea blockade of Europe prevented import of black pepper and caused a switch to chili in Hungary and probably other parts of Europe. Early European colonists reintroduced *Capsicum* to North America.

Cinnamon, Cassia *Cinnamomum* spp.
Lauraceae

The bark of various species of *Cinnamomum* is used as a spice. True cinnamon (*C. verum*) is native to Sri Lanka. Cassia has a stronger flavor, and is widely used in the United States, where it is not distinguished from cinnamon. Cassia includes Chinese, Indonesian, and Saigon cassia (*C. cassia, C. loureirii* and *C. burmannii*). Massoi bark, another cinnamon substitute, comes from the *Cryptocarya massoia* of New Guinea, while *C. camphora* is the source of camphor.

Authorities disagree on whether the cinnamon and cassia mentioned in the Old Testament and in ancient Egyptian embalming recipes were species of *Cinnamomum*. According to Herodotus, cinnamon was harvested from twigs used by birds to build nests attached to steep cliffs. The birds carried baits of donkey meat to these nests, which then collapsed, enabling the cinnamon harvesters to collect the twigs. Pliny dismissed this as nonsense, invented to increase prices, and described instead how cinnamon and cassia were brought "over vast seas on rafts" to Africa. Nero burned a year's supply of cinnamon on his wife's funeral pyre in 66 AD.

Cassia is mentioned in Chinese texts of the 4th century BC and the Emperor who conquered part of southern China in the 3rd century BC gave it the name it still bears: *kwei lin* (cassia grove). Cassia (possibly Saigon cassia) was used at the Court of Huề in such quantities that wild trees were overexploited and cultivation increased.

Cleaning cinnamon on boards.

In 15th century England, "synamome" was "for lordes," but cassia for the "commyn people." Cinnamon was one of the spices sought on European 15th and 16th century voyages. After their seizure of Ceylon in 1505, the Portugese exacted a tribute of 110,000 kg of cinnamon bark per year, obtained from wild trees. The Dutch started strictly controlled cultivation of cinnamon, and their monopoly subsequently passed to Britain in 1796. In 1771, the French introduced cinnamon to the Seychelles, where it has now run wild, and from the 19th century cinnamon was more widely cultivated.

Today the major producers of cinnamon are Sri Lanka, the Seychelles, and Madagascar, while cassia comes mainly from China, Indonesia, and Vietnam.

See: The Hunter-Gatherers, p. 3; Fragrant Plants, p. 241

Clove *Syzygium aromaticum*
Myrtaceae

Cloves are dried unopened flower buds of a tree in the Myrtaceae, native to the Moluccas (Indonesia). From the Moluccas, cloves were exported early to China and India. In China, cloves were used at court in the Han Dynasty (220–206 BC) to combat halitosis and avoid "giving offence to the broad-minded and sagacious Emperor." In India, one use was to fasten betel pepper leaves (*Piper betle*) around quids of betel nut (*Areca catechu*). "Clove" comes from the French *clou* (nail), because the long inferior ovary and "head" of other floral parts give the clove bud the shape of a nail.

Arabs probably introduced cloves from India to the Mediterranean region. Cloves reached Egypt by AD 176. In 1265, in England, the Countess of Leicester paid ten to twelve shillings per pound for cloves. Marco Polo saw planted cloves in the East Indies in 1291 to 1297. The Portuguese occupied the Moluccas in 1514 to control the valuable monopoly of cloves and nutmeg/mace, but this did not stop others acquiring cloves clandestinely. The only captain to survive Magellan's circumnavigation of the world brought twenty-six tons of cloves back to Spain, while Sir Francis Drake jettisoned three tons of cloves, *inter alia*, to refloat the *Golden Hind* when she went aground in Sulawesi.

When the Dutch captured the Moluccas (1605–21), they attempted to prevent further haemorrhaging of their monopoly by confining clove cultivation to one island. They destroyed perhaps seventy-five percent of clove trees in the Moluccas, significantly reducing genetic diversity. In the late 18th century, the French secretly collected young cloves from the Moluccas and introduced a few trees to their islands in the Indian Ocean. From there, cloves reached Zanzibar, where the Sultan required plantation owners to plant cloves or forfeit their land. Zanzibar then dominated the world clove market, until epidemics of "sudden death" (a fungal disease of cloves) on Zanzibar in the 20th century. This, combined with ever-increasing demand in Indonesia for clove cigarettes (tobacco plus thirty to forty percent by weight of shredded cloves) has led to the homeland of clove once again becoming the world's largest producer—and consumer—of cloves.

Coriander, Cilantro *Coriandrum sativum*
Apiaceae

Coriander is native to southern Europe and the Mediterranean region. The green leaves and the dried fruit have different, equally distinctive, flavors and are much used in Mediterranean and Latin American cooking. Ground coriander fruits (often termed seeds) also comprise up to forty percent of curry powder, although the spice is not native to India and it is not known when it reached the subcontinent.

Coriander was used by the Egyptians and the Israelites before the exodus of the latter from Egypt, and the *Ebers Papyrus* of around 1550 BC refers to both its medicinal and culinary uses. Its medicinal properties were mentioned also by Hippocrates around 400 BC and by the army physician Dioscorides in his *Materia Medica* of AD 65. Some early descriptions of coriander may have represented wishful thinking (for example, the aphrodisiac qualities attributed to coriander in *The Thousand and One Nights*), but coriander does have minor pharmaceutical uses. It is also employed in perfumery.

In the 3rd century BC, the Roman statesman Cato recommended coriander as a food seasoning. The epicure Apicius published the first of his ten books on the art of cooking under the Emperors Augustus and Tiberius and used coriander in a recipe involving minced oysters, mussels, sea urchins, chopped toasted pine kernels, rue, celery, pepper, coriander, sweet cooking wine, fish vinegar, Jericho dates, and olive oil. In fact, the Roman market demanded so much coriander that it was grown in bulk, mainly in Egypt. The fruits have been recovered from shops in Pompeii. Pliny gave the plant the name coriander from the Greek *koris* (bedbug), from the odor of the foliage.

The Romans introduced coriander to England, while in 812 AD Charlemagne ordered it to be grown on the imperial farms in central Europe. Coriander was one of the first European herbs and spices to be introduced to North America by the colonists. It reached Massachusetts before 1670.

Because the crop is labor-intensive and fetches a relatively low price, today most commercial production of coriander is centered in countries with low labor costs, notably India, the former Soviet Union, central Europe, and Morocco.

Cumin *Cuminum cyminum*
Apiaceae

It is now difficult to determine the natural range of cumin. It is usually said to come from the southern and eastern Mediterranean region, but some place it further east, in central Asia. Its fruits figure in indigenous medicine and are an essential ingredient of curry powder in India, but its antiquity there is less well documented than in the Middle East and Mediterranean.

Fruits of cumin have been recovered from archaeological sites from the 2nd millennium BC in Syria and the Iron Age in Jordan. Its name is possibly derived from the Babylonian *ka-mu-nu*. This became *kuminon* in Greek and *cyminum* in Latin. A basket of cumin was included in the burial of Kha, architect to the 18th dynasty Pharaoh Amenophis III.

Cumin was valued in Europe for its medicinal properties and as a spice. It was one of the thirty-six ingredients of the antidote to poisoning prepared by the physician Crateuas for Mithridates the Great in about 80 BC. This reputedly prevented Mithridates from poisoning himself to avoid capture by Pompey. In about AD 65, Dioscorides listed cumin among the fragrant herbs used in compound ointments. Greek influence on the Roman Empire is thought to have led to increased demand for luxuries such as spices. Cato the Censor (234–149 BC) included in his *De Agri Cultura* a recipe for a sweet wine cake flavored with cumin and aniseed. Cumin also figured in the recipes of Apicius the epicure. It was grown in bulk in Egypt for the markets of Imperial Rome. Pliny regarded cumin as the best appetizer of all the condiments, and it was among the products whose maximum price was fixed by Diocletian's edict of AD 301.

Cumin remained popular in mediaeval Europe. In England in 1419, its import was taxed to provide income for the Crown. After the Middle Ages, cumin was increasingly supplanted by caraway.

Cumin is much used in Latin American cooking. The increasing popularity of both Mexican and Indian food has increased world demand. It is grown commercially in southeast Europe, north Africa, the Middle East, India, and China.

Curry leaf *Murraya koenigii*
Rutaceae

Murraya koenigii is a shrub or small tree cultivated throughout India for ornament and for its leaves, which are much used to flavor curries and chutneys. Medicinal uses of the leaves include treatment of diarrhea, dysentery, and vomiting, and external application to bruises. *Murraya* belongs to the same family as *Citrus* and, like *Citrus,* has gland-dotted leaves containing an aromatic oil which, in the case of *M. koenigii,* has a spicy smell and pungent clove-like flavor.

See: Herbs and Vegetables, p. 102

Dill *Anethum graveolens*
Apiaceae

Dill is native to the Mediterranean and west Asia, where both leaves and fruit have long been used. Dill was present in Late Neolithic lake settlements in Switzerland, 7th century BC Samos, and the tomb of Amenophis II in Egypt. It figures in Theophrastus' *Inquiry into Plants* and Apicius' recipes. In the Middle Ages, burned dill seeds placed on wounds reputedly speeded healing. Dill was also used for indigestion and insomnia. It has been cultivated in England since 1570, but grown commercially in the United States only from the early 19th century. It is an ingredient of curry powder.

See: Herbs and Vegetables, pp. 102–3

Fennel *Anethum foeniculum* (syn. *Foeniculum vulgare*)
Apiaceae

Although often placed in a different genus from dill, fennel is so closely related to dill that the two will hybridize if planted close together. Like dill, both the leaves and fruits of fennel are used medicinally and as a flavoring. Fennel was known to the Classical Greeks and Romans, while to the Saxons it was one of the nine sacred herbs used to combat the nine causes of disease. India exports one to two thousand tons of fennel seed annually.

See: Plants as Medicines, p. 214; Herbs and Vegetables, p. 103

Fenugreek, Methi *Trigonella foenum-graecum*
Fabaceae

Fenugreek seeds, like those of most legumes, are rich in protein, so fenugreek adds valuable nutrients as well as flavor to the diet. The plant is native to west Asia and southeast Europe. Charred seeds from the Middle East date from about 4000 BC onwards. Fenugreek was present in Tutankhamen's tomb and was a component of the "holy smoke" used for fumigation and embalming. It was used in a medieval cure for baldness. The steroid in its seeds may be useful in oral contraceptives, and is widely used as a galactagogue (to increase milk supply). Fenugreek is an important component of curry powders, a popular ingredient of Egyptian and Ethiopian bread, and the principal flavoring of artificial maple syrup.

Greater and lesser galangal *Alpinia galanga* and *A. officinarum*
Zingiberaceae

Galangal is cultivated throughout south and Southeast Asia for its rhizomes, regarded as essential in local cooking. Its taste resembles a mixture of pepper and ginger. Its name is probably derived from the Chinese kao-liang kiang (large mild ginger), used to distinguish *A. galanga* from *A. officinarum* (liang kiang). The Roman Apicius used galangal in his recipes and galangal figured in early trade from Asia to Europe. International demand is now small and supplied mainly by India, Thailand, and Indonesia. The Netherlands is a major importer. The fruits may be used as a substitute for cardamom. Medicinal uses date back to Sanskrit writings of 600 AD. Galangal is still used locally in traditional human and veterinary medicine, where it reputedly cures all diseases of elephants.

Ginger *Zingiber officinale*
Zingiberaceae

Zingiber officinale is unknown in the wild. It may have originated anywhere from India to China and Southeast Asia. The earliest records are from China, where Confucius (557–479 BC) mentioned ginger in his *Analects* and reputedly never ate without it. In AD 406 Fa-hsien recorded in his *Travels* that pot-grown ginger was carried on Chinese ships voyaging to Southeast Asia, to help to prevent scurvy. Records from Sung times (960–1279) show that ginger was still being imported from the south, although it was also grown in China. Fresh ginger remains an important ingredient in Chinese cooking.

Ginger reached the Greek and Roman Empires via the Arabs, accompanied by the Sanskrit name *singabera* (antlered thing, referring to the shape of the rhizomes, which are the source of the spice). Greek and Roman writers referred frequently to both its culinary and its medicinal uses. Dried rather than fresh rhizomes seem to have been imported, possibly overland, possibly by sea routes around the Indian Ocean to Arabia or East Africa. Dioscorides and Pliny both associated ginger cultivation with Arabia and Trogodytica (which may correspond to present-day Somalia). In AD 150, Ptolemy listed ginger as a product of Ceylon. The supposed aphrodisiac properties of ginger figure in *The Thousand and One Nights* and, according to the Koran, a ginger drink is served to the righteous in Paradise.

In 14th century England ginger was the most common spice after pepper and one pound (0.5 kilogram) cost as much as a sheep. Henry VIII recommended it against plague, while gingerbread was one of his daughter Elizabeth's favorite confections.

By the Middle Ages both preserved ginger and living rhizomes were imported into Europe. The latter facilitated further spread of ginger cultivation. The Arabs introduced ginger to east Africa in the 13th century and the Portuguese to west Africa in the 16th century. Francisco de Mendoza took ginger to the Americas early in the 16th century and by 1547 Jamaica was exporting 1,100 tons to Spain.

Today, India produces most of the world's ginger, followed by China, west Africa, Jamaica, Thailand, and Australia.

See: Plants as Medicines, p. 226

Juniper *Juniperus communis*
Cupressaceae

Juniper is native to temperate Europe, Asia, and North America. The fleshy fruits are used to flavor drinks, particularly gin, and as a spice. The best quality fruits come from Italy, but juniper "berries" are also exported from the former Yugoslavia, the Czech Republic, Slovakia, and Hungary. The fruits have been used to adulterate black peppercorns and as a substitute for pepper. Roasted and ground, they were used in parts of Sweden instead of coffee. They were used medicinally in Pharaonic Egypt and by the Greeks and Arabs.

Juniper (*Juniperus communis*). USDA-NRCS PLANTS Database/Britton, N.L., and A. Brown. 1913. *Illustrated flora of the northern states and Canada*. Vol. 1, p. 66.

Licorice *Glycyrrhiza glabra*
Fabaceae

Dried rhizomes and roots of a legume, *Glycyrrhiza glabra*, distributed from Europe to western Asia and North Africa, are the source of licorice, used to flavor tobacco and confectionery. In India, licorice is chewed with betel nut. Medicinally, it has been valued against indigestion. Derivatives of licorice have been used to treat stomach ulcers. "Spanish" (including Italian and Greek) licorice is considered the best quality. Lower quality "Oriental" licorice is cultivated in the Middle East, the Caucasus, and north China. In the United States, roots from different regions are blended to make various brands of licorice extract.

See: Plants as Medicines, p. 220

Melegueta pepper, Grains of paradise, Guinea grains *Aframomum melegueta*
Zingiberaceae

This west African member of the ginger family first appears in European literature in 1214 AD, as Melegetae. The herbalist Fuchs suggested that this name derives from the empire of Melle, in Niger. Others think it a corruption of Malaga, centre of Portuguese trade in this spice. The Mandingo of west Africa seem to have traded it first to Tripoli, whence it was distributed further by the Italians, who called it "grains of paradise," a name that reflected its unknown origin and great value. The seeds were used as a substitute for *Piper nigrum,* and various European countries sponsored voyages to west Africa in attempts to break the 14th century Venetian monopoly of black pepper. The west African coast consequently became known as the Melegueta or Grain Coast. Melegueta pepper was much used, together with cinnamon and ginger, in hippocras, the spiced wine of 14th to 15th century Europe. In Elizabethan England, it was used to flavor many foods and drinks, but its use later decreased, aided by an Act of Parliament which imposed a heavy fine on brewers who possessed or used grains of paradise. Melegueta pepper is still used in alcoholic drinks, to give a flavor similar to ginger ale, and in vinegars and sauces. It has also retained its traditional importance in west Africa.

Mustards *Brassica* spp. and *Sinapis alba*
Brassicaceae

Mustards have been used as vegetables, oilseeds, condiments, and medicines. White mustard (*Sinapis alba*) probably originated in the eastern Mediterranean region, black mustard (*Brassica nigra*) in the Middle East, brown mustard (*B. juncea*) in central Asia, and Ethiopian mustard (*B. carinata*) in northeast Africa. For table mustard, white mustard is combined with a more pungent species, previously black mustard but now brown mustard, mainly because *B. juncea* can be harvested mechanically whereas the dehiscent fruits of *B. nigra* must be hard-harvested. The mustard gas used in the First World War was based on the chemical structure of the pungent glycoside of black mustard.

The different species of mustard, and the reasons for their use, are difficult to distinguish in the earliest records. Seeds of *Brassica* or *Sinapis* have been recovered from around 3000 BC in Iraq, 2000 BC in India, Egyptian tombs of 1900 BC, and late Neolithic lake dwellings of Switzerland and Germany. Pythagoras (c. 530 BC) advised mustard as an antidote to scorpion bites, while Hippocrates (c. 400 BC) recommended the seeds for both internal and external use. Mustard remained a home remedy, notably an emetic and a plaster for rheumatism, until recently.

Mustard increases activity of the salivary glands, hence stimulates appetite. "Mustard" comes from *mustum ardens*, a pungent paste of ground mustard seeds and fruit juice used by the Romans. *Sinapis* comes from *sinapi*, Pliny's name for mustard. In 301, Emperor Diocletian fixed a maximum price for seeds and powder. In 812, Charlemagne decreed that mustard should be grown on imperial farms in central Europe. In mediaeval Europe, mustard was valued to flavor salted meat. Vasco da Gama included it in provisions for his voyage in 1547. In 1542, Fuchs observed that mustard was planted everywhere in gardens in Germany, while in 1597 Gerard noted that it was not common in England, but that he had distributed seed.

Mustard seeds were scattered by the Spanish fathers to mark the route of the Mission Trail in California, where it is now a nuisance, competing with native plants. Today, mustard for use as a condiment is grown mainly in North America, England, and Hungary.

See: Gathering Food from the Wild, p. 30; Nuts, Seeds, and Pulses, p. 150; Roots and Tubers, p. 72 and 73; Herbs and Vegetables, p. 107

Nutmeg, Mace *Myristica fragrans*
Myristicaceae

The nutmeg tree is unique in producing two spices. Its yellow fruits split to expose a single dark seed with a scarlet aril. Mace is the dried aril, while the dried seed, without its seed coat, is the commercial nutmeg. The yield of mace is only fifteen percent that of nutmeg, so mace usually commands higher prices.

Myristica fragrans possibly originated in New Guinea, where its closest relatives occur, or in the Moluccas. Nutmeg reached India before the 6th century and Constantinople by 540. The Arabs had pinpointed the Moluccas as the source of nutmeg by the 13th century, though they hid this from European purchasers. In the East, nutmeg was valued medicinally; but in Europe nutmeg and mace were treated primarily as spices, though the reputed ability of nutmeg to cure even the plague helped to keep prices high.

By 1526, the Portuguese monopolized the Asian spice trade. The Dutch broke the Portuguese monopoly and attempted to restrict nutmeg cultivation to the Moluccan islands of Banda and Amboina. By 1760, they had accumulated stocks reputedly representing sixteen years' production of both nutmeg and mace in warehouses in Amsterdam. These stocks were burned to keep prices high. In 1770, nutmeg seeds and seedlings smuggled from Indonesia reached Mauritius, where they were planted in the garden of Pierre Poivre, the administrator of the island, and protected by an ordinance making attempts at their export treasonable.

From 1776 to 1802 the British occupied the Moluccas. The British introduced nutmeg cultivation to Penang and Singapore, ending the Dutch monopoly. In 1802, nutmegs were introduced to the West Indies. Cultivation in Grenada started in 1843, when the ending of slavery caused sugarcane cultivation to decline. Cocoa, then nutmeg, plantations replaced sugarcane. By the end of World War I, Grenada rivaled Indonesia in nutmeg production. These two nations are still the principal producers. Flavors of nutmeg and mace from the East versus the West Indies differ, so each region has its own markets.

Essential oils of nutmeg and mace differ slightly in chemical composition. Nutmeg has a sweeter, more delicate flavor than mace. Both contain myristicin, which causes fatty degeneration of the liver (two nutmegs are reputedly a lethal dose) and is supposedly hallucinogenic. (In the 1960s it was standard practice to ban nutmeg from kitchens of federal prisons lest the inmates abuse it, after inmates at the New Jersey State Prison, Trenton, were used in testing its psychotropic action in 1960). Industrial use of nutmeg and mace has increased with growth in markets for prepared foods, sauces, pickles, and chutneys, leading to increased use of extracted oleoresins rather than the intact spices. Nutmeg oil is also used in the food industry, in beverages such as Coca-Cola, in perfumes and aerosols, and in inhalants such as Vick. The fleshy fruit wall is candied and sold as a sweetmeat in Indonesia and Malaysia.

See: Psychoactive Plants, p. 203

Pepper *Piper nigrum*
Piperaceae

"Pepper" comes from the Sanskrit word for berry (*'pippali'*). It was originally applied to fruits of long pepper (*P. longum*) of northern India, which was formerly more important than black pepper (*P. nigrum*). Black pepper is a climbing vine native to southwestern India whose fruits are harvested green or ripe. Unripe fruits are preserved in brine or vinegar (green pepper), or dried to produce black peppercorns. Ripe fruits are retted to remove the layers outside the hard endocarp, then endocarps plus seeds are dried as white peppercorns.

Pepper was well known to the Greeks and Romans. Hippocrates and Aristotle mentioned it about 400 BC. Aristotle's pupil Theophrastus, the "Father of Botany," distinguished black pepper from the more valuable long pepper, while Dioscorides was the first to mention white pepper. Pliny noted that long pepper was frequently adulterated with Alexandrian mustard, and black pepper with juniper berries. After the discovery of the monsoons in 40 AD, a regular sea trade between Rome and India was established with pepper as a staple commodity. Although pepper thus became more widely available, it remained valuable. For example, the tribute paid to prevent Alaric the Goth from sacking Rome in 408 AD included three thousand pounds of pepper. In medieval Europe, peppercorns were bequeathed in wills and were accepted currency for various transactions (as in "peppercorn rent," then a payment of value not, as now, a nominal sum).

Hindu colonists probably took pepper from India to Indonesia between 100 and 600 BC. Marco Polo observed pepper in India and Java around 1280. Vasco da Gama's search for pepper and other spices resulted in a sea route round Africa and a monopoly of the spice trade for Portugal. When the Dutch displaced the Portuguese from Indonesia in the 17th century, they captured the Indonesian pepper-growing areas, but pepper was by then too widespread for the Dutch to monopolize production. In the early 19th century the British organized pepper plantations in many parts of the Far East, including Malaya and Sarawak. After the Second World War, Japanese settlers introduced pepper plantations in Brazilian Amazonia.

Today, pepper remains the world's most important spice, with India, Indonesia, Malaysia, and Brazil all being major producers.

See: Psychoactive Plants, p. 197; Plants as Medicines, p. 226

Pepper tree, Pink pepper *Schinus terebinthifolius*
Anacardiaceae

Pink peppercorns are unrelated to black peppercorns (*Piper nigrum*). They are the fruits of the wild Brazilian pepper tree. Since the nineteenth century the pepper tree has become an invasive weed in sub-tropical areas such as Florida and Hawaii. The fruits are an important export crop from the Indian Ocean island of La Réunion. They became a popular spice in the 1980s, closely associated with the "nouvelle cuisine" movement, and are also used in "Five Pepper" blends. Pink peppercorns can have an irritant effect on skin and mucous membrane, and should be used in moderation.

Poppy *Papaver somniferum*
Papaveraceae

Recent evidence suggests that the opium poppy was domesticated in the western Mediterranean region. Charred remains occur from mid-Neolithic sites in central Europe and Bronze Age sites in the eastern Mediterranean. Hippocrates mentioned opium wine around 400 BC. Theophrastus, writing in the third century BC, described how the unripe fruit is scratched to yield the opium-producing latex. Poppy seeds were used by the Greeks, and later in Republican Rome, as a condiment, especially for flavoring bread. By the time of Nero, this use was confined to country people but, in the city, Petronius described a dish of roasted or fried stuffed dormouse sprinkled with honey and poppy seeds. By the time of Mohammed, knowledge of the medicinal and narcotic properties of poppy had spread to Arabia and thence, with Islamic traders, east to China and Southeast Asia. Poppy is now cultivated for seed in many temperate countries, particularly the Netherlands.

See: Psychoactive Plants, pp. 199–200; Ornamentals, p. 278

Saffron *Crocus sativus*
Iridaceae

Saffron, from dried stigmas of *Crocus sativus*, is the world's most expensive spice. It takes seventy thousand flowers to produce about half a kilogram of saffron. Its name comes from the Arabic *za'faran* (yellow) and saffron was the Mediterranean equivalent of the Asian turmeric. In Classical times, saffron was strewed on floors as a perfume and figured in Roman trade with India. By 960 AD, the Arabs were cultivating saffron in Spain, while the Crusaders probably introduced it to northern Europe. Being expensive, saffron was often adulterated, as described by Pliny. In 15th century Germany, traders found guilty of adulterating saffron were burned or buried alive. Saffron reputedly cured everything from toothache to plague; drinking saffron tea induced optimism; and saffron tea was even added to canaries' drinking water.

Sassafras *Sassafras albidum*
Lauraceae

Sassafras is native to the eastern United States, and in the southeast it has often been used to make a popular tea. It was formerly used to flavor the popular United States soft drink root beer, and as an ingredient in gumbo filé, used in Creole cooking. Since 1960, when the oil was described as a carcinogen by the United States Food and Drug Administration, only de-safrolized sassafras oil can be used in food/beverages. Most sassafras flavoring is now synthetic. Oil from the bark is used in flavorings and perfumes. Sassafras has also had various medicinal uses. It belongs to the same family as cinnamon (*Cinnamomum* spp.) and bay (*Laurus nobilis*).

See: Fragrant Plants, p. 242

Sesame *Sesamum indicum*
Pedaliaceae

Sesame is reported as an apparently wild plant from the Indian subcontinent. About sixty milliliters of sesame seeds were found in Tutankhamun's tomb in Egypt (c. 1350 BC). In the Iron Age kingdom of Urartu (Armenia, 900–600 BC), there are abundant remains of seed and presscake, plus apparatus

Sassafras (*Sassafras albidum*). USDA-NRCS PLANTS Database/Britton, N.L., and A. Brown. 1913. *Illustrated flora of the northern states and Canada*. Vol. 2, p. 134.

for pressing and storing the oil. An Old Babylonian text refers to sesame as food for royalty, medicine, and an ingredient of soap (mixed together with alkali and juniper resin). The Greeks and Romans cultivated sesame extensively, as an edible seed since olive oil was available for cooking. Sesame cake was given to brides as an emblem of fruitfulness because sesame was considered the most fruitful of all seeds. The 1st century AD book, *Periplus of the Erythraean Sea*, describes trade in sesame oil from India to Red Sea ports in exchange for frankincense. In 1298, Marco Polo observed that it was used in place of olive oil in Persia. General Chang Chien introduced sesame to China from the west during the 2nd century BC and the oil is still much used in Chinese cooking. African slaves, who valued the oil as medicine and considered the seeds lucky, introduced sesame to America. Sesame remains an important edible oil, while sesame paste is used, *inter alia*, to make the sweetmeat *halva* and the seeds are used as a garnish for baked goods. Today China is the major producer of sesame, but it is grown on a small scale throughout the tropics.

See: Nuts, Seeds, and Pulses, p. 151

Star anise *Illicium verum*
Illiciaceae

Star anise is the only non-poisonous species of *Illicium*. The tree is native to southwest China. The fruits and seeds are used in Chinese cooking and also medicinally for colic, constipation, insomnia, and other purposes. They contain a volatile oil, similar in composition to those of dill and aniseed. The first known record of trade in the fruits is from the Philippines in 1588. Clusius bought star anise fruits in London in 1601 and later they were traded to Europe along the tea route from China via Russia as "Siberian cardamoms." Production today remains concentrated in China.

Sumac *Rhus coriaria*
Anacardiaceae

This shrub is native to the Mediterranean region and Middle East. Its acid fruits are dried, powdered and used as a condiment on foods such as kebabs and pilaffs. The sour, lemon-like flavor is particularly valued in those parts of the Middle East where lemons are rare. The leaves and stems of this species are used in tanning.

Sweet flag *Acorus calamus*
Acoraceae

Sweet flag occurs in marshes in temperate regions of both Old and New Worlds. Its fragrant rhizomes have been used for various purposes. Fresh, sweet flag is used as a substitute for ginger; candied, and as a breath sweetener and masticatory. Sweet flag is a source of oil for flavoring alcoholic drinks such as benedictine, vermouth, gin, and beer. Medicinally, it is used as a tonic and to aid digestion. Sweet flag has also been used as an insecticide against fleas and lice. It was an ingredient in the Holy Smoke dedicated to the sun by ancient Egyptians, and is mentioned in the Old Testament. The Tartars reputedly introduced sweet flag and its uses to Poland, while in 1574 Clusius recorded that it had been introduced into cultivation in Vienna from Constantinople. Soon after, it was recorded also in France and Britain, where it has been naturalized since 1660.

See: Fragrant Plants, pp. 253–5

Szechuan pepper, Prickly ash *Zanthoxylum* spp.
Rutaceae

Zanthoxylum belongs to the same family as *Citrus*, rue, and the curry leaf tree. Fruits of some Far Eastern species, notably *Z. bungei* and *Z. piperitum*, are used as a condiment in oriental cookery. A volatile oil in the wall of the fruit is the source of the flavor. Fruits sold as spices in western supermarkets are imported from China and Japan.

Turmeric *Curcuma domestica*
Zingiberaceae

Turmeric contributes a bright yellow color and distinctive flavor to food. It is a principal ingredient of curry powder and the colorant in mustard powder and some pickles.

Turmeric comes from the dried rhizomes of *Curcuma domestica*, a sterile triploid unknown in the wild. Its closest relative may be the diploid *C. aromatica*, which has an orange-red rhizome used as a dye, cosmetic and drug, but not as a spice because it smells of camphor. *C. aromatica* is native to the Indian subcontinent, and turmeric may have originated from *C. aromatica* by human selection in cultivation.

Turmeric is widely used in Southeast Asia medicinally, as a cosmetic, and in religious rituals. This suggests to some that it spread very early from India to Southeast Asia. Turmeric probably also spread later, with Hinduism, since it is involved in some Hindu rites. Buddhist robes are dyed yellow with turmeric and turmeric spread east to Polynesia primarily as a dye. It reached Tahiti, Hawaii, and Easter Island before Europeans discovered these islands.

The spread of turmeric westward has been linked to the need of sun worshippers in Persia for more yellow dye than could be supplied by saffron. Turmeric was included among the coloring plants in an Assyrian herbal of the 6th century BC. Pliny described turmeric as an Indian plant with the appearance of ginger but taste of saffron. In mediaeval Europe, turmeric was known as "Indian saffron." In 1280, Marco Polo described in China "a vegetable that has all the properties of true saffron, as well the smell as the color, and yet it is not really saffron."

Turmeric may have been introduced to Madagascar through early contacts with Indonesia. It apparently reached east Africa in the 8th century AD and west Africa in the 13th century, though as a dye, not a flavoring. It was introduced to Jamaica in 1783.

Today turmeric has declined as a textile dye, but its importance in coloring food may increase with the swing away from synthetic food colors. India is the largest producer and source of the best quality turmeric.

See: Roots and Tubers, p. 69; Natural Fibers and Dyes, p. 312; Plants as Medicines, p. 228

Vanilla *Vanilla fragrans*
Orchidaceae

Natural vanilla comes from fermented and dried fruits of an orchid. The principal source is *Vanilla fragrans*, from tropical Mexico and Central America, but *V. pompona* (West Indian vanilla) and *V. tahitensis* (Tahitian vanilla) are grown in small quantities, despite their inferior quality.

Hand Pollination of Vanilla, Mexico. Copyright Edward Parker, used with permission.

Bernal Díaz, accompanying Cortés on his conquest of Mexico in 1520, described the drink *chocolatl* of the Aztec Emperor Moctezuma, made from cacao seeds and maize, flavored with ground fruits of vanilla. Vanilla was also included in the tribute paid to the Emperor. Its name, *tlilxochtli*, has been translated "black pod," presumably referring to the dark color developed when the fruits are fermented. Fermentation produces the crystals of vanillin responsible for the characteristic smell and flavor. Our name, vanilla, comes from the Spanish *vainilla* (little pod).

By the late 16th century, factories in Spain were making vanilla-flavored chocolate. Hugh Morgan, Elizabeth I's apothecary, gave some dried fruits to Carolus Clusius, who described them in 1605. An early introduction of the plant to England was unsuccessful, but in 1807 an introduction by the Marquis of Blandford flowered in London. Cuttings were sent to Paris and Antwerp, and thence to Indonesia.

Vanilla flowers cannot self-pollinate. Their natural pollinators are stingless bees and hummingbirds, both absent in the Old World. Since pollination is required to set the fruits from which vanilla comes, Mexico monopolized production until the Belgian Charles Morren produced fruits by hand pollination in 1836. (The Totonac Indians in Mexico had discovered hand pollination during the Aztec reign). In 1841, an ex-slave in Réunion, Edmond Albius, developed a method, still in use, which enables one worker to pollinate one to two thousand flowers per day. This, plus failure of the sugar crop, led to commercial cultivation of vanilla on Réunion and other French island colonies. Most vanilla grown in the Old World remains a single clone, descended from the Marquis of Blandford's plant.

Synthetic vanilla was produced in 1874 from conifer wood and in 1896 from clove oil. It is about twenty times cheaper than natural vanilla, lacks the complexities of flavor and aroma, but satisfies over ninety-five percent of world demand. Madagascar produces over seventy-five percent of the world's exports of natural vanilla.

Additional References

Andrews, J. 1995. *Peppers: The Domesticated Capsicums.* Austin: University of Texas Press.

Bedigian, D. 1998. Early history of sesame cultivation in the Near East and beyond. In *The Origins of Agriculture and Crop Domestication,* edited by A. B. Damania, J. Valkoun, G. Willcox, and C.O. Qualset. Aleppo, Syrian Arab Republic: International Center for Agricultural Research in the Dry Areas (ICARDA), 93–101.

Burkill, I.H. 1935. A *Dictionary of the Economic Products of the Malay Peninsula.* London: Crown Agents for the Colonies.

Correll, D.S. 1953. Vanilla—Its botany, history, cultivation and economic import. *Economic Botany* 7, 291–358.

Council of Scientific and Industrial Research (CSIR). 1948–1976. *The Wealth of India: A Dictionary of Indian Raw Materials and Industrial Products.* New Delhi: Publications and Information Directorate, Council of Scientific and Industrial Research.

Dalby, A. 2000. *Dangerous tastes: the story of spices.* London: British Museum.

Dalziel, J.M. 1937. *The Useful Plants of West Tropical Africa.* Westminster: Crown Agents for the Colonies.

Davidson, A. 1999. *The Oxford Companion to Food.* Oxford and New York: Oxford University Press.

Delaveau, P. 1969. *Histoire, description et usage des différents épices, aromates et condiments.* Paris: Albin Michel.

Ecott, T. 2004. *Vanilla: Travels in Search of the Luscious Substance.* London: Michael Joseph.

Greenberg, S. and Ortiz, E.L. (1983). *The spice of life.* New York: Amaisyllis Press.

Guzman, C.C. de, and Siemonsma, J.S. 1996. (Eds.). *Spices. Plant Resources of South-East Asia Handbook 13.* Wageningen: Pudoc.

Hedrick, U.P. (Ed.). 1919. *Sturtevant's Notes on Edible Plants.* Albany, NY: New York Agricultural Experiment Station.

Higton, R.N., and Akeroyd, J.R. 1991. Variation in *Capparis spinosa* L. in Europe. *Botanical Journal of the Linnean Society* 106, 104–112.

Huntingford, G.W.B. (Ed.) 1980. *The Periplus of the Erythraean Sea.* London: Hakluyt Society.

Jacobs, E.M. 1991. *In pursuit of pepper and tea: the story of the Dutch East India Company.* Walburg: Netherlands Maritime Museum.

Jensen, P.C.M. 1981. Spices, *Condiments and Medicinal Plants in Ethiopia, Their Taxonomy and Agricultural Significance.* Wageningen: Centre for Agricultural Publishing and Documentation.

Long-Solís, J. 1986. *Capsicum y Cultura: La Historia del Chilli.* Mexico City, Mexico: Fondo de Cultura Económica.

Miller, J.I. 1969. *The Spice Trade of the Roman Empire: 29 BC to AD 641.* Oxford: Clarendon Press.

Milton, G. 1999. *Nathaniel's nutmeg: how one man's courage changed the course of history.* London: Sceptre.

Morison, S.E. 1963. *Journals and other documents on the life and voyages of Christopher Columbus.* New York: Limited Editions Club.

Norman, J. 2002. *Herb & Spice: the cook's reference.* London: Dorling Kindersley.

Parkinson, J. 1629. *Paradisi in sole paradisus terrestris.* London: Humfrey Lownes and Robert Young.

Pearson, M.N. (Ed.) 1996. *Spices in the Indian Ocean world.* Aldershot: Variorum.

Purseglove, J.W., Brown, E.G., Green, C.L., and Robbins, S.R.J. 1981. *Spices.* Vols. 1 and 2. London: Longman.

Renfrew, J.M. 1973. *Palaeoethnobotany: The Prehistoric Food Plants of the Near East and Europe,* London: Methuen.

Riviera Núñez, D. et al. 2002. Archaeobotany of capers (*Capparis*) (Capparaceae). *Vegetation History and Archaeobotany* 11: 295–314.

Rosengarten, F., Jr. 1969. *The Book of Spices.* Wynnewood, PA: Livingston Publishing Company.

Smartt, J., and Simmonds, N.W. (Eds.). 1995. *Evolution of Crop Plants* (2nd ed.). Harlow: Longman.

Somos, András. 1984. *The Paprika.* Budapest: Akadémiai Kiadó.

Sopher, David E. 1964. Indigenous uses of turmeric (*Curcuma domestica*) in Asia and Oceania. *Anthropos* 59, 93–127.

Uphof, J.C. Th. 1968. *Dictionary of Economic Plants* (2nd ed.). Lehre: J. Cramer, and New York: Stechert-Hafner.

van der Veen, M. with S. Hamilton-Dyer. 1998. A life of luxury in the desert? The food and fodder supply to Mons Claudianus. *Journal of Roman Archaeology 11,* 101–116.

van Harten, A.M. 1970. Melegueta pepper. *Economic Botany 24,* 208–216.

Watt, G. 1908. *The Commercial Products of India,* London: John Murray.

Weiss, E.A. 2002. *Spice Crops.* Wallingford: CABI.

Welch, J.M. 1994. *The spice trade: a bibliographic guide to sources of historical and economic information.* Westport, CT: Greenwood Press.

Willard, P. 2001. *Secrets of saffron: The Vagabond Life of the World's Most Seductive Spice.* Boston, MA: Beacon Press.

Zohary, D., and Hopf, M. 2000. *Domestication of Plants in the Old World* (3rd ed.). Oxford, UK: Oxford University Press.

10

Caffeine, Alcohol, and Sweeteners

HANS T. BECK

How can caffeine, alcohol, and sweeteners have anything in common? Besides being all sourced from plants, there is also a cultural connection that interweaves history, art, and music. Caffeine, for instance, the most taxonomically widespread alkaloid in the plant kingdom, can trace musical connections from Bach's "Coffee Cantata" to Ella Fitzgerald's "Black Coffee," to Dance Hall Crashers' "Java Junkie." Why should musicians from the Baroque period to modern punk rock focus on the attraction to caffeine? Part of the answer lies in the fact that the four most popular addictive substances—sugar, caffeine, tobacco, and alcohol—are all derived from plants. Caffeine is the mildest of the habit-forming drugs, and caffeine use has a side effect in what is called cross-addiction: heavy drinkers of coffee (seven plus cups a day) are often also heavy users of tobacco and alcohol. Picture in your mind's eye, for example, Edward Hopper's painting *Night Hawks*, and see the late-night party-goers retreating from alcoholic cocktails and seeking wakefulness from a cup of coffee spiked with sugar and smoking cigarettes. Or think about one typical post-theatre combination—a brandy, a cup of black coffee or tea, accompanied by a double chocolate torte or mousse.

Sometimes coffee and tea are referred to as stimulant beverages. What is a stimulant? The Merck Manual states, "Central nervous system stimulants are widely used to increase alertness, inhibit fatigue, and suppress the appetite." This definition is useful but fails to provide us with insight into the classifications of narcotic and stimulant plants. In Lewin's classic book on the subject—*Phantastica: Narcotic and Stimulating Drugs*—he organizes plant materials by their use and effect: Euphorica—mental sedatives, Phantastica—hallucination-inducing substances, Inebriantia—alcohol, Hypnotica—soporifics, and Excitantia—stimulants. Some of the plants we will consider in this chapter fall under Lewin's Excitantia and Inebrientia. In their book *Medical Botany: Plants Affecting Man's Health*, Lewis and Elvin-Lewis discriminated between the various types of stimulants, grouping coffee, tea, cola, cocoa, and other beverages as mild stimulants. In contrast, strong stimulants such as coca, tobacco, and ephedra were not included.

In order to grasp the relatedness of these plants' cultural uses, a brief review of the chemistry of mild stimulants is useful. The xanthine group of plant alkaloids contain a nitrogen in a closed ring of atoms. By substituting a methyl group on the core xanthine ring (Table 10.1), three natural methylated xanthines are derived: theophylline (found in tea, *Thea*), theobromine (found in cocoa, *Theobroma*), and caffeine (found in coffee, *Coffea*). These xanthine-related alkaloids are intimately

173

Spreading coffee beans in Brazil, ca.1940. Library of Congress, Prints & Photographs Division, FSA-OWI Collection, LC-USF344-091285-B DLC.

related to one another and are all very weak alkaloids. Although some chemists don't like to ascribe them to the alkaloid class, they are by definition alkaloids, and in physiological action they are addictive. More than sixty plant species throughout the world contain caffeine, ranging between such taxonomically distant plant families as Liliaceae to Asteraceae. Although the Liliaceae and Asteraceae are widespread, the plant families Sterculiaceae, Rubiaceae, Aquifoliaceae, Theaceae, and Sapindaceae lead the list of taxa from which the most culturally significant stimulant beverages are derived.

The physiological reactions of human bodies to the three methylated xanthines are similar. The xanthines competitively inhibit the enzyme phophodiesterase, which results in an increase of one the body's basic energy units, cyclic adenosine monophosphate (cAMP), with a subsequent release of endogenous epinephrine in the blood. This results in the direct relaxation of the smooth muscles of the lungs' bronchi and pulmonary vessels, leading to stimulation of the central nervous system and induction of diuresis, coupled with an increase digestive tract activity and gastric acid secretion, an inhibition of uterine contractions, and topped off with a weak positive

TABLE 10.1 Substituting a Methyl Group at One or More Positions R1, R3, R7 of the Xanthine Ring Produces Mild Stimulants

Chemical Name	Common Name	R1	R3	R7
xanthine	heteroxanthine	H	H	H
1, 3-dimethylxanthine	theophylline	Me	Me	H
3, 7-dimethylxanthine	theobromine	H	Me	Me
1, 3, 7-trimethylxanthine	caffeine	Me	Me	Me

chronotropic and isotropic effect on the heart. In other words, you feel energized, breathing is easier, and you can't help but to run more often to the toilet. After consumption of coffee, tea, or cocoa, demethylated products appear in the blood within three hours, but complete clearance of caffeine and its derivates takes as much as a week. Caffeine has a toxicity rating of 4 on a scale of 10, and symptoms of caffeine poisoning include tachycardia (rapid heart rate), excitement, convulsions, restlessness, tremors, frequent urination, tinnitus, nausea, and vomiting. Six percent of all Americans suffer from chronic caffeine intoxication severe enough to require professional medical treatment.

However, concerns about addiction to caffeine, alcohol, and sugar have not dominated most of man's cultural history. Indeed, we shall learn of a great range of plants serving mankind's desire for beverages and sweeteners. The positive properties of beverages derived from these plants reflect a long tradition of culinary customs, brewing heritage, and technological development. Beverage and sweetener plants have played a central role in the cultures of most peoples, past and present.

Non-Alcoholic Beverages

Tea *Camellia sinensis*
Theaceae

Camellia has eighty-two species, yet only one has a dominant role in various cultures and has risen to dominate worldwide beverage markets. *Camellia sinensis* is an evergreen shrub or tree, kept artificially small by the harvesting and plucking of top, terminal leaf shoots. These leaves, variously processed, contain caffeine (1 to 5 percent) and traces of theophylline, theobromine, and other xanthine alkaloids; however, it is the essential oils that are responsible for the flavors. Polyphenols (5 to 27 percent) are responsible for the dark brown tannin color.

The plant's origin is in China, but its geographic distribution today reflects its cultivation in China and Japan and in countries that were previously colonies of the British Empire, such as Sri Lanka, India, Kenya, and the Carolinas in the United States, as well as Uganda, Turkey, and Indonesia. The plant can grow as far north as 43°N in the Caucasus mountains and as far south as 28°S in Argentina. There are two varieties cultivated in different environments: *C. sinensis* var. *sinensis* is a dwarf tree with small leaves that is grown in the highlands, whereas *C. s.* var. *assamica* is a larger tree with larger leaves that is grown in the lowlands. Tea plants are cross-pollinated, and they are propagated by seeds

Tea pickers in the Himalayas, India, between 1890 and 1923. Library of Congress, Prints & Photographs Division, LC-USZ62-82954.

and cuttings. The harvesting of tea is very labor-intensive: in the lowlands, every 7 days; in the highlands, every 14 days. A mature plant yields about 2 lb green leaf per year, or about 0.5 lb dry weight. One acre yields 1000 lb per annum, and 2 billion lb of black tea are harvested each year.

Three main production processes can occur with the plucked leaves, each of which produces a well-known type of tea; however, the general public tends not to associate the process with the product. Green tea is made from shoots that are steamed (Japanese) or parched (Chinese), rolled, and dried. The heat halts the oxidative degradation process that turns leaves dark-colored, leaving them green and light tasting. While traditionally consumed mostly in Asia, green tea is earning a global reputation for its healthful properties. Pouchong or Oolong tea is made from shoots that are withered in the sun, stored indoors, and pan-dried at a succession of temperatures. This produces the typical curly leaves of this "half-fermented" tea. Oolong tea, typically encountered in Chinese restaurants in Europe and North America, has a relatively minor level of consumption worldwide compared to black tea. Black tea is made from fresh shoots that are first rolled and then crushed; thereafter, the mass is allowed to ferment, which allows oxidation of phenolic compounds, giving the dark, black color. The application of heat ends the fermentation, and the leaves are pan dried. Although India and Sri Lanka produce most of the black tea, it is consumed throughout world. However, there are rather striking regional differences in consumption. These differences are related directly to the story of coffee.

See: Age of Industrialization and Agro-industry pp. 369–71

Coffee *Coffea* spp.
Rubiaceae

Within the large, alkaloid-rich, tropical family Rubiaceae, *Coffea*, with approximately 90 species, shares fame with the notable medicinal and psychoactive genera *Cephaelis* (formally *Cephaelis*) (from which is derived ipecacuanha, used as an emetic and expectorant), *Cinchona* (quinine), *Pausinystalia* (yohimbine, used in a prescription drug to treat male erectile dysfunction), and *Uncaria* (cat's claw, traditionally used in South American folk medicine, and shown to be an immunostimulant). Three economically important species of the *Coffea* family are the major source of stimulant beverages: arabica coffee, *C. arabica* L.; robusta coffee, *C. canephora* (*C. robusta*); and liberica coffee, *C. liberica*. There is a large discrepancy in relative importance of the three coffee species in international commerce: *C. arabica*, with 75.5 percent of the market, is mostly cultivated in tropical America; *C. canephora*, with 24 percent, is produced mostly in Africa; and *C. liberica*, with 0.5 percent, has mostly a small regional market.

Originating in Ethiopian upland forests, these small understory trees of tropical climates produce a fleshy red fruit (technically a drupe, or stone fruit) that has been called "green gold." Coffee is best cultivated in equatorial latitudes, and its current geographic distribution reflects a colonial history. Coffee first spread to India and Ceylon in 16th and 17th centuries, and thereafter spread to the West Indies and South America.

The harvesting of coffee is dependent on the flowering flushes that occur three to four times per year. Hand-picked harvesting is necessary due to the staggered timing of fruit maturation. The icon of Colombian coffee advertisements—Juan Valdez hand-picking every coffee bean—is not just a traditional rural farmer carrying on some quaint tradition; the biological reality of coffee reproduction has not allowed for mechanical harvesting. Interestingly, the primary coffee production is outside of *Coffea's* home range of tropical Africa: Brazil has half of the world market, and Colombia is another large producer. Other smaller or specialty production centers exist in Costa Rica, Mexico, Guatemala, El Salvador, the Antilles, Hawaii, Angola, Tanzania, Uganda, Kenya, and Ivory Coast.

At this point, we must consider coffee leaf rust and why the British drink tea. *Coffea arabica* is susceptible to a serious fungal disease—coffee leaf rust (*Hemileia vastatrix*)—that is native to Ethiopia. This disease spread to Ceylon, but neglect of treatment caused an epidemic in colonial coffee plantations, and the industry collapsed. The British replanted with tea, and to this day, Ceylon teas are

TABLE 10.2 Tea and Coffee Consumption Compared

Consumption Per Capita	United States	United Kingdom
Tea	0.7 lb	10 lb
Coffee	3.0	0.16 lb

most popular in Britain. We see this reflected in consumption per capita statistics (see Table 10.2). Coffee leaf rust is now present in Brazil, and there is concern over the historical fact of coffee's very small genetic base in South America. Only six trees were originators of the massive market. The fungus is being controlled with fungicides but more importantly by hybridizing arabica stock (*C. arabica*) with robusta stock (*C. canephora*); the resulting hybrids are more resistant to rust.

The coffee "bean" is a really a seed, and there are two seeds per fruit. In order to extract those seeds, two types of processing have been developed—a traditional dry process and an industrial wet process. The local, simple dry process involves spreading the fruits in sun and removing dried husk by hand processing. The more common industrial wet process soaks the fruits in water, where the "bad" ones float and the "good" ones sink. The good ones are taken to a depulping machine, where the fruits are gently macerated; afterwards, the mass is fermented, followed by a second rinse. The green seeds are then dried and stored, to await sale to coffee houses and roasters. The roasting of green beans caramelizes the sugars that give the characteristic flavor. Two notable adulterants or substitutes for coffee are made from the ground taproots of chicory (*Cichorium intybus*, Asteraceae) and dandelion (*Taraxacum officinale*, Asteraceae).

The chemical composition of coffee varies remarkably between species and with processing. Green coffee has 0.6 to 3.2 percent caffeine, usually in the range of 1.5 to 2.5 percent, whereas roasted coffee has slightly less caffeine than green: *C. arabica* has 1.5 percent caffeine; *C. canephora*, 2.7 percent caffeine.

Cacao, Cocoa, Chocolate *Theobroma cacao*
Sterculiaceae

Theobroma cacao, one of 22 species, has been cultivated since antiquity and has complex subspecific relationships. Several subspecies have been recognized, with a great number of cultivars developed from them. The two main subspecies are Criollo, *T. cacao* ssp. *cacao* f. *cacao*, and Amazon Forastero, *T. cacao* ssp. *sphaerocarpum*. A hybrid Criollo × Amazon Forastero is called Trinitario. The Criollo

Cacao seeds drying in Chachi canoe, Ecuador. Photo by Hans Beck.

subspecies is further divided with varieties Alligator cocoa, *T. cacao* ssp. *cacao* f. *pentagonum*, and Porcelain Java Criollo, *T. cacao* ssp. *cacao* f. *leiocarpum*.

The earliest evidence of chocolate consumption is based on cacao residue in Mayan drinking vessels dated at about 480 AD. However, scientists believe the pulp was eaten fresh or blended into a beverage well before the roasting process was developed. Carved words and pictures on special pots uncovered in archeological sites suggest that they were used to produce chocolate products for the Mayan nobility. The Aztecs consumed a cold beverage made by combining corn, cacao seeds, water, chilies, and other spices. It was believed to be an aphrodisiac. Spanish colonists in Central and South America eventually refined this bitter beverage by leaving out the chilies, adding sugar, anise, and cinnamon, heating it to improve the texture, and serving it warm. As the demand for chocolate products increased after European colonization, the chocolate tree was cultivated on large plantations. In Europe, only wealthy people were able to enjoy the early chocolate products because the cacao seeds had to be imported. Evidence indicates that cacao seeds were used as currency in the Yucatan until 1850 and were still considered valuable until about 1923.

Truly a gift from the gods, these small trees produce a leathery, American football/rugby ball-shaped fruit (drupe), within which are found seeds covered by sweet, delicious, marshmallow-like arils (the thick fleshy envelope around the seed). However, it is the seeds that are the source of the international commodity. While originally from neotropical flooded rainforests, cacao's primary producers are now Ghana (30 percent), Nigeria (15 percent), and Brazil (20 percent). Interestingly, the primary consumers are the United States (25 percent), Germany (13 percent), the United Kingdom (10 percent), and the Netherlands (9 percent).

Flowering is cauliflorous; that is, flowers are borne on the trunk and stems. The small, delicate, white flowers are pollinated by midges, with a typical success rate of 1 in 500 flowers producing a large drupe. The fruits are harvested at the end of the rainy season, yielding 200 to 2000 lb/acre. Cacao harvesting involves collecting fruits by hand, shelling, seed washing to remove the aril, fermenting, and drying; the dried seeds at this stage are called raw cacao. For factory processing, seeds are roasted to facilitate seed-coat removal and hydraulically shattered into nibs, then ground under pressure and heat. This produces a thick dark paste called "chocolate liquor"—the base for all cocoa products. If the liquor is cooled and hardened, the result is "baking chocolate." If the liquor is subjected again to high pressure, an amber liquid called "cocoa butter" is extruded, and the remaining chocolate press cake is ground to "cocoa powder" (the basis for powdered cocoa beverages). If the cocoa butter is blended with more chocolate liquor, the mixture is on its way to becoming chocolate candy: bittersweet, semisweet, and milk chocolate; white chocolate is mostly cocoa butter. Chocolate is a stimulant, containing theobromine (0.5 to 2.7 percent) and caffeine (0.25 to 1.7 percent).

Carob *Ceratonia siliqua*
Caesalpiniaceae

Carob bean is the fruit of an evergreen long-lived tree with hard woody leaves, which naturally grows on barren, rocky, and dry regions of the Mediterranean basin. It is said that the "locusts" that John the Baptist lived on in the wilderness were carob pods, as "locust bean" is another name for carob bean pods; the carob tree is thus sometimes called St. John's bread. Archeobotanical discoveries in the Middle East show that carob existed in the Eastern Mediterranean basin long before the start of agriculture. Early literary sources indicate that its domestication took place relatively late (only in Roman times). The probable reason for this late date is that the carob does not lend itself to simple vegetative propagation, and its cultivation had to wait until the introduction of scion grafting into the Mediterranean basin. Carob cultivation reached its peak in this region in early Islamic times.

The carob pod is dried, roasted, and, after the seeds are removed, ground to be consumed as a beverage, as a coffee substitute. The ground pods also compete with cocoa powder, and can be made into bars of carob "chocolate." Whole carob pods, with their thick, sweet pulp, were sold as "sweets" to be chewed raw in the United Kingdom during World War II when sweets were rationed, and are also used as animal fodder in the Middle East. As carob does not contain either caffeine or theobromine, it is marketed as a healthy substitute for coffee and chocolate.

Cola, Kola *Cola* spp.
Sterculiaceae

Typically cultivated in tropical West Africa as large trees, there are approximately 125 species of *Cola* known, but only two have entered significantly into U.S. and European markets for their beverage use. The removal of the fleshy red seed coat exposes the embryo, or "kola nut," which is the caffeine-containing seed that is chewed or used as a caffeine source and flavoring for cola drinks. The history of the slave trade triangle (Africa, New World, Europe) plays into this story of the beverage: slaves brought *Cola nitida* (originally distributed from Sierra Leone to Cameroon) to the Caribbean, and *C. acuminata* (abata cola, distributed from Benin to Angola) to Brazil. Other species such as *C. anomala* and *C. verticillata* are also grown in West Africa and are locally commercially important.

Cola propagation is by seeds or cuttings, full seed production begins at 20 years, and trees will produce well until 70 years. Typical yields are about 500 pounds of kola nuts/acre. *Cola nitida* is cultivated extensively in the tropics and is the major source of commercial cola nuts. The primary producers are Nigeria (100,000 tons/year) and Ivory Coast (30,000 tons/year). The main products and use of cola is as a flavoring and stimulant source for carbonated soft drinks and as diet/energy formulas in tablet form. With just trace amounts of theobromine, cola has a good amount of caffeine (1.0 to 2.5 percent), and can be found in many soft drinks such as Coca-Cola and Pepsi-Cola, although synthetic flavorings are now supplanting this use.

See: Nuts, Seeds, and Pulses p. 137; Plants as Medicine pp. 208–9

Guaraná *Paullinia cupana*
Sapindaceae

The beverage with the highest natural caffeine content in the world is made from the roasted seeds of guaraná, a woody climber of tropical Amazonia. The fruits are orange to red capsules containing black seeds partially covered by white arils. The contrast of the colors in the split-open fruit gives them the appearance of eyeballs. Indeed, the origin myth behind guaraná's domestication is attributed to the Sateré-Maué Indians of Brazil, the first consumers of the guaraná beverage, who tell of a malevolent god who lures into the jungle and kills a beloved village child out of jealousy. The village finds the dead child lying in the forest , and a benevolent god consoles them in their grief with a gift in the form of guaraná. The good god plucks out the left eye of the child and plants it in the forest, where it becomes the wild variety of guaraná; the right eye is planted in the village garden, where it sprouts and produces fruits resembling the eye of the child, forever after a pleasant reminder of their favorite but lost child. The Sateré-Maué continue to cultivate guaraná orchards, harvest the large sprays of fruit, extract the seeds, roast them, and form them into smoke-cured sticks (bastao). In preparing traditional guaraná beverage, these sticks are rasped with the hyoid bone of the pirarucu fish, producing a powder that is mixed with water and consumed fresh.

The fame of the caffeine-rich seeds spread throughout the Amazon, and the global demand for guaraná now is primarily supplied by industrial plantation cultivation, where the plants are maintained with a shrub-like habit. The main production centers are in Maués and Manaus, Brazil. Active research and breeding programs seek, by seed or asexual propagation by stem cuttings, to develop and maintain high seed-yielding cultivars with disease resistance. Seed harvest begins after the third year, and continues for up to 80 years, with yields of approximately 125 kg per harvest. Fruit maturation occurs in October–November, and the harvest coincides with the guaraná festival in Maués. The small agricultural town sees its population swell with visitors from throughout Amazonia during one week of festivities, including parades, pageants, folkloric theatre, music, agricultural extension service, and, of course, consumption of guaraná beverages.

The varied uses and products include local hot or cold beverages, carbonated soft drinks, and energy/diet pills. Guaraná is claimed to be the national drink of Brazil, where more than 17 million bottles per day are consumed. While the limited Amazonian production restricts international exports, guaraná is showing up in international markets in tea powder, herbal mixes, and expensive, specially-crafted, high-energy/high-caffeine drinks. Its use in such liquid products is due to its high levels of caffeine (4 to 7.5 percent) and traces of theobromine and theophylline.

Yerba maté, Brazilian tea, Paraguay tea *Ilex paraguariensis*
Aquifoliaceae

The stimulant beverage yerba maté is derived from the leaves of medium-sized trees grown in subtropical South American plantations. Primary production is centered in southern Brazil, Paraguay, and Argentina. The harvest involves clipping leaves, drying them, and grinding the dried leaves in order to produce a powder. Maté production is mostly regional, and international exports are limited, restricted to herbal tea mixes like "Morning Thunder." However, maté is the national drink of Paraguay, Uruguay, and Argentina, where it enjoys a social and cultural status rivaling that of tea and coffee in its depth of custom. The dried leaves are sold in bulk under many brand names, packaged typically in half-kilogram bags.

Aficionados of this beverage have their brand preferences, beverage recipes, and special utensils. The paraphernalia needed to consume hot infusions of yerba maté are a *maté* (a traditional cup made from a gourd or horn), a *bombilla* (a spoon-shaped straw with built-in strainer), a hot water bottle or pot, and a pouch to carry one's yerba maté supply and utensils. The maté is filled with *yerba maté* and then covered with hot water. The liquid is sucked through the bombilla's strainer, leaving behind the leaves in the cup, which is filled multiple times with water. The consumption of the beverage, with its caffeine level of 2 to 2.5 percent, is a social affair, with groups of people, college students especially, often spending leisurely hours in conversation, sharing matés and passing the hot water containers. The utensils are often decorated, sometimes highly filigreed with silver, gold, and jewels.

Alcoholic Beverages and Beverage Flavorings

Absinthe, Wormwood *Artemisia absinthium*
Asteraceae

Known since 77 AD in the time of Pliny, this aromatic herb was often used for its vermifugic (deworming) and stimulant properties, the latter due to a psychoactive ketone, thujone. The genuine absinthe beverage was an expensive liqueur prepared by distilling alcohol in which the leaves of *Artemisia absinthium* were soaked. The drink is especially associated with the French distillery Pernod Fils and was popular with the Impressionist painters. A huge demand for the beverage among the wealthy and the intelligentsia of western Europe in the late 19th century led to a craze for absinthe in the lower working class. Imitation absinthe, containing harmfully high levels of neurotoxic antimony and copper sulfate that produced a syndrome labeled "absinthism" soon flooded the market. Public outcry against all absinthe use led to bans in Europe and the United States by 1912.

African grain beers, Betso, Buza
Poaceae (Gramineae)

In Africa, much finger millet (*Eleusine coracana*), pearl millet (*Pennisetum glaucum*), sorghum (*Sorghum bicolor*), and African rice (*Oryza glaberrima*) is used to make beer. Finger millet's amylase enzymes readily convert starch to sugar, having a saccharifying power second only to barley, the world's premier beer grain. The superb productivity and yield of the millets and sorghum under a variety of environmental conditions provides local peoples and industrial entrepreneurs with plant resources to brew excellent African beers.

African beers are but one of a range of cereal beverages, many non-alcoholic, that are the result of lactic acid bacteria and yeast fermentation. Fermentation has many nutritional advantages, including increased availability of protein, amino acids, minerals and vitamins (especially certain B-group vitamins).

Agave, Pulque, Mescal, Tequila *Agave* spp.
Agavaceae

Various species of *Agave*, short-stemmed succulents with a fleshy leaf-base and trunk, have been used as the source of pulque, or agave beer, made from the fermented sap. This beverage is the national drink of Mexico. The beverages mescal and tequila, distillates of fermented pulque developed in Mexico, have been produced only since the Spanish introduced the practice of distillation. Modern operations use cooked stem hearts, crushing them with water to make a mash, which is then fermented. Like the terms scotch, pilsener, münchner, champagne, port, and so forth, "tequila" is a place name applied to a beverage, in this case a mescal brandy. This town, in the Mexican state of Jalisco, contains modern factories that produce the best brand of mescal.

The maguey or *Agave americana*, indigenous to Florida, Mexico, and other parts of tropical America, is the commonest source of pulque. Other species used are *A. tequilana*, *A. angustifolia*, and *A. palmeri*. From time immemorial the maguey has been cultivated for the abundant sap, which collects in the cavity made in the heart of the plant by removal of the young central leaves. The juice is rich in sugar when the maguey is about to flower, and a natural fermentation process can take place within the plant. Pulque is thick, milky, and slightly sweet. The smell can be off-putting but the liquid is refreshing and nutritious (rich in vitamins), and usually contains three to four percent alcohol.

Anise *Pimpinella anisum*
Apiaceae

Cultivated since 2000 BC in Greece and Egypt as a flavoring for food and drink, anise is used in alcoholic beverages (anis, anisette, ouzo, pastis, raki, and absinthe).

Angostura bitters *Angostura trifoliata*
Rutaceae

The bitter, alkaloid-rich bark of this tropical American tree produces angostura, which is used in pink gins and other beverages.

Apple cider *Malus domestica*
Rosaceae

Apples, the most popular temperate fruit tree crop worldwide, have ancient and complex origins, with many cultivars today. There is evidence from the Neolithic and Bronze Age lake-dweller cultures of Switzerland that "wild" apples such as *M. sylvestris* were exploited.

In addition to being eaten fresh or cooked, the fruit is used for its juice, which is sometimes fermented to "hard" cider or distilled to apple brandy. By the 12th century cider was a popular drink in France. In the United Kingdom, the process of cider-making was introduced early from Normandy, and today 70% of apple juice is fermented to apple cider. Colonists introduced cider-making to the United States in the 1600s. Although cider itself is popular, carbon dioxide given off during fermentation may also be harnessed in a secondary bottle fermentation to give a naturally sparkling cider (*cidre bouché*); carbon dioxide may also be artificially added to keg cider. Southwestern England, northwestern France, and northern Spain are centers of cider production.

Fruit Wines, Brandies, and Liqueurs

Brandies are usually distilled wines, but can also be distilled from cider. The most famous wine brandies are those from Armagnac and Cognac in France. The exquisite flavors of those brandies are due to the wines from which they are distilled and the fact that they are carefully matured for many years after distillation. Many distilled wines are labeled as brandies with a qualifying adjective indicating the kind of fruit from which they were distilled. Some, such as kirsch (cherry brandy), have special names. Applejack or calvados are apple brandies distilled from cider.

Liqueurs and cordials differ from brandies in that sugar or syrup or both added to the distilled liquid. Liqueurs also contain characteristic flavors. The flavorings used include leaves, roots, herbs, fruits, and barks. Chartreuse, a fine liqueur that has been made since 1605 by Carthusian monks in France and Spain, is rumored to contain 130 different flavoring agents.

Barley in beer and whisky *Hordeum vulgare*
Poaceae

As in the case of wine, no one knows when people first began to brew beer, but the practice was well established by the beginning of recorded history. Beer production was well-established in Mesopotamia and Egypt from least 3000 BC. Analysis of beer from tomb vessels has shown that beer production in ancient Egypt was sophisticated. Malted emmer wheat or barly grains were mixed with ground grains that had been heated in hot water. The enzymes from the malted grains broke down the starch in the resulting mixture, which was then fermented by yeast and bacteria to give a product perhaps similar in character to traditional African beers. The idea that Egytian beer was made from bread is not supported by recent research.

Barley is now mostly used in malt-based beverages. When germinated in water and kiln-dried, barley can be used as a substrate for yeast in beer. Beer was first successfully bottled in 1736, and is now one of the world's most popular beverages. Over 29 million pints per day were consumed in the United Kingdom in 1988, equivalent to 108 liters per head per annum, more than ten times the consumption of wine.

Whisky, distilled from malted barley, was first recorded in Scotland in 1494 by a friar buying malt to make whisky. Now over four million bottles a day are made in Scotland. Barley is the preferred grain for malting. Nonetheless, grains of rye, *Secale cereale*, are rich in gluten and used to make whisky in the United States, rye beer in Russia, and gin in the Netherlands. Wheat grains, *Triticum aestivum*, are fermented to produce "weiss" or white beer, a beverage typical of Germany, and are also distilled to vodka, which is typical of Russia.

Bog myrtle, Sweet gale *Myrica gale*
Myricaceae

The leaves of this nitrogen-fixing shrub, a common plant of wet heaths of North America, northwestern Europe, and northeastern Siberia, are used to flavor and improve the foaming of beer.

Grapes in wine *Vitis vinifera*
Vitaceae

Grapes are woody vines cultivated for their fruit, which is eaten fresh, dried, or drunk as a juice or fermented. By definition, wine is any fermented fruit juice; however, in practice, the term "wine" is overwhelmingly used for the fermented juice of grapes. The simple act of collecting ripe grapes and merely bruising or crushing them can cause fermentation to occur due to the naturally occurring yeast on the fruit skin. Although some scientists think it is possible that man began making wine as early as 8000 BC, the first concrete evidence of wine making comes from residues found in clay vessels from western Iran that have been dated to 5500 years ago.

Cultivated clones probably arose in southwest Asia and have been grown since the 4th millennium BC in Syria and Egypt and since 2500 BC in the Aegean region. The Egyptians used wine primarily for religious ceremonies, but it was only between 2000 and 1000 BC that wine became a popular beverage in Greece. The cultivation of wine grapes spread from the eastern Mediterranean to France at roughly about 600 BC, and later to Spain, Portugal, and Algeria. Columbus introduced plants into the West Indies on his second voyage, and the Spanish began cultivation in California around 1769. By middle of the 19th century, viticulture had established a special foothold in California. Australia, the United States, Argentina, and South Africa are now among the top wine-producing countries.

The spread of grape vines to the New World saved the European wine industry from collapse after an outbreak of *Phylloxera* rootlouse in 1867 that devastated European orchard stock. The American species were reintroduced and used as resistant stocks, especially *V. labrusca*. There are dozens of classical cultivars of *V. vinifera*—for example, the reds pinot noir and cabernet sauvignon, and the whites chardonnay and riesling—and many are grown for wine in the major European wine-growing nations of France, Spain, Italy, and Germany.

Grape wine has been associated with religion and other ceremonial acts since antiquity. The ancient Greeks drank diluted wine, and wine is part of the Christian sacrament. The consumption of wine beverages has become an intrinsic feature of Western society. Grape wines are fortified with brandy or other alcoholic beverages in order to make sherry, port, and Madeira, and the distillation of grape wine produces brandy. With added herbs and spice flavorings, grape wine produces vermouth and martinis.

Hops *Humulus lupulus*
Cannabaceae

The female flowers of the hop plant are the most widely used flavoring agent in beer, in which it is important both for the bitter resins that balance the sweet taste of malt, and for the essential oils that enhance the aroma.

Wild hops are climbing plants found in fens and riverbanks in Europe. Archaeological evidence suggests that wild hop flowers were first used in brewing in the early Middle Ages, from about 700 AD. Cultivation of hops is well documented in historical sources in Germany from about 850 AD. Hops were not used in ancient Egypt, and were rarely used in southern Europe, which is primarily a wine-drinking region. Hops quickly became the dominant beer additive of central Europe and were widely traded, but in areas of northwest Europe where sweet gale fruits (*Myrica gale*) were collected and used, there was resistance to the use of hops. Hops were not widely used by English brewers until the late 15th century, but in time a series of beer protection laws led to the use of sweet gale being banned in Germany and other countries.

The hop plant is dioecious; that is, the male and female flowers are borne on separate plants. The female flowers are borne in dense cone-like clusters. Each cone contains numerous leaf-like bracteoles, and at the base of each bracteole there are many small lupulin glands containing resins and essential oils. The resins are responsible for the bitterness of hops, and are made up of a number of alpha and beta acids, including humulone and lupulone. The essential oils contribute to the aroma of the beer. Hops also have antimicrobial, preservative qualities, and the better traveling and keeping qualities of hopped beer were an important factor in its displacement of sweet gale as a flavoring agent.

In cultivation, hops were trained onto poles or (nowadays) onto wires. Dwarf forms of hop that are easier to harvest are becoming increasingly important. Hops are locally important crops in several temperate areas, including the western United States, the southeast and Midlands of Britain, and Germany and Czechoslovakia. Around 50 percent of the world hop harvest is used in extract form, in which the flavoring compounds are extracted with ethanol or liquid carbon dioxide.

Juniper *Juniperus communis*
Cupressaceae

The common juniper, a shrub of northern temperate zones, produces a sweet aromatic fruit used for flavorings, especially in gin and liqueurs. Over 200 tons, collected in wild places in Central Europe, are imported annually into the United Kingdom.

Maize in beer, Chicha, Bourbon *Zea mays*
Poaceae

The majority of uses of maize, a cultigen with a complex genetic history, are for food and industrial products. However, various traditional peoples in the Andes and Mexico continue to produce a beverage from the food form that is both nutritious and slightly alcoholic. Chicha, or maize beer, is produced by fermenting hydrolyzed cornstarch. Cornstarch cannot be directly fermented, and must first be enzymatically converted to sugar, which can be fermented. Central and South American Indians learned that chewing corn kernels and spitting the quids into the corn mash produced an alcoholic beverage. This action introduces a salivary amylase that hydrolyzes, or breaks down, the starch into sugar.

Maize is an important adjunct in modern brewing operations, and overcoming this problem of starch conversion is critical to industrial fermentation. Cornstarch is hydrolyzed to yield the corn syrup sugars, glucose and fructose. Sour mash made from corn is a major ingredient in the production of bourbon.

See: Grains, p. 54

Manioc, Cassava, caxiri *Manihot esculenta*
Euphorbiaceae

Known by its more popular food product names, yucca, cassava, farinha, manioc, and tapioca, the tubers of *M. esculenta* yield starch and sugar, which are produced into alcoholic beverages called kaschiri. Cassava is a shrubby tree with large tuberous roots that are rather immune to insect attack because of the high levels of cyanide in the tuber skin. There are many cultivars with differing amounts of cyanide. The poison is removed by squeezing the grated and ground tubers in water and then heating or evaporating the products. The starchy juice pressed out of ground cassava is fermented, or it is chewed in the starchy form, which aids the change into sugar.

See: Roots and Tubers, pp. 68–9

Palm wine and arrack *Arenga, Borassus, Corypha, Nypa,* and *Phoenix* spp.
Arecaceae

The inflorescences of palms, when scratched with a knife, exude a sugary sap. This sap is copious due to the large size of palms, and humans learned to gather this sweet liquid to produce jaggery sugar and alcoholic beverages. Naturally occurring yeasts in the environment start fermentation almost as soon as the cut is made. The sap of numerous palms, including date, *Phoenix* dactylifera; palmyrah, *Borassus flabellifer*; talipot, *Corypha* spp.; sugar palm, *Arenga* spp.; and nipa, *Nypa fruticans,* are harvested to make fermented toddy or palm wine. Arrack is a potable spirit distilled from toddy or the sugary sap of palms.

Rice beer, Sake, Mirin *Oryza sativa*
Poaceae

The principal carbohydrate of Asia, rice, is fermented and brewed to beer, and eventually to sake. Sake is the traditional beverage of Japan. The fermentation process is distinct from regular malting, which uses the yeast *Saccharomyces cervisae*, in that the rice starch is first converted to sugar by the fungus *Aspergillus. Aspergillus* produces enzymes to break down starch and *Saccharomyces* ferments the sugars. The final alcoholic content of sake can reach 18 percent. Before being marketed, sake is filtered and pasteurized. Most sake should be consumed within one year of its production. The consumption of sake, traditionally served heated, is unlike that of most other alcoholic beverages.

Traditional Sweeteners

Unlike high-intensity sweeteners (often referred to as low-calorie sweeteners; see the following discussion), which are proteins or other compounds, there are two traditional sweetener sources—manna and sugars—which are both natural carbohydrates and which have long been exploited for their sweetness. For alcoholic beverages, sugars (e.g., glucose, sucrose, and fructose) are the dominant source of carbohydrates for fermentation; manna cannot be fermented with yeast.

"Manna from heaven": the biblical phrase evokes an edible, usually sweet, material. Mannas are of plant origin, often exudates following insect punctures of the stem, and are composed of syrupy secretions that set into hard sugar. (Honeydew is of insect origin, secreted by aphids and scale insects onto the leaves of shrubs.) The manna of the Bible could have been from *Haloxylon (Hammada) salicornicum* (Chenopodiaceae) or a lichen. Modern commercial sources include the manna ash, *Fraxinus ornus* (Oleaceae), collected only in Sicily, and exudates collected from other plant species such as *Larix decidua* (Pinaceae) and *Tamarix* spp. (Tamaricaeae).

Date palm *Phoenix* dactylifera
Arecaceae

Fruits of these palms are an important source of sugar. Dates are a staple of nomadic peoples of Arabia and North Africa, and are dried or preserved with sugar and used as a dessert fruit. Evidence suggests that the date palm has been cultivated in Arabia and northeast-Saharan Africa since 4000 BC. The date palm *Phoenix dactylifera* is a cultigen thought to be a spontaneous hybrid with *P. reclinata* and the African wild date, *P. sylvestris.*

In India, wild date palm is an important source of palm sugar and toddy. The fruit is composed of 60–70 percent sugar. Each palm tree can produce up to 700 kg of fruit annually. Date palm is now a common crop in Iraq, North Africa, and California.

See: Fruits, pp. 83–4

Nipa palm *Nypa fruticans*
Arecaceae

With its creeping rhizomes and fronds reaching up to 10 m long, this palm, characteristic of mangrove swamps, is originally native to a region extending from India to the Ryuku and Solomon Islands; but it is now naturalized in western Africa and Panama. The sap from the flower stalks is tapped for sugar. The famous "three palms pudding" uses coconut milk, sago palm, and gula melaka, a sugar boiled from the nipa palm sap.

Palmyra palm, Borassus palm *Borassus flabellifer*
Arecaceae

From India to Burma, the Palmyra palm is cultivated for many uses: 801 uses are noted in an ancient Tamil poem. Its inflorescence is tapped for toddy from which sugar (jaggery), vinegar, and other sugar-based products are prepared. Over 120,000 liters of toddy can be produced by one palm during its lifetime.

Sugar palm, Gomuti palm *Arenga pinnata*
Arecaceae

This Malaysian palm is widely cultivated. The male inflorescences are cut off and tapped for the syrupy sap, which is evaporated to produce palm sugar (jaggery), fermented to produce palm wine (toddy, and distilled into arrack (see the preceding discussion).

Sugar beet *Beta vulgaris*
Chenopodiaceae

First grown in Medieval Europe, these forms of beet with thickened roots, derivatives of the wild sea-beet, are grouped as *B. vulgaris* ssp. *Vulgaris*. Sugar beet is a biennial with sugar reserves in the root forming up to 20% of the weight. Napoleon realized the economic value of sugar beet as a domestic sugar source, and he ordered research to be conducted. Crystallized sugar is obtained from processing harvested sugar beets. The first factory for beet-sugar extraction was set up in Silesia in 1801. Today, sugar beets provide raw material for most of the sugar production in Europe.

See: Age of Industrialization and Agro-industry pp. 368–9

Sugar cane *Saccharum officinarum*
Poaceae

The source of half the world's sugar, sugar cane is a large perennial grass that stores its photosynthetic reserves as sugar. These "noble" canes were domesticated by indigenous peoples in New Guinea. There the modern cultigen arose from hybridization with other *Saccharum* and grass species.

The process of producing unrefined sugar has been known in the Far East and India for several millennia. The Ottoman Turks produced refined white sugar in the 1300s. The Crusaders took sugar cane from Israel to various Mediterranean islands and Iberia. From this launching point, it spread rapidly as colonists traveled to Madeira and the Canary Islands, then to the West Indies and the Americas, finally arriving in the islands of the Indian Ocean. Now grown throughout the tropical world, especially in the West Indies and Hawaii, almost all cultivars are derived from *S. officinarum*, a complex aggregate of hybrids. Nearly all sugar cane is vegetatively propagated by means of stem cuttings. One of the reasons for the successful establishment of the cane industry worldwide is that this method of propagation allows for the harvesting of one crop per year.

Sugar-cane cultivation under plantation systems has done much to destroy original native vegetation and has been responsible for the transmigration of peoples from Africa to the Caribbean, from China to Central America, and from Polynesia to Australia. The famous American and British raw sugar-rum-slavery triangles of the 1700s involved New England and England, Africa, and the West Indies: raw sugar and molasses was shipped from Caribbean islands to northern cities, where it was made into rum, which was sent to Africa to buy more slaves for the sugar-cane plantations.

Stems are crushed to extract the sweet sugar-containing liquid. In many countries, the fresh juice or a simple fermented juice is consumed within 48 hours of crushing. The process of refining sugar cane involves boiling down the juice until crystallization occurs, and produces various grades of sugar. The uncrystallizable sugar by-products—molasses or treacle—are fermented and distilled into rum, aguardiente, or cachaça.

Harvesting sugarcane in Cuba, 1940. Library of Congress, Prints & Photographs Division, FSA-OWI Collection LC-USF344-091312-B DLC.

See: Age of Industrialization and Agro-industry pp. 366–8

Sugar maple *Acer saccharum*
Aceraceae

Sugar maple is a forest tree of southeastern Canada and the northeastern United States. Native Americans developed the technology for making maple syrup and sugar, and this technology was adopted by early European settlers. Every spring, as the ground thaws, the trunk is tapped by boring a hole and placing a spigot or tube to collect the rising sap. It takes about 40 gallons of sugar maple sap to boil down and produce one gallon of maple syrup. Massachusetts alone used to produce 200,000 kg per annum. The syrup is used in a variety of confections as well as a traditional syrup flavoring for pancakes and baked goods.

High-Intensity Sweeteners

The search for sweet-tasting plants and their extracts has a long history, but the most intensive search has occurred in the last 75 years. Several plant-derived compounds of the protein, terpenoid, and phenolic types have commercial use as sweeteners. Most of these new compounds are prototype "high-intensity" sweeteners ranging from 30 to 2,000 times sweeter than sucrose that may be worthy targets for chemical synthesis or for semisynthetic modification to produce substances with enhanced sweetness properties.

Curculin *Molineria latifolia*
Hypoxidaceae

The sessile palm-like plant from Indo-Malaysia has fruits that are tasteless but which sweeten the taste of drinks for one hour after consumption. It is an unstable compound, and it remains commercially useless at this time.

Hernandulcin *Phyla dulcis*
Verbenaceae

Known to the Aztecs as a sweetening agent, *P. dulcis* is a sweet herb endemic to tropical America. The intensely sweet sesquiterpenoid hernandulcin (800 times as sweet as sucrose) accounts for this characteristic.

Licorice root, Glycyrrhizin *Glycyrrhiza glabra*
Fabaceae

Many species in *Glycyrrhiza* are local sources of licorice, but the rhizome of the cultivated Mediterranean and Central Asian species *G. glabra* is the source of licorice used in confectionery, cough medicines, and lozenges, and in the brewing of stout. The oligoglycoside glycyrrhizin is 50 times sweeter than sugar, but it has a licorice aftertaste that limits widespread use.

Monellin *Dioscoreophyllum cumminsii*
Menispermaceae

The fruit (called serendipity berry) of this West African species yields a natural protein sweetener, monellin, which is nine thousand times as sweet as sucrose. Monellin has been tried in low-calorie foods and beverages.

Miraculin, Miraculous berry *Synsepalum dulcificum*
Sapotaceae

The fruits of this West African species cause sour and salt things to taste sweet by affecting the taste buds. Miraculin is a glycoprotein that also depresses the appetite.

Phyllodulcin, Amacha *Hydrangea macrophylla* ssp. *serrata*
Hydrangeaceae

The sweet taste of the leaves of this shrub derives from phyllodulcin, an organic nonnutritive sweetener that accumulates in the leaves. In Japan and Korea, the steamed leaves of *H. macrophylla* ssp. *serrata* are ingredients in the beverage amacha.

Stevioside, Caa-ehe *Stevia rebaudiana*
Asteraceae

Long used by southern South American indigenous people for sweetening beverages, the leaves of this shrub contain stevioside, a diterpene glycoside that is up to 300 times as sweet as sucrose. Today, *S. rebaudiana* is under intensive vegetative reproduction for the production of stevioside, which is much used in Japan as a sweetener, but is not approved for use as a sweetener or food additive in the European Union or the United States.

Thaumatin *Thaumatococcus daniellii*
Marantaceae

The thaumatins are a class of intensely sweet proteins (1,600 times as sweet as sucrose) isolated from the aril of the fruit of the tropical west African species *Thaumatococcus daniellii*. Thaumatin is approved for use in many countries, and it serves both as a flavor enhancer and a high-intensity sweetener. The supply of naturally occurring thaumatin is limited, which has prompted extensive research into its synthesis via transgenic organisms. The gene encoding thaumatin has been introduced into various microorganisms. The unique properties of thaumatin as a food additive could well be exploited by the food and beverage industry. Or, we might yet see the thaumatin gene engineered directly into fruit and vegetable crops to improve their flavor and sweetness.

References and Further Reading

Non-Alcoholic Beverages

Batlle, I., and Tous, J. 1997. *Carob tree*. Ceratonia siliqua *L. Promoting the Conservation and Use of Underutilized and Neglected Crops 17*. Rome: International Plant Genetics Resources Institute.
Braun, S. 1996. *Buzz: The Science and Lore of Alcohol and Caffeine*. New York: Penguin Press.
Brücher, H. 1989. *Useful Plants of Neotropical Origin and Their Wild Relatives*. Berlin: Springer-Verlag.

Chatt, E.M. 1953. *Cocoa*. New York: Interscience Publishers.
Cheney, R.H. 1947. The biology and economics of the beverage industry. *Economic Botany 1*, 243–275.
Coe, S.D., and Coe, M.D. 1996. *The True History of Chocolate*. New York: Thames and Hudson.
Coste, R. 1992. *Coffee: The Plant and the Product*. London: Macmillan.
Dicum, G., and Luttinger, N. 1999. *The Coffee Book: Anatomy of an Industry from Crop to the Last Drop*. New York: New Press.
Eden, T. 1976. *Tea*. London: Longman.
Erickson, H.T., et al. 1984. Guarana (*Paullinia cupana*) as a commercial crop in Brazilian Amazonia. *Economic Botany 38*, 275–286.
Gutman, R.L., and Ryu, B.-H. 1995. Rediscovering tea. *Herbalgram 37*, 33–48.
Ham, V. van der, and Wessel, M. 2000. *Stimulants*. Leiden: Backhuys.
Harler, C.R. 1964. *The Culture and Marketing of Tea*. London: Oxford University Press.
Hill, A.F. 1952. *Economic Botany: A Textbook of Useful Plants and Plant Products*. New York: McGraw-Hill Book Co.
Hobhouse, H. 1999. *Seeds of Change: Six Plants That Transformed Mankind*. London: Papermac.
Lewis, W.H., and Elvin-Lewis, M.P.F. 1977. *Medical Botany: Plants Affecting Man's Health*. New York: John Wiley and Sons.
Marakis, S. 1996. Carob bean in food and feed—current status and future potentials: a critical review. *Journal of Food Science and Technology (Mysore) 33*, 365–383.
Meyer, F.G. 1965. Notes on wild *Coffea arabica* from southwestern Ethiopia, with some historical considerations. *Economic Botany 19*, 136–151.
Moxham, R. 2003. *Tea: Addiction, Exploitation, and Empire*. London: Constable.
Porter, R.H. 1950. Mate: South American or Paraguay tea. *Economic Botany 4*, 37–51.
Purseglove, J.W. 1974. *Tropical Crops: Dicotyledons*. New York, Wiley.
Raffauf, R.F. 1996. *Plant Alkaloids: A Guide to Their Discovery and Distribution*. New York: Food Products Press/The Haworth Press, Inc.
Schapira, J., et al. 1975. *The Book of Coffee and Tea*. New York: St. Martin's Press.
Schivelbusch, W. 1992. *Tastes of Paradise*. New York: Pantheon.
Schultes, R.E., and Raffauf, R.F. 1990. *The Healing Forest: Medicinal and Toxic Plants of the Northwest Amazonia*. Portland, OR, Dioscorides Press.
Simmonds, N.W. (Ed.). 1976. *Cocoa Production*. New York: Praeger.
Simpson, B.B., and Conner-Ogorzaly, M. 1986. *Economic Botany: Plants in Our World*. New York: McGraw-Hill Book Co.
Sivetz, M., and Desrosier, N.W. 1979. *Coffee Technology*. Wesport, CT: AVI Publishing.
Stone, D. (Ed.). 1984. *Pre-Columbian Plant Migration*. Cambridge, MA: Harvard University Press.
Stone, T., and Darlington, G. 2000. *Pills, Potions, and Poisons*. New York: Oxford University Press.
Tippo, O., and Stern, W.S. 1977. *Humanistic Botany*. New York: W.W. Norton.
Ukers, W.H. 1935. *All about Coffee*. New York: The Tea and Coffee Trade Journal Co.
Ukers, W.H. 1935. *All about Tea*. New York: The Tea and Coffee Trade Journal Co.
Ukers, W.H. 1936. *The Romance of Tea*. New York: Alfred Knopf Co.
Uribe, C.A. 1954. *Brown Gold*. New York: Random House.
Urquhart, D.H. 1955. *Cocoa*. New York: Longmans, Green Company.
Van Hall, C. J.J. 1914. *Cocoa*. London: Macmillian.
Weinberg, B.A., and Bealer, B.K. 2001. *The World of Caffeine: The Science and Culture of the World's Most Popular Drug*. London: Routledge.
Wellman, F.L. 1961. *Coffee—Botany, Cultivation, and Utilization*. London: Leonard Hill Books Ltd.
Willson, K.C., and Clifford, M.N. 1992. *Tea: Cultivation to Consumption*. London: Chapman and Hall.
Willson, K. 1999. *Coffee, Cocoa, and Tea*. Wallingford, Oxon: CABI Publishing.
Wood, G.A.R., and Lass, R.A. 1985. *Cacao*. New York: Longman Scientific and Technical.
Wrigley, G. 1988. *Coffee*. New York: Longman Scientific and Technical.
Young, A.M. 1994. *The Chocolate Tree: A Natural History of Cacao*. Washington, D.C.: Smithsonian Institution Press.
Zohary, D. 2002. Domestication of the carob (*Ceratonia siliqua* L.). *Israel Journal of Plant Sciences 50*, S141–S145.

Alcoholic Beverages and Beverage Flavorings

Bahre, C.J., and Bradbury, D.E. 1980. Manufacture of mescal in Sonora, Mexico. *Economic Botany 34*, 391–400.
Beadle, J. 1980. The ancestry of corn. *Scientific American 242*, 112–119.
Behre, K.E. 1999. The history of beer additives in Europe—A review. *Vegetation History and Archaeobotany 8*, 35–48.
Board on Science and Technology for International Development (BOSTID)/National Research Council. 1996. *Lost Crops of Africa: Volume 1: Grains*. Washington, D.C.: National Academy Press.
Boulton, N., and Heron, C. 2000. The chemical detection of ancient wine. In *Ancient Egyptian Materials and Technology*, edited by P.T. Nicholson and I. Shaw. Cambridge: Cambridge University Press, 599–602.
Braidwood, R.J., et al. 1953. Did man once live by bread alone? *American Anthropologist 55*, 515–526.
Campbell-Platt, G. 1987. *Fermented Foods of the World: A Dictionary and Guide*. London: Butterworths.
Chang, T.T. 1995. Rice. In *Evolution of Crop Plants,* edited by J. Smartt and N.W. Simmonds. London: Longman Scientific and Technical, 147–155.
Corran, H.S. 1975. *A History of Brewing*. Newton, MA: David and Charles Abbot.
Dirar Wallingford, H.A. 1993. *The Indigenous Fermented Foods of the Sudan: A Study in African Food and Nutrition*. Wallingford: CAB International.

Hornsey, I.S. 1999. *Brewing*. Cambridge: Royal Society of Chemistry.
Hornsey, I.S. 2003. *A History of Beer and Brewing*. Cambridge: Royal Society of Chemistry.
Johnson, H. 1978. *The World Atlas of Wine: A Complete Guide to Wines and Spirits of the World*. New York: Simon and Schuster.
Johnson, H. 1989. *The Story of Wine*. New York: Simon and Schuster.
La Barre, W. 1938. Native American beers. *American Ethnologist 40*, 224–234.
McGovern, P. 2003. *Ancient Wine: The Search for the Origins of Viniculture*. Princeton, NJ: Princeton University Press.
McGovern, P., Fleming, S.J., and Katz, S.H. (Eds.). 1996. *The Origins and Ancient History of Wine*. Amsterdam: Gordon and Breach.
Murray, M.A. 2000. Viticulture and wine production. In *Ancient Egyptian Materials and Technology*, edited by P.T. Nicholson and I. Shaw. Cambridge: Cambridge University Press, 577–608.
Neve, R.A. 1991. *Hops*. London: Chapman and Hall.
Neve, R.A. 1995. Hops. In *Evolution of Crop Plants*, edited by J. Smartt and N.W. Simmonds. New York: Longman Scientific and Technical, 33–35.
Olmo, H.P. 1995. Grapes. In *Evolution of Crop Plants*, edited by J. Smartt and N.W. Simmonds. New York: Longman Scientific and Technical, 485–490.
Price, R. 1954. *Johnny Appleseed: Man and Myth*. Bloomington, IN: Indiana University Press.
Protz, R. 1995. *The Ultimate Encyclopedia of Beer: The Complete Guide to the World's Great Brews*. New York: Smithmark.
Robinson, J. (Ed.). 1994. *The Oxford Companion to Wine*. Oxford: Oxford University Press.
Rose, A.H. (Ed.) 1977. *Alcoholic Beverages*. New York: Academic Press.
Samuel, D. 2000. Brewing and baking. In *Ancient Egyptian Materials and Technology*, edited by P.T. Nicholson and I. Shaw. Cambridge: Cambridge University Press, 537–576.
Tomlan, M.A. 1992. *Tinged with Gold: Hop Culture in the United States*. Athens, GA: University of Georgia Press.
Valenzuela-Zapata, A.G., and Nabhan, G.P. 2003. *Tequila: A Natural and Cultural History*. Tucson: University of Arizona Press.
Watkins, R. 1995. Apple and pear. In *Evolution of Crop Plants*, edited by J. Smartt and N.W. Simmonds. London: Longman Scientific and Technical, 418–422.
Webb, A.D. 1984. The science of making wine. *American Scientist 72*, 360–367.

Traditional Sweeteners

Balick, M.J., and Beck, H.T. 1990. *Useful Palms of the World: A Synoptic Bibliography*. New York: Columbia University Press.
Donkin, R.A. 1980. Manna: An Historical Geography. The Hague: W. Junk.
Galloway, J.H. 1989. *The Sugar Cane Industry: An Historical Geography from Its Origins to 1914*. Cambridge: Cambridge University Press.
van Gelderen, D.M., de Jong, P.C. and Oterdoom, H.J. 1994. *Maples of the World*. Portland, OR: Timber Press.
Hora, B. 1986. *The Oxford Encyclopedia of Trees of the World*. London: Oxford University Press.
Mohldenke, H.N., and Moldenke, A.L. 1952. *Plants of the Bible*. Waltham, MA: Chronica Botanica Co.

High-Intensity Sweeteners

Compadre, C.M., et al. 1985. Hernandulcin: An intensely sweet compound discovered by review of ancient literature. *Science 227*, 417–419.
Ellis, J.W. 1995. Overview of sweeteners. a symposium paper. *Journal of Chemical Education 72*, 671–675.
Federation of American Societies for Experimental Biology. 1974. *Evaluation of the Health Aspects of Licorice, Glycyrrhiza, and Ammoniated Glycyrrhizin as Food Ingredients*. Bethesda, MD: FASEB Life Sciences Research Office, Bureau of Foods. [Available from National Technical Information Service, Springfield, VA.]
Gordon, M. 1997. Japanese shape up with naturally sweet solution. *New Scientist 154*, 28.
Kinghorn, A.D., and Kennelly, E.J. 1995. Discovery of highly sweet compounds from natural sources. *Journal of Chemical Education 72*, 676–679.
Piggott, M.S., Jr. 1991. *Handbook of Sweeteners*. Glasgow: Blackie and Son Ltd.
Wu, C. 1997. Yeast make berry sweet sugar substitute. *Science News 151*, 284.

11
Psychoactive Plants

RICHARD RUDGLEY

Although it might seem that mind-altering plants play a comparatively small role in the great drama of social history, this is not the case. Medicines, poisons, and intoxicating substances are often made from the same plant and there is no reason why the latter usage should be of lesser antiquity. Such evidence as we have concerning the paleoethnobotany of archaic peoples indicates that the use of psychoactive species goes back to a remote era. The origins of drug use itself is more a question for the zoologist than for the cultural historian, as a number of mammals—reindeer and bear among them—actively seek out plants and fungi for their psychoactive effects.

As a species, humans seem to have an innate drive to seek out altered states of consciousness. The full range of plants typically consumed by humans includes not only staple foods but herbs, spices, and the milder stimulants (like tea and coffee), and narcotics, hallucinogens, and other powerful mind-altering substances. Hallucinogens, narcotics, and the more powerful stimulants are the subject of this chapter.

In modern society drugs are often associated with marginal groups, hedonistic pursuits, and criminal activity. However, in many traditional and indigenous societies drug plants were used in social and religious contexts. Typically such plants have been associated with healing and ceremonies more concerned with social cohesion than anti-social behavior. In short, such plants have played a central role in the cultures of many peoples past and present.

Ayahuasca, Yajé *Banisteriopsis* spp.
Malpighiaceae

The tropical forest liana *Banisteriopsis* (usually *B. caapi*) is used as the central ingredient of an Amazonian psychoactive potion called ayahuasca, or yajé, which contains other plants as well. To date ethnobotanists have reported about a hundred different plants that have been used as additives, but a species of *Banisteriopsis* is always present. More than 70 different indigenous peoples have used such potions. The name *Ayahuasca* itself means "vine of the soul". The bark of the *Banisteriopsis* vine is either mashed to a pulp and then mixed with cold water, or it can be boiled for a number of hours and then drunk. The combination of *Banisteriopsis* and other plants results in the brew's potent hallucinogenic effects.

A serious scientific study of *Banisteriopsis* was first undertaken by Richard Spruce in the 1850s. Subsequently anthropologists have reported that for various native peoples it plays a very important

Thornapple, Jimson weed (*Datura stramonium*). Otto Wilhelm Thomé, *Flora von Deutschland, Österreich und der Schweiz—in Wort und Bild für Schule und Haus* (1885–1905).

cultural role, and the geometric shapes and other symbols they have seen in visionary states induced by it are used in their paintings, architecture, pottery, masks, musical instruments, necklaces, stools, weapons, and so on. Their songs and dances are said to be derived from auditory and visual hallucinations.

Ayahuasca use continues to flourish not only in the rain forest, but also in the towns and cities of the Amazon region. It is used by shamans and native healers in the treatment of a number of disorders, most of them psychological (such as depression). The use of this hallucinogen has become more widespread partly as a result of its use in religious cults like Santo Daime which now has branches in the United States and Europe.

Betel *Areca catechu*
Piperaceae

Betel-chewing is one of the world's most popular forms of stimulant drug use. Even conservative estimates suggest that more than ten percent of the human race uses betel regularly. It is used in an area stretching from East Africa to India, and Southeast Asia and in the Pacific islands of Melanesia and Polynesia. The habit goes back at least 7,500 years, as archaeological evidence from Spirit Cave in Thailand demonstrates.

Although it is commonly called betel-chewing, this is something of a misnomer as the "nut" that is used is in fact the seed of the areca palm (*Areca catechu*), which is mixed with lime and then wrapped in a betel leaf (*Piper betle*). The nut and leaf also are not really chewed, but are put between the cheek and tongue and left there, as with a coca or tobacco quid. The main psychoactive constituent is arecaidine, which has stimulating effects similar to those of nicotine.

Like tea and coffee sets, betel-related equipment is used to send out social messages. In India it is common practice to add numerous flavorings and colorings to betel. The rich assortment of paraphernalia involved in making up these complex betel mixtures includes receptacles for each of the separate ingredients, mortars, dishes, spittoons, and the especially ornate betel cutters. In New Guinea, India, and elsewhere, betel has been the inspiration for minor art forms, and in Melanesia there are many finely decorated lime spatulas, lime containers, and other objects incorporated into the betel-chewing kit.

See: Nuts, Seeds, and Pulses pp. 134–5

Man chewing betel nut, Sumba, Indonesia. Copyright Edward Parker, used with permission.

Black henbane, Stinking nightshade, Stinking Roger *Hyoscyamus niger*
Solanaceae

There are about fifteen species of henbane found in Europe, North Africa, and Asia. Black henbane (*H. niger*) is the most well-known for its traditional use as a hallucinogen. This species grows to a height of 1–2.5 feet (0.3–0.7 m) and has grayish-green leaves, white (or faintly yellow) flowers with purplish veins, and dark gray seeds. The plant has a distinctly unpleasant smell, hence its folk names Stinking nightshade and Stinking Roger. The psychoactive tropane alkaloids hyoscyamine, scopolamine, and atropine are present in all parts of the plant, but are concentrated in the seeds and roots. High doses can be toxic. It produces strong soporific effects and hallucinations (accompanied by the sensation of flight), which are typically accompanied by a number of side effects—disorientation, temporary memory loss, profuse sweating, and severe bodily discomforts.

Its hallucinogenic and medicinal properties have been well known throughout history. The Greeks may have used it to inspire prophetic utterances and visions, and also as a surgical anesthetic. In parts of the Middle East, it was known as *bang*, and it seems to have played a role in ancient Iranian religion. It has been used as a traditional hallucinogen in China, Pakistan, Kashmir, and North Africa. It is highly effective when administered directly onto the skin and may have been one of the main ingredients in the ointments made by European witches.

Coca, Cocaine *Erythroxylum coca*
Erythroxlyaceae

The coca plant is a bush or shrub that grows to a typical height of about 3 feet (1 m), and the chewing of its leaves is a traditional practice in a wide area extending from Central America throughout the Andes and into the Amazon region. There are fourteen different alkaloids (alkaline, nitrogen-containing organic substances, usually with toxic potential) contained in the leaves of the coca plant with the most famous of course being cocaine (although nicotine is also present in smaller quantities).

The "chewing" of the coca leaf is something of a misnomer. Although the leaves of the coca plant may be chewed when first put in the mouth (in order to form a quid), the mixture is subsequently left lodged between the cheek and gums. An alkaline substance is added to the quid in order to facilitate the release of the psychoactive properties of the leaves.

Harvesting coca leaves near Cusco, Peru. Copyright Edward Parker, used with permission.

The amount of the cocaine alkaloid so obtained is of course far lower than in chemically pure extracts from the plant. By introducing the coca leaf orally, the cocaine is absorbed slowly and without harming the digestive system. It is a stimulant used to suppress hunger, to increase physical endurance and, in the Andes, to help cope with high altitudes. According to archaic indigenous beliefs coca "chewing" is harmless, and this conclusion has also been reached by distinguished ethnobotanists such as Timothy Plowman and Wade Davis. Impartial scientific investigations and studies have shown that coca leaves are extremely nutritious, and that regular use of coca is not harmful, and no major social problems are known to have resulted from its traditional use in the Andes over the millennia.

There is archaeological evidence for the ancient use of coca. A frozen mummy of an Inca boy accompanied by a feathered pouch containing coca leaves was discovered at a burial site in the Chilean Andes. Peruvian pots from around 500 AD are often decorated with motifs clearly indicating the use of coca. Many of these containers show people with their cheeks bulging in the place where the coca quid is held inside the mouth. To the Incas coca was the entheogen, or sacred plant, *par excellence*. Coca use was largely restricted to priests, aristocrats, and the court orators who used it to stimulate their powers of memory while reciting incredibly long oral histories. The sacrifice of coca accompanied almost all of their rites and ceremonies. The Quechua Indians of Ecuador make offerings of it to Mother Earth and bury the dead with a coca quid in their mouth.

There are abundant accounts of the powerful stimulating effects of coca. It is so intimately connected with the indigenous lifestyle of the indians of the Peruvian sierra that they calculate long journeys in units called *cocada*. Each *cocada* represents the distance that can be covered while chewing a single quid of coca. Because the distance that can be traveled depends on the terrain, the *cocada* is as much a unit of time as of space. Reliable accounts show that intense physical labor is possible for days on end, with only a couple of hours sleep per night and no food, as long as a coca quid is chewed every two or three hours.

Coca's reception in the modern, urban world has been a mixed one. In Bolivia and Peru, where coca is legal, it is used in toothpaste, chewing gum, wine, and, most widely, in the form of a tea (*mate de coca*). Elsewhere it is often reviled because it is the botanical source of cocaine.

In 1860 Albert Niemann of the University of Göttingen isolated the alkaloid cocaine from the coca plant. It soon had many champions in the medical world including, most notably, Sigmund Freud. However, Freud and others soon realized that this powerful anesthetic also had addictive properties. It subsequently became a major street drug and the first cocaine epidemic occurred from the 1880s until the 1920s. Its use declined in the 1930s and subsequent decades until a second wave of widespread use began during the 1970s, when it was the "champagne drug" of a new media and entertainment elite. Soon the drug became inextricably linked with major crime syndicates such as the Colombian cartels, and its glamorous image was replaced by a more sinister one. Cocaine production and distribution has become one of the world's most lucrative industries, and this alone guarantees the continuing significance of the coca plant in human affairs.

Cohoba, Yopo, Yupa, Vilca *Anadenanthera* spp.
Fabaceae

A number of species of trees in the *Anadenanthera* genus (*A. peregrina, A. colubrina* among them) have hallucinogenic properties and have been widely used to make powerful psychoactive snuff in a number of indigenous South-American Indian cultures. The snuff is prepared by making a paste from the beans, then mixing in an alkaline additive made from either snail shells or a plant source. It plays a central role in shamanistic practices such as divination, trance, communication with spirits, and the diagnosis of illness. Its use is usually limited to adult males. Although it is a powerful hallucinogen in larger doses, smaller amounts are sometimes taken as a stimulant. The seeds are also used to treat urinary disorders and infertility, and as emetics, purgatives, and laxatives.

Deadly nightshade (*Atropa belladonna*). Otto Wilhelm Thomé, *Flora von Deutschland, Österreich und der Schweiz—in Wort und Bild für Schule und Haus* (1885–1905).

The earliest written report of *Anadenanthera* use dates from 1496, but it was certainly in use before then, as archaeological evidence demonstrates. More than six hundred snuffing kits have been found by excavators at San Pedro de Atacama, an oasis in the Atacama desert of northern Chile. Botanical samples dated to 780 AD were analyzed and showed the presence of the psychoactive substances DMT, 5-MeO-DMT, and bufotenine, all of which are found in *Anadenanthera*.

Deadly nightshade, Belladonna *Atropa belladonna*
Solanaceae

The perennial plant belladonna grows to a height of 3–4 feet (0.9–1.2 m). It has ovate leaves, drooping bell-shaped purple flowers, and shiny black berries. The whole plant has psychoactive and highly toxic properties, and ingestion of it can be fatal. It is native to Europe, but is also found in North Africa and Asia and has escaped from cultivation in other regions including the United States. It contains the psychoactive tropane alkaloids hyoscyamine, atropine, and scopolamine. Its effects include hallucinations, ecstasy, and severe disorientation.

It is used in folk remedies in North Africa and Nepal, and may have been an ingredient in the wine drunk at the Bacchanalian orgies. There is also speculation that belladonna was one of the most important drugs used in European witchcraft, applied (along with a myriad of other plant and animal extracts) in the form of a salve.

Today it is an important source of medical drugs, including powerful and effective antispasmodic agents. Atropine, derived from belladonna, is used to dilate the pupils. It is also used to treat ergot poisoning.

Iboga, Eboga, Eboka *Tabernanthe iboga*
Apocynaceae

The West African shrub *Tabernanthe iboga* is the botanical source of the hallucinogen *eboga* (*eboka*, *iboga*). The root (especially its bark) is commonly utilized as an aphrodisiac, and in small doses it is valued as a stimulant. Its role as a hallucinogen is restricted to ritual use. *Eboga* is the sacred psychoactive plant of the Bwiti cult. This cult developed in Gabon as a reaction to colonialism, and fuses traditional and Catholic symbolism. Initiates of the cult take *eboga* in order to communicate with ancestors and to learn to come to terms with the inevitability of death. However, large doses can be fatal and some members have died through excessive ingestion.

The kava ceremony: straining. Tungua, Ha'apai, Tonga. Photo G.B. Milner, 1950. Courtesy Royal Geographical Society, London.

Kava *Piper methysticum*
Piperaceae

Kava is the name given by Pacific islanders to both *Piper methysticum*, a shrub belonging to the pepper family Piperaceae, and the psychoactive beverage made from it. *P. methysticum* is a hardy perennial which often grows up to 10 feet (3 m) or more. The rootstocks or stumps which contain the psychoactive substances, kavalactones, are prepared by pounding, chewing, or grinding them, and soaking them in cold water. This preparation is drunk and induces a state of mild euphoria and tranquility followed by sleep. There are also reports of increased mental clarity under its influence.

Recent research suggests that *P. methysticum* may have first been domesticated less than three thousand years ago in Vanuatu in eastern Melanesia. The use of kava seems then to have diffused westward to New Guinea and parts of Micronesia, and eastward to Fiji and then Polynesia. An indication of just how important kava cultivation has been in the Pacific is the sheer number of types which the indigenous people recognize. In Vanuatu alone natives are known to classify kava into 247 types! Kava was, and still is in many regions of the Pacific, an important medicine used in the treatment of rheumatism, menstrual problems, venereal disease, and tuberculosis.

Kava is also of great religious significance and is seen as a way of communicating with the ancestors and the gods—not simply an offering to the spirits, but a way of actually gaining access to their world. It is used in healing ceremonies and as a means of divination.

Although there have been many unsuccessful attempts to stamp out kava use in parts of the Pacific its use continues undiminished and in some areas it is preferred by the local authorities to a more negative alternative—alcohol. Kava bars have sprung up in the Pacific islands and provide a modern way to consume it in an informal and sociable setting. Kava, like many other intoxicants, is not without its health risks and medical research indicates the danger of liver toxicity.

See: Plants as Medicines, p. 238

Khat, Catha, Qat *Catha edulis*
Celastraceae

Qat (*Catha edulis*) is an evergreen shrub that grows over a very large area of Africa extending from South Africa to Ethiopia. The plant is said to still grow wild in some mountainous regions of East Africa. Although the centers of cultivation and consumption are in East African countries (Kenya,

Somalia, and Ethiopia), it is also reported to be grown in Afghanistan and Central Asia. Qat is used as a mild stimulant, more akin to tea or coffee than to stronger psychoactive substances, despite its portrayal as a potentially dangerous drug in some Western medical and media reports. The main psychoactive alkaloid of qat is cathinone.

The earliest source believed to refer to qat comes from al-Biruni in the 11th century AD. It has been suggested that *hajis* (pilgrims) may have brought the plant back with them on returning from Mecca. Not all Muslims were immediately enamored of qat, but it has since firmly established itself in East Africa and parts of the Arabian peninsula. In addition to its role as a social stimulant, the plant is also used in traditional Islamic medicine.

Qat is an integral part of culture in contemporary Yemen, where qat houses abound. In these establishments, men consume coffee, tobacco, and qat, and they are one of the most important social forums for Yemenis. Although the market for qat is currently limited to East Africa and parts of Arabia, on a local scale its cultivation and production is big business, and provides employment for hundreds of thousands in Somalia and elsewhere.

The use of qat first came to the attention of Europeans toward the end of the 17th century, but unlike other exotic stimulants known at the time (such as tea, coffee, and chocolate), it never became popular. This may have been partly because the stimulating effects of the leaves begin to fade within 48 hours. It is now sold as a "legal high" in the United Kingdom, but the market for it is very small.

Marijuana, Cannabis, Hashish *Cannabis sativa*
Cannabaceae

Cannabis sativa (hemp) is the most widely used illegal drug in the world. Although the cannabis plant is now ubiquitous, it is probably native to Central Asia. Its spread all over the world is no doubt due to a combination of cultural and natural factors. Cannabis, not only a psychoactive plant, but also of major technological importance as the source of a fiber for rope and other cordage, and was probably a popular trade item from a very early date, although its mention is scarce in archaeological records until the Iron Age in Europe (ca. 400 BC). Cannabis spread via different routes: east to China, south to India and southeast Asia, and west to Europe, Africa, and eventually the Americas.

The earliest of the Chinese pharmacopoeias, the *Pên Ching* (written in the first century BC but containing much older material), makes it clear that the ancient Chinese knew of the psychoactive effects of cannabis. The Taoists used it by combining it with other ingredients, which were then inhaled from incense burners. Despite the numerous Chinese references to cannabis, it has never been as important in Chinese social life when compared to its role in that of the Middle East and India.

Cannabis (*bhang, ganja*) has been widely used in India throughout history and is still common today. In the ancient text *Artharva Veda,* cannabis is described as one of a number of herbs that "release us from anxiety." Various psychoactive preparations containing cannabis were sacred to the gods, particularly Shiva and Indra.

Artefacts containing the charred remains of hemp seeds have been found in Iron Age Europe. In the 5th century BC the Greek historian Herodotus reported cannabis use among the Scythians living near the Black Sea. Scythian burial sites from the same period (but much further east in the Altai mountains of southern Siberia) contain abundant evidence of cannabis-smoking equipment. Cannabis has been widely used in the Middle East, North Africa, and Sub-Saharan Africa, mainly for recreational purposes, but also in ceremonial contexts.

In Europe, cannabis went through a renaissance in the 19th century when French visitors to North Africa rediscovered its psychoactive effects. Back in Paris numerous members of the artistic circles began to experiment with it, including Honoré de Balzac, Alexandre Dumas, and Charles Baudelaire. However, by the end of the 19th century American newspapers were already running sensational tales of the moral and physical decline experienced by hashish users. In 1915 California became the first state to make it illegal to possess cannabis, and by the 1920s it had become a major illicit drug. Cannabis use again increased dramatically in the 1960s and became socially acceptable to millions. More recently, its medicinal applications and the sheer number of its users have prompted political authorities in the western world to seriously consider the possibility of decriminalization and legalization.

See: Natural Fibers and Dyes, pp. 296–7; Age of Industrialization and Agro-industry, pp. 361–2

Mescal bean *Sophora secundiflora*
Fabaceae

Mescal beans are the psychotropic seeds of *Sophora secundiflora* and are not associated with the peyote cactus that is also sometimes known as mescal (see following discussion). This small tree or evergreen shrub is native to Texas, New Mexico, and Mexico. The pods contain up to eight seeds, which are maroon or orange-red in color. The principal alkaloids contained in the seeds are cytisine, N-methylcytisine, and sparteine.

Despite the use of mescal beans in Native American vision quests, none of these alkaloids are known to have hallucinogenic properties. Depending on the amount consumed and the method of preparation, mescal beans can cause a range of effects, from vomiting, headaches, and nausea to intoxication, stupor, and even death. Mescal beans are usually consumed in a decoction. Some 30 Native American peoples have made use of mescal beans, almost all of them using the beans for their decorative value; less than half of them have used mescal for its psychoactive effects. Mescal beans have been found at archaeological sites dating back to 7,000 ^{14}C years ago, in Texas, New Mexico, and Mexico, where they may have been used for ornamental purposes.

Mormon tea, Ma huang *Ephedra* spp.
Ephedraceae

Ephedra is an evergreen shrub that grows up to 4 feet (1.2 m) in height and is widely distributed in Asia, Europe, and North America. Several species of *Ephedra* growing in the southwestern United States, including *E. viridis* and *E. trifurca*, are known as Mormon tea. *Ephedra* contains alkaloids with stimulant properties, including ephedrine and norpseudoephedrine (the latter is also present in khat, *Catha edulis*).

Ephedra has been proposed as a candidate for the mysterious hallucinogenic drink "soma" used by the ancient Indo-Iranians in religious ceremonies. Archaeological finds of the plant in prehistoric Iranian temples in central Asia have been claimed, but the details are not yet fully published. The Chinese have traditionally used *E. sinica* (ma huang) for respiratory disorders. It is widely used today as a dietary and sports supplement, but has recently been the source of some controversy, with suggestions of severe side effects including strokes and heart attacks. In early 2004 a ban on U.S. sales of products containing ephedra was introduced.

See: Plants as Medicines, p. 223

Opium poppy *Papaver somniferum*
Papaveraceae

The opium poppy is undoubtedly one of the three or four most important drug plants of all, both in terms of its historical and contemporary influence on human society and in the power of its properties to both cure and destroy. Opium is the congealed juice of the poppy and is extracted from the plant by slitting the seed capsule.

The origins of opium poppy cultivation remain obscure but it is likely to have been first domesticated as an oil plant 6,000 ^{14}C years ago, in the western Mediterranean. Botanical remains demonstrate that it was widely used in late-Neolithic Europe. The role of opium in funerary cults—well known in ancient Crete, Cyprus, and Greece—seems to have its roots even earlier in prehistory. The discovery of numerous opium poppy capsules at a burial site in southern Spain shows that such a function dates back to at least 2500 BC.

Opium poppy is absent from ancient Mesopotamia. Claims of opium finds from ancient Egypt are controversial. Pottery juglets, shaped like poppy heads, were imported from Cyprus in the Eighteenth Dynasty (1550-1295 BC), but analysis of their contents is inconclusive. A stronger case can be made for opium consumption in Cyprus and Crete at this time, based on the presence of ivory pipes and other smoking equipment. In Crete the ceremonial use of opium was probably part of a cult centered around the worship of a fertility goddess. Some Mycenaean signet rings depict a female deity holding what are probably opium poppies, a motif that may well be the inspiration for later classical Greek images of the goddess Persephone holding opium poppies.

In the Middle East opium was widely recognized as a medical panacea by Muslim physicians long before its value was understood by their Christian counterparts. Opium came to be used among European doctors largely through the influence of Paracelsus (1493–1541), the great alchemist and physician. Not only did he do more to popularize the use of opium in medicine than any of his contemporaries, but he was also the first to make laudanum, a mixture of opium and alcohol.

In Victorian England opium was an integral—and legal—part of daily life and was almost routinely consumed by all sectors of society, used even to pacify crying infants. However, it would soon be vilified and identified with the "yellow peril" as an unwarranted wave of anti-Chinese xenophobia swept through the country. The sickly aura of the opium den conjured up in the popular imagination was to be replaced by the even more sordid image of the backstreet junkie. As the botanical source of morphine, codeine, and heroin, opium continues to exert a powerful economic, political, and medical hold on society.

See: Ornamentals, p. 278; Plants as Medicines, p. 221; Spices, pp. 167–8

Peyote, Mescal *Lophophora williamsii, L. diffusa*
Cactaceae

Peyote is a small spineless cactus found in the desert regions of Texas and Mexico. The head of the cactus is cut off and consumed as a sacred hallucinogen by a number of Indian peoples. Peyote is typically eaten raw or dried, and is also taken in the form of a tea. It contains many alkaloids including mescaline, a powerful hallucinogen.

Archaeologists have discovered remains of the peyote cactus dating back several thousand years in the Texas/Mexico border region. Rock art in the area includes peyote images bearing a striking resemblance to those used by the Huichol Indians of Mexico today, suggesting a continuous tradition lasting through the millennia.

By the 20th century, the use of peyote had extended far beyond its natural homeland. It has been particularly associated with the Plains Indians of the United States. By 1906 there was a loose intercultural network of peyote-using native peoples from Oklahoma in the south to Nebraska in the north. In 1918, with the help of anthropologist James Mooney, the Native American Church, as it was then known, was officially incorporated in Oklahoma. It subsequently became known as The Native American Church of North America in order to accommodate peyote users north of the U.S. border with Canada.

San Pedro cactus *Echinopsis* spp.
Cactaceae

Psychoactive members of the *Echinopsis* family (*E. pachanoi, E. peruviana* ssp. *peruviana*, and others) are commonly called San Pedro cactus. They are large cacti (up to 20 feet or 6 m tall) that contain the hallucinogen mescaline, as does its more famous cousin peyote. The standard method of consumption is to boil pieces of the stem for a few hours and then drink the liquid. It is widely used among Andean peoples in Peru, Bolivia and Ecuador.

A temple carving depicting the San Pedro cactus has been found at an archaeological site in northern Peru and is more than three thousand years old. The cactus is also an important artistic motif throughout later periods, demonstrating continuous use from the distant past up to the present. Folk uses are not restricted to shamanistic vision quests, and San Pedro cactus is also deemed to be effective in the treatment of fever, hepatitis, and alcoholism.

Thornapple, Jimson weed *Datura* spp.
Solanaceae

The traditional use of the *Datura* species is well attested from Africa, Europe, Asia, and the Americas, making it one of the most widely used of all hallucinogenic plants. There are fifteen to twenty species, the most well-known of which is *Datura stramonium*. The members of this species are annual plants which typically grow to between 1 and 3 feet (0.3–0.9 m) tall. The leaves are coarse and the entire plant has an unpleasant smell. The flowers are white or purplish and the capsular fruit is typically covered with spines. The "tree daturas" (*Brugmansia* spp.), previously classed in the same genus, are now categorized as a distinct genus. Both genera contain the tropane alkaloids typically found in other psychoactive plants of the Solanaceae family (such as belladonna and henbane; see the preceding discussion)—atropine, hyoscyamine, and scopolamine. Large amounts may cause a state of intoxication that can last for a number of days.

The *Datura* plant was well known to the Greeks and Romans: Dioscorides made reference to its psychoactive properties nearly two thousand years ago. In India the *Datura* plant has a long history as an entheogen (or sacred hallucinogen) and is an integral element of the cult of the god Shiva. In East Africa it is smoked as a drug and added to beer to create a potent combination.

Datura has also been used in shamanism and initiation cults in the Americas. Excavations in the border regions of Texas and Mexico suggest use of the plant ca. 700–1000 AD, demonstrated by spiked ceramic pots that strikingly resemble the spiny fruit of *Datura*. In more recent times, Native American uses for the plant have been manifold—as a divinatory agent, an aphrodisiac, and a surgical anesthetic. Its potential for abuse is also recognized; the Navajo call it "imbecile-producer," for instance.

Tobacco *Nicotiana* spp.
Solanaceae

With the exception of Australia (which has its own indigenous *Nicotiana* that was chewed by Aborigines), the world ultimately owes its supplies of tobacco to the Americas. The cultural history of tobacco use begins in the remote prehistory of South America. According to current understanding, the lowlands of Patagonia, the Pampas, and Gran Chaco are the probable home of the tobacco plant. The early history of tobacco in South America is obscure, as few, if any, definite archaeological finds have been made. However, it is likely that the indigenous inhabitants of the region began to cultivate tobacco in their gardens several thousand years ago, growing about 12 different species altogether, *Nicotiana tabacum* and *N. rustica* being the most common. *N. rustica* had been taken by 200 AD to North America, where wild tobaccos had been harvested since at least 400 BC; *N. tabacum* probably followed ca. 1000 AD.

Although tobacco is most commonly used as a stimulant, in sufficient quantities (such as those used traditionally by some Native Americans) it can have what, for all intents and purposes, may be called hallucinogenic effects. It is certainly perceived as such among South American shamans, and although apprentices are told beforehand about the type of visions they are going to see, this effect appears to be not just the result of cultural conditioning, but also of the actual chemistry of the plant. Tobacco contains the harmala alkaloids harman and norharman, and the closely related harmine and harmaline are known hallucinogens. The levels of harman and norharman in cigarette smoke are between forty and one hundred times greater than in tobacco leaf, confirming that the burning of the plant generates a dramatic increase in concentration. The effects of nicotine on the central nervous system are still far from understood. The hallucinogenic effects of the plant become far more explicable when it is borne in mind that the strains of tobacco smoked by Native Americans were far more potent than our commercially produced varieties.

Not only were the indigenous people of South America the first to grow tobacco, they also discovered all known ways of using it, including some means of ingestion unknown in the Western world. They chewed tobacco quids, smoked it (in both cigars and pipes), took it as snuff, drank tobacco juice and syrup, licked tobacco paste, administered tobacco enemas, and even rubbed it on their skin and in their eyes.

Shamans on the Orinoco have been reported to smoke five or six three-foot cigars in a single ritual session. The toxic effects of the plant are well understood by the shamans of South America, and sometimes use severe tobacco poisoning to induce near-death experiences, according to the anthropologist Johannes Wilbert. These tobacco shamans believe that humankind's hunger is for food, while the spirit's hunger is for tobacco. Taking it is seen as a means of communicating with the spirits. Before the arrival of Europeans, tobacco seems to have been restricted mainly to ceremonial functions in both South and North America.

The 16th century physician Nicholas Monardes wrote that tobacco was of great significance to the Indian priests. He reports that when a priest was asked questions, his patients expected him to find the answer by undertaking a tobacco-induced trance. After inhaling the smoke, the priest "fell downe upon the ground, as a dedde manne, and remaining so, according to the quantitie of the smoke that he had taken, and when the hearbe had doen his woorke, he did revive and awake, and gave theim their answeres, according to the visions, and illusions which he sawe." He also noted that they would chew a quid of tobacco and coca, a combination that he said made them "out of their wits" as if drunk.

Other early accounts corroborate this description of native tobacco use. Edmund Gardiner, writing at the beginning of the 17th century, describes native "enchanters" (i.e., shamans) getting drunk on tobacco smoke and then falling into a deep sleep. On awakening they would tell of the visions they had seen and divine their meaning. Although Gardiner, in line with most of his contemporaries, interpreted Native American experiences with tobacco as delusions of the devil, it is clear that the plant was attributed with inebriating and hallucinogenic properties by early Europeans as well as by the native peoples themselves.

Reactions to the arrival of tobacco on the European stage were mixed. There were those who saw it as a gift from the gods and as a precious medicine, while others, like King James I, were highly vocal in their condemnation of the new habit of tobacco use. The monarch described it as a "stinking fume" from hell and dangerous to health. Despite such protestations, tobacco use spread rapidly across the continent. Initially it was usually smoked in a pipe, but by the 1680s taking it in the form of snuff had become all the rage. Even before this, it had started to beguile Asian peoples as easily as it had the Europeans. The Spanish introduced it to the Philippines, from which it diffused into China, Korea, and Southeast Asia, and Portuguese traders also brought it into many ports in southern Asia. The Russians spread its use across the vastness of Siberia and even traded it with Alaskan natives, thus returning it to the Americas, where it had first started on its journey across the globe.

See: Age of Industrialization and Agro-industry, pp. 371–4

Water lily *Nymphaea* spp.
Nymphaeaceae

In Europe the water lily was known by a number of names, including the water rose, Clavis Veneris, and Digitus; apothecaries knew it as Nenuphar and claimed it to have effects similar to but weaker than those of opium. Its oils and decoctions were used as soporifics and anaphrodisiacs. Water lilies were occasionally used in the psychoactive ointments made by witches.

Maya art depicts the water lily in conjunction with two other psychoactive substances(the toad and the mushroom), and ancient Egyptian iconography portrays the blue water lily in close relationship with the opium poppy and the mandrake (*Mandragora* spp., also well known in European folklore). The specialist on narcotic plants William Emboden has long held the view that both the Maya and the Egyptians utilized lillies for their psychoactive effects. Alkaloids extracted from *N. ampla* are very close in chemical structure to apomorphine, which is a synthetic derivative of morphine. Emboden tried the Old World water lily (*N. nouchali* var. *caerulea*) for himself and found that the extracts of the flower caused visual and auditory hallucinations. However, there is no ethnographic or archaeological evidence for the psychoactive use of either species, and their cultural role remains uncertain.

See: The Hunter-Gatherers, p. 9; Fragrant Plants, p. 255; Gathering Food from the Wild, p. 38

Yakee, Paricá *Virola* spp.
Myristicaceae

There are about sixty species of tree belonging to the genus *Virola* that grow in the tropical zone of the Americas. A number of them (*V. theiodora, V.* spp.) have been used for the hallucinogenic red resin of the inner bark. It is taken either in the form of a snuff, or simply made into pellets and eaten. After an initial phase of excitation, numbness and loss of coordination set in, this in turn being followed by hallucinations and eventual narcosis.

Virola is utilized by indigenous peoples in Colombia, Brazil, Venezuela, Ecuador, and Peru. In Colombia it is named *yákee* or *yáto* and used mainly by shamans, while elsewhere its use is less restricted. In Brazil it is known as *paricá, ebene* and *epéna*—general names for snuff not referring exclusively to mixtures containing species of *Virola*.

References and Further Reading

Abel, E.L. 1980. *Marihuana: The First Twelve Thousand Years*, New York and London: Plenum Press.

Adovasio, J.M., and Fry. G.F. 1976. Prehistoric Psychotropic Drug Use in Northeastern Mexico and Trans-Pecos Texas. *Economic Botany* 30(1), 94–96.

Antonil [pseudonym of Anthony Henman]. 1978. *Mama Coca*. London: Hassle Free Press.

Bennett, B.C., and Alarcón, R. 1994. *Osteophloeum platyspermum* and *Virola duckei* (Myristicaceae): Newly reported as hallucinogens from Amazonian Ecuador. *Economic Botany* 48(2), 152–158.

Beran, H. 1988. *Betel-Chewing Equipment of East New Guinea*. Shire Ethnography 8. Aylesbury: Shire.

Boyd, C.E., and Dering, J.P. 1996. Medicinal and Hhallucinogenic plants identified in the sediments and pictographs of the Lower Pecos, Texas Archaic. *Antiquity* 70, 256–275.

Brooks, J.E. 1953. *The Mighty Leaf: Tobacco Through the Ages*. London: Alvin Redman.

Brownrigg, H. 1991. *Betel Cutters from the Samuel Eilenberg Collection*. Stuttgart: Edition Hansjörg Mayer.

Brunton, R. 1989. *The Abandoned Narcotic: Kava and Cultural Instability in Melanesia*. Cambridge: Cambridge University Press.

Davis, W. 1997. *One River: Science, Adventure, and Hallucinogenics in the Amazon Basin*. London: Simon & Schuster.

De Smet, P.A.G.M. 1985. *Ritual Enemas and Snuffs in the Americas*. Dordrecht: Foris.

Dobkin de Rios, M. 1984. *Hallucinogens: Cross-Cultural Perspectives*. Albuquerque, NM: University of New Mexico Press.

Efron, D.H., Holmstedt, B., and Kline, N.S. (Eds.) 1979. *Ethnopharmacologic Search for Psychoactive Drugs*. New York: Raven Press.

Emboden, W.A. 1978. The sacred narcotic lily of the Nile: *Nymphaea caerulea*. *Economic Botany* 32(4), 395–407.

Emboden, W.A. 1979. *Narcotic Plants: Hallucinogens, Stimulants, Inebriants, and Hypnotics, Their Origins and Uses*. London: Studio Vista.

Emboden, W.A. 1981. Transcultural use of narcotic water lilies in ancient Egyptian and Maya drug ritual. *Journal of Ethnopharmacology* 3, 39–83.

Fernandez, J.W. 1982. *Bwiti: An Ethnography of the Religious Imagination in Africa*. Princeton: Princeton University Press.

Furst, P.T. (Ed.). 1972. *Flesh of the Gods: The Ritual Use of Hallucinogens*. London: Allen and Unwin.

Goodman, J., Lovejoy, P.E., and Sherratt, A. (Eds.). 1995. *Consuming Habits: Drugs in History and Anthropology*. London: Routledge.

Hansen, H.A. 1978. *The Witch's Garden*. (Translated by Muriel Crofts). Santa Cruz, CA: Unity Press-Michael Kesend.

Harner, M.J. (Ed.). 1973. *Hallucinogens and Shamanism*. Oxford: Oxford University Press.

Hawkes, J.G., Lester, R.N., and Skelding, A.D. 1979. *The Biology and Taxonomy of the Solanaceae*. London: Academic Press.

Heiser, C.B. 1969. *Nightshades: The Paradoxical Plants*. San Francisco: Freeman.

Janiger, O., and Dobkin de Rios, M. 1976. *Nicotiana* an hallucinogen? *Economic Botany 30*, 149–151.

Kalix, P. 1991. The pharmacology of psychoactive alkaloids from *Ephedra* and *Catha*. *Journal of Ethnopharmacology 32/1–3)*, 201–208.

Krikorian, A.D. 1984. Kat and its use: An historical perspective. *Journal of Ethnopharmacology 12(2)*, 115–178.

La Barre, W. 1980. *Culture in Context*. Durham, NC: Duke University Press.

La Barre, W. 1989. *The Peyote Cult* (5th edition). Norman, OK: University of Oklahoma Press.

Lebot, V., Merlin, M., and Lindstrom, L. 1992. *Kava: The Pacific Drug*. New Haven, CT: Yale University Press.

Lewin, L. 1964. *Phantastica: Narcotic and Stimulating Drugs, Their Use and Abuse*. London: Routledge and Kegan Paul.

Li, H-L. 1974. The origin and use of cannabis in eastern Asia: Linguistic-Cultural implications. *Economic Botany 28/3*, 293–301.

Li, H-L. 1974. An archaeological and historical account of *Cannabis* in China. *Economic Botany 28(4)*, 437–448.

Lindstrom, L. 1987. *Drugs in Western Pacific Societies: Relations of Substance*. New York: University Press of America.

Litzinger, W.J. 1981. Ceramic evidence for prehistoric *Datura* use in North America. *Journal of Ethnopharmacology 4*, 57–74.

Lockwood, T.E. 1979. The ethnobotany of *Brugmansia*. *Journal of Ethnopharmacology 1(2)*, 147–164.

Martin, R.T. 1970. The role of coca in the history, religion, and medicine of South American Indians. *Economic Botany 24(4)*, 422–438.

Merlin, M.D. 1972. *Man and Marijuana: Some Aspects of Their Ancient Relationship*. Cranbury, NJ: Associated University Presses.

Merlin, M.D. 1984. *On the Trail of the Ancient Opium Poppy*. London: Associated University Presses.

Merlin, M. 2003. Archaeological record for ancient Old World use of psychoactive plants. *Economic Botany 57(3)*: 295–323.

Merrill, W.L. 1977. *An Investigation of Ethnographic and Archaeological Specimens of Mescalbeans (Sophora secundiflora) in American Museums. Research Reports in Ethnobotany 1*. Ann Arbor, MI: Museum of Anthropology, University of Michigan.

Merrillees, R.S. 1962. Opium Trade in the Bronze Age Levant. *Antiquity 36*, 287–292.

Myerhoff, B.G. 1974. *Peyote Hunt: The Sacred Journey of the Huichol Indians*. Ithaca, NY: Cornell University Press.

Ott, J. 1993. *Pharmacotheon: Entheogenic Drugs, Their Plant Sources and History*. Kennewick, WA.: Natural Products Co.

Pope, H.G. 1969. *Tabernanthe iboga*: An African narcotic plant of social importance. *Economic Botany 23(2)*, 174–184.

Prance, G.T. 1970. Notes on the use of plant hallucinogens in Amazonian Brazil. *Economic Botany 24(1)*, 62–68.

Prance, G.T. 1972. Ethnobotanical notes from Amazonian Brazil. *Economic Botany 26(3)*, 221–237.

Rätsch, C. 1992. *The Dictionary of Sacred and Magical Plants*. Bridport: Prism Press.

Reichel-Dolmatoff, G. 1975. *The Shaman and the Jaguar: A Study of Narcotic Drugs Among the Indians Of Colombia*. Philadelphia: Temple University Press.

Reichel-Dolmatoff, G. 1978. *Beyond the Milky Way: Hallucinatory Imagery of the Tukano Indians*. Los Angeles: University of California, Los Angeles, Latin American Center Publications.

Rubin, V. (Ed.). 1975. *Cannabis and Culture*. The Hague: Mouton.

Rudgley, R. 1993. *The Alchemy of Culture: Intoxicants in Culture*. London: British Museum Press.

Rudgley, R. 1998. *The Encyclopaedia of Psychoactive Substances*. London: Little Brown.

Safford, W.E. 1922. Daturas of the Old World and New: An account of their narcotic properties and their use in oracular and initiatory ceremonies. In *Annual Report of the Smithsonian Institution 1920*. Washington, DC: The Smithsonian Institution, 537–567.

Schultes, R.E., and Hofmann, A. 1980. *The Botany and Chemistry of Hallucinogens* (2nd ed.). Springfield, IL: Charles C. Thomas.

Schultes, R.E., and Hofmann, A. 1980. *Plants of the Gods: Origins of Hallucinogen Use*. London: Hutchinson.

Sharon, D.G., and Donnan, C.B. 1977. *The Magic Cactus: Ethnoarchaeological Continuity in Peru*. Archaeology 30(6), 374–381.

Sherratt, A.G. 1987. Cups that cheered. In *Bell-Beakers of the West Mediterranean*, edited by W.H. Waldren and R.C. Kennard. Oxford: British Archaeological Reports (International Series 331), 81–106.

Sherratt, A.G. 1991. Sacred and profane substances: The ritual use of narcotics in Later Neolithic Europe. In *Sacred and Profane: Proceedings of a Conference on Archaeology, Ritual and Religion*, edited by P. Garwood et al. Oxford: Oxford University Committee for Archaeology, Monograph 32, 50–64.

Torres, C.M. et al. 1991. Snuff powders from pre-Hispanic San Pedro de Atacama: Chemical and contextual analysis. *Current Anthropology 32(5)*, 640–649.

Weir, S. 1985. *Qat in Yemen: Consumption and Social Change*. London: British Museum Publications.

Wilbert, J. 1987. *Tobacco and Shamanism in South America*. New Haven: Yale University Press.

Wilbert, J. 1991. Does pharmacology corroborate the nicotine therapy and practices of South American shamanism? *Journal of Ethnopharmacology 32(1–3)*, 179–186.

12
Plants as Medicines

MICHAEL HEINRICH, ANDREA PIERONI, AND PAUL BREMNER

Preparing a cultural history of medicinal plants is a challenging task, especially when attempting worldwide coverage. Each region has its own centuries-old, continuously changing, traditions. This is exemplified in Table 12.1, which shows the geographical origin of medicinal plants currently included in the European and the German pharmacopoeias. The latter includes a particularly large number of medicinal plants. It is of particular note in these pharmacopoeias that they still rely very heavily on European (including Mediterranean) plant resources, which are part of the tradition of Europe, and on the plants introduced by the various Arabic scholarly works.

How does one go about selecting plants to be included in this chapter? D. Moerman recorded at least one medicinal use for 8.2 percent, or 2,600 species, of a total of 31,600 North American species of higher plants (Moerman 1998), while in the case of Mexico, R. Bye has estimated that about 17 percent (5,000 species) of the total flora (30,000 species) are used medicinally (Bye 1993). From this overview it becomes obvious that the selection of just a few medicinal plants from any of the continents will have to highlight particularly interesting or typical examples and that such a selection has to be arbitrary. Consequently, we have selected six species from Africa, twenty from Europe (including the Mediterranean and the Near East), seven from northern Asia, fourteen from South and southeastern Asia, eleven from the Americas, and three from Australia/Oceania. Some species have been selected because they are of particular importance as medicinal plants and/or because they provide particularly interesting case studies that highlight exciting developments in the use of medicinal plants.

Africa

Very little information about plant use in Africa has been written down. In African thought, all living things are believed to be connected to each other, to the gods, and to ancestral spirits. If harmony exists between all of these, then good health is enjoyed, but if not, misfortune or ill health will result. Forces can be directed at humans by displeased gods, ancestors, and witches, resulting in disharmony which must be resolved before good health can be restored. Treatment may also involve much more than medicine. Practices such as divination and incantation may be carried out to help with diagnosis, and sacrifices may need to be made in order to placate the supernatural entity.

The traditional healer is also likely to be a religious leader, since health and spirituality are closely intertwined in Africa. Traditional healers have existed throughout Africa since prehistoric

Birthwort (*Aristolochia clematitis*). USDA-NRCS PLANTS Database/Britton, N.L., and A. Brown. 1913. *Illustrated flora of the northern states and Canada*. Vol. 1, p. 645.

times, for example, the *ifas* and *juju* men of West Africa or the *inyanga* (*sangoma* is the term used for diviners) of South Africa. Various methods have been used to identify "healing plants," such as trying to find a plant that possesses a stronger spirit than the one causing the disease, or by using the "law of signatures." This system existed in other parts of the world, including Europe in the Middle Ages (Grier 1937), and is based upon the concept that nature has provided a cure for every

TABLE 10.1 Geographical Origins of Medicinal Plants Contained in the European and the German Pharmacopoeia[§]

Geographical Region	EurPh 1997 & 98	DAB 1997 & 98	Total	DAB1926
Eurasia (incl. Mediterranean)	31	40(11)	71	76
• Central, Northern Europe	7	15(2)	22	19
• Central, Northern Europe and Mediterranean	3	3(2)	6	8
• Mediterranean only	8	7(3)	15	16
• Northern Asia and Near East	13	15(4)	28	33
Asia (Central, North, East)	3	2(2)	5	4
South and Southeast Asia	8	5(−)	13	24
Africa	5	1(−)	6	7
Australia	0	1(−)	1	1
North America	3	2	5	4
Mexico, Central and South America	9	3(3)	12	19
Oceans	1	0	1	1
Total	60	54	114	136

[§] If more than one species are accepted for a botanical drug, they are included only once.

• The data refer to the German Pharmacopoeia (DAB) only (i.e. botanical drugs not listed in the EurPh); the numbers in parentheses are the species which yield botanical drugs, which are included in the EurPh.

disease. Matching a plant to a disease for a cure involves finding a plant with a similar form to the diseased organ. For example, plants with a red flower would be used in blood disorders, and yellow-flowered plants in cases of jaundice.

The use of medicinal plants in Africa also provides the observer with timely reminders of how sustainable cultivation can protect a resource (*Aloe ferox*, Southern Africa), but a species may likewise move toward regional extinction through over-exploitation (*Prunus africana*, Cameroon). Modern treatises and texts of African *Materia Medica* include the following: for West African medicinal plants, Oliver-Bever (1986); the ethnobotanical studies of Neuwinger (1994, 2000); for North Africa, Bellakhdar (1997); and for South Africa and East Africa, Watt and Breyer-Brandwijk (1962); van Wyk et al. (1997).

African devil's claw, Grapple plant *Harpagophytum procumbens* Pedaliaceae

Devil's claw or grapple plant derives its name from the formidable "claw," the dried hooked thorns of the fruit used in seed dispersal, which are a hazard to any passing cloven-hoofed animal or careless human. The plant is native to southern and eastern Africa and it is collected in regions bordering the Kalahari Desert. It thrives in clay or sandy soils and is often found in parts of the South African veldt. The tubers are traditionally used as a tonic, for "illnesses of the blood," fever, problems during pregnancy, and kidney and bladder ailments. Since the mid-1980s and with considerable research effort, African devil's claw has been developed into a very successful and relatively well-characterized phytomedicine for the treatment of pain relief in joint diseases, back pain, and headache. Most pharmacological and clinical research has been conducted on standardized extracts. The secondary storage tubers are collected and, while they are still fresh, they are cut into small pieces and dried. The main exporters are South Africa and Namibia. Attempts are currently under way to cultivate the species because over-harvesting has led to conservation concerns.

The dried and powdered root of the plant is now included in the *European Pharmacopoeia* (2002) and some national pharmacopoeias, such as that of Switzerland. Several constituents are known, including iridoids and phenylethanoids, but the active constituents have not been identified with certainty.

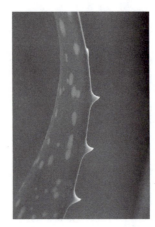

Aloe plant. Copyright Edward Parker, used with permission.

Aloe *Aloe vera*
Aloeceae

Aloe vera (also known as *A. barbadensis*) has long been used because of its medicinal properties and its mythical significance. Depictions of a plant similar to *A. vera* appear in wall carvings and paintings in ancient Egypt (2000 BC), but its use prior to the Roman period is speculative. Alexander the Great learned of aloe after conquering Egypt in 332 BC and, on Aristotle's advice, had his army capture the island of Socotra where *Aloe* grew, thus ensuring that his soldiers, rather than the opposition, would benefit from the healing powers of this plant. The first-century Greek herbal of Dioscorides gives a description of the healing powers of aloe and noted it was the sap, and not the gel, that was the healing agent. It should be clearly stated here that confusion does exist about which component of aloe is to be used. The gel, from the parenchyma cells in the inner leaf pulp, is the part that has been used medicinally for centuries and should be clearly distinguished from the exudates (or latex). This latex originates from the bundle sheath cells just beneath the outer skin of the leaves, is characterized by its yellow color and bitter taste, and is used as a strong purgative.

The plant is native to Africa. Its former scientific name, *A. barbadensis* (from the Caribbean island of Barbados), originates from the time Jesuit priests exported the plant from South America to Spanish territories in the Caribbean (Jamaica and Barbados) and subsequently imported it to England in the 17th century. *Aloe ferox*, or Cape aloe, is native to southern Africa and is widely applied in phytotherapeutic preparations. Aloe extracts incorporated into over-the-counter cosmetic and medicinal products are now widespread throughout the world. Aloe gel is used widely in many preparations including those for wound treatment, especially burn wounds and inflammation, as extensively reviewed by Reynolds and Dweck (1999) and Reynolds (2004). All *Aloe* species are protected by the Convention on International Trade in Endangered Species (CITES).

Cola nut, Sudan cola nut *Cola acuminata* and *C. nitida*
Sterculiaceae

These two taxa of evergreen trees are sometimes considered to be one species. They are endemic to the lowland forests of western Africa, especially Sierra Leone, Liberia, Ivory Coast, and Ghana, and have been introduced to Nigeria, Mali, and Guinea. *Cola nitida* is widely cultivated in the Caribbean. The medicinally used part of the plant is the cola nut (the tree's seed kernel) which is chewed to produce saliva. The seeds are rich in caffeine and related alkaloids and have consequently been used as a stimulant. However, cola's importance in African traditional societies goes beyond this: it is a symbol of friendship, and guests receive a cola nut on arriving at a house. Also many ceremonies are ended by jointly eating cola nuts. The first drawing of a cola nut was published by Clusius in 1591, but it was only in the 19th century that the nut became more widely known in Europe and North America. Travelers' stories of using it to ease the hardship of expeditions and travel highlighted the stimulating effects of *C. nitida*, but it has no modern biomedical use.

See: Nuts, Seeds, and Pulses, p. 137; Caffeine, Alcohol, and Sweeteners, p. 179

Myrrh *Commiphora myrrha*
Burseraceae

Myrrh is indigenous to Northeast Africa, particularly Somalia and Ethiopia, and is one of the oldest and most highly esteemed botanical drugs. The drug used is the gum resin (an exudate), which is excreted by the trees through the bark. In present-day Morocco, for example, it is used as a balsam and for nervous disorders, and it is applied in cleansing ceremonies such as ritual fumigations (Bellakhdar 1997). *In vivo* anti-inflammatory effects have been reported for its extract (Duwiejua et al. 1992).

See: Fragrant Plants, p. 250

Red stinkwood *Prunus africana*
Rosaceae

The bark of *Prunus africana* is obtained from a tree that grows in the high (at elevations of 1000–2500 m) mountain forests of Africa and Madagascar). It has been developed into a phytomedicine and used to treat benign prostate enlargement, with good clinical evidence pointing to the extract's efficacy. Within Africa, different decoctions of the plant are used to treat various conditions, including fevers, urinary tract infections, and inflammation (bark tea), and to prepare wound dressings (leaves), and the leaf sap is drunk for insanity. In the regions of origin a high-quality extract is produced from the botanical drug, which is mostly exported to Italy, France, and Spain. This extract could provide a sustainable and valuable income for many of the African countries, especially for Cameroon and Madagascar. However, the bark is currently harvested from this slow-growing tree in an unsustainable manner. Overharvesting has resulted in serious damage to plant populations and it is now CITES-listed. Consequently, there has been a drive to stop the collection of the bark or, alternatively, to increase efforts to cultivate the species. These have been hampered by shortage of seed stocks and the long unproductive period as the tree matures.

See: Ornamentals, pp. 265–6, 269, 270, 274, and 279–80; Wood, p. 321; Nuts, Seeds, and Pulses, p. 134

Rosy periwinkle *Catharanthus roseus*
Apocynaceae

A small, erect shrub, the rosy periwinkle (Madagascar periwinkle) is native to Madagascar but can now be found as a widely cultivated plant in warm climates around the world. The plant reaches a height of 3 feet (1 m) and has smooth oblong leaves; the pink or white flowers are borne all year round on the upper leaf axils. The plant is used in South Africa to treat diabetes and rheumatism, but its pervasive growth has now caused it to become a weed, especially in Kwa-Zulu Natal and Mpumalanga.

In Madagascar the plant is widespread, reflecting its multiple uses in herbal treatments. The striking floral display of the plant has promoted its spread around the world and has led to variations in its medicinal applications. By the early 20th century it was being used as an oral hypoglycemic agent (to lower blood sugar levels) in South Africa, southern Europe, and the Philippines; to treat diabetic ulcers in the West Indies; and to control hemorrhages and scurvy in Brazil. The role

Rosy periwinkle. Copyright Edward Parker, used with permission.

of the plant as an antidiabetic agent in the Caribbean led to the discovery of its effective anticancer activity. Diabetics in Madagascar were taking a periwinkle tea in preference to the more expensive insulin treatments. This prompted a small scientific study which found that rats given the tea had a significantly lowered white blood cell count. Although fatal to the rats, this finding prompted the researchers to investigate the effect of the Madagascar periwinkle against leukaemia, a disease caused by an abnormal increase in white blood cells. The active agent was identified as a new alkaloidal compound termed vinblastine which was licensed in the United States and approved for use in cancer treatments in 1961. A related alkaloid from the same plant, vincristine, was licensed as a drug two years later. Rosy periwinkle is now widely grown for extraction of these alkaloids.

Europe and the Mediterranean/Near East

Europe differs from the other regions under discussion, in that there the historical knowledge about medicinal plants is rarely a popular tradition or based on knowledge handed down from one generation to the next. Instead herbal traditions form part of mainstream pharmaceutical traditions and relatively well-researched botanical medicines are often available. Most of this information is now transmitted via the mass media, especially through popular books about herbal medicines and in countries such as Switzerland, Italy, France and Germany. Other countries, like the United Kingdom, consider many herbal medical products to be a form of alternative and complementary medicine. However, in both cases medicinal plants have become a commodity.

The Mediterranean and Near East region has been characterized by mutual influences among popular medical practices and schools of medicine. The oldest written information in the European-Arabic traditions originates from the Sumerians and Akkadians of Mesopotamia, and another region for which we have documents that have survived through the millennia is Egypt. There are many problems in translating ancient texts. Because medicinal plants are precious and used in small quantities, they are rarely found in the archaeological record. Translation of ancient names therefore relies on internal evidence—for example, descriptions of the plant concerned—and on philological techniques. Many plant names from ancient Mesopotamia and Egypt are of uncertain meaning or cannot be translated at all. The Egyptians documented their knowledge, including medico-pharmaceutical knowledge, on papyruses (e.g., the Papyrus Eber), using paper made from the sedge *Cyperus papyrus*.

Greek medicine has been the focus of pharmaceutical historical research for many decades. The Greek scholar Pedanius Dioscorides from Anarzabos (first century AD) is considered to be the father of (Western) pharmacy. The doctrine presented in his works governed pharmaceutical and medical practice for more than 1,500 years and has heavily influenced European pharmacy. He was an excellent pharmacognosist and described more than six hundred medicinal plants. A number of Greek and Roman scholars were influential in developing other fields of health care practice and the natural sciences, including the following:

- The Greek medical doctor Hippocrates (ca. 460 to 375 BC) who heavily influenced European medical traditions and was the first of a series (of otherwise largely unknown) authors which produced the so-called Corpus Hippocraticum (a collection of works on medical practice).
- The Graeco-Roman medical doctor Claudius Galen (Galenus) [130 to 201 (?) AD] who summarized the complex body of Graeco-Roman pharmacy and medicine and whose name survives in the pharmaceutical field of "galenics" or pharmaceutical technology.
- Pliny the Elder [23 or 24 AD to 79 AD (killed at Pompeii while observing the eruption of Vesuvius)], who was the first to give a "cosmography" (a detailed account) on natural history, which included cosmology, mineralogy, botany, zoology, and medicinal products derived from plants and animals.

After the conquest of the southern parts of the Roman Empire by Arab troops, the Greek medical texts were translated into Arabic and adapted to the needs of the Arabs. Many of the Greek texts survived only in Arab transcripts. *Ibn Sina* or Avicenna from Afshana (980–1037) wrote the monumental treatise *Canon of Medicine* (Qânûn fi'l tibb; ca. 1020), which was heavily influenced by Galen and which in turn influenced the scholastic traditions, especially of southern Europe. This five-volume book remained the most influential work in the fields of medicine and pharmacy for more than 500 years (together with direct interpretations of Dioscorides' work). While many Arab scholars worked in Eastern Arabia, the Arab-dominated parts of Spain became a second center of classical Arab medicine. An important early example is the "Umdat at-tabîb" (The medical references) by the "Unknown Sevillan botanist." Thanks to the tolerant policies of the Arab regimes, many of the most influential representatives of Arab scholarly traditions, such as Maimoides (1135–1204) and Averroes (1126–1198), were Jews.

In the Christian parts of Europe the texts of the classic Greeks and Romans were copied and annotated during the Middle Ages, especially in the versions that had been passed on by the Arabs. The Italian monastery of Monte Cassino is one of the earliest examples of such a tradition, and others developed around the monasteries of Chartres (France) and St. Gall (Switzerland). A common element of the monasteries was a medicinal plant garden, which was simulateneously used for growing medicinal (and aromatic) herbs required for treating patients and for teaching the knowledge about medicinal plants to the younger generation. The species included in these gardens were common to practically all monasteries, and many of the species are still important medicinal plants of today.

Of particular interest in this context is another text, the *Capitulare de villis* of Charles the Great (Charlemagne, 747–814), who ordered that medicinal and other plants were to be grown in the King's gardens and in the monasteries. He specifically listed twenty-four species of medicinal plants. Walahfried Strabo (808 or 809 to 849), abbot of the monastery of Reichenau (Lake Constance), deserves mention because of his *Liber de cultura hortum* (book on the growing of plants), the first textbook on (medical) botany, and the *Hortulus*, a Latin poem on the medical plants grown in the district (Vogellehner 1987). This poem is not only famous not only because of its poetry, but also because of its vivid and excellent descriptions of the appearance and virtues of medicinal plants.

The climax of the medieval medico-botanical literature was reached in the 11th century with *De viribus herbarum* [On the virtues of herbs] of "Macer Floridus," a Latin poem written in approximately AD 1070 and attributed to Odo of Meung. In this educational poem, 65 medicinal plants and spices are described. Another frequently cited source was *Physica, Causae et curae* by the Benedictine nun, early mystic, and abbess Hildegard of Bingen (1098–1179). She described the medical benefits of plants and included many remedies popularly used during the 12th century. On the one hand, it is an early botanical work, and on the other hand her writings focus on prophetic and mystical topics. The botanical works of both scholars are only available in later copies and in prints. Unfortunately these copies only give a rather distorted idea of the original documents and are heavily reinterpreted texts.

For over 1,500 years the classical and most influential medical book in Europe had been Dioscorides' five-volume *De materia medica* (published ca. 50–70 AD). Until the Europeans' (re)invention of printing in the mid-15th century, these texts were handwritten codices that were almost exclusively used by the clergy and scholars in monasteries. Information about medicinal plants in Europe became much more widely available following the writing of the early herbals which rapidly became very popular and which made information about medicinal plants accessible in the languages of the lay people. Important authors of herbals published during the 16th century include Otto Brunfels, Eucharius Rösslin/Adam Lonitzer, Leonhard Fuchs (German), Nicolás Monardes (Spanish), William Turner, John Gerard (English), Pietro A. Mattioli (Latin/Italian), Garcia ab Horto (Orto) (Portuguese), and Antoine Constantin (French). The early texts were still strongly influenced by Graeco-Roman concepts, but other influences from many sources became increasingly important during the 16th century. The herbals rapidly became available in the various European languages and many of the later authors copied, translated and reinterpreted the earlier books or manuscripts.

The use of medicinal plants has always been an important part of all medical systems, including those in Europe during the various historical periods. Little is known about the popular traditions of Medieval and early modern Europe, since these were generally not written down. Our knowledge starts with the availability of written (printed) records on medicinal plant use by common people (see the previous discussion). As pointed out by Barbara Griggs (1981) a woman in the 17th century was a "superwoman," capable of administering "any wholesome receipts or medicines for the good of the family's health." A typical example of such a remedy is foxglove (*Digitalis purpurea*), reportedly used by an English housewife to treat dropsy and then used more systematically by the physician William Withering (1741–1799). He transformed British herbalism from knowledge simply passed by word of mouth into a science used by medical doctors (see *Digitalis purpurea*, following).

In the 17th and 18th centuries, knowledge about plant-derived drugs expanded, but the attempts to "distillate" the active ingredients from plants were unsuccessful. The main developments during these centuries consisted of detailed observations on the clinical usefulness of the medicinal products recorded in previous centuries or of those brought over from non-European countries. The next main shift in emphasis came in the early 19th century when it became clear that the pharmaceutical usefulness of plants was due to specific molecules, which could be isolated and characterized. This led to the development of a field of research now called natural-product biology or—in the specific case of plants—phytochemistry. Pure chemical entities were isolated and their structure determined. Some of these were then developed into medicines or were chemically modified and then used as medicines. In the second half of the 20th century, micro-organisms became another important source of biologically active molecules and many of these have been developed into medicines. However, the following brief accounts are exclusively on plants that have been of considerable pharmaceutical importance in Europe during the last centuries.

North temperate Europe

Arnica *Arnica montana*
Asteraceae

It is well known that the German poet, philosopher, and natural historian J.W. Goethe (1749–1832) highly valued *Arnica montana*, and that he received a tea prepared with arnica after he had suffered a heart attack in 1823. Today, arnica is still an important medicinal plant, but pharmaceutical uses are exclusively external, for the treatment of bruises and sprains, and as a counterirritant. However, the task of establishing uses for the plant in the Middle Ages and the Renaissance has proven to be a difficult one. Arnica was hardly known in Greek, Roman, and Arabic medicine, and the first reliable evidence dates back to the 14th century (Matthaeus Silvaticus) and the 15th century. The situation was made even more complicated when this species was confused with water plantain, *Alisma plantago-aquatica*. In Jacobus Theodorus Tabernomontanus' *New vollkommentlich Kreuterbuch* (1588), there is a picture of *Arnica montana*. However, the text refers to water plantain (*Alisma plantago-aquatica*). Hence it comes as no surprise that the reported uses of these botanically completely different species are often very similar (especially during the 16th and 17th centuries). In the 16th century *Arnica montana* became an outstanding wound remedy.

Chamomile, German chamomile *Matricaria recutita* (syn. *M. chamomilla*)
Asteraceae

Chamomile is one of the most important medicinal plants of Europe, and has been in use for more than 2,000 years. It is used internally for gastrointestinal and respiratory complaints, as well as topically for inflammatory skin conditions (eczema, wounds, haemorrhoids). Hippocrates called it *euanthemos* (the real or good flower), Dioscorides called it *anthemis* and *anthyllis*, and Galen called it *anthemis* and *chamaimelon* (apple growing on the ground). It continues to be an important medicine in academic as well as popular medicine. The species has been exported to many areas and is today an essential element of medical systems all over the world. *Chamaemelum nobile*, or Roman chamomile, has very similar uses, especially as a (bitter) tonic and a gastrointestinal remedy, but it is chemically rather different. Historically it has been an important medicinal plant in England, France, and Italy.

Chamomile (Matricaria recutita). USDA-NRCS PLANTS Database/Britton, N.L., and A. Brown. 1913. *Illustrated flora of the northern states and Canada*. Vol. 3, p. 521.

European birthwort or snakeroot *Aristolochia clematitis*
Aristolochiaceae

Roots of the European snakeroot or birthwort, *Aristolochia clematitis*, and *A. rotunda* had long been used in medicine as an emmenagogue (to promote menstrual discharge), an abortifacient (to induce abortions), a diuretic, and for the treatment of arthritis. *A. serpentaria* (Virginia snakeroot) was similarly used in the United States, and was also used by Native Americans to treat snakebites. Until the early 1980s several phytomedicines from *A. clematitis* were widely sold in Europe. Its main active constituent, aristolochic acid, was also formerly used in Germany to treat abscesses, eczemas, and other long-lasting skin diseases and as a nonspecific stimulant of the immune system. However in 1981 both the extract and the pure compound were withdrawn due to serious carcinogenic and nephrotoxic (i.e., damaging to kidney cells) effects. A traditional Chinese medical herbal preparation which was used for its supposed weight-reducing effects, and which contained *Aristolochia fangchi* (Chinese Snake Root), known to possess large amounts of aristolochic acid, was used extensively in Belgium during the early 1990s. *A. fangchi* had inadvertently been substituted for another Chinese drug (*Stephania tetrandra*). Kidney failure owing to a progressive form of renal fibrosis due to the use of *A. fangchi* was observed eight to ten years later in many of the patients, and the import of this species is now widely banned.

Fennel *Foeniculum vulgare*
Apiaceae

Fennel and fennel seeds are a classic household remedy used as a cure for flatulence. It was important to Greeks and Romans and included in the *Capitulare de villis* as well as in the *Hortulus* of Walahfried Strabo (809 AD). It was reputed to ward off evil spirits. In his 16th century herbal, Leonhard Fuchs recommends the aerial parts and the seeds for stimulating lactation and as a diuretic. The seeds are also reported to be useful if one has a "weak stomach." Another of Fuch's uses is for the macerated root: "if one puts it in honey it protects one against angry dogs" (sic) (Fuchs 2001).

See: Gathering Food from the Wild, p. 40; Herbs and Vegetables, p. 103

Feverfew *Tanacetum parthenium*
Asteraceae

Feverfew has been used as a bitter tonic and an antipyretic (i.e., to reduce fever) for many centuries; there was renewed interest in the plant in the 1990s as a potential treatment for migraines. *Tanacetum parthenium* is known in South America as Santa Maria and is an important species in

Feverfew (*Tanacetum parthenium*). J.S. Peterson @ USDA-NRCS PLANTS Database.

Foxglove. Copyright Dave Webb, used with permission.

many indigenous medical systems, especially in South and Central America, where it is used for menstrual disorders.

Foxglove *Digitalis purpurea* and *Digitalis* spp.
Scrophulariaceae

Foxglove, *Digitalis purpurea*, is one of the best-known examples of a remedy developed from close observation of the pharmacological effects of the species. Digitoxin from this and related species is still extracted from several members of the genus and is used today in the treatment of coronary disease. It was reportedly used by an English housewife to treat dropsy, and then used more systematically by the physician William Withering (1741–1799). Prior to William Withering, herbalism in Britain spread simply by word of mouth; his endeavors transformed it into a form of science used by medical professionals. Herbalism during this period was more of a clinical procedure concerned with the patient's welfare, than a systematic study of the virtues and chemical properties of medicinal plants.

Horse chestnut *Aesculus hippocastanum*
Hippocastanaceae

The seeds of the horse chestnut are used topically and orally for chronic varices (circulation problems in veins) and other inflammatory circulatory disorders. The active ingredient is a mixture of saponins collectively called aescin. The species originated in the Balkan peninsula or Anatolia, but is now cultivated in many regions of the world. In Europe the first trees were planted in Vienna in 1576 (Clusius), and the species was also introduced into North America. in the middle of the 19th century. The horse chestnut is commonly used in Continental Europe, and the German medical control agency published a "positive" (i.e., approved) monograph, or legally binding analysis of the accepted medical uses of the species, that was adopted by the health authorities (BFA 1994). However, the horse chestnut is of lesser importance in North America.

See: Wood, p. 323

Mint *Mentha* spp.
Lamiaceae

Peppermint, *Mentha × piperita*, is a very widely cultivated medicinal and aromatic plant, and is particularly important for the production of menthol, which is used as a flavouring agent. Its medical uses are to treat flatulence and gastrointestinal cramps. The species is also used to flavour chocolate and in liqueurs. It is a hybrid of *M. aquatica* × *M. spicata*, which spontaneously originated in 1696 in an English garden. It can only be propagated via its runners. Today peppermint is one of the most important medicinal and aromatic plants, and is widely grown in some parts of Europe (southeastern Europe, Spain, Germany) and especially in the United States. It was reportedly first brought to the United States in the early 19th century and cultivated in the state of Massachusetts. From there the production spread to other East Coast states and to the northern Midwest (Michigan in 1835), and eventually to the West Coast (Oregon in 1919). In the 1940s production started in Wisconsin. Today the major production centers of the world are the Columbia River Basin, Washington, the Willamette Valley, Oregon, and the states of Indiana and Idaho. Another important representative of the genus, *M. spicata*, or spearmint, is menthol-free and commonly used to flavor toothpaste. The production of spearmint started later and its extension followed that of peppermint. It is mainly grown in the state of Washington.

Mentha arvensis, or corn mint, originally from Eurasia, is another prominent member of this genus used medically in numerous cultures of the world. The closely related *M. canadensis* is the Japanese peppermint used for a large variety of illnesses because of its high menthol content.

See: Fragrant Plants pp. 249–50; Herbs and Vegetables, pp. 106–7

Scotch marigold *Calendula officinalis*
Asteraceae

Scotch or pot marigold is used pharmaceutically, for example, to treat chilblains, chronic wounds (including ones which heal very slowly), and other inflammatory conditions of the skin. In some rural regions of Europe (e.g., Switzerland and the German Black Forest) the species is still widely used externally as a veterinary anti-inflammatory remedy. Normally the flower heads are mixed with lard or another animal fat, heated for a short period in order to extract the active lipophilic constituents, and the resulting product is stored for usage. The flower heads are also used to color butter and to thicken soups.

The species originated in southeastern Europe and is today widely cultivated (most commonly as a garden plant). It is uncertain whether the species was already used by the Greeks and Romans. The core problem is the lack of proper botanical identification of plants with names similar to the ones used for *Calendula* in much later periods. The first certain evidence is from the *Physica* (first printed 1533) of Hildegard of Bingen (1098–1179). According to Mayer and Czygan (2000b), this points to use in the popular (peasant) medicine of this period and later became a constant element of the various medieval texts (especially the later Circa Instans). This species is an example of the continuous usage of a medicinal plant for similar medical indications.

See: Ornamentals, p. 262

St. John's wort *Hypericum perforatum*
Hypericaceae

St. John's wort is a perennial which plays a primary role in modern evidence-based phytotherapy. The aerial parts of the plant have been used as an anti-inflammatory, healing application for wounds and as an antimelancholia remedy since the time of the ancient Greeks. In the Middle Ages many believed it to have magical powers to protect one from evil. Early Christian mystics named the plant after John the Baptist and it is traditionally collected on St. John's Day, June 24, and soaked in olive oil for several days in order to produce a red anointing oil to be used externally. This practice is still very common in many folk medicines of eastern and southern Europe, and some pharmacological evidence in support of the usage exists.

Several controlled studies have shown positive results when using standardized St. John's wort extract to treat patients with mild depression. Improvements were shown with no reported

St. John's wort (*Hypericum perforatum*). Jennifer Anderson @ USDA-NRCS PLANTS Database.

side effects. The active constituents in the plant (there are over fifty) include antraquinones (hypericin and pseudohypericin), prenylated phloroglucinol derivatives (hyperforin), and flavonoids.

However, the side effects of St. John's wort include stomach complaints, fatigue, and especially allergic reactions (photodermatitis). Most authorities warn strongly that *Hypericum* extracts should not be taken alongside other anti-depressants; doing so can result in a syndrome called serotonin syndrome.

See: Ornamentals, p. 282

Valerian (root) *Valeriana officinalis*
Valerianaceae

The combined root and rootstock of the (common) valerian with its characteristic "old socks" smell is used pharmaceutically to treat sleep disorders and other related conditions. It is a licensed medicine in many European countries and is currently grown commercially in Russia, Japan, the United States, Belgium, Holland, and Germany. The species has been used through the centuries in many European regions. L. Fuchs lists a variety of uses including as a diuretic and gynecological aid, but he does not include uses for sleep disorders.

White willow *Salix alba* and *Salix* spp.
Salicaceae

The genus *Salix* includes numerous trees and shrubs common in alpine ecosystems and along the margins of streams. The white willow, *Salix alba*, is a tree that commonly grows in areas periodically flooded along streams and lakes. Willow bark (known to pharmacists as *Salicis cortex*) is a European phytomedicine with a long tradition of use for treatment of chronic pain, rheumatoid diseases, fever, and headache, and one of its main compounds, salicine, served as a lead substance for aspirin (acetyl<u>salic</u>ylic acid). Leonard Fuchs devotes a chapter in his *New Kreütterbuch* (1543), illustrated with three drawings of different species, to the various "classes" of willow. The leaves are reported to be good for treating some gastrointestinal complaints, and the bark to be useful for treating warts and corns. Acetylsalicylic acid is today used in a similar way. Fuchs's use of willow as a treatment for podagra (i.e., gout, especially of the big toe) mirrors modern uses in the treatment of a variety of chronic inflammatory conditions. A positive monograph for its use in fever-related diseases, rheumatic complaints, and headache was published by the German medical control agency (BFA). The drug is also monographed in the European Pharmacopoeia. Clinical evidence, including two double-blind studies, points to the effectiveness of willow bark for the treatment of chronic pain and rheumatoid diseases. This potency as a treatment is the result of more than one substance (most of them so far unknown). The extract is effective at a dosage of salicylates (a group of compounds chemically related to salicylic acid) which is much lower than the one reported for aspirin. However, the mechanism of action of this complex mixture and the individual compounds responsible for this activity remain unknown.

See: Ornamentals, p. 284; Materials, p. 348

Wormwood *Artemisia absinthium*
Asteraceae

Wormwood or absinthe has been used for many centuries not only as a bitter tonic, stomachic (strengthener of the stomach), and stimulant, but also in the production of liqueurs (e.g., vermouth) and it remains a useful phytotherapeutic option for mild gastrointestinal disorders. It was known to Wallafried Strabo in the 9th century (Vogellehner 1987) and is the first species mentioned in Fuchs' *New Kreüterbuch* (1543). There it is recommended for a number of gastrointestinal problems, and for a large variety of other disorders. The species is an example of how systematic exploration of its clinical effectiveness and pharmacological effects over the last few centuries has helped to reduce the uses of wormwood to therapeutically validated ones. One of its constituents, the neurotoxic monoterpenoid (i.e., a terpenoid with ten carbons in the skeleton) thujone, is concentrated in vermouth wines, especially if they are distilled, but the use of such products has been outlawed in Europe since the 1920s. Consequently, thujone-free species and varieties are used in the production of such wines.

See: Caffeine, Alcohol, and Sweeteners, p. 180

Yarrow *Achillea millefolium*
Asteraceae

Yarrow has been used as a medicine for many centuries. The species is native both to temperate Europe and to North America. Particularly common in Europe were uses associated with inflammatory conditions (e.g., in the form of an externally applied poultice). Even though the species had been an important medicinal plant in many European countries and, for example, is mentioned by Fuchs as a useful treatment for a variety of inflammatory conditions such as wounds and abscesses, today it is of limited importance and is only used in some regions, as a popular medicine.

See: Ornamentals, p. 276; Herbs and Vegetables, p. 106

Mediterranean and the Near East

Alexandra senna *Senna alexandrina* and Tinnevelly senna *S. angustifolia*
Fabaceae

Both species are of desert origin: Tinnevelly senna, *Senna angustifolia*, is native to Arabia, West Africa and Asia, as far as Punjab, while Alexandra senna, *S. alexandrina*, grows naturally in northeastern Africa and it is harvested and cultivated in Sudan, China, and India.

About 1,000 years ago the Arabs introduced the use of dried leaves and especially fruits of senna into Western pharmacopoeias as a laxative. Senna was mentioned in detail by Ibn al-Baytar (1197–1248), one of the most important Arabian scholars of the Middle Ages and the author of the famous medical treatise *Jami' al-mufradat*. Over the centuries senna has proved its worth as an herbal drug and today represents one of the most widely used herbal drugs in the classical pharmacy.

Artichoke *Cynara cardunculus*
Asteraceae

Formerly known as *Cynara scolymus*, the artichoke is the best example of a food-medicine in the whole of European phytotherapy. Artichokes originated in the Mediterranean region and numerous diverse cultivars were subsequently developed. Many Mediterraneans used artichokes by soaking them in wine, then drinking the liquid as a digestive and a reconstituent for various illnesses. Today the culinary use of the fleshy flower receptacles and the base of the bracts is known worldwide. However, the utilization of the leaves in modern phytotherapy is quite recent and focuses on their cholekinetic (increasing the flow of bile acids), antihyperlipidemic (lowering the level of lipids in the blood), and hepatoprotector (liver-protecting) properties. Artichoke leaves are the main ingredients of the Italian bitter liqueur Cynar, which has contributed significantly to changes in Italian social habits. During the 1960s and 1970s Cynar was very popular and drinking it after a meal became a "must" in many urban milieus.

See: Herbs and Vegetables, pp. 119–20

Chaste tree *Vitex agnus-castus*
Lamiaceae

This shrub is native to the Mediterranean and northwestern India. Dried fruits of *Vitex agnus-castus* seem to be one of the oldest phytomedicines, dating back to the beginning of European civilization. The Greek name of this plant, "lygos" (pliable branch), hints at its usage in viticulture for staking vines and in livestock farming for pasture fences. The other Greek expression, "agonos,"meaning chaste and pure, distinguishes it as a feminine plant of goddesses like Hera, Demeter, and Artemis (Diana), a cult plant of womanliness. Agonos is the source of the medieval name "agnus-castus" (the chaste lamb). The ancient Greek physicians Hippocrates, Theophrastus, and Dioscorides all made reference to it, as did the Greek-Roman historian Pliny The Elder. Both Dioscorides and Pliny reported its use in suppressing the libido. The Greeks' use closely resembled modern practices: they recommended it as an aid in healing of external wounds and complaints of the spleen, and for use in child birth. The English believed it would suppress the libido, as did the Catholic church, which had it placed in the pockets of novice monks in order to help them in fulfilling their vow of chastity. Adam Lonicer (1679) wrote that the aerial parts of the species were macerated into wine and honey, and early American physicians used the fruits to stimulate lactation and as an emmenagogue.

In the Mediterranean the fruits are sometimes used as a diuretic, to induce production of milk, and as an antirheumatic treatment. Recent studies have confirmed the pharmacological validity of its use in relieving complaints associated with premenstrual syndrome. Specifically, it is effective in relieving insufficiency of the corpus luteum (luteal phase defect). These fruits, and also the aromatic leaves, have been occasionally used as a condiment and pepper substitute in southern Europe and northern Africa, and both are part of the legendary Moroccan spice mixture "ras el hanout."

Garlic *Allium sativum*
Alliaceae

Garlic represents one of the oldest medicinal plants in the Mediterranean. Garlic was first discovered in the wild, probably in the Iranian region, and is well documented by textual and archaeological records in Mesopotamia and Egypt from 3000 BC. The early Sumerian diet included garlic as a mainstay, and garlic is mentioned in the *Shih Ching* (The Book of Songs), a collection of ballads said to have been written by Confucius. Garlic was highly prized in ceremony and ritual, and it is said that lambs offered for sacrifice in China were seasoned with garlic to make them more pleasing to the gods. Herodotus wrote that the Egyptians fed it to slaves building the pyramids in order to increase their stamina. Workers deprived of their ration of garlic went on strike. An Egyptian papyrus from 1500 BC recommended garlic for twenty-two ailments. It represented a kind of combination food–medicine.

In the fourth book of the Bible, it is reported that the Jews returning to Sinai had nostalgia for the eating of garlic, which they had known and appreciated in Egypt. In ancient Greece and Rome, it was claimed to have additional uses, such as repelling scorpions, treating dog bites and bladder infections, and curing leprosy and asthma. By 1000 AD garlic was grown in virtually the entire known Medieval world, and was universally recognized as a valuable plant. Many cultures elevated garlic beyond a dietary staple, and suggested that it had medicinal and even spiritual uses. Philosophers and scholars credited garlic with many virtues. Aristophanes suggested that athletes and those going into battle should eat garlic to enhance their courage. Pliny wrote about garlic's ability to cure consumption and numerous other ailments. Virgil commented that garlic enhanced and sustained the strength of farm workers. Celsius recommended garlic as a cure for fever. Hippocrates thought it was a good medicine for many health problems, and Mohammed, the prophet, claimed that if garlic was applied directly to a sting or a bite wound, it would ease the pain. For many centuries there was a widely held belief that garlic would keep evil at bay. Wreaths of garlic hung outside a door or dwelling were believed to ward off witches or vampires. Bull fighters may well wear cloves of garlic around their neck during a bullfight in order to protect themselves from the vicious horns. In the Middle Ages it was thought to prevent the plague.

In the Middle Ages the use of garlic as a medicine and food was spread to central Europe mainly by the Jews; the Jewish habit of frequently eating garlic became one of the characteristics reviled by the Nazi regime during the 1930s and 1940s. Later, in the 1960s, garlic was spread further by immigration of the southern Europeans. Louis Pasteur first documented that garlic had an anti-bacterial property. in 1858, and nowadays garlic preparations play a primary role in modern phytotherapy as a treatment for high blood pressure and to reduce high levels of lipids in the blood. Some clinical evidence points to garlic's effectiveness in treating these conditions.

See: Herbs and Vegetables, p. 104

Licorice *Glycyrrhiza glabra*
Fabaceae

This shrub originated in the semi-arid areas of the Eastern Mediterranean, Near East and Central Asia. The oldest report on the use of the roots of this species comes from a Sumer tablet of Mesopotamia (2000 BC), and shortly thereafter the use of *Glycyrrhiza uralensis* in China is documented. By then the sweet taste of the licorice root was already known (its botanical name, *Glycyrrhiza*, sweet root, comes from Greek). Alexander the Great, the Scythian armies, Julius Caesar, and even India's great prophet, Brahma, were known to have endorsed the benefits of licorice. Arab physicians used licorice to treat coughs and relieve side effects of laxatives.

Licorice began to be cultivated intensively in Syria, southern Italy, France, and Spain, particularly from medieval times onward. The tradition of licorice use then spread to reach central Europe, where St. Hildegard of Bingen first wrote about its use in the 12th century, and England, where the roots were used to aromatize beer in the 14th century. For over 3,000 years, licorice root has been used by Mediterranean and Near Eastern populations as a remedy for sore throats and coughs, and sometimes to heal ulcers. These two medical applications remain the basis on which *G. glabra* continues to be important in modern evidence-based phytotherapy. Licorice, which is now mainly cultivated in Turkey, the former Soviet Union, and China, is also used in the food industry. A decoction of its washed roots, filtered and concentrated, continues to be one of the favorite snacks for children and adults.

See: Spices, p. 165

Poppy *Papaver somniferum*
Papaveraceae

Papaver somniferum has long been popular in the Near East and in the Mediterranean as a remedy and source of pharmaceuticals. It is one of the prime examples of a medicinal plant. Illustrations of the Greek and Roman gods of sleep, Hypnos and Somnos, show them wearing or carrying poppies. Classical Greek physicians either ground the whole plant or used opium extract. Galen lists its medical properties, noting how opium "resists poison and venomous bites, cures chronic headache, vertigo, deafness, epilepsy, apoplexy, dimness of sight, loss of voice." Dioscorides described both the latex of the capsules (*opos*) and the extract of the whole plant (*mekonion*). By the 8th century AD, opium use had spread to Arabia, India, and China. The Arabs both used opium and organized its trade. Thomas Sydenham, 17th-century pioneer of English medicine, wrote, "among the remedies which it has pleased Almighty God to give to man to relieve his sufferings, none is so universal and so efficacious as opium." In 1874, English pharmacist Alder Wright had boiled morphine and acetic acid to produce diacetylmorphine, which was synthesised and marketed commercially by the German pharmaceutical giant Bayer. In 1898, Bayer launched the best-selling drug of all time, heroin.

Opium, the latex obtained by making a cut in the unripe capsules of this species, was imported into England from Iran as early as 1870. Today it still represents the industrial source used for isolating the opium alkaloids, which play a central role in the modern pharmaceutical chemistry as analgesics (morphine-derivates) and as a treatment for coughs (codeine-derivates). Today morphine is isolated from opium in large quantities—over 1,000 tons per year—although most commercial opium is converted into codeine by methylation (a simple chemical process which replaces an –OH group by an –O–CH$_3$ group). On the illicit market, opium gum is filtered into morphine base and then synthesized into heroin.

In the popular folk medicine of southern Europe and in the Near East, low doses of *P. somniferum* fruits, as well as the flowers of *P. rhoeas*, are still occasionally used as a tranquillizer.

See: Psychoactive Plants, pp. 199–200; Spices, pp. 167–8

Sage *Salvia officinalis*
Lamiaceae

Sage is native to the eastern Mediterranean region. The plant was already known in the medical practices of the Greek and Roman period, mainly as a digestive and cough remedy. At that time, its role as a culinary herb had yet to be discovered. In his recipe books, Apicuius refers to the seeds of sage, but not the leaves. Sage became an important medicinal plant for Mediterranean monks of the Middle Ages who dispersed it to central Europe. The medical treatises of the Tuscan physician Pietro Andrea Mattioli (16th century) and the German physician Adam Lonicer from Marburg describe the use of sage leaves as a mouth antiseptic, to heal bronchial diseases, as an appetizer and an emmenagogue (to promote menstrual discharge). The essential oil of sage contains thujon, the compound that makes the dried leaves of sage an important herbal drug today, to heal mild infections of the mouth and upper respiratory tract. It is widely used in over-the-counter phytotherapeuticals.

See: Herbs and Vegetables, p. 109

Northern Asia

In the history of medicinal plant use in eastern Asia and Siberia, a very important school of medical practice, traditional Chinese medicine, links practices from a number of traditions that have been handed down by word of mouth (as in Siberia or northern China) and for which written historical sources are very rare and poorly investigated (e.g., Mongolian traditional medicine and the Tibetan school).

The Chinese *Materia Medica* has been growing throughout the last 2,000 years (Benski and Gamble 1993). This increase results from the integration of drugs into the official tradition from China's popular medicine as well as from other parts of the world. The first major *Materia Medica* after Tao Hong Jing was the *Xin xiu ben cao* 659 AD, also known as Tang Materia Medica, which was the official pharmacopoeia of the Tang dynasty. It contained 844 entries and was China's first illustrated *Materia Medica*. *Zheng lei ben cao*, 1108 AD, was the major medical treatise during the Song dynasty and contained 1,558 substances. However, China's most celebrated medical book is represented by Li Shi-Zhen's *Ben cao gang mu*, posthumously printed in 1596 AD, with 1,173 plant remedies, 444 animal-derived drugs and 275 minerals. This tradition has continued into the modern era with the publication in 1977 of the Jiangsu College of New Medicine's monumental 25-year project entitled *Encyclopedia of the Traditional Chinese Medicinal Substances* (*Zhong yao da cin dian*) containing 5,767 entries. Traditional Chinese medicine and Chinese plant remedies are nowadays used throughout Western countries, whereas other botanico-medical traditions remain largely restricted to their regions of origin.

Chinese foxglove *Rehmannia glutinosa*
Scrophulariaceae

Although it belongs to the same family as European foxglove and is sometimes known as Chinese foxglove, *Rehmannia glutinosa* has a completely different usage in traditional Chinese medicine.

R. glutinosa represents the most widely used drug of the Pharmacopoeia of China, where the plant, called *di huang*, is one of the most popular tonic herbs and is considered to be one of the fifty fundamental herbs. The drug is found in the root, which is used in different ways (raw, dried, charcoaled). Charcoaled *R. glutinosa* root is used to stop bleeding and to tonify (a form of balancing the body) the spleen and stomach. The fresh root is used to treat thirst, the rash from infectious diseases, and bleeding due to pathological heat. The dried root is used to treat bleeding due to blood deficiency and to nourish the vital essence, and the prepared root is used to treat chronic tidal fever, night sweats, and dizziness and palpitations due to anaemia or blood deficiency.

The root of Chinese foxglove is one of the ingredients of the "Four Things Soup," the most widely used woman's tonic in China, the other three ingredients being *Angelica sinensis*, *Ligusticum wallichii*, and *Paeonia lactiflora*. *Rehmannia* root is another example of a food–medicine and it is eaten after having been boiled "nine times."

Chinese rhubarb *Rheum officinale, R. palmatum,* and their hybrids
Polygonaceae

Chinese rhubarb originally grew wild in northwestern China, the best varieties coming from mountainous regions of Kansu Province, and in eastern Tibet. The dried roots of rhubarb have been used as a laxative for thousands of years, and it is known in the Chinese Pharmacopoeia by the name *dà huàng*.

Chinese rhubarb is mentioned in historical sources from 2700 BC and was cultivated for medicinal purposes. Wu, an emperor of the Liang dynasty (557–579), was advised to cure his fever, but only after being warned that rhubarb, being a most potent drug, had to be taken with great moderation. During the Yuan dynasty (1115–1234) a Christian sentenced to hard punishment was pardoned after using rhubarb to heal some soldiers. Marco Polo, who knew the Chinese rhubarb rhizome very well, talked about it at length in the accounts of his travels in China, while Chinese rhubarb was already widely used in European pharmacy, especially in the School of Salerno, possibly introduced through Arabic medicine. Later on, the Chinese emperor Guangzong (reigning from 1620 to 1621) was miraculously cured with rhubarb from a severe illness he contracted after having had a joyful time with four "beautiful women." In 1828 the Daoguang emperor sent out an edict stating no more tea and rhubarb could be sold to Western merchants. Rhubarb has played an important role in the historical Western Pharmacopoeias of the last centuries and it is still used today for its peculiar phytotherapeutical properties: it acts as an astringent at low doses (0.1–0.2 g of dried drugs) and laxative at high doses (1.0–2.0 g).

Ephedra, Ma huang *Ephedra sinica*
Ephedraceae

Ephedra sinica represents one of the oldest medicinal plants in China, where it is known by the name of *ma huang*. It is estimated that its use began 4,000 years ago, particularly in northern China and Mongolia. Ephedra was used in ethnomedicine as a stimulant, to increase perspiration, and as an anti-inflammatory. In the Chinese school of medicine, a preparation called *mimahuang*, containing roasted honey and chopped dried aerial parts of this species, is claimed to be an effective treatment for flu and respiratory tract inflammations. Ephedra, which contains ephedrine and similar alkaloids, has been used extensively in the ancient pharmacy as an antihistaminic in the treatment of asthma and as a natural decongestant. It has become a very popular ingredient in herbal combinations for allergies and hay fever. Since it is a central nervous system stimulant and increases the metabolism and increases body temperature, it has been used to control weight and to help prevent sleep, and by athletes in order to improve their performance. One of the side effects of increased metabolism is an increased pulse rate and a potentially dangerous rise in blood pressure. Occasionally, *ma huang* has been mixed with other stimulants or food supplements and used as a narcotic stimulant. Dangers stemming from lack of knowledge about the pharmacological effects of the drug caused several American states to ban ephedra in 2003, followed by a federal ban in 2004 on sales of products containing ephedra. It is currently on the doping lists of banned substances and ingredients of all the international sport associations. *E. equisetina* in China and *E. gerardiana*, *E. intermedia*, and *E. major* in Pakistan have also been used in a similar way.

Dried aerial parts of several Ephedra species, including *E. trifurca*, are known in the US as "Mormon tea," a very popular stimulant beverage made by Native Americans and the early settlers in the southwestern United States and Mexico.

See: Psychoactive Plants, p. 199

Ginkgo *Ginkgo biloba*
Ginkgoaceae

The name *ginkgo* is thought to come from the Chinese word *sankyo* or *yin-kuo*, meaning "hill apricot" or "silver fruit." This refers to the seeds, produced by female trees, which resemble apricots, but have a smell like rotting flesh or rancid butter (due to their content of butanoic and hexanoic acid). *Ginkgo biloba* was originally found in China and Japan, where it was cultivated as a temple tree. Fossilized leaf material from the Permian is remarkably similar to the modern *Ginkgo biloba*. Ginkgo had survived in China, where it was mainly found in monasteries in the mountains and in palace or temple gardens, where Buddhist monks cultivated the tree from about 1100 AD. It was spread by seed to Japan (around 1192 AD, linked with Buddhism) and Korea. Ginkgo is considered a sacred tree of the East and it is associated with longevity. In the 11th century (Sung dynasty) gingko is referred to in literature as a plant native to eastern China. Prince Li Wen-ho (first half of the 11th century), who came from the south, transplanted it to his residence, after which it became well known and was spread through propagation. Ginkgo has been depicted on Chinese paintings and appeared in poetry from that time onward. Scientists thought that it had become extinct, but in 1691 the German Engelbert Kaempfer discovered *G. biloba* trees in Japan, and it was brought to Europe (Utrecht) in 1730.

The earliest known medicinal use dates back to 2800 BC and is ascribed to the pseudofruits of *G. biloba*. They have been used as both food and medicine: the pseudofruit, prepared by fermentation and cooking, have been considered a delicacy during weddings and feasts, while ginkgo seeds are used in traditional Chinese medicine as a kidney "yang" tonic, to increase sexual energy, halt bedwetting, and soothe bladder irritation. The seeds are boiled as a tea used to treat lung weakness and congestion (especially asthma), wheezing, coughing, vaginal candidiasis, frequent urination, cloudy urine, and excess mucus in the urinary tract. In Malaya, the seeds are popularly used for making desserts and are recommended for their beneficial effect on the brain, circulation, and eyes. Interestingly, the leaves are much less frequently used in eastern Asia and include the treatment of chilblains (reddening, swelling, and itching of the skin due to frostbite), and as a throat spray for asthma. Today, ginkgo leaf extracts have enjoyed worldwide popularity and have stepped into the "herbal spotlight." This has been due to heavy media coverage and also to a number of very interesting clinical findings, recently supported by many pharmacological studies of its main constituents (diterpenes [ginkgolides], sesquiterpenes [bilobalide], and biflavonoids). *G. biloba* leaf derivatives have shown to be very effective in the treatment of intermittent limping due to chronic thickening of the arteries, senile dementia, vertigo, tinnitus, and hearing loss caused by severely reduced blood supply.

See: Nuts, Seeds, and Pulses p. 137

Qing hao *Artemisia annua*
Asteraceae

Artemisia annua has been used for hundreds of years in traditional medicine in China. The leaves were harvested in the summer, before the flowers appear, and dried for later use. The dried leaves are generally used in the treatment of fever, malaria, colds, diarrhea, as a digestive, and, externally, as a wound remedy. As recently as 1971 a Chinese research group isolated the active principle, the sesquiterpene lactone artemisinin, which proved to be very effective against the malarial parasite *Plasmodium falciparum*, and particularly towards chloroquine-resistant malaria.

In an attempt to overcome the frequent problem of many patients having a recurrence of the illness one month after the treatment, a number of derivatives of artemisinin have been developed (ethers, such as artemether and arteether, and esters, such as sodium artesunate and sodium artenlinate). Although artemisinin has been synthesized, this process is complex and not economically viable, and it is currently still extracted and isolated from the aerial parts of *A. annua*. Today artemisinin and its derivatives represent a promising group of antimalarial agents.

Korean ginseng *Panax ginseng*
Araliaceae

Korean ginseng is a small herbaceous plant growing wild in mountainous areas from Nepal to Manchuria, and from eastern Siberia to Korea. The earliest mention of it comes from a book of the Chien Han Era (33–48 BC), but by the 3rd century AD China's demand for ginseng created international trade of the root which allowed Korea to obtain Chinese silk and medicine in exchange for wild ginseng. Long before Vasco da Gama opened up a sea route to Cathay (China) in 1497, the "Three Kingdoms" (the Koreas and part of Manchuria) had a thriving trade selling ginseng to China. Soon after, word began to filter through to Europe about the northern woodland plant that had miraculous healing and restorative powers. The first reference to ginseng in Europe dates from 1643 and is a traveller's report published in Rome. In 1653, Hendrick Hamel of Holland and his fellow seamen were sailing from Formosa to Japan; during a lengthy storm they were blown off course and ran aground on Cheju Island, Korea. They were taken into custody by the Korean government and held on what was meant to be a permanent basis. In Hamel's diary, he noted that Korea paid "tribute" to China entirely in the form of ginseng. Three times a year an envoy would arrive from China to collect it. In 1709, a Jesuit named Father Jartous returned to Europe from an assignment in China. In the Memoirs of the Royal Academy in Paris he wrote of the amazing medicinal ginseng root he had learned about. Later it was translated into English for the *Philosophical Transactions of the Royal Society of London*.

Although probably originally used as food, it quickly became revered for its strength-giving and rejuvenating powers, and its human shape became a powerful symbol of divine harmony on earth. Ginseng roots often resemble a human-like figure, which is why they are referred to as "human-root." The more human-like the root appears, the higher its value. The age of the root plays an important factor in assuming this human-like shape—thus, older roots are more valuable than younger ones. For customers in China, the shape of the root is very important: a 'skinny' root does not have the beauty that a larger one has and, hence, is not as valuable. Chinese herbalists have prescribed ginseng root for centuries to treat problems of the digestive and pulmonary systems, nervous disorders, diabetes, and a low sex drive in men. It is said to increase energy and improve memory. Ginseng is mainly used in modern evidence-based Western phytotherapy as an "adaptogen" (a substance which helps the body to deal with, or adapt to, stress or adverse external conditions): to stimulate the central nervous system, increase resistance to fatigue and stress, and to improve memory.

Interestingly, in 1711, Father Joseph Francois Lafitau, a Jesuit, was sent from France to a mission near present-day Montreal, Canada. Jartous' writings eventually reached him in 1716. Realising that the local latitude was about the same as the area in China where ginseng grew, he wondered if some might grow in his vicinity. When he showed a drawing of the ginseng plant to the Indians, they immediately took him to a similar plant nearby, which was American ginseng, *Panax quinquefolium,* growing wild in the eastern half of North America. The Indians used it in ways somewhat different to those of the Chinese: as a tonic, for healing wounds, as a headache cure, to soothe eyes and muscular cramps, and to cure croup in children. Nowadays, American ginseng, together with other ginseng species (Japanese ginseng, *P. japonicus*; Himalayan ginseng, *P. pseudoginseng*; and San-chi ginseng, *P. notoginseng*) are sold widely in the United States and Europe, and are prescribed for similar medical conditions as the Korean ginseng.

Siberian ginseng *Eleutherococcus senticosus*
Araliaceae
Siberian ginseng is a thorny bush common in western Siberia, from the Amur to Sakhalin Island, and is commonly used in the popular medicine of Siberian populations. Russians in particular have used its roots as a cheaper substitute for the expensive Korean ginseng. Cosmonauts have relied on this nourishing herb to help them manage their rigorous training, while Olympic athletes have often used Siberian ginseng to maximize their potential. Siberian ginseng is considered by modern phytotherapy to be an "adaptogen" (a substance which helps the body to deal with, or adapt to, stress or adverse external conditions), as is Korean ginseng.

Southern and Southeastern Asia

India

The current practices within traditional Indian medicine reflect an ancient tradition that can be traced back to at least 900 BC, to written Ayurvedic records (Kapoor 1990; Mukherjee 2001). These practices, all holistic in nature, are divided into three principal systems: Ayurveda, Siddha, and Unani. The most ancient is Ayurveda, literally meaning the "science of life," and has a basis in the spiritual as well as the temporal. The practice of Ayurveda is aimed at the intrinsic whole of the patient and involves the administration of medicinal preparations of complex mixtures containing animal, plant, and mineral products. Siddha can be considered similar to Ayurveda and is governed by the understanding that everything, including the human body, is made up of the five basic elements: earth, water, fire, air and space. In addition, 96 major elements are considered to constitute human beings and include the constituents of physiological, moral, and intellectual elements. An imbalance among any of these is believed to result in disease. Siddha medicine is based more on a psychosomatic system in which treatments are based on minerals, metals, and herbal products. The Unani medical system can be sourced to the writings of the Greek philosopher–physician Hippocrates (460–377 BC). The system was introduced by Arabic practitioners seeking refuge in India from the Mongol hordes invading Persia and central Asia. Unani medicine is based on the medical history of a patient and still includes a major element of plant drug administration that compliments dietary and physical regimes. However, in this case the drugs are usually single herbs administered for specific diseases, rather than the herbal mixtures commonly found in Ayurveda or Siddha.

Ginger *Zingiber officinale*
Zingiberaceae

Ginger is one of the world's most commonly used culinary spices. The medicinal use of ginger has an ancient history and can be traced back to Greek and Roman times. In Indian medicinal practice the plant has been mentioned in religious scriptures dating back to 2000 BC. The rhizome of the plant is used and is best recognized in Ayurvedic medicine as an aid to digestion and to treat rheumatism/inflammation. Trikatu ("three spice") is an example of an Ayurvedic remedy containing ginger, used to improve digestive and respiratory function, which also contains black pepper (*Piper nigrum*) and Indian long pepper (*Piper longum*). Across the world ginger is a popular food additive to aid digestion. Some of its more unusual uses are to promote hair growth (Japan), treat nerve disorders (India), and as an aphrodisiac (Cuba, Yemen). Ginger is acknowledged in Ayurveda to be effective against nausea and neurological dysfunction. Ginger consumption has also been reported to have a beneficial effect in alleviating the pain and frequency of migraine headaches.

Studies on the effectiveness of ginger in rheumatic conditions have shown a moderately beneficial effect, but falls short of proving ginger to be a positive, nontoxic remedy for such conditions.

See: Spices, pp. 163–4

Indian pennywort *Centella asiatica*
Apiaceae

Indian pennywort is a creeping herbaceous plant that prefers a moist habitat and is found near reservoirs and streams. This pantropical species is commonly found in India and throughout Sri Lanka, Madagascar, China, South Africa, the southeastern United States, Mexico, Venezuela, and Colombia. It is possible that *Centella asiatica* is identical with the medicinal plant called mandukpani in the Sanskrit text of Susruta (ca. 1200 BC), and today it is used not only in Ayurvedic medicine, but in a large number of medical systems, for a great variety of diseases, and throughout its native range. The major application of *C. asiatica* has been in the treatment of leprous lesions, first reported in 1887. The active triterpenoid principles were identified 70 years later, but in clinical trials have only been found to be effective in aiding wound healing, vein problems, and striae gravidarum (stretch marks of pregnancy). Standardized extracts of *C. asiatica* are today incorporated into various ointments and tablets. Other potential pharmaceutical uses are also currently being investigated.

Lemon Grass *Cymbopogon citratus*
Poaceae

Lemon grass, named after the prominent and relatively strong odor of the crushed leaves, can grow up to 3 feet (1 m) high, with long slender leaves. The base of the stalk is used extensively in oriental culinary dishes and as a tea. Its center of origin is unknown. Today it is widely cultivated throughout the tropics and is an important element of popular medicine in many countries. For example, in Cuba it was shown in a poll conducted in the 1980s to be one of the most widely used medicinal plants. The pharmaceutical uses are extremely diverse, but its use as a carminative (to treat flatulence) and for other gastrointestinal disorders seems to be most relevant, owing to the presence of essential oil.

See: Fragrant Plants, pp. 247–8

Makandi *Plectranthus barbatus*
Lamiaceae

Formerly known as *Coleus forskohlii*, this is an aromatic herb whose roots are the source of the drug used in medicinal preparations. The roots are a dull orange color and in India they are used to improve the appetite and in the treatment of anemia and inflammation. The roots are the only source of the important diterpene (a terpene with 20 carbon atoms) forskolin. Research was carried out on the possible use of forskolin to reduce high blood pressure and it was shown to be a potent inhibitor of the adenylate cyclase. Although this compound could have been developed into an antihypertensive drug, its development was halted and the compound is today mostly used as a model substance in experimental biochemistry as an inhibitor of adenylate cyclase.

Mishmi, Gold thread *Coptis teeta*
Ranunculaceae

Coptis teeta is a Ayurvedic herb found in the eastern Himalayan regions, particularly the Mishmi hill range of Arunachal Pradesh in northeast India. The rhizome of the plant is a prized medicinal commodity and is used to treat gastrointestinal complaints (WHO 1999) and malarial infections. However, the plant has been brought close to extinction by deforestation and overexploitation for its medicinal properties. Conservation schemes have recently been proposed and these may lead to the recovery of *C. teeta*. *Coptis* species found in other continents are also used medicinally by local populations and include *C. trifolia* (gold thread, North America) *C. chinensis* (China), and *C. japonica* (Southeast Asia). The principal active constituent of *Coptis* is the alkaloid berberine and it is known to display antidiarrheal and antimicrobial activity (WHO 1999).

Neem *Azadirachta indica*
Meliaceae

Azadirachta indica or neem is a principal species used in the Ayurvedic medicine of India and today is a tree grown throughout the (sub)tropics. The species is drought resistant and thrives in arid conditions with annual rainfall of between 15–46 inches (400–1200 mm). It can grow between 0–1500 m above sea level, but it is intolerant to freezing, extended periods of cold, and waterlogged soils. Neem trees can reach a height of 80–100 feet (25–30 m) and provide valuable shade with their dense canopy of pinnate leaves.

Neem is thought to have originated in the northeastern region of India (Assam) and in Myanmar (formerly Burma). The exact location of origin is uncertain and some attribute it to the whole of the Indian subcontinent, others to dry forest regions throughout South and Southeast Asia. Neem has also been established in China, Nepal, Australia, West Africa, central America, and the United States. The introduction of neem to East Africa is thought to have arisen during the construction of the Kenya–Uganda railways. Indian migrant workers are believed to have brought neem seed with them in order to cultivate this important medicinal plant. However, the tree is most widely used in India (Williamson 2002).

The neem tree possesses many medicinal uses derived from all parts of the plant. As part of Ayurvedic medicine, the leaves are chewed for 15 days in late winter in order to maintain a healthy body. Tonics prepared by boiling the leaves, often with other herbal constituents such as ginger (*Zingiber officinale*), are useful against intestinal worms, fevers, and internal ulcers. Externally, the juice of the leaves is applied to the skin for the treatment of boils and eczema. The twigs are used extensively in dental hygiene to brush the teeth and are incorporated into pastes or mouth washes. The fruits of neem are used against leprosy, intestinal worms, and urinary diseases. Neem oil (Margosa) is a chemically diverse mixture that includes the isoprenoids azadirachtin and nimbidin, plus numerous fatty acids such as oleic and palmitic acids. The oil is used to treat chronic skin complaints, leprosy, and ulcers, and it is also sold as a natural botanical insecticide. Azadirachtin is the major insecticidal principle of neem.

Much controversy surrounds the development of this traditional insecticide and medicine. In 1992 the U.S. company W.R. Grace submitted a patent application in which a simple extraction procedure from the seeds of the neem tree was described. The plant material is extracted with a lipophilic solvent (e.g., ethyl ether), instead of with a watery one, as has been done for many centuries in India. This results in an increased stability. However, is this an innovation? American patent law does not recognize oral traditions like the Indian ones and approval of such a patent would have, for example, resulted in the exclusion of Indian companies from the U.S. market. This patent and some related ones have been revoked, but the conflict continues.

See: Materials, p. 342

Turmeric *Curcuma longa*
Zingiberaceae

Turmeric is endemic to peninsular India, especially the provinces of Tamil Nadu, West Bengal, and Maharashtra. The plant's growth habit is erect with large, pale green elongated/ribbed leaves; the flower develops a spike 4–6 inches (10–15 cm) with a cylindrical inflorescence of yellow flowers. *Curcuma longa* has a small, branched rhizome that is bright yellow on the interior.

The rhizome is the source of turmeric, widely used in Indian cuisine, the dyeing of cloth, and traditional medicine. At the close of the 19th century, turmeric was used in laboratories in "turmeric paper" to test for alkalinity before litmus paper was produced. Medicinally, turmeric is considered to be a strong antiseptic and is used to heal wounds, infections, jaundice, urinary diseases, ulcers, and to reduce cholesterol levels. Turmeric, in the form of a paste, has been used to treat external conditions such as psoriasis (i.e., as an anti-inflammatory) and athlete's foot (i.e., as an antifungal). A major medicinal compound from *C. longa* is the phenolic constituent curcumin. This compound has been shown to be effective against some forms of cancer and has been intensively examined as a possible anti-inflammatory drug. In the United Kingdom an extract is currently being developed as a veterinary medicine for use in canine arthritis.

See: Spices, p. 170; Natural Fibers and Dyes, p. 312

Psyllium *Plantago* spp.
Plataginaceae

The seeds, which have a seed coat particularly rich in mucilage, are used medicinally for treating constipation, dysentery, irritable-bowel syndrome, and a variety of skin conditions. Certain types of mucilage are well known to have beneficial effects, on the gastrointestinal tract including antidiarrheal and anti-inflammatory effects. A number of species are used, including blond psyllium (*Plantago ovata*, Asia), black psyllium (*P. afra*, Asia), *P. asiatica* in Indochina, and great plantain (*P. major*, widespread throughout Asia).

Winter cherry *Withania somnifera*
Solanaceae

Winter cherry, also known as Indian ginseng, has similar alleged rejuvenating properties to that of ginseng in Chinese medicine. The plant is endemic to India, particularly in the sub-Himalayan (1000 m) tracts of Himachal Pradesh, Punjab, and the drier parts of India. Its use can be traced back to Assyrian sources, and the drug was already used in Mesopotamia as a narcotic. Ancient sacred writings of Hinduism from India praise the plant as a wonder drug, and it was used as a charm and as an aphrodisiac. In Europe it has been known since the 16th century and it is included in many herbals. In Ayurvedic medicine the roots are used to treat ulcers, fever, breathing difficulties, cough, tuberculosis, dropsy, and a variety of nervous disorders.

Southeast Asia

Java tea *Orthosiphon aristatus*
Lamiaceae

Java tea, or Indian kidney tea, is native to Southeast Asia and Australia and has a long tradition of use in India and neighbouring regions. It was introduced into European phytomedicine only toward the end of the 19th century, and today the leaves are widely used in Europe in the treatment of kidney and bladder problems, especially infections. Other uses include treatment of gout and rheumatism.

The Americas

This is the only geopolitical region which extends from the Arctic circle to the Antarctic circle. This, in combination with other geographic factors, results in an impressive biological diversity—more than 100,000 species of higher plants occur naturally on these continents. At the same time it is or was the home to numerous indigenous groups speaking a multitude of languages. It is estimated that about 1,200 ethnolinguistic groups existed in 1492, but today about 420 (i.e., only a third) remain. Most of these belong to the poorest sections of society in their respective countries. Recent attempts to strengthen indigenous traditions have been diverse and it is to be hoped that these attempts succeed in improving the generally appalling living conditions and strengthening the local traditions. The Amazon basin and the Central American region are particularly diverse botanically. Historically, some regions of the Americas have distinguished themselves for the development of dominant cultures that left impressive religious and civil monuments, like the Maya, Zapotecs/Mixtecs, and Aztecs (Nahua) in Mesoamerica, and the Inca in South America. In the case of the Aztecs, some written manuscripts or codices are available which record the use of plants for food, medicine, and many other purposes. The most important and oldest source is a herbal written in Nahuatl by Martin de la Cruz and translated into Latin by Juan Badiano. It was given to the King of Spain, Carlos V, in 1552. It was written rather hastily and has numerous color illustrations of medicinal plants. There have been several attempts to identify plants from this herbal and most of the identifications seem to be botanically sound. The major problem with this source is the European influence that being felt only 30 years after the conquest of México-Tenochtitlan. Also the Nahuatl author attempted to show "European sophistication."

Another important source is Fray Bernadino de Sahagún's work. It is certainly the best source available for the early historical period. This Franciscan missionary arrived in Mexico in 1529 and worked there until his death in 1590. The methods he used compare favorably with modern ethnographic techniques. He posed questions in Nahuatl to a group of ten to twelve elderly informants, and their answers were recorded in Nahuatl by trilingual (Nahuatl, Latin, Spanish) student scribes. The questions were then developed into a more extensive questionnaire used in communities around México-Tenochtitlan. He left several codices (among them the *Codex Florentino*, compiled ca. 1570) and on the basis of these documents he wrote the *Historia General de las Cosas de Nueva España* (published 1793). From an ethnobotanical point of view, the source is somewhat more difficult to use than the Codex Cruz/Badianus because there are fewer botanical identifications and these are less certain. The strength lies more than anything in its description and analysis of medicinal concepts.

Let us now turn to North America. Most of our knowledge about medicinal plant traditions on the continent is due to anthropological and ethnobiological research conducted mostly since the second half of the 19th century. An excellent summary of these data is provided by Moerman (1998), who summarizes fieldwork by ethnographers who recorded such oral traditions in the late 19th and early 20th centuries.

America North of the Rio Grande

American mayapple and Podophyllotoxin *Podophyllum peltatum*
Berberidaceae

American mandrake or American mayapple is a poisonous weed which was commonly used in many regions of North America for many centuries. Its main use (e.g., by the Cherokee, Delaware, Iroquois) was as a laxative and the resin had been included in the American pharmacopoeia of 1820 for this purpose. Another use for the resin is the treatment of warts. It is one of the main sources of podophyllotoxin, a lignan which has resulted in semisynthetic derivatives essential in the chemotherapy, for example, of leukemia, especially teniposide, which was introduced into clinical use in 1967. It is well known that this substance revolutionized the chemotherapy of leukemia and has saved untold numbers of young lives.

Californian yew, Pacific yew *Taxus brevifolia*
Taxaceae

Taxus brevifolia or Californian yew has been used by a variety of West-American Indian groups in the United States and Canada as a medicine, and also for producing a variety of other useful products (canoes, brooms, combs). Very diverse pharmaceutical uses of the root and the bark are recorded and include several reports of the treatment of stomachache and, in the case of the Tsimshian (British Columbia, Canada), the treatment of cancer.

However, this ethnobotanical information was not the basis for the development of a pharmaceutical product. In 1962 several samples of *Taxus brevifolia* were collected at random for the

Scientists are trying to find alternative ways of producing taxol, one being from the yew needles rather than the bark of the Pacific yew tree. Source: Dr. Gorden Cragg. National Cancer Institute.

National Cancer Institute (NCI, United States) and the United States Department of Agriculture. These samples were included in a large screening program at the NCI. A potent cytotoxic effect was documented in one *in vitro* system. However, the *in vivo* effects (especially in cases of leukemia) were less promising and demonstrated a *general toxicity* of the samples. Bioassay-guided fractionation led to the isolation of 0.5 g of an active substance from a total of 12 kg of bark. The substance was named taxol. It took from 1961 until 1971 to establish the compound's structure. It is a diterpenoid with fifteen carbon atoms forming a complex ring system—a pentadecene ring system.

Clinical studies started 13 years later in 1984. Prior to this, studies on the compound's toxicology and the pharmacological mechanism of its action were conducted. It took a further ten years before taxol was approved in the treatment of anthracyclin-resistant metastasis-forming breast cancers. In the meantime the compound has been approved for a variety of other cancers, and semisynthetic derivatives are now also employed.

Some fascinating questions arise out of the use of this species and the compound isolated from it:

- *Taxus brevifolia* is a very slow-growing species and produces the relevant active ingredients only in very small amounts. Because taxol was isolated from the bark for many years, the trees had to be felled in order to obtain the botanical drug. The amount of taxol required to cover the annual therapy requirements of patients with ovarian cancer in the United States is estimated to be 15–20 kg. If other cancers common in the United States are to be treated with this compound, around 200 to 300 kg will be required per year. This amount can only be isolated from approximately 1.45 thousand tons of bark. Collecting such amounts would have been completely unsustainable and would have resulted in the extinction of the species within a few years. In the 1990s the semisynthetic production of taxol from natural products in other *Taxus* species (10-desacetylbaccatin II isolated from the European yew, *Taxus baccata*) enabled the production of large amounts of taxol. Up to this point an enormous conflict of interests between conservation and medicine/pharmacy was unavoidable. An alternative proposal is the production of taxol in *in vitro* plant cell cultures Goodman and Walsh (2001) have discussed the economic and political rise and fall of the Californian yew as a source of taxol in great detail.

This example again raises a series of important questions:

- Californian yew is native to a region in the main industrial power of the temperate zones. What sort of consequences would these studies have had if the species had been from a "developing" country? And specifically, what would have happened with respect to this conflict between conservation and medicine/pharmacy?
- Today the drug is produced from the needles of another species of *Taxus*—*T. baccata*—and more and more is produced biochemically in large-scale fermenting processes. The Californian yew is no longer needed for the production of the drug and this has also meant that the local economy no longer profits from this drug first developed from a native tree. What means are there to assure adequate benefit-sharing between the regions of origin and the industrial partners?

Echinacea *Echinacea angustifolia*
Asteraceae
Species of echinacea have been used widely in the indigenous medical systems of North America. Particularly common is the use of *E. augustifolia*—indigenous groups in North America frequently used it in the treatment of pain and, topically, of inflammatory skin conditions. The

Echinacea angustifolia. Clarence A. Rechenthin @ USDA-NRCS PLANTS Database.

following uses of *E. augustifolia* by the Winnibago tribe (modern Wisconsin, United States) were recorded.

- The juice serves as a remedy for pain and for treating burns (several reports).
- The species is used, in the form of incense, in the treatment of headache.
- It is an antidote for snakebites.
- Topical applications of a compress are used for treatment of enlarged lymph nodes (mumps).
- It is applied topically for treatment of toothache.
- It is used as a veterinary medicine for horses suffering from distemper.

According to Bauer (1998), the various species of this genus were used ethnopharmaceutically all over their natural area. In the second half of the 19th century, settlers of European origin started to use echinacea. The "medical doctor" H.G.F. Meyer is particularly well known, who widely advertised "Meyer's Blood Purifier" as a remedy for rheumatism, headache, digestive problems, tumors and boils, open wounds, vertigo, scrofula (tendency to get tuberculosis), bad eyes, intoxications resulting from plants or snakes. The plant used in this preparation was subsequently identified as *E. angustifolia*. Once larger amounts of echinacea were required, *E. pallida* with its larger rootstock became popular. In 1916 the rootstock of echinacea was included in the National Formulary of the United States. Both species were allowed for pharmaceutical use. At the beginning of the 20th century, echinacea became popular in Europe and has recently become an important immunostimulant, and an enormous number of (mostly ill-defined) herbal preparations are on the market both in Europe and in the United States. For some standardized extracts, sound clinical data are available that indicate that these extracts are clinically superior to placebo in the treatment of some common respiratory problems. However, the compounds responsible for the effect have not yet been identified.

Two aspects of the cultural history of these species deserve special mention:

- It is a phytomedicine, i.e., not a single compound but an extract, which hopefully is standardized to a series of lead compounds (in order to assure reproducible quality) as the active "drug."
- The modern phytomedicines with echinacea as an active ingredient were developed on the basis of the indigenous uses in North America. If such a product were to be developed

based on modern ethnobotanical studies, appropriate steps would have to be taken in order to adequately compensate the traditional bearers of this knowledge.

Mesoamerica

Guava *Psidium guajava*
Myrtaceae

A brief mention has to be made of guyava or guava, a Mesoamerican tree which is now pantropically cultivated especially for its fruit. As is the case with many food plants, it is also used medicinally. The leaves and bark are commonly used in the treatment of diarrhea and some other gastrointestinal disorders. While it is not used biomedically, it is used in popular medicine and dates back to at least the Aztecs of the 16th century.

See: Fruits, p. 93

Passionflower *Passiflora incarnata*
Passifloraceae

In the days following their conquest of South America, the Europeans became interested in the passion flower. Its unique and very conspicuous flower reminded the conquistadors of the Passion of Christ, and they saw all the symbols of his suffering in the flower: the crown of thorns in the corona, the five wounds in the five anthers, the nails of crucifixion in the three stigma, the ten apostles (less Peter and Judas) in the five plus five lobes of the calyx and corolla, and the hands and whips of Christ's persecutors in the tendrils and lobed leaves. The aerial parts (*Passiflora herba*) are today used in the treatment of nervousness and unrest.

See: Fruits, p. 93

Quinine tree *Cinchona* spp.
Rubiaceae

Cinchona, the quinine tree, raises a series of fascinating questions about indigenous plant use and drug development. It is uncertain whether species of this genus were used pharmaceutically by the native populations of tropical South America prior to or during the 17th century. According to Schneider (1974) and Tschirsch (1910), the bark of this species was not used medically to a great extent, and it was known to specialists as a remedy in the treatment of "fever" (malaria) only in a very limited region. Since it is still doubtful whether malaria was endemic to the Americas prior to the conquest by the Europeans, it would seem unlikely that the bark would have been of any relevance. The bark and its use was popularized by the Jesuits ("Jesuits' powder"), and since 1687 it has been recorded in several lists of medicines (first in the "Arzneitaxe" of Frankfurt-on-the-Main, Germany). Its use as a remedy for malaria, and also for other forms of fever, spread very rapidly across Europe (Schneider 1974). In the mid-19th century the tree was introduced to Java and is now grown in many regions of the tropics. Quinine was isolated from the bark in the early 19th century and was obtained, for example, from the bark of *C. pubescens* and *C. officinalis* in relatively large amounts. The structure of quinine was not established until 1951.

However, the bark was used for various ill-defined forms of "fever" until in 1880 LaVeran identified the organism responsible for the illness—eukaryotes from the genus *Plasmodium*. In 1897–1898 Ronald Ross and Battista Grassi demonstrated the life cycle of the parasite and its dependency on the *Anopheles*-mosquito were. Only at this stage was it possible to understand the mechanism of action of Jesuits' powder, and this in turn allowed a rational phytotherapy using the isolated active ingredient, quinine. Since then a large number of quinine derivatives have been isolated. Another core ingredient is quinidine, a classic anti-arrhythmic drug.

This example demonstrates the long development process from initial observations (whether they were by the Jesuits or indigenous Americans) to several natural products used biomedically.

Wormwort, Epazote *Chenopodium ambrosioides*
Chenopodiaceae

The wormwort or wormseed is also known under its alternative Latin name, *Teloxys ambrosioides*, or its Aztec name, *epazotl* (modern mexican Spanish: *epazote*). It is another species with a long tradition of uses. Fascinatingly, it is used both as a spice for a variety of dishes, especially ones with Mexican black beans (frijoles), and as a medicine for gastrointestinal parasites. The name seems to be derived from the Nahua term for skunk, *epatl*, and relates to the rather unpleasant smell of the plant (some liken it to the urine of a skunk). As long ago as the 16th century, Fray Bernardino de Sahagún mentioned "epázotl" as a food. Today it is one of the most popular spices and is used medicinally as a vermifuge (to treat worms) as well as to reduce flatulence. It was included in many pharmacopoeias, including the ones of Mexico, the United States, and many European countries, but because of the toxic side effects (mostly of the essential oil) and a lack of evidence in support of its vermicidal effects, it has now been substituted by synthetic vermifuges. Once used worldwide as a medicine, it is today largely restricted to its region of origin, especially Mexico, where epazote is an essential part of the local cuisine and medical tradition, and also a powerful symbol for Mexican identity.

See: Herbs and Vegetables, pp. 118–9

Zoapatle *Montanoa tomentosa*
Asteraceae

Zoapatle or cihuapatli (*cihuatl*, woman and *patli*, medicine) is a classic example of a women's medicine and the main uses have been retained by the indigenous Mexican populations at least since the Aztec period. Sahagún described the use of its root for inducing labor ("which has the virtue to bring the new-born out") and in combination with the Mesoamerican sweatbath temazcal. It was so famous with the midwives of this period that it was rapidly accepted as a medicine for regulating "menstrual irregularities" by the Spanish conquistadors. However, the very nature of its use made it a very controversial remedy, and even though it is still commonly used in Mexico, it is not currently used in biomedicine.

Australia and Oceania

Australian aboriginal societies have for millennia been based on a migratory lifestyle, and at the time of the conquest in the 17th century about 250 languages were spoken among them. Many of these are under considerable threat or have already disappeared, and the cultures have been forced to change extremely rapidly. Until fairly recently many societies still had this migratory lifestyle, but increasingly the Australian Aborigines have settled especially in the central deserts and in the north of the continent. In the south of Australia much of the local indigenous knowledge has disappeared forever, but in the north communities are once again gaining autonomy through "native title" to their lands, and are eager to salvage what remains of their culture. Statistics show that Aboriginal and Torres Strait Island people are the most disadvantaged group in Australia, with high unemployment rates, poor housing conditions, low life expectancy, and high infant mortality and morbidity rates.

Surprisingly, few medicinal plants have been adopted by the white conquerors or have made their way to other continents. One of the most noteworthy exceptions is *Eucalyptus globulus*, which is now one of the most widely cultivated trees of the world.

Australian tea tree *Melaleuca alternifolia*
Myrtaceae

In Australia, Europe, and North America, tea tree oil (*Melaleuca aetheroleum*) from a tree native to subtropical coastal regions of New South Wales has become very popular in the last decades. Today it is obtained almost exclusively from cultivated material. The use of this species in biomedicine is based on the medical traditions of the Australian Aborigines and includes the treatment of infectious skin conditions (acne). It is also used in cosmetics.

See: Fragrant Plants, p. 245

Eucalyptus *Eucalyptus globulus*
Myrtaceae

Today, leaves from *E. globulus* (blue gum, fever tree) are used as a medicine and for the production of commercial oil of eucalyptus, which has 1, 8-cineole (= eucalyptol) as the main component (up to 85 percent in the case of *E. globulus*). However, several other species are used in the production of the essential oil, including *E. smithii* and *E. polybractea*. The oil and pure cineole have antiseptic effects and are mildly irritant to the mucosa and the skin. Inhalations to treat inflammatory infections of the upper respiratory tract (bronchitis) and creams/balsams to treat a variety of mild to moderate respiratory problems (including problems of the nose and throat) are the most common applications. Commercial material for the pharmaceutical industry is today produced in the Mediterranean (Spain, Morocco) and around the Black Sea (Ukraine).

See: Fragrant Plants, pp. 244–5; Wood, p. 328

Conclusions

Human cultures have for millennia found medicinally useful plants in their environment and the this knowledge has passed from generation to generation mainly by word of mouth. These practices began to be recorded in ancient scripts of India, China, Mesoamerica, and Arabia. This chapter provides an overview of medicinal plant usage around the world and although it highlights only some examples, it clearly shows the diversity in cultural history of human uses of medicinal plants. However, the sociocultural and industrial context within which these practices take place demands a closer analysis, for example, highlighting how overexploitation of medicinal plants has occurred in African countries through factors including the following:

1. Loss of natural habitat and dramatic reduction in the distribution and availability of medicinal plants due to competing uses of either the land (logging) or the medicinal plants themselves (timber logging; commercial harvest for export and extraction of pharmaceuticals; use for building materials and fuel; collections from the wild, e.g., *Prunus africana*).
2. A high rate of urbanization, in which a large proportion of the population seek traditional health practitioners to heal ailments that are often perceived to result from the action of individuals or ancestral spirits.
3. The change of traditional medical practices in the urban centers from a small-scale, specialist activity to one that is driven by wider economics. This produces a need to obtain as much medicinal plant material as possible without regard for the traditional conservation ethos found in rural communities.
4. Traditional practitioners moving to the urban centers where business is economically rewarding. In addition, the stresses and strains produced by urban living—such as finding

and maintaining employment, or nurturing and keeping a partner—all increase the demand for psychosomatic or symbolic symbols, all willingly provided by traditional practitioners using medicinal plants.

The opportunity to make money, whether from wild collected herbs sold from a market stall or a plant-derived drug marketed the multinational company, can skew the attitudes of the culture toward its plants to one of profit-making. Researchers studying the relationship between people and plants (ethnobotany) have much to contribute to our understanding of this change in order to highlight the need for cultural and botanical conservation. This requires a proper resolution of the intellectual property rights that local peoples have in respect to the sharing of their knowledge. Generally this is dealt with through legislative frameworks based on international treaties such as the Convention on Biological Diversity (CBD), but the implementation of such policies has proved difficult for financial, political, and social reasons.

A particular problem has been to ensure that benefits return to indigenous communities rather than governments. This is particularly true where medicinal plants are concerned. One current and successful case study deals with the small Samoan tree *Homalanthus nutans*—the source of the anti-HIV-1 drug prostratin (Cox 1997). The tree had long been a staple medicinal plant of local peoples in the treatment of hepatitis. Ethnobotanical studies and cooperation with local communities have been conducted for many years and the drug is now classified by the National Cancer Institute and the AIDS Research Alliance as a candidate anti-AIDS drug. A significant portion of the license income is to be returned to the Samoan communities.

In terms of conservation, new technologies can aid in the regeneration of medicinal plants through such techniques as plant tissue culture. Ongoing efforts include regeneration of a number of medicinal plants and production of biologically active compounds.

Many of the examples given in this chapter show the value of approaches with an interdisciplinary focus, integrating biological and social science. Medicinal plants are not just a commodity, but an integral element of cultures all over the world. Ethnopharmaceutical approaches may lead to improved primary healthcare in marginal societies, as well as the discovery of new drugs. Thus a core task for future research on medicinal plants is to integrate the cultural history of medicinal plants and our current biological knowledge of these resources, and to use this information for sustainable development of this exciting interaction between people and plants.

Bibliography

General references

Barnes J, Anderson, L.A. and Phillipson, J.D. 2002. *Herbal Medicines. A Guide for Healthcare Professionals* (2nd ed.). London: Pharmaceutical Press.

Blumenthal, M., et al. 2000. *Herbal Medicine: Expanded Commission E Monographs*. Newton, MA: Integrative Medicine Communications.

Blumenthal, M., et al. 1998. *The Complete German Commission E Monographs: Therapeutic Guide to Herbal Medicines*. Philadelphia: Lippincott Williams and Wilkins.

Chevallier, A. 1996. *The Encyclopedia of Medicinal Plants*, London: Dorling Kindersley.

Collin, H.A. 2001. Secondary product formation in plant tissue cultures. *Plant Growth Regulation 34*, 119–134.

Cox, P.A., and Heinrich, M. 2001. Ethnobotanical drug discovery: Uncertainty and promise. *Pharmaceutical News 8* (3), 55–59.

Cunningham, A.B. 1997. *Medicinal Plants for Forest Conservation and Health Care. Non-Wood Forest Products 11*. Rome: Food and Agriculture Organization of the United Nations.

Etkin, N. 1988. Indigenous patterns of conserving biodiversity: pharmacologic implications. *Journal of Ethnopharmacology 63*, 233–245.

European Pharmacopoeia Commission. 2002. *European Pharmacopoeia* (4th ed.) Strasbourg: European Directorate for the Quality of Medicines, Council of Europe.

Evans, W.C. 1996. *Trease and Evans' Pharmacognosy* (14th ed.). London: WB Saunders.

Grier, J. 1937. *A History of Pharmacy*, London: The Pharmaceutical Press.

Griggs, B. 1981. *Green Pharmacy*, London: Jill, Norman and Hobhouse.

Heinrich, M. 2001. *Ethnobotanik und Ethnopharmazie: Eine Einführung*. Stuttgart: Wissenschaftliche Verlagsgesellschaft.

Heinrich, M., and Gibbons, S. 2001. Ethnopharmacology in drug discovery: an analysis of its role and potential contribution. *Journal of Pharmacy and Pharmacology 53*, 425–432.

Hiller, K., and Melzig, M. F. 1999. *Lexikon der Arzneipflanzen und Drogen*. Heidelberg: Spektrum Akademischer Verlag.

Joyce, C. 1994. *Earthly Goods: Medicine-hunting in the Rainforest*. London: Little, Brown and Company.

Mills, S., and Bone, K. 1999. *Principles and Practice of Phytotherapy*. Edinburgh: Churchill Livingstone.

Nunn, J.F. 1996. *Ancient Egyptian Medicine*. London: British Museum Press.

Plotkin, M. 2000. *Medicine Quest: In Search of Nature's Healing Secrets*. London: Viking.

Rout, G.R., Samantaray, S., and Das, P. 2000. *In vitro* manipulation and propagation of medicinal plants. *Biotechnology Advances 18*, 91–120.

Schneider, W. 1974. *Lexikon zur Arzneimittelgeschichte* Bd. V/1:3. *Pflanzliche Drogen*. Frankfurt: Govi Verlag.

Schultes, R.E., and Hofmann, A. 1992. *Plants of the Gods: Their Sacred, Healing and Hallucinogenic Powers* (2nd ed.). Vermont: Healing Art Press.

Schulz, V., Hansel, R., Tyler, V.E. and Tegler, T.C. 2001. *Rational Phytotherapy: A Physician's Guide to Herbal Medicine* (4th ed.). Berlin: Springer.

Sneader, W. 1996. *Drug Prototype and their Exploitation*. Chichester: Wiley.

Tschirch, A. 1910. *Handbuch der Pharmakognosie* (1st ed.). Leipzig: Tachnitz.

Weiss, R., and Fintelmann, V. 2000. *Herbal Medicine*. Stuttgart: Thieme.

World Health Organization. 1999. *WHO Monographs on Selected Medicinal Plants*. Volume 1. Geneva: WHO.

Europe

Allen, D.E. and Hatfield, G. 2004. Medicinal plants in folk tradition: an ethnobotany of Britain and Ireland. Portland, Or.: Timber Press.

Bundesinstitut für Arzneimittel und Medizinprodukte. 1994. *Hippocastani semen* (Roßkastaniensamen)/Trockenextrakt. *Bundesanzeiger 71*. Bonn: Bundesanzeiger Verlagsgesellschaft. [Monograph, 15.04.1994.]

Bork, P., et al. 1999. Hypericin as a non-anti-oxidant inhibitor of NF-κB. *Planta Medica 65*, 297–300.

Chrubasik S., et al. 2000. Treatment of low back pain exacerbations with willow bark extract: A randomized double-blind study. *American Journal of Medicine 109*, 9–14.

Fuchs, L. 2001. *Das Kräuterbuch von 1543*. Cologne: Taschen Verlag. [Originally published as *New Kreütterbuch*, Basel, 1543.]

Hatfield, G. 1999. *Memory, Wisdom and Healing: The History of Domestic Plant Medicine*. Stroud: Sutton.

Houghton, P.J. 1997. *Valerian: The Genus Valeriana*. Amsterdam: Harwood Academic Press.

Isaac, O. 1993. *Chamaemelum nobile* (L.) Allioni—Römische Kamille. *Zeitschrift für Phytotherapie 14*, 212–222.

Lonicer, A. 1679. *Kreuterbuch*. Reprinted 1962. Grünwald b. München: Verlag Kölbl.

Mayer, J.G., and Czygan, F.-Ch. 2000b. Die Ringelblume—*Calendula officinalis* L. *Zeitschrift für Phytotherapie 21*, 170–178.

Mayer, J.G., and Czygan, F.-Ch. 2000a. *Arnica montana* L., oder Bergwohlverleih. *Zeitschrift für Phytotherapie 21*, 30–36.

Schmid, B., et al. 2000. Wirksamkeit und Vertraeglichkeit eines standartisierten Weidenrindenextraktes bei Arthrose Patienten: Randomisierte, placebo-kontrollierte Doppelblindstudie. *Zeitschrift für Rheumatologie 59*, 314–320.

Swerdlow, J.L. 2000. *Nature's Medicine—Plants that Heal*, Washington, D.C.: National Geographic Society.

Vogel, G. 1989. *Aesculus hippocastanum* L.—Die Rosskastanie. *Zeitschrift für Phytotherapie 10*, 102–106.

Vogellehner, D. 1987. Les jardins du haut Moyen Age (VIIᵉ–XIIᵉ siècles). *Flaran 9*, 11–40.

Mediterranean and Near East

Bellakhdar, J.J. 1997. *La pharmacopoée marocaine traditionelle. Médecine Arabe ancienne et savouirs populaires*. Ibis Press: Paris.

Mattioli, P.A. 1568. *I Discorsi di M. Pietro Andrea Matthioli. Appresso Vincenzo Valgrifi*. Reprinted 1966. Venice.

Lenz, H.O. 1859. *Botanik der alten Griechen und Römer*. Gotha: Thienemann Verlag.

Miller, J.I. 1998. *The Spice Trade of the Roman Empire*. Oxford: Oxford University Press.

Africa

Duwiejua, M., et al. 1992. Anti-inflammatory activity of resins from some species of the plant family Burseraceae. *Planta Medica 59*, 12–16.

Neuwinger, H.D. 1994. *African Ethnobotany: Poisons and Drugs*. London: Chapman and Hall.

Neuwinger, H.D. 2000. *African Traditional Medicine. A Dictionary of Plant Use and Application*. Stuttgart: Medpharm Scientific Publishers.

Noble, R.L. 1990. The discovery of the vinca alkaloids—Chemotherapeutic agents against cancer. *Biochemistry and Cell Biology 68*, 1344–1351.

Oliver-Bever, B. 1986. *Medicinal Plants in Tropical West Africa*. Cambridge: Cambridge University Press.

Reynolds, T., and Dweck, A.C. 1999. *Aloe vera* leaf gel: a review. *Journal of Ethnopharmacology 68*, 3–37.

Reynolds, T. (ed.). 2004. *Aloes: the genus Aloe*. Boca Raton, FL: CRC Press.

Van Wyk, B-E., van Oudtshoorn, B., and Gericke, N. 1997. *Medicinal Plants of South Africa*. Pretoria: Briza Publications.

Watt, J.M., and Breyer-Brandwijk, M.G. 1962. *The Medicinal and Poisonous Plants of Southern and Eastern Africa* (2nd ed.). London: Livingstone.

Americas

Bauer, R. 1998. The Echinacee story. In *Plants for Food and Medicine*, edited by H.D.V. Prendergast, N.L. Etkin, D.R. Harris, and P.J. Houghton. Richmond: Royal Botanic Gardens, 317–332.

Bye, R. 1993. The role of humans in the diversification of plants in Mexico. In *Biological Diversity of Mexico: Origins and Distribution*, edited by T.P. Ramamoorthy, R. Bye, A. Lot, and J. Fa. New York: Oxford University Press, 707–731.

Eiden, F. 1998/1999. Ausflug in die Vergangenheit: Chinin und andere China-Alkaloide. *Pharmazie in unserer Zeit 27*, 257–271; *28*, 11–20; and *28*, 67–73.

Heinrich, M., et al. 1998. Ethnopharmacology of Mexican Asteraceae (Compositae). *Annual Review of Pharmacology and Toxicology 38*, 539–565.

Heinrich, M. 1992. Economic botany of American Labiatae. In *Advances in Labiatae Science*, edited by R.M. Harley and T. Reynolds. Richmond: Royal Botanical Gardens, 475–488.

Goodman, J., and Walsh, V. 2001. *The Story of Taxol: Nature and Politics in the Pursuit of an Anti-cancer Drug*. Cambridge: Cambridge University Press.

Lizarralde, M. 2001. Biodiversity and loss of indigenous language and knowledge in South America. In *On biocultural diversity. Linking language, knowledge and the environment*, edited by L. Maffi. Washington, D.C.: Smithsonian Press, 265–281.

Moerman, D.E. 1996. An analysis of the food plants and drug plants of native North America. *Journal of Ethnopharmacology 52*, 1–22.

Moerman, D.E. 1997. Heilpflanzen aus Nordamerika. *Zeitschrift für Phytotherapie 18*, 20–33.

Moerman, D.E. 1998. *Native American Ethnobotany*. Portland, OR: Timber Press.

Mors, W.B., Toledo Rizzini, C., and Alvares Pereira, N. 2000. *Medicinal Plants of Brazil*. Algonac, MI: Reference Publications.

Ortiz de Montellano, B. 1990. *Aztec Medicine, Health and Nutrition*. New Brunswick, NJ: Rutgers University Press.

Suffness, M. 1995. *Taxol: Science and Application*. Boca Raton: CRC Press.

Wolters, B. 1996. Agave bis Zaubernuβ: Heilpflanzen der Indianer Nord- und Mittelamerikas. Greifenberg: Urs Freund Verlag.

Asia

Benski, D., and Gamble, A. (Eds.). 1993. *Chinese Herbal Medicine. Materia Medica*. Seattle: Eastland Press.

Brinkhaus, B., et al. 2000. Chemical, pharmacological and clinical profile of the East Asian medical plant *Centella asiatica*. *Phytomedicine 7*, 427–448.

Carbajal, D., et al. 1989. Pharmacological study of *Cymbopogon citratus* leaves. *Journal of Ethnopharmaclology 25*, 103–107.

Duke, J.A. and Ayensu, E.S. 1985. *Medicinal Plants of China*. Algonac, MI: Reference Publications.

Padua, L.S. de, Bunyapraphatsara, N., and Lemmens, R.H.M.J. (Eds.). 1999. *Medicinal and Poisonous Plants. Plant Resources of South-East Asia (PROSEA) Handbook 12* (1). Leiden: Backhuys.

Kapoor, L.D. 1990. *Handbook of Ayurvedic Medicinal Plants*. Boca Raton: CRC Press.

Mukherjee, P.K. 2001. Evaluation of Indian traditional medicine. *Drug Information Journal 35*, 623–632.

Marcus, D.M., and Suarez-Almazor, M.E. 2001. Is there a role for ginger in the treatment of osteoarthritis? *Arthritis and Rheumatism 44*, 2461–2462.

Jarvis, A.P., et al. 1997. Identification of azadirachtin in tissue-cultured cells of neem (*Azadirachta indica*). *Natural Product Letters 10*, 95–98.

Kadidal, S. 1998. United States patent prior art rules and the neem controversy: a case of subject matter imperialism. *Biodiversity and Conservation 7*, 27–39.

Mueller, M.S., et al. 2000. The potential of *Artemisia annua* L. as a locally produced remedy for malaria in the tropics: agricultural, chemical and clinical aspects. *Journal of Ethnopharmacology 73*, 487–493.

Pandit, M.K., and Babu, C.R. 1993. Biology and conservation of *Coptis teeta* Wall.—an endemic and endangered medicinal herb of Eastern Himalaya. *Environmental Conservation 25*, 262–272.

Parrotta, J.A. 2001. *Healing Plants of Peninsular India*. New York: CABI Publishing.

Rani, G., and Grover, I.S. 1999. *In vitro* callus induction and regeneration studies in *Withania somnifera*. *Plant Cell Tissue and Organ Culture 57*, 23–27.

Schulick, P.1996. *Ginger: Common Spice and Wonder Drug*. Brattleboro, VT: Herbal Free Press.

Sharma, S.K, Satyanarayana, T., Yadav, R.N.S., and Dutta, L.P. 1993. Screening of *Coptis teeta* Wall. for antimalarial effect: A preliminary report. *Indian Journal of Malariology 30*, 179–181.

Williamson, E.M. 2002. *Major Herbs of Ayurveda*. Edinburgh: Churchill Livingstone.

Surh, Y-J., et al. 2001. Molecular mechanisms underlying chemopreventive activities of anti-inflammatory phytochemicals: down-regulation of COX-2 and iNOS through suppression of NF-κB activation. *Mutation Research 480–481*, 243–268.

Oceania

Collocott, E.E.V. 1927. Kava ceremonial in Tonga. *J. Polynesian Soc. 36*, 21–47.

Cox, P.A. 1997. *Nafanua: Saving the Samoan Rain Forest*. New York: Freeman.

13
Fragrant Plants

BY SUE MINTER

The fragrance of flowers derives from volatile oils known as essential oils. More than three thousand essential oils have been identified from over eight-seven plant families worldwide. These oils are not found only in the flowers, but also frequently in the leaves, roots, rhizomes, seeds, rind, or bark. Plants usually produce these oils as a defense against insect attack, plant diseases, or extreme heat. The bioactivity of essential oils is the key to their potency as plant defenses and to their therapeutic properties. White flowers are disproportionately represented in scented genera, because the oil-producing glands compete for location with cells producing pigmentation.

The use of essential oils, gums, and resins from plants dates back thousands of years and was particularly important in ancient Egypt and the civilizations of the Middle East with their incense trade routes. The Old World proved to be a rich source of aromatic plant materials, whereas relatively few originate from the New World. The term *perfume* derives from the Latin (and, from thence, the Italian) "per" (through) and "fumare" (to smoke), reflecting the religious origins of fragrances used for incense and for mummification. The Arabs made great advances in distillation techniques, with the physician Avicenna developing steam distillation in the 11th century.

In medieval times, perfume was believed to be important in the prevention of sickness. Before germ theory of disease was developed in the 19th century, "foul and pestilential airs" were believed to cause infection, hence the strewing of herbs on floors to release supposedly protective fragrances and the use of "tussie mussie" bouquets and pomanders as prophylactics. These may have been indirectly effective in repelling insect carriers of disease-causing organisms. The production of oils and scented waters was a domestic industry, and manor houses had "still" rooms. In southern France, a major farming industry developed around Grasse during the early 17th century, growing rose, violet, jasmine, tuberose, myrtle, cassie, and narcissus, among other fragrance crops. This industry declined after the Second World War due to rising land prices and labor costs, although the town does retain the production of lavender and has several perfume museums. Today, most world perfume production is concentrated in China, Brazil, Turkey, Indonesia, Morocco, and Egypt, and the most valuable products are the mint, citrus, and rose oils.

The development of "aromatherapy" using plant oils was an offshoot of the fragrance industry. This term was coined in 1928 by a French perfume chemist, Rene Gattefossé, when he healed his burnt hand in a flagon of lavender oil. Further 20th-century developments in scent additives

Frangipani (*Plumeria rubra*). Copyright Dave Webb, used with permission.

include increased use of pharmaceutical flavorings and a massive increase in demand for fragrances and flavorings in the processed-food industry from the 1960s onward, coupled with the demand for highly fragranced household products. Though many of these products have been sourced from synthetics, there is now cachet value to the use of "natural botanicals." The fragrance industry has traditionally used cultivated crops, and some plant genera, for example, sweet peas and lilacs, have been subjected to breeding for use in perfume. Occasionally wild harvesting raises conservation issues, but there is much less of a problem with the harvesting of fragrance plants than with the widespread harvesting of a great variety of wild medicinal plants to support a much-expanded herbal industry.

The use of scented plants for decoration in the home and garden is likewise ancient, ranging from the rose petals at Roman banquets to the use of plants with religious connotations (such as the Madonna lily) in medieval gardens and the development of "florists' flowers" (such as clove pinks) in the 17th century. New species were first introduced by the Romans, then via the Eastern spice routes, and subsequently by successive waves of plant hunting. In the 18th century, plant hunters explored the Americas, the Cape, the Middle East, and Australia in their search for new species, and in the 19th century they turned their attention to China and Japan.

Twentieth-century developments in ornamental plant use include changes in the distribution of the cut-flower industry according to global economics. For example, it is now cheaper to import carnations to Britain from Colombia than to produce them in the Channel Isles. Designers creating "sensory" gardens, targeted for therapeutic use in hospice and hospital grounds and to augment the experience of those with disabilities, have given a boost to the use of scented plants. Increased overseas travel has also introduced growing numbers of consumers to tropical fragrances and stimulated development of ever more exotic perfumes and air freighting of cut flowers. In other cases, scents have been lost through breeding (e.g., in roses).

Agarwood, Eaglewood, Gharuwood *Aquilaria malaccensis*
Thymelaeaceae

This is the most important of a group of several species of trees from Malaysia and western China that produce oil similar to sandalwood oil, with the strongest oils coming from fungally infected heartwood. Agarwood has been used for thousands of years, especially for incense in eastern religious customs, and the high-grade wood is very valuable. Agarwood is mentioned in Tamil texts from the third century AD, in which a reference to an actress drying her hair over the smoke of agarwood can be found, for example. The trees are now highly endangered and are included in Appendix Two of the Convention on International Trade in Endangered Species (CITES) in order to allow control and monitoring of trade. Agarwood is possibly the aloe wood referred to in the Old Testament.

Bay rum *Pimenta racemosa*
Myrtaceae

This evergreen tree is native to and cultivated in the Caribbean and adjacent geographic areas. In the Virgin Islands, the leaves are distilled with rum to produce bay rum, which is a popular toilet water and hair tonic. Bay leaf oil is also distilled by more conventional means for use in perfumery.

Benzoin, Gum Benjamin *Styrax benzoin*
Styracaceae

Benzoin has been traded in Southeast Asia for at least 1,000 years, and is mentioned by Chinese writers of the Sung Dynasty (AD 960–1279) as an import from Sumatra and Cambodia. The fragrant resin from several species is collected, mainly in Indonesia and Laos, by wounding the bark of the tree. This resin has medicinal uses, for example in cough treatments, and ceremonial uses, particularly in incense. It is also used for food and cigarette flavoring. A resin is also obtained from *S. officinalis*, which grows in southern Europe and Turkey.

Brown boronia *Boronia megastigma*
Rutaceae

This western Australian species is cultivated for its flowers, from which fragrance is extracted by washing with organic solvents such as hexane, which is then evaporated to leave a waxy residue known as a "concrete." Chilled alcohol is then used to remove the wax, leaving a perfume "absolute." Cultivation remains small scale, concentrated in Tasmania and New Zealand. The warm, woody, rather spicy oil is used to flavor food and also in perfumery, and brown boronia is grown especially as a niche Australasian crop for tourists. Plantations are established from cuttings of selected stock material.

Buchu *Agathosma betulina*
Rutaceae

This is a heath-like shrub native to South Africa and thought to have been introduced to Britain in the late 18th century, along with other *Agathosma*, often generically termed "buchus." The plant is valued for its aromatic foliage, which has a resinous peppermint aroma, and is sometimes steam-distilled for the oil. The genus name means "pleasant smell" in Greek.

Camphor *Cinnamomum camphora* and *Dryobalanops aromatica*
Lauraceae and Dipterocarpaceae

Common or "true" camphor is derived from *C. camphora*, an evergreen tree native to the forests of eastern China and Japan, but now grown widely in eastern Asia. The aromatic timber has long been used for making moth-proof chests. Camphor oil is traditionally obtained by steam distillation from wood chips or leaves. Camphor is also produced in *D. aromatica* and other species, as crystal-line deposits in the wood and as an oily exudate. The native distribution of the large evergreen trees (up to 196 feet/60 m tall) is in Malaya, Borneo, and Sumatra. Oil is also distilled to a lesser extent from the leaves of *Blumea balsamifera* (Asteraceae), also from Southeast Asia and known as ngai camphor.

The use of camphor from *C. camphora* is documented in Sanskrit texts and Chinese texts dating to about 400 BC. Traditional Asian uses of camphor include a skin powder, appreciated for its cooling and odor-controlling properties, and incense, burned in Hindu and Buddhist religious ceremonies. Camphor from both major species reached Europe in the Medieval period via Arab traders, and was important in perfumery and medicine by the 13th century. Large-scale plantations of *C. camphora* in Taiwan, Japan, and China dominated twentieth-century production. However, production has steeply declined owing to the manufacture of synthetic camphor.

Chinese sassafras oil *Cinnamomum camphora* and Brazilian sassafras oil *Ocotea* spp.
Lauraceae
Chinese sassafras oil is obtained from *C. camphora* (see previous discussion), and Brazilian sassafras oil from several wild species of *Ocotea* (also in the Lauraceae family) in Brazil, including *O. pretiosa*. Sassafras oil contains 80 percent safrole, an important raw material in the production of pyrethroid insecticides. Sassafras oil should not be confused with the spice, *Sassafras albidum*.

Cassie, Popinac *Acacia farnesiana*
Fabaceae
A tropical American species of acacia growing to 23 feet (7 m) in height, introduced to Europe in the early 17th century, and now widely cultivated in warmer parts of the world. Production of cassie is still concentrated on the French Riviera and around Grasse, where it is extracted from the flowers. Cassie is used in perfumes classified as "floral" or "oriental," particularly in violet perfumes, but annual production is low, on the order of 100 kilograms. The species was named after Rome's Farnese Palace.

Cedarwood oil, Cedar of Lebanon *Cedrus libani, C. atlantica*
Pinaceae
The Cedar of Lebanon is the national tree of Lebanon, where it is highly endangered, but is also native to northwest Syria and south central Turkey. Its wood was probably burned as incense in ancient Egypt and oil distilled from the wood was used in cosmetics. The tree became a symbol of country estates in Britain following its introduction there in the 1640s.

Cedarwood oil was possibly the first essential oil to be extracted from a plant and is now obtained from various species of cedar and juniper (*Juniperus*), particularly in China, the United States and, on a small scale, Morocco (from *C. atlantica*). It is used in fragrances.

See: Wood, p. 330

Champac *Michelia champaca*
Magnoliaceae
This species, native to Southeast Asia, has white or yellow scented flowers sold locally for floating on water in dishes in order to perfume the domestic environment. An absolute is derived for use in perfumery by extraction using volatile solvent and alcohol. The fresh flowers are also used to make garlands, sprinkle on bridal beds, and compound hair lotions. This species is favored by the Hindus and Jains for a variety of traditional ceremonies, and is cultivated around temples.

Cherry pie, Heliotrope *Heliotropium arborescens*
Boraginaceae

Heliotropium arborescens is a tender Peruvian shrub introduced to Britain in 1757 for the Chelsea Physic Garden, and is generally grown from seed or cuttings as an annual for use in bedding displays or conservatories. It has intensely fragrant purple flowers that smell of cooked cherries, hence its common name. It can also be grown as a "standard," in a similar way as fuchsias are.

Christmas box, Sweet winter box *Sarcococca* spp.
Buxaceae

The *Sarcococca* genus includes some of the most useful dwarf winter-flowering scented evergreens. All come from Southeast Asia, western China, or the Himalayas and prefer partial shade. The fragrant species (all except *S. saligna*) were introduced to Britain in the first decade of the 20th century. The white flowers seem insignificant, but their perfume is very powerful when planted en masse.

Citrus *Citrus* spp.
Rutaceae

The citrus oils are second in value only to menthol and are produced worldwide from a number of species. Various parts of the plant are used, from the fruit peel to the wood and the flowers. With the exception of grapefruit, which is from the New World, most citrus species originated in Southeast Asia and were introduced to the West on trade routes. (For example, the lemon was first grown in Europe in the 12th century and then carried by Columbus to the West Indies in the 15th century.) In terms of oil production for use in perfumery and as a flavoring, orange and lemon are the most significant species, followed by lime, grapefruit, and mandarin. Generally the oil is produced as a byproduct of canning and juicing of the fruit. The southern United States, Central America, and the Mediterranean account for most of the world's production, with ten percent from Australia (which received its first lemons in 1788) and eastern Asia, and ten percent from South America.

Lemon (*Citrus limon*), grapefruit (*C. × paradisi*), and sweet orange (*C. sinensis*) oils are generally cold-pressed to express the oils contained in the glands of the peel, usually as part of the juicing process. Lime fruit (*C. aurantiifolia*) is usually steam distilled in order to remove phototoxic coumarins, especially when used in perfumery (these compounds can thin the blood and, in this case, sensitize the skin to the effects of sunlight and produce rashes). Uses for citrus oils are numerous and expanding, and include food and drink, flavorings, perfumery, pharmaceuticals, confectionery, and adhesives.

The most important citrus oil in perfumery is Neroli, distilled mainly from the flowers of the Bitter or Seville orange (*C. aurantium*), and named after the wife of the 16th century Flavio Orsini, Prince of Neroli. Two subspecies are known, the Bergamot orange, *C. aurantium* subsp. *bergamia*, whose oil is used to flavor Earl Grey tea, and Petitgrain, *C. aurantium* subsp. *amara*. The oil of the latter is distilled from leaves and twigs. In perfumery the nomenclature reflects the source of the oil: flower oils are often known as Neroli; expressed oils as peel oils, and leaf oils as Petitgrain, often qualified by country of production. The finest Nerolis come from southern France and Tunisia, and the peel oils come from Jamaica, the Dominican Republic, and Haiti. Petitgrain oils are used to extend other oils. Not all lemon oil used in perfumery is from lemon; lemon scent is also obtained from lemongrass (*Cymbopogon citratus*).

See: Fruits, pp. 80–1

Copaiba balsam *Copaifera* spp.
Fabaceae

Several species of *Copaifera*, an Amazonian tree, are tapped for resin. Distillation results in copaiba oil, used medicinally in Brazil, and internationally by the fragrance industry.

Damask rose, Rose oil (Rose otto, Attar of roses) *Rosa × damascena*
Rosaceae

The rose is one of the most ancient of garden plants and the practice of extracting its oil for use in anointing dates to at least the 9th century BC. The use of rose petals at Roman feasts was chronicled by the painter Alma-Tadema in the 19th century. The supply came from special gardens at Paestum, near Naples.

Damask rose is not known in the wild, and is thought to have originated by hybridization of wild species, perhaps in Turkey. The so-called Damascus rose became familiar to Europeans through the Crusades. Cultivation began in the Ottoman Empire in the 15th century around Kazanlik, now part of Bulgaria, and today this remains the center of production of rose oil (also known as rose otto or attar of roses), based on locally raised cultivars, particularly *R. damascena* var. *triginti-petala*. Bulgaria has nationalized and de-nationalized the industry following the fall of communism, but it still remains important. Turkey is second in world production; smaller quantities are produced in Russia, India, Morocco, and Egypt. Rose oil is also produced on a smaller scale from other species of rose, including *R. alba*, *R. centifolia*, and *R. gallica*.

Rose oil is produced by twice steam-distilling flowers from plantations, hand-picked in the early morning. Rose water from the steam distillation is valuable in itself and sold directly. Concretes and absolutes are also produced by solvent extraction. Rose oils are extremely expensive and valuable in perfumery, though synthetics often substitute for them. They are also used in sweetmeats, especially in Turkey (in Turkish delight) and in India.

See: Ornamentals, pp. 279–80

Daphne *Daphne odora*
Thymelaeaceae

Daphne odora is an evergreen from China and Japan, introduced to Britain in 1771 and now usually seen in the variegated form *D. odora* "Aureomarginata." It bears intensely fragrant pink flowers in early spring. There are many other *Daphnes* (practically all fragrant), ranging from prostrate shrubs (*D. blagayana* and *D. jasminea*) to alpine shrublets (*D. cneorum*), to semi-evergreen hybrids (*D. × burkwoodii*). Several are winter-flowering, including the Himalayan *D. bholua* and its selected forms, and the European *D. mezereum*. All are excellent shrubs for the scented garden provided virus-free stock can be obtained.

Eucalyptus *Eucalyptus* spp.
Myrtaceae

Eucalyptus originates from Australia, but nowadays the majority of its oil is produced elsewhere. Eucalyptus is grown in regions between 45° of latitude north and south of the equator for timber, paper pulp, hydro- or steam-distillation in oil production, or a combination of these uses. Europeans colonizing New South Wales began to distill the so-called peppermint gums, particularly *E. piperita*, in 1788. Commercial production of eucalyptus oil for medicinal use started in 1852 and Australia was the world's largest supplier between 1900 and 1950 until Brazil, Spain, South Africa, and, above all, China established production at lower cost.

Eucalyptus globulus subsp. *globulus* (the Tasmanian Blue Gum) and *E. polybractea* (the Blue-leaved mallee) are the two main species used for commercial medicinal production of the so-called "eucapharma oils." The former is found particularly in Brazil, Argentina, Chile, Columbia, India, Russia, and Spain, where its oil is distilled from waste foliage from timber trees for use under the British Pharmacopoeia (BP) for coughs and chest conditions, as well as in household cleaning products and for balsam notes in perfumery. *Eucalyptus polybractea* oils are produced mainly in New South Wales and used in pharmaceuticals and dental products.

Corymbia citriodora (formerly *E. citriodora*) is endemic to Queensland, Australia, with large plantations in São Paulo, Brazil; in Russia, around the Black and Caspian Seas; and in China. In India (Punjab, Uttar Pradesh, and Andhra Pradesh) it is a small-farm crop. It is distilled for citronellal and hydroxycitronellal, which are lemon-scented oils widely used in perfumery.

Other species are *E. dives*, used in menthol and thymol synthesis, and *E. radiata*, used for antiseptics.

In Southeast Asia, cajeput oil obtained from *Melaleuca cajuputi* (a closely related genus, also in the Myrtaceae) has very similar medicinal uses to eucalyptus oils. *Melaleuca quinquenervia* is grown on a small-scale in the south Pacific for niaouli oil, and is an invasive, weedy plant in North America. Tea tree, *M. alternifolia*, is an increasingly well known source of antiseptic essential oil with medicinal value.

See: Wood, p. 328; Plants as Medicines, p. 235

Frangipani *Plumeria rubra*
Apocynaceae

These are deciduous shrubs from tropical America bearing waxy fragrant flowers on leafless stems in the dry season. The flowers range from red to pink, yellow, white, or a combination of these colors, and are extensively used in Hawaii in garlands. Frangipani is now frequently used in hotels and landscape plantings throughout the tropics but flowers best where there is seasonal drought. It is the source of an Italian perfume called frangipane.

Frankincense, Olibabum *Boswellia* spp.
Burseraceae

Frankincense is obtained as a resin from various species of *Boswellia* in Arabia and the Horn of Africa, specifically *B sacra*, *B. carteri* (doubtfully a distinct species from *B. sacra*), *B. frereana*, and *B. papyrifera*. *Boswellia* are small trees preferring dry, rocky habitats in the mountains, and the resin is obtained as teardrops oozing from cuts on the bark. Its use for incense probably dates from ancient Egypt, although no archaeological material has yet been definitively identified by chemical analysis. Its symbolism as a gift of the Three Wise Men was surely a reference to the holiness of Christ. There is still a heavy demand for incense in Christian churches, whereas medicinal uses are important in China. Olibabum is used as a chewing gum in the Near East.

Gardenia, Cape jasmine *Gardenia jasminoides* (formerly *G. augusta*)
Rubiaceae

This is an evergreen from China, Japan, and Taiwan that was introduced to Britain in 1754 and produces waxy white flowers with a very heavy scent. It is usually grown in selected double-flowered forms and is popular as a pot plant, though it requires considerable heat and acid soil. The flowers are very popular for bridal bouquets and as the source of an exotic perfume in soaps.

Geranium *Pelargonium* spp.
Geraniaceae

"Geranium" oils are derived from *Pelargonium* species, which are generally of South African origin. *P. fragrans* was first introduced to Europe by William A. van der Stel in 1794. The use of the rose-scented leaves of the *P. capitatum* × *P. radans* cross, called "Rose," for commercial oil production commenced in Grasse in southern France in the early 19th century. The oil is used in most rose-scented perfumes and soaps. French colonies such as Algeria developed a geranium oil industry in the 1840s, as did Réunion 40 years later with material sent from Grasse. In Réunion it remains as a small-farm crop that is grown on steeply sloping land unsuitable for other crops. Réunion oil is sometimes called Bourbon geranium oil.

The French also took Réunion plant material to Tamil Nadu in India at the turn of the 20th century, and later took Algerian plants to the Nilgiri Hills, where they are now important and successful crops. Geranium oils are also produced in China and in Australia, Egypt, the United States, Cuba, and Israel from crops grown as perennials from stem cuttings or through micro-propagation. The crop normally comes from small farms, where it is cut eight months after planting (or is mechanically harvested on larger plantations) and then hydro- or steam-distilled.

See: Ornamentals, p. 227

Ginger lily *Hedychium* species
Zingiberaceae
The ginger lilies are rhizomatous plants from the forest edges of tropical Asia and the Himalayas and have very sweetly scented yellow to orange or white flowers. One species, *Hedychium coronarium* (with white flowers), is truly tropical, almost aquatic, whereas the yellow *H. gardnerianum* is suitable for conservatory use, or outdoor use in Mediterranean climates, in order to achieve a tropical effect. Both are used in Polynesian garlands. Several other Himalayan species are nearly as hardy, but none are as fragrant. *Hedychium coronarium* was introduced to Britain from India in 1791, and *H. gardnerianum* was introduced in 1819.

Jasmine *Jasminum* spp.
Oleaceae
Jasmine oil and flowers have an ancient tradition. In Iran, the name means "fragrant." Most species come from India and Southeast Asia. The most important species in India is *J. auriculatum,* and the most commonly cultivated is *J. sambac,* or Arabian jasmine or sambac. The latter was introduced to Europe by the Duke of Tuscany in 1691. The double form of this species is named after him and its flowers are used for threading into garlands, as well as for perfume preparation.

Jasmine flowers were used in ancient Egypt in garlands and for feasting and bathing. Jasmine reached China via the silk routes and was used there to scent tea (using either *J. paniculatum* or the double form of *J. sambac*), a practice noted by the plant hunter Robert Fortune in the mid-19th century.

Jasmine is also associated with the development of Grasse as the French perfume center in the 17th century, based on the cultivation of *J. grandiflorum*. This was dependent on cheap labor, since all jasmine oil production was by enfleurage, that is, the hand placing of hand-picked flowers into animal fats in order to absorb the perfume, which is then washed free by solvents. Jasmine cultivation in the Mediterranean is now based in Algeria, Morocco, Sicily, Spain, and Egypt, and very little production remains around Grasse.

Jasmine concretes and absolutes are very high in value and coming under increasing demand. It is one of the commonest constituents in perfume formulations, valued for its long-lasting base notes, although synthetics often substituted for it.

Lakawood *Dalbergia parviflora*
Fabaceae
The aromatic heartwood of this Indonesian tree is used today in joss sticks in China, India, and Malaysia, and was widely burned for magical and medicinal purposes in T'ang China (618–907 AD).

See: Wood, pp. 317, 327–8

English lavender (*Lavandula angustifolia*). *Flora von Deutschland Österreich und der Schweiz.*

Lavender *Lavandula* spp.
Lamiaceae

Thirty species of lavender are native to Southern Europe, North Africa, and India. *Lavandula angustifolia*, known as common or English lavender, is an important crop, as is the hybrid lavandin (*L. ×intermedia*), which originates in the wild from the crossing of *L. angustifolia* and spike lavender, *L. latifolia*. There are numerous cultivars of both, some selected for perfumery, others for garden use. French lavender, *L. stoechas,* and the less hardy *L. dentata* have also been grown since the late 16th century.

Lavender has been used in perfumery and medicine since at least Roman times, with small-scale cultivation in European gardens since the Medieval period. Large-scale cultivation of *L. angustifolia* began in the 19th century in England, centered on the Surrey village of Mitcham, now a suburb of London. Cultivation shifted to its current center in Norfolk in the 1930s. In North America, lavender cultivation is strongly associated with Shaker communities. However, the major global center for lavender oil production has been around Grasse in France. Initially dependent on gathering naturalized wild lavender (both *L. angustifolia* and *L. latifolia*) from the surrounding hills, production has been based on cultivated fields of *L. angustifolia* and the hybrid *L. × intermedia* since the 1920s.

True lavender oil is obtained from *L. angustifolia,* and ranks alongside citrus, rose, and mint oils as one of the most important essential oils in trade, with an annual production worldwide of 250 tons, mostly from France and Bulgaria. It is used in luxury perfumery, whereas the somewhat harsher lavandin oil, from *L. × intermedia*, is used in cheaper cosmetics and soaps and as a food additive. These and other species such as *L. stoechas* are also grown as a source of dried leaves and flowers, for use in potpourri and aromatic sachets.

Lemongrass and Citronella *Cymbopogon* spp.
Poaceae

Cymbopogon citratus (West Indian lemongrass) and *C. flexuosus* (East Indian lemongrass) are grasses cultivated throughout the tropics for the oil in their leaves. Distillation in Kerala, in southern India, began in the 1880s but commercial plantations boomed in South and Central America during the Second World War. The citrus content is released by steam distillation of the leaves and used as a cheap fragrance for household polishes, waxes, and toiletries. It can also be used as a base for the production of ionones, which are used to produce violet-like fragrances, and is a powerful insect repellent. Production in China and Vietnam is currently increasing. West Indian lemongrass is widely used in Southeast Asia as an ingredient of spicy sauces. The related *C. martinii* grows in India (both wild and cultivated) and produces palmarosa oil, used in soaps and as an aromatherapy oil.

Citronella oil is obtained from *C. nardus* and *C. winterianus*. Both species are thought to have originated in south Asia. *C. nardus* is grown in Sri Lanka, while *C. winterianus* is cultivated in Indonesia, China, India, and Central America. Citronella oil is distilled from the leaves, and has a wide range of uses in perfumery and cosmetics. The oil is increasingly popular as an insecticide and insect repellant.

See: Plants as Medicines, pp. 226–7

Lemon verbena *Aloysia triphylla* (syn. *Aloysia citriodora, Lippia citriodora*)
Verbenaceae

A slightly tender deciduous shrub from Chile and Argentina, introduced to Britain in 1784. Generally grown for home use as a wall shrub in protected gardens or in conservatories, its leaves are used in lemon tisanes or to imbue lemon flavoring to cakes. However, it is not grown commercially as a source of lemon oil: pressed *Citrus* peel or steam-distilled lemongrass (*Cymbopogon citrates*, see preceding discussion) is used for this purpose.

See: Herbs and Vegetables, p. 105

Lilac *Syringa* spp.
Oleaceae

Syringa is a genus of thirty species, native to eastern Asia and southeastern Europe (it is not related to the sweet syringa or mock orange, *Philadelphus* spp.). Not all species are sweet-scented. The most fragrant species is the common lilac, *S. vulgaris*, a south European species cultivated since the 16th century and extensively hybridized by the French Lemoine family at the beginning of the 20th century.

Lily *Lilium* spp.
Liliaceae

The principal *Lilium* species that are grown for their fragrance are the Madonna lily, *L. candidum*, from the Balkans and eastern Mediterranean; the Easter lily, *L. longiflorum*, from southern China and Japan; and *L. regale*, introduced to Britain from Sichuan in China by plant collector Ernest Wilson in 1903. There are many lily hybrids; sadly, however, many have little or no perfume.

See: Ornamentals, p. 275

Lily of the valley *Convallaria majalis*
Liliaceae

This is a rhizomatous perennial, native to woodlands in northern temperate regions, which bears intensely fragrant white flowers in May. The flowers are used for cut flowers and in bridal bouquets. The crowns can be forced to flower at several other times and are a high-value crop. All parts of the plant are rich in cardioactive glycosides and are poisonous. In the past, lily of the valley was used as a milder substitute for digitalis in treating heart problems.

Madagascar jasmine *Stephanotis floribunda*
Asclepiadaceae

This is an evergreen climber from Madagascar that is valued for its waxy, white, jasmine-scented blooms. The flowers are valued for table display as a pot plant or in bridal decorations where the flower clusters or "pips" are often used in headdresses. This species is sometimes called "bridal wreath," and is not to be confused with *Spiraea arguta*.

Magnolia *Magnolia* spp.
Magnoliaceae

The magnolia is a forest tree, predominantly of the temperate Northern Hemisphere, with centers of distribution in the Himalayas, China, Japan, and the United States. Many species are highly fragrant and as much valued for this as for their dramatic blooming. Among the most fragrant are *Magnolia obovata* (formerly *M. hypoleuca*), citrus-scented *M. denudata* from eastern and southern China, *M. grandiflora* from the southeastern United States, *M. virginiana* from the eastern United States, and *M. wilsonii* from western China. Some of these have been the result of plant hunting expeditions, and the latter is named for Ernest ("Chinese") Wilson and was introduced to Britain in 1908. In addition there are a number of recent introductions specially bred for scent, for example *M.* "Heaven Scent," one of the Gresham hybrids.

Manila elemi *Canarium luzonicum*
Burseraceae

The essential oil, with an aroma similar to fennel, from this multipurpose tree is produced from resin obtained by tapping the bark. The main center of production is the Philippines, for domestic and export use in soaps and perfumes.

Mastic *Pistacia lentiscus*
Anacardiaceae

Mastic resin is an important export of the Aegean island of Chios, just off the west coast of Turkey. Extensive plantations of the mastic tree cover the southern part of the island. Mastic exudes from incisions cut into the bark each summer, and the resulting drops are collected from the ground. After sorting and cleaning, the resin is exported by farmers' cooperatives for use in chewing gum and picture varnish, and for a flavored spirit known as mastica.

Although the mastic tree grows throughout the eastern Mediterranean, it is only on Chios that resin yields are high. Although mastic cultivation on Chios is only documented from Medieval times, it is likely that the island was the source of the mastic gum widely used in the ancient world. One ton of mastic gum was found during excavation of the Late Bronze Age Ulu Burun shipwreck (c. 1300 BC) on Turkey's south coast, and chemical analysis has shown that mastic gum was extensively burned as incense in the New Kingdom period of ancient Egypt (1500–1100 BC).

May Chang *Litsea cubeba*
Lauraceae

Cultivated in Southeast Asia, this tree has many uses, including extraction of may chang oil from the fruits. The oil is rich in lemon-scented citral and is used in cheaper fragrance products.

Mignonette *Reseda odorata*
Resedaceae

A Mediterranean annual or short-lived perennial, bearing greenish-yellow flowers with a delicate musk fragrance. The plant was very popular in the 19th century and is occasionally seen nowadays as a cottage garden plant.

Mint *Mentha* spp.
Lamiaceae

Mint oils are the most important of all essential oils in terms of value in world trade, with most production coming from cornmint (*Mentha arvensis*), peppermint (*M.* × *piperita*, a hybrid of *M. aquatica* and *M. spicata*), and spearmint (*M. spicata* from south and central Europe). Spearmint was introduced to western European gardens in Roman times, and is well documented as a Medieval garden plant. All three species are important culinary and medicinal herbs, as well as major sources of essential oil. Cornmint, native to Europe and North Asia, is the most widely cultivated species, with oil production concentrated in China, Brazil, and India. The essential oils are used in a wide range of products, such as ointments and cough lozenges, as flavorings, and in perfumes.

See: Herbs and Vegetables, pp. 106–7; Plants as Medicines, pp. 215–6

Mock orange, Sweet syringa *Philadelphus* spp.
Hydrangeaceae

Philadelphus is a shrub native to a wide area, from northern Africa to the West and South of the United States and Mexico. *P. coronaries*, known as mock orange due to its characteristic orange-blossom fragrance, grows up to 10 feet (3 m) high. *P. microphyllus* is pineapple-scented. Many species are distributed through North America to the Himalayas, China, and eastern Asia. Various significant selections have been made, for example, *P. coronarius* (from southern Europe) and its hybrid with *P. microphyllus,* known as *P. × lemoinei.*

See: Ornamentals, pp. 276–7

Mother of the evening, Sweet rocket *Hesperis matronalis*
Brassicaceae

This is a useful biennial or short-lived perennial for the wild garden where fragrance is wanted. It will also act as a food source for butterflies. Possibly originating in Siberia, it has now become naturalized throughout Europe and is a popular cottage garden plant in either its single- or its double-flowered forms. In North America, the plant is invasive and is classified as a noxious weed.

Musk mallow, Ambrette *Abelmoschus moschatus* (syn. *Hibiscus abelmoschus*)
Malvaceae

Abelmoschus moschatus is an annual or biennial plant, cultivated for its seed oil in India, Madagascar, Java, and parts of the Americas. The oil and the distilled essential oil have a rich, floral, musky aroma, used in luxury perfumery and as a food and liquor flavoring. The oil has potential as a substitute for musk from the endangered musk deer.

Myrrh *Commiphora* spp.
Burseraceae

Many *Commiphora* species produce oleoresins, but the main source in commerce is *Commiphora myrrha.* Myrrh is excreted as "tears of myrrh" from the stems and trunk of this shrub from Arabia, Somalia, and eastern Ethiopia. The name bdellium refers to myrrh from *C. wightii* (formerly *C. mukul*), from East Africa and India, while opopanax refers to *C. kataf* from east Africa. The taxonomy and distribution of *Commiphora* species is poorly known and needs further fieldwork to ascertain it. Myrrh features frequently in Sanskrit literature from the 7th century BC onward. Possibly used in ancient Egypt for incense and for the embalming of bodies, it was also a wound healant. It is thought this was the purpose of the gift from the Three Wise Men—prefiguring the sufferings of Christ. *C. gileadensis* is native to southwestern Arabia, but is probably the balm of Gilead said by ancient writers to have been cultivated at Jericho and elsewhere in Palestine in Roman times.

See: Plants as Medicines, p. 209

Orris *Iris × germanica* var. *florentina*
Iridaceae

Orris is obtained from the rhizome of this variety of the bearded iris. It has long been cultivated around Florence for its violet scent (irone), and is important in its own right and as a perfume fixative.

Orris was used frequently in Elizabethan times to scent snuff. Fresh root and juice can cause severe skin irritation.

See: Ornamentals, pp. 273–4

Pandang, Kewda, Keora *Pandanus tectorius*
Pandanaceae

A popular Hindu perfume is distilled from the white male flowers of this species of screwpine from coastal areas of India and the Far East. It is unknown in the West except as an architectural plant for warm conservatories, but, along with champac, *Michelia champaca* (see preceding discussion), and sambac, *Jasminum sambac,* it forms the triumvirate of Hindu perfumery.

Patchouli *Pogostemon cablin*
Lamiaceae

Patchouli is a musky oil normally solvent-extracted from the downy leaves of the Southeast Asian *Pogostemon cablin.* Though very important in 20th-century perfumery (it became a hallmark of the "hippy" era), patchouli was known in ancient China at least 2,000 years ago, where it was added to writing ink. The Indian textile industry used it like vetiver (see following discussion), to perfume textiles, and the Arabs used it to scent carpets, from whence its odor would become known in Europe as "malabathron." Today patchouli is used in perfumery for its musky basal notes, in soaps, in breath-fresheners, and in low tar tobacco.

Patchouli oil can be detected upon crushing the leaves, but it is normally produced by drying or by lightly fermenting the foliage for between 2 and 14 days. Dried leaves were imported into London in 1844 but the industry based on cultivation of the plant and extraction of its essential oil began in Malaysia in the 19th century, and had become established in Java and Sumatra by the 1920s. Indonesia is the largest producer, with some production in India, the Seychelles, Taiwan, and Brazil. Dried leaves are no longer exported due to the transport costs.

Patchouli is an herbaceous shrub, needing plenty of water and good drainage, high soil fertility, and high light levels. It is frequently grown by small farmers, often on recently cleared land, but occasionally intercropped with rubber or oil palm, or under coconuts. As a plantation crop it is renewed every 3–5 years.

Pinks, Carnation *Dianthus* spp.
Caryophyllaceae

Pinks have been cultivated for their fragrant flowers for centuries, with 300 species distributed throughout Europe and Asia. *Dianthus barbatus* (sweet William) and *D. caryophyllus* (carnation) from Central Europe have been cultivated since the 16th century or earlier. More recent introductions include *D. chinensis* grown as an annual, *D. alpinus* and *D. plumarius*. These are classic cottage garden plants grown as fragrant flowers for cutting. Breeding of carnations and pinks became fashionable in the 19th century in Europe, and around 30,000 cultivar names have been registered. Carnations have also become an important part of the international cut-flower trade.

Red gum, styrax *Liquidambar styraciflua*
Hamamelidaceae

This is a tree native to eastern North America and Central America, valued for its sweet balsam-bearing foliage and rich amber autumn color. Liquid storax, of use in the perfume trade, is obtained on a small scale from the inner bark of *Liquidambar* in Central America and from its slow growing relative *L. orientalis* from Asia Minor. The essential oil from the resin is used in the fragrance industry.

Rock rose, Sun rose *Cistus* spp.
Cistaceae

The gum obtained from leaves of species of *Cistus* (principally *Cistus creticus* subsp. *creticus*, *C. laurifolius*, and *C. ladanifer*) is known as ladanum or labdanum. It has been collected in the eastern countries of the Mediterranean, Crete, and Cyprus since biblical times, either by trimming the beards of goats grazing on *Cistus* or by using a wooden implement laden with long leather, wood, raffia, or (more recently) plastic strings. In Crete these implements are known as "ladanistiri," but the scraping of the gum from the thongs has generally been replaced by wholesale cutting and baling of the shrub, followed by steam distillation of the leaves. Ladanum from Crete and Cyprus was used in incense and in holy oils and in medicine, especially to treat rheumatism and bronchitis.

In perfumery, ladanum is used as a fixative and as an ingredient in perfumes classified as "oriental" or "amber" having a rich exoticism.

Rosewood oil, Bois de rose, Cayenne rosewood *Aniba rosaeodora*
Lauraceae

Once an important source of linalool, a raw ingredient used in the fragrance industry, the production of this oil has declined ever since synthetic linalool became available in the 1960s. Rosewood oil is still produced from several species of *Aniba* found in the Amazonian rainforest of Brazil. Wild trees are felled and the wood oil is distilled from wood.

Sandalwood *Santalum album*
Santalaceae

Sandalwood oil is obtained from the roots and heartwood of this semiparasitic evergreen tree, which is native to the drier parts of Indonesia. The oil is normally hydro- or steam-distilled uprooted plant material harvested from the wild.

Sandalwood is described in pre-Christian Vedic texts and became important in Hinduism and Buddhism, particularly through the use of sandalwood sawdust, which was molded using plant gums into "agarabatti" or incense sticks, to be used in temples and palaces, after the scented wood was carved into sculptures. There are extensive stands of sandalwood in southern India, probably introduced from Southeast Asia some 2,000 years ago.

The trade in sandalwood products was significant from the 3rd century AD onward, extending from Indonesia and India to China, where it was adopted into daily and religious life, and along the Arab trade routes to the Middle East. Payment in sandalwood for cargoes of tea from China in the 19th century led to severe overexploitation of closely related species in Polynesia and led to the substitution and adulteration common in the trade today.

Sandalwood oil is used in high-quality perfumery, and in today's aromatherapy it is thought to improve self-esteem. It is used in high-value soaps and incense, and the wood is still used for inlay and high-value woodcarving and to scent the funeral pyres of members of the highest castes of India. In parts of Tamil Nadu and Karnataka, it has become an endangered species; although harvesting is government-controlled, supplies are also stockpiled by bandits in order to control the price.

See: Wood, p. 328

Spikenard *Nardostachys grandiflora*
Valerianaceae

This herbaceous perennial of the Himalayas has stout rhizomes, from which spikenard oil is distilled. The oil has medicinal and cosmetic uses that have led to overharvesting, and the species is now threatened. The pounded roots are used for incense throughout South Asia. Spikenard is referred to in the New Testament as a very costly oil, and it was popular as a perfume in the Roman Empire.

Star jasmine *Trachelospermum jasminoides* and *T. asiaticum*
Apocynaceae

Members of *Trachelospermum* are scented evergreen self-clinging climbers suitable for warm walls or conservatories. *Trachelospermum asiaticum* from Korea and Japan bears yellow flowers and is hardier than *Trachelospermum jasminoides* with pure white, jasmine-like flowers—a favorite in Mediterranean gardens. The latter was introduced to Britain from China by the plant hunter Robert Fortune in 1844, a product of his first expedition in quest of Chinese plants for the Horticultural Society, now the Royal Horticultural Society.

Stock *Matthiola* spp.
Brassicaceae

The night-scented stock, *Matthiola longipetala,* originating from Greece and the Ukraine, is an annual highly valued for the fragrance it produces in the evening. The Brompton stock, *M. incana*, is a biennial from southwest Europe whose the fragrance is best in the double-flowered forms used for cut flowers.

Sweet flag, Calamus root *Acorus calamus*
Acoraceae

This rhizomatous aquatic plant originates from Southeast Asia and has been known in Europe since classical times. Its fragrant leaves were used as strewing material to freshen the air at a time when "foul, pestilential airs." was thought to cause disease. Production of the citrus-scented roots continues in India and Southeast Asia for medicinal purposes, but the potentially carcinogenic properties of the essential oil restrict its current use. It has potential as a natural insecticide.

See: Spices, p. 169

Sweet myrtle *Myrtus communis*
Myrtaceae

Myrtle is an evergreen shrub from the Mediterranean and Middle East and figures in the mythology of several peoples, having been sacred to the ancient Greeks, Persians, and Romans. In Egypt, it was used in garlands and it is still a popular garden plant today, seen as a symbol of love and constancy. Its leaves are collected from the wild in Tunisia, Morocco, and Algeria, and are distilled to yield myrtenol, which has woody base notes and is used in perfumery.

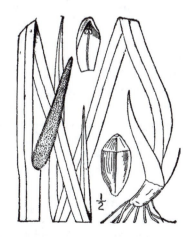

Sweet flag (*Acorus calamus*). USDA-NRCS PLANTS Database/Britton, N.L., and A. Brown. 1913. *Illustrated flora of the northern states and Canada.* Vol. 1, p. 446.

Sweet osmanthus *Osmanthus fragrans*
Oleaceae
Sweet osmanthus comes from China, Japan, and the Himalayas, and produces very small white flowers with an intense apricot fragrance. In China it is used to scent tea and is planted around temples. Its oil is extracted using solvents in order to produce an absolute that is used in very high value cosmetics and perfumes, within China and for export. Other species such as *Osmanthus delavayi* are valued as garden plants.

Sweet violet *Viola odorata*
Violaceae
This is a species native to south, south central, and northern Europe, which was cultivated in Greece before the time of Christ, and which flowers from late winter to spring. Sweet violet oil is extracted by solvent from the leaves and flowers for use in perfumery and scenting confectionery. "Parma violets" are less hardy hybrids.

See: Ornamentals, pp. 283–4

Tacamahac, Balsam poplar *Populus balsamifera*
Salicaceae
This is a tall tree, easily reaching 98 feet (30 m) in its habitat, native to the northern United States, the former Soviet Union, and Canada. The buds bear a fragrant yellowish resin or gum, which scents the air on warm spring days.

Tolu balsam, Peru balsam *Myroxylon balsamum*
Fabaceae
This is an evergreen tree from tropical zones of Mexico and central and northern South America, the bark of which is tapped for a sweet balsam. Tolu balsam (var. *balsamum*), introduced to Britain in 1735, is lemon-scented and its oil is used in pharmaceuticals as a flavoring and in perfumery. The related Peru balsam (var. *pereirae*) is darker, with a "balsamic" smell, and is used for treating skin disorders and as a fixative for perfumes.

Tuberose *Polianthes tuberosa*
Agavaceae
A Mexican half-hardy bulb grown for its spikes of waxy white flowers, as intensely fragrant as gardenias. It is generally grown in the double-flowered forms called "The Pearl" or "Excelsior Double Pearl," but single forms are also known. It was a Victorian favorite grown in cool glasshouses or on hotbeds, and was also used by other cultures: by the Aztecs as a flavoring for cocoa, the food of the gods, and by Hindus in garlands. It was also grown for the cut-flower and perfume industry, especially in southern India. Cold enfleurage (oil extraction performed by pressing the flowers into fat and washing out the oil with solvent) requires 1150 grams of flowers to produce 1 gram of essential oil. Tuberose is used in heavy floral perfumes, and some people find it overpowering.

Vetiver *Vetiveria zizanioides*
Poaceae
Vetiver is an Indian grass reaching 10 feet (3 m) in height, with roots containing the essential oil of the same name. The oil had ancient uses in Hindu cultures as the oil of tranquility, and has been taxed since the 12th century. Traditionally the rhizomes and roots were woven into screens, mats, and fans, which, when wetted, release a delicious woody fragrance. This practice is still seen today; roots are also hung inside wardrobes. Vetiver has insecticidal and insect-repellant properties.

The equivalent material in Europe was calamus (sweet flag) from *Acorus calamus*. Vetiver is used today in perfumery and soapmaking, and as a food flavoring, for example, in canned peas.

Extraction of vetiver oil (normally by steam) was developed in Java by the Dutch before World War II, and small quantities have been distilled in China since the 1950s. The French introduced vetiver to Réunion Island in the 1950s to join the many other oil crops grown there. It is also grown in the Seychelles, Haiti, Brazil, and Japan.

Vetiver is also widely grown as a soil stabilizer throughout the tropics, as its roots can reach 9 feet (3 m) in length, and its distribution throughout the wet tropics is due to this as much as to its use in perfumery. It is normally grown from splits as a plantation crop for its oil, using the southern Indian, non-flowering form. For maximum yield, the roots are harvested when 18–30 months old, either by hand or using mechanical harvesters, and can be stored prior to steam extraction in order to obtain the best price at market.

Viburnum *Viburnum* spp.
Caprifoliaceae

The *Viburnum* genus includes 150 species of shrubs, mostly from the temperate Northern Hemisphere. Many of them are fragrant, especially the species and hybrids flowering in winter and early spring. European species like *Viburnum opulus* and *V. tinus* have long been cultivated. In the 18th century various North American species were introduced, but these remain uncommon in European gardens. The introduction of species from northeast Asia, notably *V. betulifolium*, *V. × carlesii*, and *V. farreri*, was far more significant. The variety has been further enhanced by numerous hybrids raised in cultivation.

The most fragrant are *V. farreri* and its hybrid *V. × bodnantense*, both of which bear almond-scented pink flowers on leafless stems in the winter. *Viburnum × burkwoodii* is semi-evergreen and flowers in early spring, closely followed by one of its parents, the deciduous *V. carlesii*, and its offspring *V. × carlcephalum* and *V. × juddii*. All are excellent shrubs for the scented garden and for cutting.

Wallflower *Erysimum cheiri* (syn. *Cheiranthus cheiri*)
Brassicaceae

There are about eighty *Erysimum* species in Europe, western Asia, and North America. The common bedding wallflower *Erysimum cheiri* is by far the most important in gardens, and has been cultivated since the 16th century or earlier. Native to southern Europe, this short-lived sub-shrub has a heavy and very rich perfume that was much appreciated by the Victorians.

Water lily *Nymphaea* spp.
Nymphaeaceae

The native European *Nymphaea alba* has probably been cultivated for ornament since the 16th century. However, the genus consists of 35 species, and others have been brought into cultivation as well, although many of these species are not hardy in temperate regions.

An early introduction from North America was *N. odorata*, a day flowering hardy water lily native to the eastern United States with more than twenty hybrids or cultivars. The flowers range from deep pink to rose to yellow, and have been used as cut flowers, even though they usually close in the afternoon.

Hybridization of water lilies has been the specialty of the French, particularly Marliac in the 19th century, who was reputed to have taken his secrets to the grave, and of Americans, whose work includes fragrant tropical species.

See: The Hunter-Gatherers, p. 9; Gathering Food from the Wild, p. 38; Psychoactive Plants, p. 203

Wax flowers *Hoya* spp.
Asclepiadaceae

These are evergreen tropical and subtropical climbers or epiphytes (plants which live in the forest canopy perched on others for support, and absorbing nutrients from debris or the air), and are used for warm glasshouse or house decoration. The most fragrant species, with honey-like perfumes, are *Hoya carnosa* from Burma, southern China, and India; *H. lanceolata* subsp. *bella* from the Himalayas; *H. carnosa* from Japan; and *H. multiflora* from China.

Wintergreen, Checkerberry *Gaultheria procumbens*
Ericaceae

A prostrate shrub from boggy acid habitats of eastern North America. Its foliage can be steam-distilled to produce oil of wintergreen, which is rich in methyl salicylate and is used in inhalation remedies, including Olbas oil. Although synthetic methyl salicylate is used in perfumery, natural extracts are still important as flavorings, for example, in cola soft drinks.

Wintersweet *Chimonanthus praecox* (syn. *Chimonanthus fragrans*)
Calycanthaceae

A deciduous shrub reaching 9–13 feet (3–4 m) and producing spicy-scented yellow flowers on naked twigs in winter. The plant was introduced to Britain from China in 1766 and, though ungainly and insignificant throughout the summer, it is a valuable addition to the winter garden. Outside of very sheltered areas, it is best planted against a wall.

Ylang ylang, Perfume tree *Cananga odorata*
Annonaceae

Ylang ylang is a fast growing, elegant tree native to monsoon areas of Southeast Asia, where it is widely grown as an ornamental under the name perfume tree. The essential oil is borne in the petals of the yellow-green flowers and carried in the axils of the leaves. The species is related to the familiar tropical fruits custard apple and soursop.

First described by Blume in 1829 (in *Flora Javae*), the tree was subsequently introduced throughout tropical Africa, the Americas, and the islands of the Indian Ocean, principally by the French. The main producer is now Madagascar, followed by Indonesia, with small amounts coming from Réunion Island, Comoro, and the Philippines, where the industry was once much greater. Two types of oil are produced, ylang ylang and the harsher cananga. True ylang ylang is reminiscent of freesias, though much sweeter, and is purportedly an aphrodisiac. The perfume description is floral with woody leathery notes. Locally the flowers are used as temple offerings, in garlands, as house fresheners, and to adorn bridal beds.

The flowers are normally steam-distilled. The first to do so was Albertus Schwenger in the Philippines, closely followed by Steck and Sartorius, two Germans who started a commercial distillery based on plantations around Manila at the end of the 19th century. Another industry in Réunion developed in parallel to the Philippines-based production, starting from an original introduction of ylang ylang by the French captain d'Etchevery in 1770. The industry in both countries was destroyed by the First World War. The postwar industry in Madagascar was encouraged by a priest, the Reverend Raimbault. The industry on the adjacent Comoros Islands developed throughout the 20th century, although recent increases in tourism and in forced migration from Madagascar have led to a reduction in cultivation.

Ylang ylang is hand picked in the early morning, from trees in plantations that are topped at 9 feet (3 m) and have their naturally pendant branches pegged down. Now popular in soaps and as an exotic aromatherapy oil, ylang ylang would have been a well-known fragrance to the Victorians in Macassar hair oil, the use of which spawned "antimacassar" cloths to prevent staining on upholstery.

Zdravetz *Geranium macrorrhizum*
Geraniaceae

Zdravetz oil is the only essential oil produced from the *Geranium* (as opposed to *Pelargonium*) species. Well-known to gardeners as an herbaceous *Geranium*, this species is in fact endangered in its native Bulgaria, where its leaves and rhizomes are still collected for local distillation from managed natural stands. Zdravetz oil is a sweet, woody, resinous material used in perfumery for its basal notes.

References and Further Reading

Bauer, K., Garbe, D. and Surburg, H. 1997. *Common Fragrance and Flavor Materials: Preparation, Properties and Uses*. Weinheim: Wiley-VCH.

Boer, E., and Ella, A.B. (Eds.). 2000. *Plants Producing Exudates. Plant Resources of South-east Asia 18*. Leiden: Backhuys.

Bown, D. 2001. *Herbal*. London: Pavilion Books.

Bygrave, P. 2001. *Cistus*. London: Chelsea Physic Garden Company.

Coppen, J.J.W. 1995. *Flavours and Fragrances of Plant Origin. Non-wood Forest Products 1*. Rome: Food and Agriculture Organization.

Coppen, J.J.W. 1995. *Gums, Resins and Latexes of Plant Origin. Non-wood Forest Products 6*. Rome: Food and Agriculture Organization.

Genders, R. 1994. *Scented Flora of the World*. London: Robert Hale.

Genders, R. 1994. *Scented Flora of the World*. Bury St. Edmunds: St. Edmundsbury Press.

Hay, R., and Waterman, P.G. (Eds.). 1993. *Volatile Oil Crops*. New York: Wiley.

Lawless, J. 2001. *The Aromatherapy Garden: Growing and Using Scented Plants*. London: Kyle Cathie.

Macmillan, H.F. 1991. *Tropical Planting and Gardening* (6th ed.). Kuala Lumpur: Malayan Nature Society.

Manniche, L. 1999. *Sacred Luxuries: Fragrance, Aromatherapy and Cosmetics in Ancient Egypt*. London: Opus Publishing.

Minter, S. 1996. *Thinking with Your Nose*. London: Chelsea Physic Garden Company.

Nicholson, P.T., and Shaw, I. (Eds.). *Ancient Egyptian Materials and Technology*, Cambridge: Cambridge University Press.

Ohloff, G. 1994. *Scent and Fragrances: the Fascination of Odors and their Chemical Perspectives*. Berlin: Springer-Verlag

Oyen, L.P.A. and Nguyen Xuan Dung (Eds.). 1999. *Essential Oil Plants. Plant Resources of South-east Asia 19*. Leiden: Backhuys.

Miller, J.I. 1969. *The Spice Trade of the Roman Empire 29 BC to AD 641*. Oxford: Clarendon Press.

Weiss, E.A. 1997. *Essential Oil Crops*. Wallingford, Oxfordshire: CAB International.

14
Ornamentals

PETER BARNES

Any selection of the most important decorative plants in the gardens of Europe and North America will unavoidably be subjective. The topic is vast, and it is made much more complex by the dominance, in many genera, of variants selected or bred for their horticulturally desirable properties—cultivars—that are often of complex hybrid origin. Whereas elsewhere in this book individual species are discussed, that approach is scarcely possible under this heading. No one could deny roses a place, for example, but it would be unrealistic to pick out a single cultivar as representative of the entire genus; although many of the species are cultivated, they have a much less prominent place in gardens than the cultivars. It is the important role of certain of the species in the ancestry of the cultivars that underpins the whole subject.

For this reason, the approach taken here has been to review briefly a number of genera of indisputable importance to Western gardens, trying at the same time to achieve a balance of plant types. The subject matter would merit a book to itself, so the entries are necessarily brief, with longer treatments for about a dozen of the most important genera. In addition, where appropriate, the development of cultivars or cultivar groups derived from these species is briefly reviewed.

Significant plant introductions to European and North American gardens are discussed, from both geographical and historical viewpoints, including dates of introduction where known. It is surprising how little-documented many important plants are, in this respect. For the most part, these dates refer to introductions to Western Europe, and mostly to the British Isles, which has, probably, a longer and deeper involvement in the development of garden plants than any region other than China and Japan.

Until the establishment of botanic gardens in the United States from the mid-19th century, it is probable that virtually all cultivated plants were introduced to that country via the gardens of Europe. From the late 1850s, however, the growth of American-sponsored plant exploration was rapid, especially after the founding of the Arnold Arboretum of Harvard University in 1872.

Early gardening is, by definition, largely shrouded in history, but it is generally agreed that the Chinese have the longest tradition of gardening for ornament and relaxation, rather than purely for edible plants. The 2nd millennium BC saw the creation of lavish gardens by a succession of emperors, laying the roots of a tradition that has influenced gardening in many other countries. This influence reached Japan in the 6th century AD, and since that time that country has developed along

Columbine (*Aquilegia*). Copyright Dave Webb, used with permission.

its own course. Another thread in the pattern is traced to Mesopotamia, where gardens were planted initially perhaps primarily for shade and for edible fruit. The famed hanging gardens of Babylon, apparently an elaborate roof garden, are said to have been built in the 6th century BC.

In Europe the Romans were early gardeners, again, largely with edible or otherwise useful plants in mind, but also undoubtedly of ornamental flower beds. Curiously, in England gardening seems almost to have died out with the end of the Roman occupation, only to resurface in Medieval times, along with a growing interest in medicinal plants in monastic gardens. Only in the 11th century did purely ornamental plantings reappear, but their major development in the English style (in reality, a plurality of styles) occurred from the 16th century onwards. From this period, the development of the garden was increasingly closely tied to the plant collecting activities sponsored within a growing empire.

Plant hunting has inevitably tended to follow geographical exploration, often simultaneously, but the earliest plant introductions followed the available trade routes. Hence the earliest nonindigenous plants to be cultivated in the UK—up to the mid-16th century—were those native to Europe and Western Asia. The Silk Road from Western Asia also provided a corridor by which a small number of ornamental plants of Chinese origin, notably the peach, *Prunus persica*, arrived in the West. Edible and medicinal plants were of more interest than ornamentals at this period.

The 16th century saw the entry of European adventurers into South America, but the very few plant introductions made then were of economic, rather than ornamental, plants. Not until the visits of William Lobb in the 19th century were significant ornamental introductions made. Plants from the more accessible parts of southern Africa, notably the Cape area, started to reach Europe late in the 17th century. At about the same time, North America began to be opened up by explorers such as John Tradescant, father and son, and later by John Bartram, but plant hunting expeditions there on a large scale peaked about 150 years later, most notably with the indefatigable John Douglas.

Many authorities agree that the greatest impact on gardens in the West has come from the introduction of plants from the temperate parts of Eastern Asia. The Himalayan mountains and forests were extensively explored by collectors such as J.D. Hooker in the mid-1800s, contributing most importantly to the rise in interest in the genus *Rhododendron*. For centuries, Japan and China were, to say the least, difficult to access by Westerners. The Treaty of Nanking in 1842 allowed some access to China, and Robert Fortune was one of the earliest and most successful to work this area. The glory days of plant hunting in China, however, occurred towards the end of that century and in the first three decades of the 20th, with plant collectors such as Ernest 'Chinese' Wilson, George Forrest, and the Austrian Joseph Rock often working in collaboration with resident French missionaries. Both Wilson and Rock did much work for the Arnold Arboretum of Harvard University.

Japan "opened up" only in the 1860s, following the Meiji Restoration. Phillip von Siebold, resident in the Dutch trading base of Dejima, Nagasaki, had managed to make some earlier introductions, but it was to the likes of Robert Fortune, J.D. Veitch, Charles Maries, and, later, E.H. Wilson and Charles Sprague Sargent that we owe many of the fine garden plants that this relatively small country has yielded.

African daisy, Cape daisy, Freeway daisy, Osteospermum *Osteospermum*
Asteraceae

Osteospermum is a genus of about seventy species native to southern Africa through to Arabia. The plants are about a foot tall with daisy-like flowers. Early introductions from southern Africa to European gardens include *O. acutifolium* in 1774, followed by *O. jucundum* in 1862 and *O. ecklonis* in 1897. The genus became fashionable towards the end of the 20th century, when considerable hybridizing and selection occurred. Currently, over sixty cultivars are available.

African violet *Saintpaulia*
Gesneriaceae

The twenty-one species of *Saintpaulia* are native to East Africa, especially Tanzania. Most of the species have flowers in shades of blue, purple, or lavender, and are grown indoors as houseplants. *Saintpaulia ionantha*, introduced in 1893, is the most important species in cultivation, although a few other species are grown by specialists. Many cultivars have been raised in North America and Europe. Recently, miniature African violets have been developed from *S. pusilla*, *S. shumensis*, and *S. magungensis* var. *minima*.

Alumroot, Coral bells, Heuchera *Heuchera*
Saxifragaceae

About fifty species of *Heuchera* are native to temperate North America. *Heuchera americana* reached Europe as early as 1656, but it was in the 19th century that a number of species spread to Europe, including *H. pubescens* and *H. villosa* in 1812; *H. glabra*, *H. micrantha*, and *H. richardsonii* in 1827; and *H. sanguinea* in 1885. The closing years of the 20th century saw the genus increasing in popularity, with many new cultivars being introduced. Reaching from 12 to 18 inches (30 to 45 cm) high, some types are grown for their evergreen foliage, while some have masses of tiny bell-shaped flowers.

African and French marigold *Tagetes*
Asteraceae

There are thirty species of marigold, all native to tropical America. Of these, two species have led to a range of important annual bedding plants. French marigolds are derived from *T. patula* and African marigolds (sometimes called Aztec marigold) from *T. erecta* (actually native to Mexico). Intensive breeding, especially in the 20th century, has introduced diversity in flower form and color. Afro-French marigolds represent a fusion of characteristics from both species.

Anemone, Lily-of-the-field, Windflower *Anemone*
Ranunculaceae

This genus of about 120 species, native to both northern and southern temperate regions, has long been important in gardens. European species, mainly of Mediterranean origin, include *Anemone blanda*, *A. apennina*, *A. narcissiflora*, *A. sylvestris*, *A. coronaria* (lily-of-the-field, poppy-flowered anemone), *A. pavonina*, and *A. hortensis*. All have been cultivated since the 16th century. The last three have been much developed to produce the large-flowered St. Bavo Group and the St. Brigid Group with double flowers, both important to the cut-flower trade. From the Himalayas, *A. polyanthes* (ca. 1885) and *A. obtusiloba* (1843) are cultivated, but are less common than the autumn-flowering Japanese anemones (*A. × hybrida*). These are derived from *A. hupehensis* var. *japonica* (Chinese, in spite of the name; it has long been naturalized in Japan) and the Nepalese *A. vitifolia* (1829). Several cultivars survive, most apparently raised in France in the mid-19th century.

Asters, Michaelmas daisy *Aster*
Asteraceae

Asters are also known as Michaelmas daisies, but this name is specifically for the commonly grown *Aster novi-belgii*. Widespread in the northern temperate region with about two hundred species, European asters such as *A. amellus* (Italian aster) have been grown in gardens since around 1650. The introduction of *A. novi-belgii* and *A. novae-angliae* (New England aster) to Europe from the United States in 1710 greatly increased the range, soon to be followed by other American species and Asiatic species such as *A. thomsonii* (1887) and *A. yunnanensis* (ca. 1900). Nowadays, the wild species have largely been supplanted in gardens by the many cultivars, selected for improved flower colors, disease resistance, or other attributes. This is especially true of those derived from American species.

Balsam, Busy Lizzy, Jewel weed *Impatiens*
Balsaminaceae

This is a genus of over eight hundred species ranging from the tropics to the northern temperate region, but only a relatively few species are widely cultivated. *I. walleriana*, of East Africa origin, and hybrids derived from it, are among the most widely grown bedding plants in temperate regions. Another common annual is *I. balsamina* (introduced from Southeast Asia, before 1810 and possibly as early as the late 16th century); many selections of both are listed by seedsmen in Europe, North America, and Japan. The Himalayan *I. glandulifera*, introduced in 1839, has naturalized aggressively in parts of Europe and North America. *Impatiens hawkeri*, from New Guinea, has recently yielded a number of cultivars with spectacularly colored foliage.

Barberry, Hollygrape *Berberis*
Berberidaceae

Over five hundred species of *Berberis* occur in the temperate regions, mostly in Asia and South America. Although long cultivated, the European *B. vulgaris* is a secondary host of a wheat rust disease, and so its cultivation is undesirable in wheat-growing areas. Evergreen shrub species from South America, notably *B. empetrifolia* and *B. darwinii*, quickly became popular, as did their hybrid, *B.* × *stenophylla*. *Berberis thunbergii*, Japanese barberry, was brought from Japan in 1864 and has spawned numerous cultivars. The early 20th century saw the introduction of several important Chinese species, notably *B. julianae*, *B. wilsonii*, and *B. gagnepainii*. Although commonly grown for habit of growth and foliage, all species have small flowers, frequently quite showy.

Cornflower, Bachelor's buttons, Knapweed *Centaurea*
Asteraceae

There are five hundred species of *Centaurea* in Europe and western Asia ranging from perennials to annuals. The cornflower, *C. cyanus*, grown since the 16th century, and sweet sultan, *C. moschata* from Iran, are popular annuals with varieties in a wide range of colors. The perennials include yellow-flowered *C. macrocephala* from the Caucasus, the European *C. montana* (16th century), and the silvery-gray-leaved *C. cineraria* (1710).

Begonia *Begonia*
Begoniaceae

This genus includes over one thousand species—all with attractive flowers or foliage. Most are native to South America; a few are from other parts of the subtropics of both hemispheres. In the early 19th century, Central and South American species such as *B. metallica* and *B. scharffii* were introduced. Various tuberous-rooted species from Central America arrived in the late 19th century in both Europe and North America, although precise dates of introduction are not known; these species rapidly gave rise to a large number of hybrid cultivars selected for desirable flower or foliage characteristics. The Brazilian *B. semperflorens* is the main ancestor of the popular bedding begonias and *B. sutherlandii* from South Africa has recently become popular for hanging baskets. Important groups include the Rex-Cultorum, Semperflorens-Cultorum, and Cane-stem groups. In all, possibly ten thousand cultivars have been raised, mostly of hybrid origin.

Bellflower *Campanula*
Campanulaceae

About three hundred species of *Campanula* occur in the northern temperate region. West European species such as *C. poscharskyana*, *C. portenschlagiana*, and *C. glomerata* have been grown in flower gardens since the 16th century or, in the case of *C. medium*, since the Medieval period or earlier. More recent introductions include *C. barbata*, *C. isophylla* (Falling Stars), and *C. punctata*, and smaller species such as *C. carpatica* and *C. cochlearifolia*. Although introduced only in the 1980s, the Korean *C. takesimana* has rapidly become widely grown. Flowers are commonly bell-shaped, but may have a flatter, star shape, and may be borne on tufted, sprawling, or tall erect stems.

Bleeding heart, Dutchman's breeches *Dicentra*
Papaveraceae

Nineteen species of *Dicentra* are native to North America and East Asia. Small heart-shaped flowers with protruding inner petals occur in sprays along the stems, above fern-like foliage. The American species reached Europe in the 18th century, namely *D. cucullaria*, *D. formosa*, and

D. eximia. More important, however, was the introduction in 1841 of the Chinese *D. spectabilis* from Japan, where it had long been cultivated. This was considered to be "one of the most brilliant hardy plants added to our collections for many years" (Wright and Deward 1910).

Broom *Cytisus*
Papilionaceae

Cytisus is a genus of about fifty species, mostly native to Europe. Various species have been introduced since the 18th century, valued for their showy, often yellow flowers in summer, and ranging from small trees to creeping rock-garden shrubs. Numerous hybrid cultivars have been raised subsequently, mostly in the UK in the late 19th and 20th centuries, with the objectives of increasing the range of flower color and flowering season.

Butterfly bush *Buddleja*
Buddlejaceae

The one hundred species of this genus are distributed throughout eastern Asia, Africa, and North and South America. The highly fragrant flowers are attractive to insects, especially butterflies (hence the name). *Buddleja salviifolia* from South Africa has been cultivated since 1783, and *B. colvilei* since 1849. However, these early introductions were all tender. The Chinese *B. davidii* (1890) was the first truly hardy species, soon yielding many cultivars. Other Chinese species, notably *B. fallowiana* have subsequently been grown and hybridized.

Camellia *Camellia*
Theaceae

Camellia is a genus of about two hundred species, occurring in eastern Asia, especially China. The first to be introduced was *C. japonica*, from China (where it has been cultivated for centuries, although probably not indigenous) in 1739. Since then, camellias, more than most other plants from the Far East, have made a major contribution to Western gardens. For many years *C. japonica* was assumed to be tender and was grown as a greenhouse plant, although it has proven to be quite hardy. It remains the most popular species, and many cultivars are now available. These include a wide range of flower colors, single and double flowers ranging from anemone-centered to formal double.

The free-flowering Chinese *C. saluenensis* was soon hybridized with *C. japonica* to give a range of excellent hybrid cultivars known collectively as *C.* × *williamsii*, considered the best for cooler areas. *C. reticulata*, again from China, is a tender species that is perhaps the most spectacular of all, with handsome foliage and very flamboyant flowers. It is hardy only in warm temperate regions, but is frequently grown as a conservatory shrub.

The Japanese *C. sasanqua*, which flowers in winter and early spring, has smaller flowers, but these are pleasantly fragrant. Very distinct is *C. cuspidata*, a hardy species with narrow leaves and small white flowers; it reached Europe and North America in 1900. Several other species are in cultivation, mostly rather tender, and their main significance has been for use in breeding, especially to introduce new colors such as yellow to the large-flowered garden hybrids.

Cultivars of *C. japonica* and *C. reticulata* have been bred for centuries in China and Japan, and some early introductions were of native cultivars that were often renamed in the West. In the 19th century, many cultivars of *C. japonica* were raised in Western Europe, especially in France, Belgium, and England. In the second half of the 20th century, camellias became popular in North America, Australia, and New Zealand, and many of the finest new cultivars have been raised here.

Ceanothus, Californian lilac *Ceanothus*
Rhamnaceae

Most of the fifty-five species of *Ceanothus* occur in the United States Southwest, with a few in the eastern United States and others in Central America. *Ceanothus* first reached the Old World in 1713, with the introduction of the white-flowered *C. americanus,* known as New Jersey Tea or redroot. The leaves of this species were used in New England as a substitute for tea during the American War of Independence. The Western species were brought to Europe from the 1840s onwards, and these have had the greater impact, with their showy blue flowers carried over a long period. Some hybridizing took place in France in the 19th century, and a trickle of new cultivars appeared in Europe and the United States in the 20th century.

China aster, Annual aster *Callistephus*
Asteraceae

The single species of this genus, *Callistephus chinensis*, was introduced to Europe from China in 1731, and probably reached North America early in the 19th century. Many cultivars are now listed, some developed for bedding, others for cutting for flat daisy-like or large chrysanthemum-like flowers. In recent years, Japan has become important in breeding these plants.

Chrysanthemum *Chrysanthemum*
Asteraceae

Formerly a large and diverse genus, *Chrysanthemum*, as currently recognized, is restricted to about twelve species of herbaceous perennials, all native to northeast Asia, mainly China and Japan. Several of these species are believed to be ancestors of the familiar garden chrysanthemum and the florists' "mum," *C. grandiflorum*. A few closely allied species are also cultivated. Other species formerly included in *Chrysanthemum*, both annual and perennial, have been reassigned to a number of small genera that more closely reflect their natural relationships.

The garden chrysanthemums, known as *C. grandiflorum*, are complex hybrids involving several species, the result of perhaps 2000 years or more of breeding in the Far East. Chrysanthemums still have great significance in Japan, and not only as an ornamental flower—the symbol of the Japanese Imperial dynasty is a stylized chrysanthemum. In autumn, many chrysanthemum exhibitions are held, featuring life-size figures created from the blossoms and leaves of chrysanthemums, as well as competitive displays of the cultivars of this highly developed flower. In addition, the garland chrysanthemum, *C. coronarium*, is commonly eaten in Japan, where it is known as shungiku.

Chrysanthemums first reached France in 1789, a separate introduction arriving in England in 1795. In the intervening years, they have become some of the most popular garden flowers, and a major element in the commercial cut-flower trade. The range of cultivars is enormous and diverse in both form and color. In general, the larger-flowered varieties require protection, mainly from the rain, but there are numerous small flowered varieties that are excellent border perennials, as are many of the Korean chrysanthemums.

From the late 19th century onwards many cultivars have been raised in North America and Europe (notably the British Isles), either from seed or as bud-sports, in other words, spontaneous mutations. In addition, Japan and China continue to introduce new cultivars in a range of distinctive styles, and some of these also reach the West. A classification system for the many cultivars is in general use in the British Isles, where about five hundred cultivars are currently available.

Cinquefoil, Buttercup shrub *Potentilla*
Rosaceae

Over five hundred species of *Potentilla* are found throughout the northern temperate region. Low-growing shrubs with small flowers, they are usually grown as individual specimens. European species, several native to the British Isles, have been cultivated since the 17th century, such as *P. fruticosa* and *P. recta*. From outside Europe, significant introductions include the Himalayan *P. atrosanguinea* var. *argyrophylla* (formerly *P. argyrophylla*), *P. atrosanguinea*, and *P. nepalensis* (both cultivated by 1822). There remain many attractive species from East Asia rarely grown in gardens.

Clematis, Virgin's bower, Leather flower *Clematis*
Ranunculaceae

A genus of over two hundred species of woody climbers, shrubs, and herbaceous perennials, *Clematis* is native to the temperate regions of both the Northern and Southern Hemispheres. *Clematis viticella* and other European species were cultivated in the 16th century, but it was the arrival of *C. florida* in 1776, and *C. patens* and *C. lanuginosa*, all from China or Japan, that allowed the development of the large-flowered hybrid cultivars so familiar today. Most of these have been raised in Europe, and a few in North America, Australia, and Japan.

Columbine, Granny's bonnet *Aquilegia*
Ranunculaceae

About seventy species of *Aquilegia* are found in the northern temperate region. Long stems have nodding flowers with a spur at the bottom. The European *A. vulgaris* has been a common cottage garden plant since the 16th century or earlier. Several North American species, notably *A. chrysantha*, introduced about 1873, have contributed to the development of modern hybrid strains. The late 19th century saw the introduction of *A. flabellata* from Japan, the fragrant *A. viridiflora* from China, and others from East Asia, but these are still to be exploited by plant breeders.

Cotoneaster *Cotoneaster*
Rosaceae

There are over two hundred species of *Cotoneaster* in Europe and Asia east to China. Several European species have been cultivated since the 16th century for their flowers, berries, and, in some cases, foliage that turns purplish-red in autumn. The influx of species from eastern Asia in the late 19th and early 20th centuries greatly enriched the range available, and also resulted in the raising of some fine hybrids. Notable among these are *C. frigidus*, and *C. henryanus*, *C. salicifolius*, and *C. horizontalis*, all from western China.

Crabapple, Apple *Malus*
Rosaceae

About twenty-five species of *Malus* are found across the northern temperate region. Although orchard apples have long been cultivated, ornamental crabapples did not appear in gardens until the 18th century, with the introduction of *M. coronaria*, *M. prunifolia*, and *M. baccata*. The last two were hybridized about 1815 to produce the familiar Siberian crabapple, *M.* × *robusta*. An influx of species from East Asia in the late 19th century raised the profile of crabapples considerably including *M. floribunda*, *M. halliana*, *M. sieboldii*, and *M. tschonoskii*, all now well established in cultivation.

See: Fruits, p. 85

Crocus *Crocus*
Iridaceae

With over eighty species in Europe, Western Asia, and North Africa, crocuses have been cultivated since Medieval times, initially for medicinal and culinary purposes, as with saffron, *C. sativus*. Plant hunting in the Middle East during the early 20th century resulted in the introduction of several newly described species, but *C. chrysanthus*, *C. speciosus*, *C. tomassinianus*, and *C. vernus* now represent the bulk of the cultivated plants grown for spring flowers. These four species have produced many cultivars in a range of flower colors, mostly raised in western Europe in the late 19th and early 20th centuries.

See: Natural Fibers and Dyes, p. 310; Spices, p. 168

Cupressus, False cypress *Chamaecyparis*
Cupressaceae

Chamaecyparis nootkatensis (Alaska cedar, or Nootka or Sitka cypress) and the familiar hedging plant *C. lawsoniana* (Lawson or bird-nest cypress) reached Europe from the western United States in the 1850s, to be followed in 1860 by the Japanese species, *C. obtusa* and *C. pisifera*, and a number of native Japanese cultivars. The Nootka cypress is significant as a parent of the fast-growing Leyland cypress, × *Cupressocyparis leylandii*, sometimes a problem when neglected hedges outgrow their space. There are numerous cultivars of these species.

Cyclamen, Sowbread, Persian violet *Cyclamen*
Primulaceae

Of the fifteen species of *Cyclamen*, all native to southern Europe and western Asia, several have been cultivated since the 16th century. *Cyclamen hederifolium* (1583), *C. coum* (1596), and *C. purpurascens* (1596) are among the hardier species grown. The more tender *Cyclamen persicum* was introduced from the eastern Mediterranean region in 1731. It was subsequently developed extensively by plant breeders, initially in Europe and North America, more recently also in Japan. There are now many cultivars with large flowers in a wide range of colors.

Cypress *Cupressus*
Cupressaceae

The twenty species of *Cupressus* occur in North and Central America, Asia, and Mediterranean Europe. *Cupressus lusitanica* arrived from Central America as early as 1683, but Himalayan, Chinese, and North American species were introduced only in the mid- and late 19th century. The Californian *C. macrocarpa* (1838) is important as the parent of a ubiquitous intergeneric hybrid × *Cupressocyparis leylandii* (see previous entry, *Cupressus*).

Daffodil *Narcissus*
Amaryllidaceae

The importance to gardens of the genus *Narcissus* is quite disproportionate to its size. Opinions vary as to the number of species, but it is generally considered to be between twenty-five and fifty. These are native to Europe, North Africa, and West Asia, with the center of distribution lying in the western Mediterranean area. Most of the species are in cultivation, some having been cultivated for several centuries.

Narcissus pseudonarcissus, native to much of Europe including the British Isles, has been grown in gardens since the 16th century or earlier. Some of the first cultivars date back to the 17th century, and this species has contributed to many newer cultivars.

Dwarf species such as *N. bulbocodium* and *N. triandrus*, both from southwest Europe, and *N. cyclamineus* are grown in their own right, and have also been used by plant breeders.

It is the cultivars of *Narcissus* that make by far the most significant contribution to gardens. The international cultivar register is maintained by the Royal Horticultural Society and some 24,000 names are recorded. A formal system of cultivar classification has been developed, based to some extent on the dominant ancestral species. Traditional centers of daffodil breeding, mostly in Europe, such as England, Ireland, and the Netherlands continue to be important, but there are also important breeders in North America, Australia, and New Zealand. Current trends in breeding aim to refine the shape of the flowers, both single and double sorts, as well as to extend the color range, with good pink-cupped cultivars now appearing. Other species important in daffodil breeding include *N. tazetta*, *N. poeticus*, and *N. jonquilla* all native to Southern Europe. As well as cultivation in borders and containers, many daffodils, both cultivars and species, are suitable for naturalizing in grass. Daffodils are a major item in the cut flower trade, with certain cultivars having been selected especially for this purpose.

Dahlia *Dahlia*
Asteraceae

About twenty-seven species of *Dahlia* occur in Central America, from where *D. coccinea* and *D. pinnata* spread to Europe in the late 18th century. Other species arrived later, but these two are the main ancestors of the 20,000 or more cultivars raised since the early 19th century. The main plant breeding activity is in Europe, the United States, and Australia. An enormous variety of bloom shapes and sizes is available.

Daylily *Hemerocallis*
Hemerocallidaceae

Fifteen species of daylily occur in northeast Asia. *H. lilioasphodelus* (syn. *H. flava*) and *Hemerocallis fulva* have been cultivated, and the latter naturalized, in Europe since the 16th century, but are presumed to originate in China. From China, *H. minor* arrived in 1759, and the Japanese species *H. dumortieri* in 1833. Hybridizing began in the 19th century, becoming a major activity in both North America and Europe in the early 20th century. Over 20,000 cultivar names have been registered to date, representing a far wider range of flower colors and forms than is found in their wild ancestors.

Delphinium *Delphinium*
Ranunculaceae

There are about 250 species of *Delphinium* in the temperate Northern Hemisphere. The familiar tall perennials with spikes of blue to purple flowers are largely derived from the southern European *D. elatum*, cultivated since the 16th century. The arrival in 1816 of *D. grandiflorum* led to the Belladonna hybrids, and some Chinese species have also been used. Red-flowered species from western North America have recently been used to develop the red- and pink-flowered University hybrids, the name referring to the University of Agriculture, Wageningen, The Netherlands, where much of their development was carried out by Dr. R.A.H. Legro. Larkspurs, sometimes placed in the genus *Delphinium*, are annuals of the related genus *Consolida*, native to southern Europe.

Deutzia *Deutzia*
Hydrangeaceae

There are seventy species of *Deutzia* native to the Himalayas, China, and Japan; all have proved hardy and reliable shrubs. The Himalayan *D. corymbosa* was the first introduction in about 1830, but was soon eclipsed by the Japanese *D. crenata* and *D. gracilis* (Japanese snowflower), and by *D. longifolia*, *D. purpurascens*, and *D. setschuenensis*, from China. Several interspecific hybrids and many cultivars have been raised subsequently, with the majority of such development taking place in France in the early 20th century.

Dogwood *Cornus*
Cornaceae

Cornus sanguinea (blood-twig dogwood), native to Europe including England, has never been widely cultivated. More important in gardens are North American species such as *C. alternifolia* and *C. stolonifera*, both grown for their colorful stems in winter, and the flowering dogwoods *C. florida* and *C. nuttallii*. These were joined by *C. capitata* from the Himalayas and *C. kousa* from Japan. The name dogwood is a corruption of the old name, dagwod or daggerwood, applied because daggers for skewering meat were made from the wood of some native forms. All dogwoods flower, but most are grown for individual form specimens and colorful stems in winter.

Escallonia *Escallonia*
Escalloniaceae

First introduced in the early 19th century, *Escallonia* quickly became valued as flowering shrubs and for hedging, especially in seaside areas. There are about forty species native to South America, from southern Brazil to Chile and Argentina. The Patagonian *E. rosea* (introduced 1847) and Chilean *E. rubra* (1827) have yielded numerous fine hybrids, many raised in Northern Ireland by Slieve Donard Nursery in the early 1900s. Several other species are grown in specialist collections. Mostly evergreen, the trees and shrubs have tiny leaves and an abundance of colorful, small flowers.

European marigold, Marigold *Calendula*
Asteraceae

Although there are fifteen *Calendula* species in Europe, western Asia, and North Africa, just two annuals, *C. arvensis* and *C. officinalis* (pot marigold) have become important garden plants. Both reached the British Isles from southern Europe before 1200, the petals of the latter being used as a food colorant as well as an ornamental. Selective breeding has resulted in a wide range of modern cultivars.

See: Plants as Medicines, p. 216

Firethorn *Pyracantha*
Rosaceae

Although there are only seven species, native to Europe and Asia, the firethorns are widely cultivated for their highly decorative winter berries. The European *P. coccinea* has been grown since 1629. Important introductions from China include *P. rogersiana* (1911) and *P. crenatoserrata* (1906). *Pyracantha koidzumii*, imported from Taiwan to the United States in 1937, brought a valuable degree of disease resistance to later hybrids.

Flowering cherry, Almond, Peach *Prunus*
Rosaceae

Prunus is a genus of four hundred species, distributed throughout the temperate Northern Hemisphere. Numerous species are of economic value, especially as fruit trees: plum, apricot, peach, damson, almond, and cherry. The timber of many species is also valuable. *Prunus* vies with *Magnolia* as a genus of outstanding flowering trees, the native species having been enjoyed as ornamentals for centuries.

Ornamental cherries have come into cultivation from three main directions: long-cultivated European and Middle Eastern species, later arrivals from North America, and most recently those from northeast Asia.

The first foreign species to be introduced to western Europe came from Asia, such as *P. cerasifera* (16th century, probably from Central Asia), *P. tenella* (west Asia, 1683) and *P. laurocerasus* (west Asia, 1576).

North American species arrived in Europe as early as 1629 (*P. serotina*), but neither this nor other North American species, such as *P. americana* (1768) or *P. virginiana* (1724), have proved important. *Prunus besseyi* (central United States, 1892), although little cultivated, provided the popular dwarf flowering cherry, *P.* × *cistena*.

It is fair to say that the world of ornamental cherries has been transformed by the introduction of various species from East Asia, in particular, Japan. Following the opening up of Japan to trade and exploration around 1860, some outstanding introductions were made, often simultaneously in Europe and North America. Among these, *P. mume* (1841), *P. sargentii* (1890), *P. subhirtella* (1894), and *P.* × *yedoensis* (1902) stand out. This period also saw the first introductions of the Sato-zakura, the cultivars regarded in the West as Japanese flowering cherries. These are often complex hybrids involving species such as *P. serrulata* var. *spontanea*, *P. lannesiana*, and *P. serrulata*. Many are now grown in the West, the great majority being old Japanese cultivars.

From China came *Prunus glandulosa* in 1774, *P. triloba* in 1884, and *P. cerasoides* as late as 1931, but their impact has been far less than that of the Japanese cherries.

See: Wood, p. 321; Fruits, pp. 85–6, 86–7, and 87–8

Flowering quince, Japonica *Chaenomeles*
Rosaceae

Chaenomeles speciosa and *C. cathayensis,* both from China, reached Europe by 1800, the former soon becoming widely grown, with several cultivars raised in England and France. *Chaenomeles japonica* was brought from Japan in 1869, but is not widely grown in the West. There are hybrids between all three species, notably *C.* × *superba* (*japonica* × *speciosa*) with numerous cultivars. Although the fruits of *Chaenomeles* are used in preserves, the genus should not be confused with true quince, *Cydonia oblonga*, originating in the Caucasus, and grown for its fruits.

Forget-me-not *Myosotis*
Boraginaceae

Although the fifty species of *Myosotis* occur throughout both northern and southern temperate regions, only a few European species have had a major impact on gardens. The widely cultivated bedding forget-me-nots with small blue flowers are derived from *M. sylvatica* and *M. scorpioides,* both native to Europe, and many cultivars are now available. A few Australasian species, with white or yellowish flowers, are cultivated in specialist collections.

Forsythia, Golden-bells *Forsythia*
Oleaceae

A genus of seven species, one native to the Balkans, the others native to East Asia, *Forsythia* species are yellow-flowering shrubs, mostly have arching or spreading branches. The first species to be introduced to gardens were the Chinese *F. suspensa* in 1833, followed by *F. viridissima* in 1844. The European *F. europaea* was brought into cultivation in 1899, *F. giraldiana* from China in 1910, and the compact *F. ovata* from Korea in 1918. There are several garden hybrids and numerous cultivars of *F.* × *intermedia* (*suspensa* × *viridissima*, raised around 1880).

Fuchsia, Lady's eardrops *Fuchsia*
Onagraceae

Fuchsia is a genus of about a hundred species. Most occur in Central and South America from Mexico to southern Argentina, but there are also a few in New Zealand and Tahiti. Although *F. triphylla* was the first species described (by Linnaeus in 1753), the first living material to reach European gardens was *F. magellanica*, about 1789. Other South American species followed in the first quarter of the 19th century, and *F. excorticata*, a large shrub with small flowers, came from New Zealand in 1824. Another New Zealand species arrived in 1854; this was the very distinctive *F. procumbens*, a creeping shrub with upward-facing flowers combining yellow, green, red, and blue tints.

Hybrids were soon appearing in the larger collections, and in the mid-19th century nurserymen such as Standish of Bagshot, England, and various individual enthusiasts were producing a number of cultivars. Their activity increased as more species came into cultivation. In particular, *F. fulgens,*

Fuchsia. Copyright Dave Webb, used with permission.

which was in cultivation in England by 1840, came to be regarded as especially important in the development of modern fuchsia cultivars.

F. triphylla, introduced to the United States in 1873, and to Europe shortly after, resulted in the distinct group of Triphylla hybrids, many of which were raised by Bonstedt and others in Germany.

Forms and hybrids of *Fuchsia magellanica* are among the most frost-tolerant, and are widely used for informal hedging in seaside areas, sometimes becoming naturalized as in southwest England, parts of Ireland and the southwestern United States. In colder areas they may behave as herbaceous perennials, making new growth from the roots each year.

Breeding of new cultivars of fuchsia continues enthusiastically, especially in the United Kingdom, where about two thousand cultivars are commercially available. The great majority of these require greenhouse protection in cool temperate areas, and they are among the most popular of flowering shrubs for a frost-free conservatory.

Garden sage, Ramona, Chia, Sage *Salvia*
Lamiaceae

A genus of at least nine hundred species, salvias are found throughout the tropical and temperate regions, especially in the Americas, and Asia. Early decorative imports, with spiked flowers growing above the foliage, include *S. sclarea* (in cultivation before 1562) and *S. viridis* var. *comata* (syn. *S. horminum*) by 1596, *S. sylvestris*, and *S. verticillata*, all from southern Europe. *Salvia farinacea* came from the United States in 1842, and Mexico provided *S. involucrata*, *S. fulgens*, *S. microphylla*, and *S. patens*. More recent introductions include *S. × superba* from Europe (1900) and *S. guaranitica* from South America (1925).

See: Plants as Medicines, p. 221; Herbs and Vegetables, p. 109

Geranium, Cranesbill *Geranium*
Geraniaceae

Three hundred species of cranesbill (the name refers to the shape of the developing fruit) are widely distributed throughout the temperate regions, with several native species in both Europe and North America. *Geranium robertianum*, *G. pratense*, *G. sylvaticum*, and other European species have been cultivated for medicine or decorative flowers since the 16th century or earlier. North American species began to reach Europe in the 18th century, and the South African *G. incanum* arrived in 1701, but these have had little impact. The 19th century saw the introduction of several important species from eastern Europe and Asia, notably *G. wallichianum*, *G. lambertii*, *G. farreri*, *G. dalmaticum*, and

G. sessiliflorum. In recent years, many cultivars, often of hybrid origin, have been introduced. Note that pelargoniums, commonly referred to as geraniums, are not in the *Geranium* genus, and are treated separately later.

Gladiolus, Sword lily, Corn flag *Gladiolus*
Iridaceae

There are about 180 species of *Gladiolus*, a few native to Europe and western Asia, but most from southern Africa. European species such as *G. communis* and *G. byzantinus* have been cultivated since the 17th century or earlier for their stunning flower spikes. However, their cultivation developed greatly with the introduction to Europe of the South African *G. cardinalis* (18th century), *G. carneus* (1796), *G. dalenii*, *G. tristis* (1745), and a few others. Modern cultivars are numerous, and most are complex hybrids between these and other South African species. Plant breeding originally developed in Belgium, but there is now extensive breeding elsewhere in Europe, North America, and Australia. They are an important element in the international cut-flower trade.

Heath, Cape heath (South African species) *Erica*
Ericaceae

This large genus has over 450 species of evergreen shrubs in southern Africa and others in Europe—several native to Britain. Several European species have been cultivated since the 18th century and there are now many cultivars, most raised in Europe, especially Britain and more recently Germany. The south African species, which are all fairly frost-tender, began to reach European gardens in the late 18th century. Whilst a few can be grown outdoors in the warmer parts of Europe and North America, the great majority of the Cape heaths require greenhouse protection. In Europe heath or heather gardens planted almost entirely with species and varieties of *Erica* and *Calluna* are popular.

Heather, Scotch heather, ling *Calluna*
Ericaceae

A monotypic genus native to much of Europe, including the British Isles, *Calluna vulgaris* only became popular in gardens in the 19th century when the selection of cultivars began. There are now many cultivars, with single or double flowers, ranging from white through pink to crimson and lilac, as well as some with brightly colored foliage.

Hellebore *Helleborus*
Ranunculaceae

A genus of twenty species, *Helleborus* is confined to Europe and western Asia with the exception of *Helleborus thibetanus*, recently introduced from Western China. *Helleborus foetidus* (stinking hellebore) and *H. viridis*, both native to the UK, have been cultivated since at least the 17th century. The Christmas rose, *H. niger*, was introduced from Central Europe in the 16th century, and *H. argutifolius* (Corsican hellebore) and *H. lividus* arrived about 1710. Most other species did not come into cultivation until the 19th century. *H. orientalis* (the Lenten rose) from eastern Europe and western Asia has given rise to a large number of cultivars that are hardy and easy to grow. Much breeding of hellebores has taken place in recent years, with a wide color range, and even double flowered cultivars, now available. Most kinds bloom in winter and early spring.

Holly *Ilex*
Aquifoliaceae

With about four hundred species distributed through the temperate and subtropical regions of both hemispheres, relatively few holly species have been cultivated until recently. The European *I. aquifolium* has been grown for centuries and its cross with the Madeiran *I. perado* (1760) gave the popular *I.* × *altaclerensis*. Although some North American hollies were introduced as long ago as the early 1700s, they remain uncommon in European gardens. The late 19th century saw the introduction of some fine hollies from China and Japan, notably *I. cornuta*, *I. crenata*, *I. pernyi*, and *I. latifolia*. Recently, several clones of *I.* × *meserveae*, hybrid between *I. aquifolium* and the Japanese *I. rugosa*, have been raised in the United States, and are proving excellent plants.

Honeysuckle *Lonicera*
Caprifoliaceae

The 180 species of *Lonicera*, ranging from the familiar climbers, often with fragrant flowers to shrubs, are native to Europe, Asia, and North America. Some European species such as *L. periclymenum* and *L. caprifolium* have been cultivated since the 16th century or earlier; later arrivals include the Mediterranean *L. etrusca*, *L. splendida*, and *L. tatarica*. The American *L. sempervirens* arrived in Europe in 1656; other American introductions include *L. ciliosa* and the robust shrub, *L. ledebourii*. Asian species include *L. japonica*, the shrubby *L. fragrantissima*, *L. maackii*, and the evergreen *L. nitida*.

Hosta, Plantain lily, Funkia *Hosta*
Hostaceae

Hosta is a genus of about fifty species, native to Japan, Korea, and China. Long cultivated in those countries, the delimitation and origin of some species is uncertain. The Chinese *H. plantaginea* reached Europe in 1790, but it was Japanese species such as *H. fortunei*, *H. sieboldiana*, *H. lancifolia*, and *H. undulata* that brought the genus to prominence. There are now many cultivars, mostly of hybrid origin, that are grown for their bold foliage. Some came from Japan, but many have been raised, especially in the United Kingdom and in North America, in the late 20th century.

Hyacinth *Hyacinthus*
Hyacinthaceae

In spite of containing only three species, the genus *Hyacinthus* has become a major element in the horticulture industry. Native to the northeastern Mediterranean region, *H. orientalis* reached Western Europe late in the 16th century. Selective breeding, especially in the Netherlands in the 18th century, produced a number of more robust cultivars with highly fragrant, single or double flowers in an ever-widening range of colors.

Hydrangea *Hydrangea*
Hydrangeaceae

With about twenty-three species, *Hydrangea* has an unusual distribution, with representatives in East Asia and North and South America. However, Asian species such as *H. anomala* subsp. *petiolaris*, *H. paniculata*, *H. aspera*, and especially *H. macrophylla* are dominant in gardens. *Hydrangea arborescens* from the eastern United States, *H. quercifolia* from the southeastern United States, and *H. serratifolia* from Chile, are the main cultivated species from the Americas. Numerous cultivars derived from *H. macrophylla* have been raised, many in France and Germany.

Iris, Flag iris, Fleur-de-lis *Iris*
Iridaceae

About two hundred species of iris occur in Europe, North America, and Asia. Some have been cultivated for centuries for medicinal purposes. Decorative species grown for their brief but magnificent blooms have come from all three continents. The English, Spanish, and Dutch irises derive from *I. xiphium* (Pyrenees, 1596). Other bulbous species grown include *I. histrioides* (Turkey, 1821) and *I. reticulata* (western Asia, 1879). From North America came *I. cristata* (1756), *I. douglasiana* (1873), and *I. innominata* (1935); the last two are important as ancestors of the Pacific Coast hybrids. Asia has greatly increased the range in gardens, with *I. japonica*, *I. laevigata*, *I. ensata*, *I. sibirica*, and many cultivars of the last three. Bearded irises are complex hybrids, first developed in the 19th century from the European *I. lutescens*, *I. variegata*, and others, and are perhaps the most important group in the genus.

See: Fragrant Plants, pp. 250–1

Ivy *Hedera*
Araliaceae

Twelve species of *Hedera* are recognized, distributed from Europe and North Africa to East Asia. The European *H. helix* has been grown for several centuries, and many cultivars with variously shaped or variegated leaves have been named. Closely allied, but recently recognized as a distinct species, is *H. hibernica*. The large-leafed *H. canariensis* of gardens, sometimes distinguished as *H. algeriensis* (North Africa, 1833), and *H. colchica* (Caucasus, 1850) are also widely grown.

Jacob's ladder *Polemonium*
Polemoniaceae

About twenty-five species of *Polemonium* are found in North America, Europe, and Asia. The European *P. caeruleum* has been cultivated since the 16th century or earlier. North American species such as *P. reptans* (1758), *P. pulcherrimum* (1826), *P. foliosissimum* (1880s), and *P. pauciflorum* (1889) are fairly widely cultivated. The common name refers to the ladder-like, pinnately divided leaves.

Japanese cedar *Cryptomeria*
Taxodiaceae

Both Chinese *Cryptomeria japonica* var. *sinensis* and Japanese var. *japonica* arrived in Europe in 1842. An important timber tree in the Far East, it has only decorative value in the West, for its tall pyramidal form. Japanese cultivars 'Lobbii' and 'Elegans' were imported in 1850 and 1854 respectively and, from the 1860s, numerous other cultivars were introduced from Japanese nurseries, and some selected in Europe.

Kerria, Japanese rose, Jew's mallow *Kerria*
Rosaceae

This monotypic genus occurs wild in China and Japan. The double-flowered cultivar *Kerria japonica* 'Plena,' which was introduced to England from China in 1804, remains very popular. The single-flowered wild type reached Europe from Japan, thirty years later. A cultivar with variegated leaves, 'Picta,' was in cultivation before 1883; the large flowered 'Golden Guinea' is more recent in origin. There is no variation from the golden-yellow flower color of the wild plant.

Laburnum, Goldenchain, Bean tree *Laburnum*
Fabaceae

This genus contains just two species, both native to southern Europe and both brought into cultivation by the late 16th century. *Laburnum anagyroides*, despite having toxic seeds, is very popular in gardens for its long chains of yellow flowers. It has however been partly superseded by the almost sterile hybrid *L. × watereri*, which produces few of the toxic seeds and also has longer chains of flowers.

Lilac *Syringa*
Oleaceae

Syringa is a genus of thirty species, native to eastern Asia and southeastern Europe. The most fragrant species is the common lilac *S. vulgaris*, a south European species cultivated since the 16th century and extensively hybridized by the French Lemoine family at the beginning of the 20th century. The range was enhanced by species from China, notably *S. pinnatifolia* and *S. reflexa* in 1904, and the dwarf *S. meyeri* in 1908.

Lily *Lilium*
Liliaceae

Showy flowered bulbous perennials, with about one hundred species occurring from western Europe to East Asia and North America, lilies are important in both commercial horticulture and domestic gardens. Although the great majority of lilies cultivated now, either for the cut-flower trade or in gardens, are of complex hybrid origin, most species are also very beautiful and deserve to be kept in cultivation. Three major influences in the cultivation of lilies can be traced: those native to Europe, those imported from North America, and the Asian species.

European species such as *L. pyrenaicum*, *L. martagon*, and *L. bulbiferum*, together with the Madonna lily, *L. candidum*, from the eastern Mediterranean, have all been cultivated since the 16th century or, in the case of Madonna lily, since the 13th century or earlier. Until the 19th century, most lilies in gardens were species such as these. Although several other species reached western European gardens, notably *L. chalcedonicum* (Balkans, 1796) and *L. monadelphum* (Caucasus, 1820), a major step forward came with the introduction of various North American species.

Important among these were *L. superbum* (eastern United States, 1738), *L. canadense* (eastern North America, 1829), *L. columbianum* (western North America, 1872), *L. humboldtii* (California, 1872), and *L. pardalinum* (western United States, 1875). Although a few hybrids existed between the European and West Asian species, it was the influx of lilies from North America that really prompted the beginning of serious hybridization in gardens. Among these, the Bellingham hybrids, derived from *L. humboldtii* and *L. pardalinum*, remain important.

The third strand in the cultivation of lilies began in the late 18th century when the first species from East Asia began to reach Europe. *Lilium maculatum* (1745), and *L. tigrinum* (1804), *L. longiflorum* (1819), and *L. brownii* (1835) all came from China; from Japan came *L. speciosum* (1832), *L. auratum* (1862), *L. japonicum* (1873), and *L. rubellum* (1898).

Initially, the species were cultivated for their own sake—indeed, the late 19th century saw a brief but substantial wholesale trade in wild-collected bulbs from Japanese to European and North American nurseries.

The hybridizing of lilies became very popular in the 20th century, stimulated in part by the wide range of species available, but also by the difficulties in the cultivation of some species. The Royal Horticultural Society maintains the international register of lily cultivar names, and has helped to develop a system of classification for cultivars. At the present time, North America is probably the most active area in lily breeding, with significant contributions from many western European countries, especially the Netherlands, and increasingly from eastern Europe.

See: Fragrant Plants, p. 248

Lobelia *Lobelia*
Lobeliaceae

Lobelia is a genus of 375 species, predominantly subtropical and most abundant in the Americas. The bedding lobelia, *L. erinus*, was introduced from South Africa in 1752, and had become a popular bedding plant by the 1860s, with a large number of cultivars. From North America came the herbaceous perennials, *L. cardinalis* (1629) and *L. siphilitica* (1665), and from Mexico, *L. fulgens* (1809) and *L. splendens* (1814). The hybridization of these three species led to many cultivars. The distinctive but tender *L. tupa* came from Chile in 1824.

Lupin, Lupine *Lupinus*
Fabaceae

Lupins have a curious distribution, occurring in western North America, South America, and the Mediterranean region. Few species are grown, other than as fodder plants, but *L. arboreus*, introduced from California in 1793 is an exception. In 1826, it was followed by *L. polyphyllus* from western North America. Modern hybrid lupins are typified by Russell Lupins, the product of many years of selective breeding by the Englishman George Russell. It is thought that these are derived primarily from the two species just mentioned, probably with some genetic influence from some annual species. These are considered a traditional "cottage garden" plant, although developed only in the 19th and early 20th century.

Maple, Box-elder *Acer*
Aceraceae

The maples form a distinctive genus of over one hundred species, widely distributed in the temperate Northern Hemisphere. They are valued in the garden for their foliage, especially autumn color in some species, and in a few instances also for their flowers or winged fruits. Indigenous European species, including the sycamore (*Acer pseudoplatanus*), Norway maple (*A. platanoides*), and field maple (*A. campestre*), have long been cultivated, the first having been introduced to the UK possibly in Roman times. Norway maple, brought to the UK in 1683 from Continental Europe, has given rise to numerous cultivars, some with colored or variegated leaves, others of distinctive habit.

North American species cultivated for ornament include the sugar maple (*A. saccharum*), silver maple (*A. saccharinum*), red maple (*A. rubrum*), and box-elder (*A. negundo*), the first three being notable for their autumn color. These species were introduced to Europe in the late 17th and early 18th centuries, during the major age of plant exploration in North America, when the Tradescants and Bartram were major players. Numerous cultivars have been selected in the last 120 years, especially from *A. rubrum*. The narrow-crowned 'Scanlon' is well suited for street planting; others with outstanding autumn color include 'October Glory.'

However, species native to China, Japan, and the Himalayas, introduced between 1820 and 1890, have had the greatest impact. Among these, *Acer palmatum* and *A. japonicum* are especially important. The former, introduced to England from Japan in 1820, has become exceptionally popular, and many cultivars, mostly bred or selected in Japan, are now available in Europe and North America. These vary in leaf color, in growth habit, and especially in the degree of dissection of the leaves. Many also color well in the autumn. The "snake-bark" maples, so called because of the attractive silvery markings on the young bark, are valued by connoisseurs, and include *A. pensylvanicum* from eastern North America, *A. rufinerve* from Japan, and the Chinese *A. davidii* subsp. *Grosseri*.

See: Wood, p. 325

Milfoil, Yarrow *Achillea*
Asteraceae

About one hundred species of *Achillea* are widely distributed throughout Europe and Asia. *Achillea millefolium*, native to most of western Europe and cultivated since at least 1400, has given rise to numerous cultivars with pink and red flower-heads. The east European *A. clypeolata* and *A. aegyptiaca* (formerly *A. taygetea*) (introduced in the early 20th century), and *A. filipendulina* from Asia Minor (1803), have large yellow flower-heads borne above attractively divided foliage. Various other European alpine species are grown in specialist collections.

See: Plants as Medicines, p. 218

Mock orange, Philadelphus, Syringa *Philadelphus*
Hydrangeaceae

Sixty-five species of *Philadelphus* are distributed from Central and North America to Europe and Asia. In gardens, they are valued for their showy white flowers, borne over a long period and often very sweetly scented. The European *P. coronarius* (type species of mock orange) was in cultivation in England by

1596, and North American species began to arrive in the mid-18th century (*P. inodorus* and *P. lewisii*). Two important introductions from Mexico were *P. microphyllus* (1883) and *P. mexicanus* (1839), followed by Asian species like *P. satsumi* (Japan, 1851) and *P. purpurascens* (China, 1911). In the early 20th century, many cultivars of complex hybrid origin were raised, especially by Lemoine in France.

See: Fragrant Plants, p. 250

Oregon grape, Holly grape, Mahonia *Mahonia*
Berberidaceae

Seventy species of *Mahonia* occur in North and Central America, the Himalayas and East Asia. Many are not fully hardy in temperate regions, but *M. aquifolium*, also known as Oregon grape or grape holly (western United States, 1823), *M. pinnata* (California, 1838), and *M. japonica* (China, 1849) quickly became established in European gardens. Another important introduction was *M. lomariifolia* (Burma, 1931): although not fully hardy, it is important as a parent of the highly successful *M. × media*, originating in Northern Ireland about 1950.

Pelargonium, "Geranium," Storksbill *Pelargonium*
Geraniaceae

The majority of this genus's roughly 280 species—and the most important in gardens—are native to southern Africa. Since their introduction to European gardens, they have become among the most popular of plants, especially for indoor and greenhouse cultivation, and for summer bedding and containers. Although many species are cultivated, the cultivars dominate the trade, and they fall into three main groups: Zonal, Regal, and Ivy-Leaf pelargoniums. Although commonly referred to as "geraniums," they are quite distinct from the genus *Geranium*.

The earliest introductions from South Africa arose from the trading activities of the Dutch in the Cape, and the tuberous-rooted *Pelargonium triste* is thought to have been the first species cultivated in Europe, early in the 17th century. Around 1700, *P. peltatum* and *P. zonale* were brought to the Netherlands, and *P. inquinans* was in England by 1714. The last two are the main ancestors of the familiar Zonal pelargoniums, *P. × hortorum*. From *P. peltatum* were developed the Ivy-leaf pelargoniums. Regal pelargoniums are complex hybrids including in their ancestry *P. cucullatum* and *P. grandiflorum*.

A further informal group, popular since the 17th century, is the scented-leaved pelargonium. This comprises many wild species and a number of hybrids between them, grown more for the attractively fragrant leaves than for the usually small flowers. Some, such as *P. graveolens*, are grown commercially for the extraction of essential oils.

The development of garden hybrids began in the early 19th century, becoming a significant activity from the 1850s. At the present time, pelargoniums are again enjoying a surge in popularity, with many active breeders in Europe, Australia, and North America. Well over two thousand cultivars are currently in the UK trade.

See: Fragrant Plants, p. 257

Peony *Paeonia*
Paeoniaceae

The European *Paeonia officinalis* and the less common *P. mascula* have been cultivated since at least the 12th century, initially for medicinal purposes, and later for their large showy flowers. *Paeonia tenuifolia* came from the Caucasus in 1765 and *P. lactiflora* from Siberia at about the same time. The first tree peonies arriving in the West in 1789 were ancient Chinese cultivars of *P. suffruticosa*. More recent introductions include *Paeonia delavayi* (China, 1886; this includes what was formerly called *P. lutea* and var. *ludlowii*), *P. obovata* (China, 1899), and *P. mlokosewitschii* (Caucasus, 1907).

In both herbaceous and shrubby groups, the first cultivars grown were classical Chinese and Japanese ones. These formed the basis of subsequent breeding in the West, especially in France and England in the late 19th century. More recently, much breeding work has been done in the United States, latterly using species such as *P. emodi* and *P. mlokosewitschii*.

Petunia *Petunia*
Solanaceae

Petunia is a genus of about thirty species, native to South America, especially Brazil. Two species are the ancestors of the modern bedding petunia cultivars. The annual *P. axillaris* (formerly *P. nyctaginiflora*) was introduced in 1823, and the subshrub *P. integrifolia* from Argentina in 1831. Many seed-raised cultivars are now available, bred in Europe, North America, and, increasingly, Japan.

Poppy *Papaver*
Papaveraceae

Fifty species of poppy are recognized, mostly in Europe, with some in North America, Africa, and Australia. The opium poppy, *P. somniferum*, has long been cultivated as a narcotic, but ornamental strains developed in the 19th century remain popular. The European *P. rhoeas* gave rise to the Shirley poppies in the late 19th century. However, it is *P. orientale* (Southwest Asia, before 1714) and allied species that have been most developed for gardens, with well over one hundred cultivars listed, most raised in European nurseries. The Iceland poppy, *P. nudicaule* (subarctic, 1730), has also been carefully bred to give a wide color range. Although botanically quite distinct, the widely cultivated blue poppy, *Meconopsis betonicifolia*, was introduced from Tibet in 1925, by Frank Kingdon-Ward, soon followed by related species from China and the Himalaya.

See: Psychoactive Plants, pp. 199–200; Plants as Medicines, p. 221

Primrose, Polyanthus *Primula*
Primulaceae

One of the most uniformly ornamental genera, the four hundred species of *Primula* are distributed throughout the temperate Northern Hemisphere. European species such as the primrose, *P. vulgaris*; cowslip, *P. veris*; and *P. auricula* have been cultivated since the 16th century or earlier. Other European alpine species were brought into cultivation rather later, *P. marginata* in 1777 and *P. × pubescens* in 1800.

Although several species occur in North America, they are little grown either there or in Europe. More important have been Asian primulas such as *P. sinensis*, *P. capitata*, *P. sieboldii*, *P. japonica*, and *P. denticulata*.

However, the vigorous exploration of western China at the turn of the 19th century, mentioned in the introductory paragraphs, had the most dramatic impact on gardens, with the introduction of such important species as *P. obconica*, *P. malacoides*, and *P. florindae*. The same period also saw the arrival of various species of Candelabra primula, many of which have subsequently hybridized in gardens.

Although double-flowered and other variants had been cultivated since the time of Gerard (1597), serious breeding of primulas developed in the 18th century when selections of both polyanthus (*P. × polyantha* and allied hybrids) and of *P. auricula* became very fashionable. Recent years have seen a renewed enthusiasm for such plants, although most remain plants for the specialist collection. During the 20th century many hybrids between Chinese species in the Candelabra section, such as *P. beesiana*, *P. bulleyana*, and *P. pulverulenta* were produced. These Candelabra hybrids have proved well-suited to naturalizing in a wild garden setting. *Primula sinensis*, *P. malacoides*, and *P. obconica* are all tender species, which have been developed extensively in Europe and North America as house and greenhouse plants. Many seed strains of the last two are available.

Privet *Ligustrum*
Oleaceae

Privets grow wild from Europe and North Africa to eastern Asia, from where most of the fifty species originate. *Ligustrum vulgare*, native to western Europe including the United Kingdom, has long been cultivated, but was eclipsed by the introduction of various East Asian species. Of these, *L. ovalifolium* from Japan is the most important, now a ubiquitous evergreen hedging plant. The Japanese *L. japonicum*, *L. lucidum*, and *L. sinense* from China, and *L. compactum* from Himalaya and China are fine specimen shrubs, although less commonly cultivated.

Rhododendron, Azalea *Rhododendron*
Ericaceae

Rhododendron is a genus of over eight hundred species generally distributed in the north temperate region, but also extending into Southeast Asia and (a single species) Australia. Its influence on gardens is consequently varied.

The first to be cultivated was *R. hirsutum* (before 1656), a small alpine shrub from central Europe, but it remained little grown. It was an influx of species from eastern North America that brought the genus to prominence: *R. maximum* (1736), *R. catawbiense* (1809), and the azaleas, *R. calendulaceum* (1806) and *R. occidentale* (western United States, 1851). *Rhododendron ponticum* came from Turkey in 1763, becoming invasive in certain areas, and *R. caucasicum* in 1803. At much the same time, rhododendrons from the Himalayas were also reaching Europe, especially the British Isles, with *R. arboreum* (1810), *R. thomsonii*, and *R. griffithianum* in 1850, and the Chinese *R. fortunei* in 1855.

Soon, hybridization, both accidental and deliberate, occurred and *R. catawbiense*, *R. maximum*, *R. caucasicum*, and *R. ponticum* in particular gave rise to those cultivars now known as Hardy Hybrids, greatly developed in the 19th and early 20th century. Later, *R. griffithianum* and *R. fortunei* enriched the gene pool in this group.

Japan contributed several more species, especially azaleas: *R. japonicum* (1861), *R. indicum* (1877), and *R. kaempferi* (1892 to the United States). Each has given rise to a line of hybrid azaleas: Mollis, Indian, and Kurume, respectively.

However, the greatest impact on gardens resulted from plant exploration in western China by George Forrest, Joseph Rock, E.H. Wilson, and Frank Kingdon-Ward, among others, around the turn of the 20th century. Collectors found an unimagined richness of species, for example, *R. sutchuenense* (1900), *R. neriiflorum* (1910), *R. sinogrande* (1913), *R. forrestii* (1914), and many others.

Modern rhododendron breeding continues apace, especially in Europe and North America. Although traditional style Hardy Hybrids are still appearing, the emphasis has changed to developing more compact plants better suited to small modern gardens, many derived from the Japanese *R. degronianum* (including *R. yakushimanum*), introduced to England in 1934. In the United States (and in Japan) a further aim is to produce cultivars with increased heat-tolerance.

See: Invasives, p. 383

Rose *Rosa*
Rosaceae

Roses are one of the most ubiquitous features in the gardens of Europe and North America. Both regions have a number of indigenous species but, although several of these are cultivated, they have largely been eclipsed by the influence of species from Asia. The genus consists of around three hundred species, and many thousands of cultivars, many no longer in cultivation.

The so-called "Old roses"—the Damask, Gallica, and Alba groups—came from the Middle East via ancient trading routes and reaching Western Europe around the time of the 12th century

Crusades. For several centuries, they were cultivated primarily for medicinal purposes, notably attar of roses, rose-water and various products derived from the hips.

It was not until the 16th century, when the Centifolia Group was developed, that roses were grown also for their ornamental value. Roses native to western Asia offered the prospect of various desirable characteristics that were not available from European or North American species or their hybrids. An important development occurred in the mid-18th century, with the introduction of the first China roses. These were ancient Chinese cultivars derived from *R. chinensis*, and they brought the prospect of repeat-flowering roses to the West. A probably accidental cross with a Damask rose in about 1820 resulted in the more-or-less repeat flowering Bourbon group. Back crosses between these new hybrids and the China rose gave Hybrid Perpetuals, the most important roses of Victorian gardens.

Another important influence was *Rosa foetida* 'Persiana,' introduced from central Asia in 1837. This was responsible for all the bright yellow and orange-flowered rose cultivars first raised about 1900, which are now very popular.

The Chinese *R. × odorata*, introduced by Robert Fortune in 1845 and crossed with Hybrid Perpetuals, produced the truly repeat-flowering Hybrid Tea roses, the first of the modern types. Cluster-Flowered or Floribunda roses arose in the 1870s, from hybridizing the Japanese *R. multiflora*, the China rose, and Hybrid Tea varieties. Another Japanese species, *R. wichuriana*, was introduced in 1891, and soon led to the development of the rambler roses. Similarly, Climbing Hybrid Tea roses owe their habit to the Chinese *R. gigantea*.

The last two hundred years have seen a number of wild rose species from China and Japan come to prominence in their own right, rather than as hybrids. The first was *Rosa rugosa* (1796), valued for its fine foliage, large hips, and toughness. The early 20th century saw *Rosa moyesii*, *R. sericea*, *R. xanthina* f. *hugonis*, and *R. filipes*, among others, all now highly regarded for their flowers and fruits.

See: Fragrant Plants, p. 244

Shrubby veronica *Hebe*
Scrophulariaceae

At least one hundred species of *Hebe* are known, the great majority in New Zealand, with a few in Australia and southern South America. *Hebe elliptica* was the first introduced, from Tierra del Fuego in 1776. Introductions from New Zealand to Europe began with *H. speciosa* and *H. salicifolia* in the early 1840s. Their hybrid *H. × andersonii* arose in Scotland by 1849 and by the late 19th century numerous species and cultivars were grown, mainly for landscaping shrubs and often for their small flowers born on long spikes. Other significant species include *H. cupressoides*, *H. × franciscana* (*elliptica × speciosa*), *H. pinguifolia*, and *H. brachysiphon*. In garden conditions they hybridize freely, and "true" species and cultivars are frequently difficult to distinguish from hybrid offspring.

Silverberry, Elaeagnus, Oleaster *Elaeagnus*
Elaeagnaceae

There are about forty species of these shrubs and trees, predominantly Asian in origin, with one species each in North America and southern Europe. The European *E. angustifolia* (Russian olive) has been cultivated since the 16th century, and *E. commutata* (American silverberry) arrived from North America in 1813. In gardens, however, the evergreen species *E. pungens* and *E. macrophylla* from Asia are more important, grown for their foliage, often variegated.

Skimmia *Skimmia*
Rutaceae

Skimmia is a genus of about four species, ranging from the Himalaya to Japan. *Skimmia japonica* was introduced to Kew from Japan in 1838, but was little noticed until Robert Fortune introduced it to the Standish nursery in 1861. In 1841 the Himalayan *S. anquetilia* (confused with *S. laureola* in gardens, the true *S. laureola* being seldom cultivated) was introduced, followed by *S. japonica* subsp. *reevesiana*.

Snapdragon *Antirrhinum*
Scrophulariaceae

Of the forty species of *Antirrhinum* native to southern and eastern Europe and western Asia, *A. majus*, introduced from the Mediterranean region in the 16th century, is by far the most important in gardens. Although perennial, it is grown as an annual or biennial, with many cultivars bred in Europe, North America, and Japan. Recently, trailing plants derived from the Spanish *A. hispanicum* have become widely used in containers.

Snowdrop *Galanthus*
Amaryllidaceae

Galanthus nivalis is native to much of Europe including parts of the British Isles, and has been cultivated since the 16th century or earlier for its dainty white flowers borne in early spring but sometimes in mid to late winter. The other fifteen or so species are native to Europe and western Asia, and species such as *G. elwesii*, *G.* × *allenii*, and *G. fosteri* were brought into cultivation from the 1880s. The early years of the 20th century saw considerable interest in the breeding and selection of cultivars, primarily in England, and this continues to the present day.

Spiderwort, Wandering Jew *Tradescantia*
Commelinaceae

A genus of about twenty species native to North and South America, the most important representatives in gardens are the spiderworts, often listed under *T. virginiana*, but these are most likely to be complex hybrids of several North American species known as the Andersoniana Group, or *T.* × *andersoniana*. Tender species such as *T. albiflora*, *T. fluminensis* (both known as Wandering Jew), and *T. sillamontana*, all from Central or South America, are commonly grown as houseplants. Dates of introduction are unrecorded, but it is likely that some arrived in Europe in the middle of the 19th century. The generic name commemorates the two John Tradescants, father and son, among the earliest plant collectors in North America.

Spirea, Bridal-wreath, Foam-of-May *Spiraea*
Rosaceae

About seventy species of *Spiraea* occur throughout the northern temperate region as deciduous shrubs. European species such as *S. salicifolia* (1586), *S. media*, and *S. chamaedryfolia* (1789) were the earliest in cultivation. Early imports from China include *S. trilobata* and *S. cantoniensis*, and later *S. thunbergii* and *S. japonica* came from Japan. *Spiraea douglasii* arrived from western North America in 1827. Spirea is grown mainly for its small but copious flowers, often borne on arching branches.

Spurge, Poinsettia *Euphorbia*
Euphorbiaceae

Cosmopolitan and diverse in habit, the 1,500 species of spurge are especially prevalent in the subtropics. Currently fashionable in gardens, important species include the annual *E. marginata* (snow-on-the-mountain) from the United States, *E. polychrome* (cushion spurge, early 20th century) from eastern Europe, *E. sikkimensis* from the Himalayas (introduced c.1970), and *E. characias* and *E. mellifera* from the Canary Islands (ca.1910). The poinsettia, *E. pulcherrima* (Mexico, 1834) is an important house and greenhouse plant, whereas several succulent species such as *E. fulgens* (Mexico, 1836) are to be found in specialist collections.

Poinsettia, a popular flowering potted Christmas plant. Photo by Scott Bauer, ARS/USDA.

Squill *Scilla*
Hyacinthaceae

Ninety species of *Scilla* are found in Europe, Asia, and southern Africa. The Mediterranean *S. peruviana* was introduced in 1607, followed by *S. bifolia* (cultivated by 1790) and *S. siberica* (Russia, 1796). Several species arrived later, including *S. litardieri* (formerly *S. pratensis*), *S. messeniaca*, and *S. mischtschenkoana*.

St. John's wort, Rose of Sharon *Hypericum*
Clusiaceae

Nearly four hundred species of *Hypericum* occur in the northern temperate region, ranging from annuals to shrubs. A few Mediterranean species such as *H. calycinum* and *H. olympicum* have been cultivated since 1675, originally for medicinal properties, and the North American *H. prolificum* reached Europe about the same time. More important in gardens for ornamental berries are several related shrubby species from the Far East, notably the Himalayan *H. hookerianum* (1853), *H. beanii*, *H. forrestii*, and *H. pseudohenryi*, all introduced from western China between 1904 and 1908. These species have also produced some fine garden hybrids, notably 'Hidcote' and 'Rowallane.'

See: Plants as Medicines, pp. 216–7

Tobacco plant, Flowering tobacco *Nicotiana*
Solanaceae

Mostly native to South America, with a few species in Africa and Australia, a small number of the sixty-seven species of *Nicotiana* have become important decorative plants. Early introductions, all from South America, include the shrubby *N. glauca* (1827), as well as the annuals *N. acuminata* (1840), *N. forgetiana* (1901), and *N. alata* (1881). The last two were hybridized in England in the early 20th century to produce *N. × sanderae*, to which all the modern annual cultivars belong. The genus also includes *N. tabacum* and *N. rustica*, cultivated commercially for tobacco.

See: Age of Industrialization and Agro-industry, pp. 371–4; Psychoactive Plants, pp. 201–2

Tree mallow *Lavatera*
Malvaceae

The twenty-five species of *Lavatera* are distributed throughout Europe, Asia, the southwestern United States, and Australia. The tree mallow, *L. arborea*, has been cultivated since the 16th century or earlier as a medium-height shrub with large pink or white flowers. The annual *L. trimestris* (Spain, 1633) is widely cultivated, with numerous cultivars. Although long cultivated, the shrubby *L. olbia* (southern Europe, 1570) and *L. thuringiaca* (Germany, 1731) have only recently become popular, and several cultivars and hybrids have been named in recent years.

Tulip *Tulipa*
Liliaceae

Tulipa is a genus of perhaps 150 species, mostly in western Asia, with a few in Europe and East Asia. *Tulipa gesneriana* from Asia Minor was cultivated in England by 1577, and is the main ancestor of most modern tulip cultivars, whose breeding developed especially in the Netherlands from the early 17th century. At the height of excitement over the genus, single bulbs acquired absurd monetary value and there are instances of bulbs changing hands for as much as three times the average annual wage in the Netherlands. *Tulipa fosteriana*, *T. kaufmanniana*, and *T. greigii*, all introduced from Turkestan in the mid to late 19th century, have each contributed to modern dwarf cultivars.

Vetchling, Sweet pea, Wild pea *Lathyrus*
Papilionaceae

Mostly climbing plants, about one hundred species of *Lathyrus* occur throughout the temperate Northern Hemisphere, also in Africa and South America. Most important is *L. odoratus*, the sweet pea, introduced from Italy about 1700, now highly developed with many cultivars, and grown in all temperate countries. Work before 1900 by the breeder Henry Eckford preserved the scent which was so valued; this has been lost in the more modern Spencer hybrids. A traditional cottage garden annual, recent breeding has led to the development of perennial strains of sweet pea. Other garden species include *L. vernus* (central Europe, 1629), *L. latifolius* (Europe, 17th century or earlier), and *L. grandiflorus* (southern Europe, 1814).

Viburnum, Arrowwood, Black haw *Viburnum*
Caprifoliaceae

The *Viburnum* genus includes 150 species of shrubs, mostly of the temperate Northern Hemisphere, many of which have fragrant flowers, especially the winter and early spring flowering species and hybrids. European species like *V. opulus* and *V. tinus* have long been cultivated. In the 18th century various North American species were introduced, but these remain uncommon in European gardens. Far more significant was the introduction of species from northeast Asia, notably *V. betulifolium*, *V. × carlesii* and *V. farreri*. The range has been further enhanced by numerous hybrids raised in cultivation.

See: Fragrant Plants, p. 255

Violet, Pansy *Viola*
Violaceae

This genus of perhaps five hundred species occurs in the temperate zones of both hemispheres. European species such as *V. odorata* and *V. tricolor* (heartsease) have been cultivated since mediaeval times for their beautiful flowers. The 18th and early 19th centuries saw the introduction of other European species including *V. biflora*, *V. cornuta*, and *V. calcarata*, as well as the North American

V. obliqua (formerly *V. cucullata*), *V. pedata*, and *V. selkirkii*. Breeding began in the early 19th century, resulting in the three main groups, pansies (bred mainly from the wild forms *V. tricolor* and *V. lutea*), violets, and violettas. Curiously, the great potential of the many East Asian species has not been exploited in Western gardens to date.

See: Fragrant Plants, p. 254

Virginia creeper, Woodbine, Boston ivy *Parthenocissus*
Vitaceae

Ten species of *Parthenocissus* grow in North America and in Asia to the east of the Himalayas. Virginia creeper or woodbine, *P. quinquefolia*, was introduced to Europe from the eastern United States in 1629. Asian species followed, notably *P. tricuspidata* (Boston ivy) from Japan, and *P. himalayana* and *P. henryana* from China. Grape woodbine, *Parthenocissus vitacea*, is native to much of North America.

Weigela *Wegela*
Caprifoliaceae

This genus has ten species in China, Japan, and Korea. *Weigela florida* was introduced from China in 1845, and Japan yielded *W. coraeensis* in 1850, *W. japonica*, and *W. praecox* around 1894, and the yellow-flowered *W. maximowiczii* in 1925. Weigela is represented in gardens mostly by cultivars of hybrid origin, many raised in France in the late 19th century.

Willow, Sallow, Osier *Salix*
Salicaceae

With over three hundred species in both northern and southern temperate regions, willows have been grown for timber and other craft uses since very early times. Their use as ornamentals may have begun as late as 1730, when the weeping willow, *S. babylonica*, was introduced from China. European species like *S. hastata* (1780) and *S. reticulata* (1789) show the diversity of the genus. *Salix irrorata*, native to North America, reached Europe in 1898; *S. gracilistyla* came from China about the same time.

See: Plants as Medicines, pp. 217–8

Wisteria *Wisteria*
Papilionaceae

Wisteria is a predominantly Asian genus of about six species, most native to China and Japan, but also occurring in southeast United States. The American *W. frutescens* was the first introduced to European gardens, in 1724, but it remains rare in cultivation. More important are the Asian species: *W. sinensis* arrived from China in 1816, *W. floribunda* from Japan in 1830, and *W. venusta* from Japan about 1912. Several cultivars have been named, mostly from *W. floribunda*, the majority originating in Japan.

References and Further Reading

Beckett, K. (Ed.). 1993. *Alpine Garden Society Encyclopedia of Alpines*. Pershore, Worcestershire: AGS Publications Limited.
Campbell-Culver, M. 2001. *The Origin of Plants: The People and Plants That Have Shaped Britain's Garden History Since the Year 1,000*. London: Headline.
Curtis's Botanical Magazine, London.
Desmond, R. 1994. *Dictionary of British and Irish Botanists and Horticulturists* (rev. ed.). London: Taylor & Francis.
The Garden (formerly *Journal of the Royal Horticultural Society*), London: Royal Horticultural Society.
Glattstein, J. 1996. *Enhance Your Garden with Japanese Plants*. New York: Kodansha International.
Hillier Nurseries. 1991. *The Hillier Manual of Trees and Shrubs* (5th ed.). Newton Abbot, Devon: David & Charles and New York: Van Nostrand Reinhold.

Hobhouse, P. 1992. *Plants in Garden History.* London: Pavilion.

Huxley, A. (Ed.). 1992. *The New Royal Horticultural Society Dictionary of Gardening.* London: Macmillan.

Jellicoe, G.S., et al. 1986. *The Oxford Companion to Gardens.* Oxford and New York: Oxford University Press.

Kashioka, S., and M. Ogisu. 1997. *Illustrated History and Principle of the Traditional Floriculture in Japan.* Osaka: Kansai Tech Corporation (privately distributed).

Lord, Tony (Ed.). 2002. *RHS Plant Finder 2002–2003.* London: Dorling Kindersley.

Rehder, A. 1940. *Manual of Cultivated Trees and Shrubs Hardy in North America.* New York: Macmillan.

Sutherland, D.C., and J. Stuart. 1987. *Plants from the Past.* London: Viking.

Thomas, G.S. 1992. *Perennial Garden Plants* (3rd ed.). London: Dent and Portland, Oregon: Sagapress.

Wright, C.H., and D. Dewar. 1910. *Johnson's Gardener's Dictionary.* London: G. Bell & Sons.

15
Natural Fibers and Dyes

FRANCES A. WOOD AND GEORGE A.F. ROBERTS

Plant fibers can be subdivided into three groups based on where they occur in the plant: the seed or fruit, the stem, or the leaf. In the first group the fibers are present as hairs attached to the seeds or the fruit, each fiber consisting of a single, long, narrow cell. Examples are cotton, coir, and kapok. The second group comprises the bast fibers: flax, hemp, jute, ramie, kenaf, sunn, and urena. These fibers, which are made up of overlapping long cells, may be up to several feet in length and are found in the inner bast tissue beneath the bark of the plant stem where they are bound together by gummy materials that must be removed to obtain the fibers. The third group contains fibers that are found as part of the fibrovascular system of the leaf. The most important of these fibers are sisal, abacá, henequen, and phormium. They are frequently referred to as "hard fibers" to distinguish them from the bast, or "soft fibers".

Since the earliest written record of the use of dyestuffs in China, ca. 2600 BC, the use of plants to impart coloration to textiles has been extensively used up until, and even after, the production of synthetic colors. The emergence of the major dye plants and their trading throughout the world has been of great importance to the political, social, and economic welfare of the countries involved.

Although the number of dye plants is immense, this work has only sought to include those that have been of major importance on an international basis. However, the spread of knowledge in the use of these plants and, consequently, the use of synthetic dyes and the exploitation of locally grown dyestuffs have benefited the dyer since ancient times until the present day.

Seed and Fruit Fibers

Cotton *Gossypium*
Malvaceae

Cotton is the most important vegetable fiber in terms of both volume and value. The quality and quantity of fiber obtained depend greatly on the climate, particularly a warm, humid climate with high moisture content during the growing period and warm dry weather during harvesting. Originally a perennial, it is now an annual in the United States, Egypt, and India, but in more tropical regions it may still be a perennial, although even in these regions it is frequently treated as an annual. The annual plant is a shrub growing to a height of 4 to 6 feet while the perennial has a tree-like form.

There are about 50 species of *Gossypium*, but only four are cultivated: two New World species, both tetraploid (with 52 chromosomes), *G. barbadense* and *G. hirsutum*, and two Old World species,

Since the 1960s there has been increasing interest in the U.S. in kenaf as an annually renewable alternative to wood pulp for papermaking. Infrastructure for significant utilization of kenaf fiber is beginning to develop in the southern U.S. Photo by Scott Bauer, ARS/USDA.

Cotton pickers, Pulaski County, Arkansas. Ben Shahn, 1935. Library of Congress, Prints & Photographs Division, FSA-OWI Collection LC-USF3301-006220-M4 DLC.

both diploid (with 26 chromosomes), *G. arboreum* (tree cotton) and *G. herbaceum*. For commercial purposes cotton fibers are normally classified into three groups according to fiber length and fineness. The longest staple cottons come from *Gossypium barbadense* and include the well-known Sea Island cotton and some Egyptian cottons. The most important source of intermediate length fibers is the Uplands variety, which come from the species *Gossypium hirsutum*, while short staple cottons, which form the bulk of the commercial cottons of India and China, come from either *Gossypium arboreum* or *Gossypium herbaceum*. The major cotton growing countries are the United States, the former USSR, China, India, Egypt, and Brazil.

The origins of the two Old World cotton species remain obscure: no wild ancestor is known for either species. Although the earliest archaeological records are from the Indian subcontinent, dating to ca. 2000 BC, one or both cotton species may in fact have been domesticated in Africa, home to many wild *Gossypium* species. Cotton cultivation spread slowly from South Asia, reaching central Asia by the late Sasanian period (600 AD).

The earliest Old World written record of cotton is in the Asvaláyana Sranta Seitra, dating from about 800 BC, which refers to a law requiring the sacred thread of Brahmans to be made of cotton. The use of cotton wall coverings by King Solomon is described in the Bible in the Book of Esther. In the 5th century BC Herodotus wrote: "The Indians have a wild-growing tree which instead of fruit produces a species of wool similar to that of sheep, but of finer and better quality." Later reports on the Indian campaign of Alexander the Great (356 to 323 BC) described a wool-bearing tree some 3 to 4 feet high, from the bolls of which the seed was removed to give a fiber, the fabric woven from this fiber being finer and whiter than anything else. Proof that the Egyptians were also cultivating cotton by the 1st century AD comes from the writings of Pliny the Elder: "The upper part of Egypt, facing Arabia, produces a shrub which is called gossipion. . . . The fruit of this shrub resembles a bearded nut, which contains a soft wad of fine fibers which can be spun like wool and which is unsurpassed by any other substance in respect of whiteness and delicacy. It supplies the Egyptian priests with their best and most comfortable clothing." The parallels between cotton and wool drawn by these writers are reflected today in the German name for cotton, *Baumwolle*.

These factual reports contrast strongly with a number of later travellers' tales, dating from the Middle Ages, which claimed the existence in India of trees bearing small lamb-like animals. These creatures, "vegetable lambs" or "cotton sheep," were pictured variously as growing in fruit or as growing on the top of tall rigid stalks withone lamb per stalk (see illustrations).

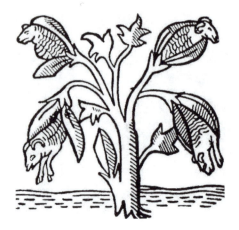

"Vegetable lambs" growing in fruit on trees (14th century woodcut).

The Romans imported considerable quantities of cotton fabrics from India, Egypt, and Greece but did not develop an indigenous cotton industry. Cotton cultivation was probably an early Islamic introduction to the Near East, c. 800 AD. It was the Moors, during their conquest of southern Europe, who introduced the cotton plant to Europe, beginning around the 10th century AD in Spain and Sicily whose climates and soil suited its cultivation.

Much of the fabric produced in Europe was fustian, a mixed fabric having a flax warp and a cotton weft. During weaving the warp yarns are under considerable tension, and European spinners were unable to spin cotton yarns of sufficient strength to be used as warp yarns. It was not until the end of

c planta mirabili opiniones
im Zoophytorum cam num
c videamur, cam plantam f

ix quomodo phytomagnetic
imus) vt fanguinaria, huiufi
mum certi generis referat. c

"Vegetable lamb" growing on a solitary plant (17th century woodcut by Kircher).

the 18th century, with the invention of Arkwright's water frame and Crompton's mule, that they could produce sufficiently strong cotton yarns. Until then all-cotton fabrics were either imported from India or woven in Europe using cotton warps imported from India, whose spinners were able to convert the short staple cotton fibers into more compact, tightly-twisted yarns than could their European counterparts. Most of the cotton processed in Europe was imported from India, Egypt, Syria, Armenia, Cyprus, and other countries of the region with much of this lucrative trade passing through Venice.

Despite the spread of cotton production through Europe in the Middle Ages, high-quality cotton goods were still imported from India. Following the setting up of the British East India Company in 1664 cotton fabrics from India became highly prized items, the craze for calico, muslin, and particularly chintz reaching such extremes that in 1700 King William III banned the import of printed fabrics from India. This legislation, although designed to protect the British wool and linen industries, only encouraged the development of calico-printing in Britain, using white calico fabric imported from India. So in 1720 a new law prohibited the wearing of any printed calico, even material printed in Britain. This law was partially repealed in 1730 to the extent that printing on fustian was permitted, a restriction that was only removed in 1774. A similar craze for these fabrics developed in France, and again attempts were made to suppress their importation. In Prussia King Frederick William I decreed in 1721 that after an eight-month grace period no one would be allowed to wear "printed or painted chintz or calico in the provinces of Churmark, Magdeburg, Halberstadt and Pomerania." All clothing made from these fabrics was to be destroyed by the end of the eight-month period, and severe penalties were imposed for any infringements.

One of the New World species, *G. hirsutum* now accounts for 90% of the world's current production of cotton. This species was domesticated, probably more than once, in central America and the Caribbean. Archaeological finds of cotton seed in Mexico date to 3500 BC. *G. barbadense*, which now accounts for a little under 10% of world prodution, was domesticated in South America. Early finds of cotton seed, yarns and fishing nets from Peru and Chile, dating to 4000-3000 BC, belong to this species. Columbus found the native inhabitants at his landfall wearing garments made of woven cotton, Cortez sent gifts of cotton fabric to Charles V of Spain, and Pizarro noted in 1522 that the inhabitants of Peru wore cotton clothing.

The first attempt at planting cotton in the American colonies was made in 1621 by the then governor of Virginia, Sir Francis Wyatt. Some one hundred years later its cultivation was started in the Carolinas (1733) and Georgia (1734) followed by Louisiana in 1758. However it was not regarded as a major crop, and almost all the cotton grown was for home consumption. Several factors were to change this: (a) the development of spinning technology by Hargreaves, Arkwright, and Crompton (see preceding), which resulted in an almost insatiable demand for cotton fiber for the spinning mills in England; (b) the invention of the cotton gin in 1793 by Eli Whitney, which in two years increased the rate at which cotton fiber could be separated from the cotton seed by a factor of several hundred; (c) the large numbers of slaves available within the southern states, brought there initially to work on the tobacco, rice and indigo plantations; and (d) the decline in the quality of the cotton fiber from India available to the English spinning mills.

In 1793, the year the cotton gin was invented, America exported 600,000 pounds (lbs) of cotton; in 1795 this rose to 6 million lbs; and by 1800 the figure reached 18 million lbs. Most of this was exported to the textile mills in England, particularly to Manchester, which was the center of the English cotton industry. During the period 1770 to 1870 Manchester underwent a massive expansion with the number of textile-related companies increasing from approximately 1,400 to 30,000. This expansion was helped by the invention of the power loom by Dr. Edmund Cartwright, a clergyman and former professor of poetry at Oxford University. It was the final technical development required for complete mechanization of what, only fifty years previous, had been essentially a cottage industry. Such expansion in manufacturing required a very large increase in population with all the attendant problems of inadequate housing, poor sanitation, and low life expectancy. Indeed it was a visit to Manchester that inspired Frederick Engels to write his classic Marxist book, *The Condition of the Working Class in England.*

The dependence of England's cotton industry on a steady supply of fiber from America led to great hardship among mill workers during the American Civil War. At the start of the war the Union navy established a blockade of major ports of the Confederacy, and as a result shipments of cotton to England virtually stopped. In the space of twelve months the number of full-time workers in the Lancashire cotton industry dropped by about sixty-five percent, and the workers faced appalling poverty, receiving very little help from the mill owners. Attempts were made to revive imports of fiber from India, but less than fifty percent of the short-fall was covered, and the problem of the poor quality, which had led to cotton fiber being sourced in America in preference to India in the first instance, still remained.

The 20th century has seen a gradual reversal of the historical spread of cotton processing from Europe back to India and other Eastern countries. Due to the great importance of cotton as an agricultural crop in America it still retains a large cotton spinning and weaving yarn and fabric industry, but in this it is unusual among Western industrialized countries.

See: Industrial Age

Coir *Cocos nucifera*
Arecaceae

Coir is the name of the fiber that envelops the hard shell, which contains the kernel, of the coconut palm (*Cocos nucifera*). The word derives from the Malayan "kayar" meaning cord. The coconut palm grows in the tropics, occurring rarely outside that zone. This plant has many uses to mankind including edible fruits, useful oils, and copra (the dried kernel of the coconut). But the coir fiber has only ever really been produced on an industrial scale in two main areas: Sri Lanka (formerly Ceylon) and southwest India.

As the coconut palm grows (up to a height of 60 to 100 feet) the terminal cone produces a continuous succession of leaves. The leaves drop after about three years, leaving scarring round the trunk. Age can be calculated from the number of scars. A fully-grown leaf can reach 20 feet in length. The inflorescence grows from buds in the leaf axils, flowering after five years at the earliest and eight years at the latest. The fruit only ripens after more than a year. Full production occurs after about fifteen years but continues for a very long time.

It is a fact that coconuts have frequently been washed up on uninhabited islands where they germinate, after drifting for many weeks. Charles Darwin noted considerable "stands" on what were then the uninhabited Cocos-Keeling Islands in the Indian Ocean. The tree is mentioned in a medical work of the Indian physician Susruta. The Greek Ktesian of Knidos, writing about 400 BC, stated there were palm trees in Persia from whose large nuts oil was obtained. However Marco Polo (1254 to 1323) stated that "the planks of Indian ships are tied together with yarn, not secured with nails . . . and the yarn is produced from the husks of Indian nuts, placed in water, and turned into bristle-like threads used for sewing, and resistant to salt water."

Records of planting in Ceylon go back to the Maharawansa Chronicle when King Agga Bodhi, in 589 BC, ordered a thirty-mile plantation to be laid out on the south coast. Under later kings plantings were increased as appreciation of the plant was recognized. The Portuguese, who established a foothold in Ceylon in 1571, took little notice of the palm. However, when the Dutch occupied the island in 1638/39 they realized they could draw revenue from the palms and hence encouraged planting. The English took over the island in 1802, by which time the entire southeastern coast was fringed with a continuous belt of coconut, an estimated 10 million trees. At the beginning of the 20th century it was estimated there were over 52 million palms. The thickest fibers (from the outside of the husk) are used as brush bristles. This thick fiber is also used for door mats. The finer (inner) fibers are spun into yarns for fish nets and cordage. During the middle of the 20th century the United States imported 6 million tons of this yarn, principally from India.

See: Materials, p. 349

Kapok (Java kapok) *Ceiba pentandra* and Indian kapok *Bombax ceiba*
Malvaceae

Java kapok is the seed fiber from the seed of the silk-cotton or kapok tree (*Ceiba pentandra*), which grows mainly in Southeast Asia, Indonesia, and the tropical regions of South America, and is believed to be pan-tropical in origin. It is an evergreen that grows to a height of about 100 feet and begins growing pods after five to seven years that can be harvested. Its productive life may be fifty years or more. Indeed it is claimed that the largest tree in Villabra, Puerto Rico, a *C. pentandra* which now measures 20.7 feet in diameter at the base, could well be the same tree noted by Columbus in 1493.

Indian kapok is the floss secured from the seeds of *Bombax ceiba*, a large indigenous tree to India, which has been used for many years as a source for stuffing pillows. In 1919 the Imperial Institute of London undertook a series of tests to determine the comparative merits of Java (ceiba) kapok and Indian (bombax) floss.

The Indian kapok fiber is known as vegetable down or bombax cotton. There are several varieties of plants from which bombax cotton may be obtained. Indian kapok probably originated in India. Main areas of production today are India, Thailand and Myanmar.

Kapok has a very important use due to its buoyancy. Compressed Java kapok is said to be able to support up to thirty-six times its weight in water, uncompressed Java kapok twenty to thirty times its weight, and Indian kapok ten to fifteen times its own weight. This characteristic has proved invaluable in lifebelts, lifeboats, and other safety equipment at sea. Java kapok was specified in the U.S. Navy requirements for safety. It is also used in upholstery and padding materials, but to date this fiber is unable to be spun into yarn.

Bast or Stem Fibers

Bast fibers are present in a large number of plants, but only a few of these are of historical or commercial importance. Bast fibers occur in the phloem or bast layer, which lies between the stiff outer cuticle and the woody core of the stems of dicotyledenous plants. They are present in the form of fibrous bundles of long thick-walled cells, which overlap each other and are bound together by pectins and other non-cellulosic materials. Thus, although the ultimate fiber lengths may be very short, in practice the fibers as isolated from the plant may be up to several feet in length.

All the bast fibers require a similar sequence of processes to extract them from the plant, although different names may be attached to them depending on the fiber being processed. These processes are: (a) retting, a fermentation process in which micro-organisms attack the gummy materials binding the fibers to the plant stem and also attack the outer cuticle; (b) breaking, a mechanical process in which the dried, retted plant stalks are crushed to break the cuticle and woody core without damaging the fibers; (c) scutching, a process in which the crushed straw is mechanically beaten to remove the broken fragments of cuticle and core; (d) hackling, a combing process that separates the coarse bundles of fibers into finer bundles and also arranges them in a more parallel fashion for subsequent processing.

Flax *Linum usitatissimum*
Linaceae

Flax, the second most important fibrous plant, is the most important of the bast fibers. In English-speaking countries the name is applied to the fiber as well as to the plant, while the name linen is given to the yarns and fabrics produced from flax fiber. However in other countries the name—lin, lino, linnen, lein, lyen, or similar—is used for the fiber as well and may also be used for the plant itself. Only one species, *L. usitatissimum*, is cultivated to any extent for fiber production. It is an annual, and when grown for fiber the seeds are sown close together, 2,500 to 3,000 plants per square meter, so that the stalks are straight and slender with branches only at the top. When grown for oil the plants are shorter and more branched, and the seeds are sown further apart, 1,000 to 1,500 plants per square meter, so as to encourage these characteristics and hence increase the yield of seed and oil.

Flax is the most ancient of the cultivated fibre crops. Its wild ancestor is *Linum bienne*, native to the fertile crescent of the Near East. Domesticated flax, identifiable by its larger seeds, first appears in the Near East about 6000 ^{14}C years BC. Perhaps first domesticated as an oilseed, flax was used early on as a fibre plant. However it appears to almost completely disappear from Central and Southern Europe in the Bronze Age (2500 to 800 BC). This may have been due to climatic changes or an increase in the use of wool due to the considerable increase in the importance of sheep breeding as the Stone Age gave way to the Bronze Age.

Flax cultivation and linen production, based on *L. usitatissimum*, were well-established in Assyria, Babylonia, and Egypt by 3400 BC, and some of the fabric dating from this period is as fine as any that can be produced today. Egypt was the major flax-producing country at that time, supplying the ancient world with a wide range of products. Its importance as a crop to Egypt is evident from the fact that its invention was ascribed to the goddess Isis, whose priests were required to dress in white linen, and also that the failure of the flax crop is one of the ten plagues referred to in the Bible. The many other references to linen in the Bible indicate its importance in the surrounding cultures, and to this day linen has retained its symbolic significance in many religions.

Although the Romans and Greeks used linen textiles these were mainly imported from Egypt and the surrounding areas, and hence there was little effort made to develop home-based flax production although some flax was cultivated to supply flax seed for food use. However its cultivation spread rapidly through Europe, into modern-day Spain, France, Belgium, Holland, and Germany, and flax from both France and Germany was being imported into Italy as a raw material by the end of the 1st century AD. It is thought that the spread of knowledge of improved manufacturing methods was due to the expansion of both the Phoenicians' trading routes and the Roman Empire.

It can be seen from monastic records and legal decrees that by the Middle Ages flax cultivation was widespread in Europe where it was the most important vegetable textile fiber, a position it retained up to the end of the 18th century. One reason for its dominance is that its growth is favored by a temperate climate typical of northern Europe. A second reason is that until the end of the 18th century, with the invention of Arkwright's water frame and Crompton's mule, European spinners were unable to spin cotton yarns of similar strengths to those from flax. From early in the 16th century Belgium was noted for the manufacture of "fustein" or "fustian", fabric with a linen warp and a cotton weft. The industry was centered on Bruges and used flax from Holland and cotton imported from the Mediterranean area, principally Cyprus. By the end of the 16th century Antwerp was a major textile center, exporting linen goods to Scandinavia, Germany, and a number of Mediterranean countries. Although its production in Germany was set back considerably during the Thirty Years' War (1618 to 1648), it recovered rapidly in the immediate aftermath and from there spread to the central plains of Russia where vast tracts of land were available for its cultivation and the conditions were very favorable. Flax was also seen as an important crop in the Austro-Hungarian Empire by both the Emperor Charles VI and his daughter the Empress Maria Theresa. Both rulers encouraged the flax industry, and a number of government publications were issued describing the best methods for growing, harvesting, and processing flax.

In Ireland the development of the flax industry was brought about by two main factors. During the 17th century the wool trade was perhaps the most important sector of the English economy, but the English manufacturers began to feel threatened by the lower-priced woollen goods produced in Ireland. To protect English merchants the English parliament passed an act in 1699 prohibiting the export of woollen goods from Ireland to any country except England itself. Because the English market was effectively closed to Irish woollen goods due to the high duties imposed this was in effect a total export ban. Textile manufacturers in Ireland therefore turned to flax production, an action that was encouraged by England. The second factor was the revision of the Edict of Nantes, which resulted in a considerable influx of skilled Huguenot refugees into Ireland. A large number of these refugees were from the Flanders area, which had been noted for the production of fine linen materials for several centuries. The Huguenots gave a major injection of expertise and, in several cases of capital also, to the Irish linen industry, and one of their leaders, Louis Crommelin, was appointed overseer of the royal linen manufacture of Ireland and granted a pension of £200 a year by the King. Because the majority of Huguenots in Ireland settled in the area around Belfast, linen manufacturing was also concentrated in this region and by the end of the 19th century Belfast, or 'Linenopolis' as it was sometimes called, was the most important linen manufacturing center in the world. Large quantities of flax fiber and yarn were imported from Europe to supplement that grown in Ireland, with fiber produced in Belgium being particularly prized for high-quality goods.

Since the end of the First World War the importance of flax cultivation and linen manufacture has declined worldwide. Since flax requires much more mechanical and chemical processing in order to obtain a fully-bleached yarn or fabric it cannot compete with cotton on a cost basis. Nowadays most flax is grown in the former Soviet Russia and exported as fiber, yarn, or fabric for further processing.

Jute *Corchorus capsularis* and *C. olitorius*
Tiliaceae

Corchorus capsularis and *C. olitorius* are the principal species grown for commercial fiber production. Each species has a large number of varieties, but these are utilized for textiles and cordage on more of a local basis. Both species are annual plants and attain heights of 12 feet or more. The shapes of their fruits differ: *C. capsularis* having globular capsules flattened at the top, *C. olitorius* having elongated pods. The *C. olitorius* does not tolerate excessive humidity and is therefore restricted to elevated sites, whereas *C. capsularis* is able to grow in low-lying and occasional flooded country.

C. capsularis probably originated in South Asia, while *C. olitorius* may have been first domesticated in Africa. The history of jute as a fibre crop is obscure before the first European contact with India, but *C. olitorius* may have had a much older role as a leafy vegetable.

The first shipments of jute reached western Europe from India in 1795. In 1820, jute was spun experimentally at Abingdon near Oxford. This gave rise in interest to the flax and hemp spinners at Dundee in Scotland. However, one of the difficulties of production was to overcome the handle of the fiber: jute having a toughness that gave cause for complaint. Only after a method was found of pretreating the jute with water and whale oil and eliminating the toughness did the fiber become acceptable. Fortunately Dundee was a whaling station, and whale oil was available in large quantities and at low cost, which proved to be advantageous.

The Scottish jute industry was further helped by several historical events. First the Dutch used jute bags as a cheap packaging material for the coffee exported from their colonies. Then the Crimean War (1854 to 1856) followed by the American Civil War (1861 to 1864) were the chief reasons for the prosperity of Dundee's industry. During the first conflict Russia's heavy export of flax and hemp ceased; in the second, the Federal armies of the North were denied cotton supplies from the South and had urgent need of containers for moving stores and supplies. Meanwhile the jute processing centers developed in France (1857), Germany (1861), and by 1890 had also been established in Italy, Austria, Belgium, Spain, Poland, Czechoslovakia, and Russia.

The jute industry expanded through the First World War, trebling the population of Dundee in the space of fifty years. During the Second World War there was a rising demand for sandbags. After the attack on Pearl Harbor the demand for jute parachutes was high as well as for wrapping supplies that were to be dropped. However the Bengal Provincial Government restricted jute growing in favor of rice cultivation so that jute production declined steeply between 1939 and 1945.

The first jute mill in the United States was in Ludlow, Massachusetts, in 1848 for weaving yarn shipped from Dundee. Other mills were established, but the low price of Asian jute made it more profitable to spin specialized materials, and so the weaving equipment fell into disuse. In the United States Census of Manufacturers, 1947, it was shown that there were thirty-five establishments in the United States producing jute fibers.

The partition of India and Pakistan in 1947 affected jute production as east Bengal was awarded to Pakistan, the new state that produced eighty-five percent of the world's raw jute. This meant that Pakistan held much of the production while India held the mills for the processing. Friction between the two countries led to low production, high cost, and fluctuating supplies.

Brazil plays a major part in jute production, but production is still low in comparison with the Asian markets.

While jute growing has been attempted in many countries throughout the world, the climatic conditions and the labor force required has kept this useful crop as a minor player.

Hemp *Cannabis sativa*
Cannabaceae

There are two main subspecies of *Cannabis*: subsp. *sativa* and subsp. *indica*. It is *sativa* that is used for the production of textile fibers. Subspecies *indica* (known as "Indian" hemp) is the plant that yields the narcotic but is not used for commercial textiles. It has been suggested that the two subspecies have no real specific difference, but when the plant is grown in a hot climate it gives rise to the intoxicating resinous extract within the plant.

The name "hemp" is often used in connection with other plant fibers, especially with some of the leaf fibers such as abacá (from *Musa textilis*), which is often called Manila hemp. The reason may well be that they were used in the same way as the true hemp (*C. sativa*). In fact there are forty-nine hemp plants listed in Matthews Textile Fibers that represent fibers from plants other than *Cannabis sativa*.

Hemp field, Britain. Copyright Edward Parker, used with permission.

Hemp is a dioecious annual, producing male and female seeds on separate plants causing difficulties in achieving simultaneous ripening of the two sexes. The fibers from the two sexes also differ in thickness so that harvesting them is usually carried out at the same time despite differing maturity times, blending the thin female fiber together with the coarse male fibers giving rise to a more uniform product. However development of monoecious forms has been carried out in the latter half of the 20th century with some stability. The height of the hemp plant depends on the variety, with 10 to 20 feet being the norm. However individual specimens reaching 26 to 33 feet or more have been known. Giant and dwarf specimens arise from biological and environmental causes, rather than genetic characteristics. However, early maturing varieties tend to yield shorter, coarser fibers. It is also possible that some varieties, such as some grown in Italy, resemble flax fiber and it is difficult to distinguish between hemp and flax. The method of hemp fiber production is similar in nature to that of flax.

Hemp probably originated some 3000 years ago in central Asia and spreading to China, where it has been cultivated for more than 6000 years. It is also thought to be the oldest textile fiber used by the Japanese, although it is likely they used wild nettle hemp up until around the 3rd century.

The plant spread to many of the Mediterranean countries of Europe, notably Italy, during the early Christian era and extended to the rest of Europe during the Middle Ages. Hemp is an important crop in Italy to this day, due to the fineness of the variety grown there.

The cultivation of the plant has been carried out in eastern Europe and nothern Asia from very early times, in the United States since early colonial times, in Chile for more than 400 years, and in Argentina in lesser quantities for a long time.

The largest use of hemp is in cordage of all types (ropes, twines, and cables) and also for sailcloth, canvas, and tarpaulins. The need for hemp by the armed services during World War II was considerable and could not be satisfied either due to transport disruption or insufficient supply to meet these needs, which necessitated the use of other fibers such as the stronger fibers of sisal, jute, or abacá instead.

Hemp production has declined in eastern Europe and elsewhere in the world. The reasons for this are due to migration of villages to the towns of developing areas and to speculative price structures. The dwindling supply during the latter half of the 20th century has forced cordage manufacturers to turn to the use of stronger, less biodegradable synthetic fibers.

See: Age of Industrialization and Agro-industry, pp. 361–2

Ramie *Boehmeria nivea*
Urticaceae

The plant, indigenous to China, is also known as china grass, grass linen, grasscloth, and rhea. The genus *Boehmeria* comprises about two hundred species. The plant is a perennial stingless nettle with stems 3 to 8 feet high, originating from a rootstock, and having few branches.

In a Chinese book dating from the 5th century BC mention is made of a hemp-like plant that is perennial which is understood to be ramie. Later a French Jesuit, Louis Lecomte, in his work "Nouveaux Mémoires sur l'état présent de la Chine" (Paris, 1697) stated that "Apart from cotton fabrics, which are very widely used, the Chinese in summer employ nettle cloth for their long robes." His description of the plant is not very clear but he refers to the white underside of the leaves as a distinguishing characteristic.

More information about ramie reached Europe from the naturalist Georg Eberhard Rumphius (1627 to 1702) who described the plant in his "Herbarium Amboinense." He derived the name "rami" or "rame," the general term for fiber in the Malay, Javanese, and Makassar languages, which was later taken by all the European languages as the designation for the chu plant.

Linnaeus classified the ramie plant as *Urtica nivea* in 1737. For several years there was some confusion over the botanical classification of the ramie plant. In 1808 William Roxburgh, superintendent of the botanical gardens at Calcutta, brought ramie seedlings from Sumatra, where the plant was cultivated by natives, to Calcutta where it was previously unknown. These plants exhibited characteristics distinguishing them from *Urtica nivea* (the species of which the leaves have silvery-white undersides), and he proposed a new species *Urtica tenacissima* (of which the leaves have green undersides). Ramie's place in the genus Boehmeria was first recognized in 1826 by the French botanist Charles Gaudichaud-Beaupré (1789 to 1854). He had found plants on Guam in the Marianas (an archipelago in the NW Pacific). The correct scientific name of the ramie plant thus is *Boehmeria nivea*.

Farel of Montpelier (1815) and Poppenheim of Combes la Ville were the first in France to plant ramie commercially. The first ramie spinning mill in England started operating at Leeds in 1840, and mills on the outskirts of Paris started later. German spinners of Zittau in Saxony also took an interest in the new fiber. Joseph Decaisne (1807 to 1882) of the Jardin des Plantes in Paris ensured that ramie material for planting went to Algeria and Guyana.

In 1889, G. Watt in his "Dictionary of the Economic Products of India" identified 18 wild species of *Boehmeria*, all of these being known as rhea, which had become the accepted term for ramie in many areas. However it is disputed whether all the material from this area was indeed ramie, but probably nettle cloths from other nettles, or possibly other fabrics imported from China.

Commercial use of ramie is limited by the fact that it cannot be spun to counts finer than 40s and is also more expensive that cotton, flax, or hemp. However it was an important fiber for many types of goods including sailcloth and cordage. Originally, ramie fiber served primarily for the production of gas mantles. It has also been used by the paper industry for the production of bank notes.

Kenaf *Hibiscus cannabinus*
Malvaceae

Hibiscus cannabinus has been cultivated in India since prehistoric times and is said to be indigenous to Africa, occurring in Sudan as early as 4000 BC. It was also mentioned in early Sanskrit literature as having been used for barkcloths. Kenaf is the fiber obtained from this plant, which is similar in appearance to the hollyhock. It takes its name from the Persian word for flax or hemp. In India it is also known as bimli jute or mesta. However like many other plants it is incorrectly identified as being hemp, jute, or flax. Because it is so similar in character to jute and is processed in a similar way, it is often misreported in world production figures.

It is a fast growing plant that is able to grow in any tropical or subtropical climate free from heavy rain or strong winds. The stalks average 8 to 12 feet in height, usually over a growing season of 120 days. The color varies from purple to green depending on variety and conditions. Leaves occur on the upper part of the stalk and are light green and five-lobed. Single yellow flowers are borne on short stalks in the axils of the leaves. It is a relative newcomer to the Western Hemisphere having only been introduced within the last 200 years. The East India Company brought samples of kenaf cordage and sacking back to England and other parts of Europe. Cultivation has also taken place in the Philippines, South Africa, and, since the Second World War, in the northeast region of China. It is grown in eastern Europe and nothern Asia, and, since about 1940, it has been cultivated in the Americas, especially the United States (early research being in Alabama then moving to Florida on a larger scale), and in Cuba and Mexico. Since Fidel Castro came to power in Cuba (1959) the industry has declined and those personnel who had managed kenaf production in the Cuban kenaf industry also transferred to Florida to aid production there.

Sunn *Crotalaria juncea*
Fabaceae

This plant is also commonly known as Indian, Sann, Bombay, Benares, Conkanee, Brown, Madras, Itarsi, Sunn, and Juddulpore hemp and has also been called Travancore flax, although the terms "hemp" and "flax" are incorrectly used. It is a plant that has grown in India and Pakistan since prehistoric times. *C. juncea* grows to about 8 to 10 feet tall with stalks up to ³/₄ inch thick. Leaves appear only at the top of the stalk when they are grown closely together, are bright green, pointed, and about 2 to 3 inches long. Its flowers are small, yellow, and borne in racemes from the axils of the leaves. It is cultivated and harvested in a similar manner to jute.

C. juncea was introduced to the Western Hemisphere during the latter part of the 18th or early part of the 19th century by the East India Company. Exports from the Indo-Pakistan subcontinent were about one-third of total annual production during the 1930s, with approximately a quarter of that going to the United Kingdom, the balance going mostly to Belgium and France. The plant was introduced into Uganda on an experimental basis in 1931 for cultivation as a fiber plant.

Imports of the fiber from Asia into the United States were around 1,000 tons annually. It is also grown in the United States for fodder and as green manure. Since the Second World War exports from India were, until 1950, approximately a quarter of the total production. However, since 1950 exports have dropped to twelve percent of total production due to the Asian jute mills supplementing the supply of jute fiber available to them. World production of sunn was reported to be about 71,000 tons in 1964. Although sunn is not as strong as hemp, it is second to ramie in wet strength, resistance to salt-water organisms, and to mildew and rot. Consequently this fiber is used in products where these characteristics are beneficial (e.g., fish nets, marine ropes, etc.). It is also used in sacking materials and paper-making, but is now far more important as a green manure.

However it is claimed in Matthews Textile Fibers that "Lack of care in fiber production has been one of the factors deterring wider use of the fiber."

Urena (*Urena lobata*)
Malvaceae

The urena fiber is obtained from the herbaceous perennial *Urena lobata*, a bast fiber that has been used in Brazil since prehistoric times. It is also known as aramina and ibaxama meaning plant rope. It grows wild in many tropical regions of the world, the weed assuming a more shrub-like growth, only attaining a height of up to 7 feet. The cultivated plant reaches a height of 10 to 12 feet, maturing in 120 to 150 days. It is harvested when it flowers in a similar manner to jute, except that the stalks are cut higher from the ground to avoid the high lignin content at the base of the plant.

Another species, *U. sinuata* (the species known as Cuban "jute" and in Bengal as kunjia), grows wild and is cultivated on a small scale in the West Indies and in tropical Asia. The fiber is similar to that of *U. lobata* and is often classified as such instead of as a separate species.

Commercial urena was first attempted in experimental trials in the Belgian Congo in 1926 and more successfully in 1929, where it compared favorably against *Hibiscus sabdariffa* (roselle) and indigenous Tiliaceae such as *Cephalonema polyandrum* and *Triumfetta cordifolia*. By far the best results were obtained by *Urena lobata*. Urena-growing was entrusted to the natives who were directed to clear land of primeval forestry. During the 1930s it became much more important, and cultivation was started in French Equatorial Africa. The British Government sponsored experimental growing trials in Guyana (formerly British Guiana), Trinidad, and the Gambia. Some experimental work has been carried out in the United States. World production reached about 6,000 tons in 1964, mostly from the Congo.

Again, interest in this fiber intensified during the Second World War for cordage, sandbags, and other uses as a substitute for the unavailable jute.

Leaf Fibers

Human ingenuity has created the ability to make textiles from plant leaves since early history. The many species and varieties that have proved capable of being used for this are immense. However, many of these have not been found commercial enough for the plants to be exported to other continents, or even other countries, but are used locally for domestic use. Examples of these are from the family *Bromeliaceae*, which includes the pineapple plant (*Anana comosus*) which has over 1,500 American species alone. The major leaf fibers that have been used commercially have been included in this work, and although some of them differ in their appearance there are also similarities between many of the species. The most important leaf fibers from a commercial standpoint come from the Order Iridales. These include sisal, abacá, and henequen, which account for eighty percent of world production of leaf fibers.

The method of processing the fibers is similar for all the leaf fibers with some slight differences according to the country of origin, the labor availability, and the proposed usage. The function of the leaf fibers is to give strength and rigidity to the leaf. Fiber removal consists of two operations: (1) separating the fibrous outer layer; and (2) removing the pulpy material, releasing the fiber strands by stripping or cleaning. This is normally carried out in the field. These latter operations are done quickly to avoid the hardening of the pulpy material, which could then not be removed. Several machines to decorticate the sheaths have been invented to mechanize this difficult process. The American spindle stripping machine, the hagotan, was superseded by the raspador or other machines invented in various countries. Once the fibers have been washed free from pulp they are dried, sometimes combed, and are then ready for use.

Sisal *Agave sisalana*
Agavaceae

The fiber derives its name from the Yucatan port of Sisal on the Gulf of Mexico where the genus *Agave sisalana* is indigenous and is known as yaxci. Sisal is a seed sterile crop of unknown origin, but probably from southern Mexico.

There are many species of *Agave* which are not easily distinguishable from each other. The plants brought to Europe (*A. americana*) became characteristic of the Mediterranean scenery. The true or white sisal plants (*Agave sisalana*) grow to about 3 feet sending up huge spiked leaves almost from ground level. The leaves are firm and fleshy, forming a rosette on a short trunk. After 6 to 7 years of growth the sisal plant sends out a flower stalk to a height of about 20 feet. After flowering the plant produces tiny buds, which fall to the ground and take root while the parent plant dies. Leaves are harvested after about four years and this continues until the plant dies. Plants may yield up to 250 leaves during their lifetime, each leaf containing up to 1,000 fibers.

The plants was introduced to Florida and grown there without any attempt at commercialization until about 1888 when cultivation began in the Bahamas. In 1893, Dr. Richard Hindorf imported 1,000 plants from Florida to East Africa but only sixty-two plants survived the journey. The surviving plants were planted in Tanganyika where they thrived, boosting the number of plants to 63,000 by 1898. Prior to the First World War cultivation had begun in Kenya and Uganda. During the 1920s plantations were also established in Mozambique and Angola. Many other African countries were to start sisal production about this time.

Indonesian production began in Sumatra in 1913 from plants imported from East Africa, and production rapidly increased. The Indonesian sisal has become one of the more desirable fibers due to careful and efficient preparation. During the First World War there was strong demand and high prices encouraging commercial production. Haiti had begun cultivation and, like the Indonesian sisal, the sisal was in much demand for its quality. The first commercial plantings in Venezuela and Brazil were in 1938 and 1939 respectively, although the plants were introduced in Brazil in the late 19th century.

Sisal is used extensively for cordage and, although thought to be of limited use for marine cordage (deteriorating in salt water too rapidly), it proved itself suitable for this use during the Second World War when other similar fibers were in short supply. It is also used for agricultural and industrial cordage, matting, and rugs.

See: Age of Industrialization and Agro-industry, pp. 362–3

Abacá *Musa textilis*
Musaceae

Musa textilis gives us an important fiber used for cordage, called abacá, but sometimes referred to as Manila hemp although it is not related to the true hemp, *Cannabis sativa*. Abacá is an herbaceous plant that grows to up to 15 to 25 feet with a diameter at its base of 16 to 20 inches. From the spreading rootstock shoots tightly packed false stems grow containing overlapping leaf-sheaths. The leaf blades can be 16 to 20 inches wide and up to 7 feet long, formed at the top of the false stem. The fruits of *M. textilis* are inedible, thick, 2 to 3 inches long and, unlike the edible banana, contain small black viable seeds. The fiber in the leaf sheath varies in color and quality depending on the position of the sheath in the stalk. The color becomes whiter and the fiber softer closer to the center of the plant. New plantings are taken from rhizomes or from suckers while seeds are only rarely used as seedling growth is slow.

Abacá is indigenous to the Philippines where ninety-five percent of the world output is still produced. The native islanders were making textiles from its fibers when Magellan (ca.1480 to 1521) reached the Philippines in 1521, where he was killed in battle after having circumnavigated the world.

Abacá was introduced into other tropical regions such as Indonesia, India, Guam, and Central America. Its success in these regions was due to the abacá from the Philippine being cut off during the Second World War. During and after the war the Japanese abandoned the plantations, and the ground was cleared for growing food plants, and some were even left fallow. NAFCO (National Abacá and Other Fibres Corporation), founded early in 1947, endeavored to restore production to pre-war levels. However due to demand during the war years, plantations in other tropical countries had been established. Also after the war, the U.S. Department of Agriculture successfully started cultivation in Panama, Costa Rica, Honduras, and Guatemala, all with suckers taken from the Philippines, specifically from Davao.

Most of the 67,000 tons from the Philippines during the 1960s annual harvests were exported into the United States, Japan, and the United Kingdom.

Due to the excellent strength and water resistance of abacá this fiber is much sought after. It is the strongest of all the hard fibers, followed by sisal, phormium, and henequen. It is used for marine cordage, twines, and hammocks, et cetera. It has also been used in the paper industry for making tea bags, mimeograph mats, and other specialty materials.

Henequen *Agave fourcroydes*
Agavaceae

Agave fourcroydes is another major species of *Agave* that is indigenous to Mexico and has greyish green leaves with a sharp point and marginal spines. It is the most important fiber plant of Mexico and yields the commercial fiber henequen. The name henequen refers to both the plant and the fiber. It is also sometimes referred to as maguey fiber, although strictly speaking this term applies to *Agave cantula* which is similar to henequen but is softer and more supple.

The henequen leaves are cut after they are six to seven years old. They can be up to 7 feet long and about 4 inches wide and 1 ½ inches thick at the base. They are cut two at a time, and then for twenty to thirty years (the lifetime of the henequen plant, compared with six to eight years, the lifetime of the sisal plant) a few leaves are removed twice a year until the plant flowers and dies. As the leaves are cut a trunk is formed. Both *A. fourcroydes* and *A. sisalana* produce roughly the same number of leaves during their lifetimes.

It has been cultivated for hundreds of years in Yucatan by the Mayan Indians. Since the 1880s most of the henequen exported from Mexico has been sent to the United States. Up until the Second World War twenty percent of production went to Europe. A Yucatan government department handled all sales of fiber. Henequen producers operated independently until 1912, but by 1915 a producers' association was authorized. By 1925 a co-operative was organized and replaced by the present organization, the Asociacion de Henequeneros de Yucatan, in 1938.

The total annual output of henequen during the middle of the 20th century was reported as about 150,000 tons. Mexico produced about four-fifths of this figure, mostly from the State of Yucatan although Brazil is now also a major producer. The majority of the remainder came from Cuba where it was introduced during the 19th century. Small quantities were grown in Jamaica, other areas of the West Indies, and in Costa Rica, Guatemala, and Honduras.

Phormium *Phormium tenax*
Phormiaceae

Phormium fiber, often termed New Zealand flax or hemp, is obtained from *Phormium tenax*. The plant is indigenous to New Zealand where it is known as harakeke. Captain Cook, who had observed the plant during his visit to New Zealand on his first voyage to the Antipodes in 1769, gave the first description of the plant to the rest of the world in "A Voyage to the South Pole and Round the World." "Phormium" is derived from a Greek word meaning basket, one of the first uses for the fiber that was observed by Captain Cook and his crew.

The plant has a creeping rootstock and sends up a collection of tough, sword-shaped leaves. The most common variety has leaves about 2 feet long, 4 inches wide, folded lengthways along the midrib, and often split at the tip. Eventually, a long inflorescence rises from the center of the cluster where many waxy red or yellow flowers appear during the summer months (November to January). In a single inflorescence as many as one hundred capsules may ripen, each containing 60 to 150 shiny black seeds. However propagation is more usually by rhizomes.

Phormium grows spontaneously throughout New Zealand and the Norfolk Islands to the northwest. The Maori took the plant to the South Island of New Zealand and also to the Auckland Islands in the south. The densest populations are found in Manawatu Valley in the North Island of New Zealand. However it was claimed that the Maori did not cultivate the plant but exploited the natural resources.

The plant was introduced to southern Ireland in 1798 where it grows well but usually as an ornamental plant. Subsequently it was introduced into parts of Europe, St. Helena, the Azores, Australia, South Africa, and Japan. It was planted experimentally in the United States, but commercial production did not succeed, and plants are grown principally in gardens as ornaments.

Exports from New Zealand rose steeply from 1831 to 1907 (increasing from 1,062 to 28,547 tons). During the 1930s cultivation of the plant was started in Chile, Argentina, and Brazil. However, after the First World War production dropped, presumably due to the increase in competition from sisal and abacá production. Export from New Zealand was discontinued during the Second World War. Small quantities of phormium are still produced outside New Zealand. Almost all production from St. Helena is exported to Europe. Production from Chile and Argentina is for domestic use only, but efforts to increase the Brazilian phormium for export have only been partly successful. The major drawbacks of phormium are the long periods between harvesting, susceptibility to moisture content of the soil, and irregular employment of labor.

Dyes

Woad, C.I. Natural Blue 1 *Isatis tinctoria*
Brassicaceae

Woad is a temperate herbaceous biennial, which produces a basal rosette of leaves during its first year followed by a single stem that eventually bears yellow flowers during the second year. The leaves are used to produce the dyestuff indigotine by fermenting out the sugar-bearing indican. Woad is not pleasant to handle since it develops a putrid smell during fermentation due to sulfur-containing chemicals in the leaves.

Woad was a native of southeastern Europe, around Greece, Italy, southwestern Russia, and North Africa and spread quickly throughout Europe in prehistoric times. Ancient Britons called it glastrum from the Celtic word glas, meaning blue, where it is thought the place name Glastonbury was derived. It was used for many centuries in the British Isles commonly as a body paint to frighten enemies. The Romans referred to these people as Picts (Celtic for "painted"). It is also possible that the name Briton derived from the Celtic "britho," which means "paint."

By the end of the 14th century woad was being cultivated in the German Dukedom of Jülich (north Rhine-Westphalia). According to old tax records during 1391 AD this plant is supposed to have yielded revenue of 1,270 gold marks, collected by the city of Aix-la-Chapelle (Aachen) from the woad cultivators. Peasants were only allowed to grow a certain amount of woad. No dyer was allowed to go out to the villages and buy up the precious product ahead of his colleagues. In Erfurt the right to purchase was, in 1480, expressly limited to citizens of the town. The peasants brought the woad in to market where it was often tested before being sold. "No one may sell untested woad" was one of the rules laid down by the Cloth-weavers' Regulations of Schweidnitz in 1335. The struggle between home-grown woad and its foreign rival, indigo, forced an edict from Henry IV (1399 to 1413) condemning to death any dyer who used the "pernicious drug, indigo" which was at the time called "the devil's dye."

The smell produced during fermentation so offended the nose of Queen Elizabeth I (1533 to 1603) that she passed an order prohibiting the cultivation of this plant within five miles of her residence. However Queen Elizabeth I also prohibited the use of indigo (and also logwood) and authorized searchers to burn both valuable dyestuffs when found in any dyehouse. This Act remained in place for a hundred years until the reign of Charles II. This helped to ensure the continued cultivation of woad despite the success of indigo abroad. During the Thirty Years' War (1618 to 1648) the woad fields of Germany were destroyed and woad cultivation was no longer possible.

The decline of woad gradually came about because of the increased importation of the higher indigoid-containing, though more expensive, *Indigofera tinctoria*. Both plants—woad and indigo— are based on the same dye, ester-like indican or glucose-like indoxyl. Woad contains only one-third as much dye, compared to indigo. The similarities in the two species however were not proven until 1778 to 1779 by the German chemists Trommsdorf and Plauer from Erfurt.

Woad was given another chance when Napoleon I banned the import of English and, consequently, Indian goods throughout Europe. He offered a reward of 425,000 French francs to find a replacement for indigo or to improve the dyeing method of woad and other dyestuffs. A decree of 1811 ordered the recultivation of woad in France, while from 1813 importation of Indian goods was strictly prohibited, as they were regarded as English. Even German chemists, including Trommsdorf, tried to invent an improved dyeing method. Austria also tried to reward research work improving woad dyeing. But this was all in vain as the indigo import figure of the 19th century reached several million kilograms, and the synthetic indigo proved itself so successful.

Indigo, C.I. Natural Blue 1 *Indigofera tinctoria*
Fabaceae

The most widely cultivated indigo species are *I. tinctoria* (Indian indigo), originating in Asia, *I. arrecta*, from east Africa, and *I. suffruticosa* from the tropical Americas. Indigo (mainly *I. tinctoria*) has been cultivated in India for at least 3000 years, but increasing trade led to a great increase in its cultivation, there and in Europe, in the sixteenth century. Other genera of plants contain similar dye chemicals and are locally used instead of indigo as blue dyes: wood (*Isatis tinctoria*) in Europe, knotweed (*Persicaria tinctoria*, Polygonaceae) in Japan and China, *Strobilanthes cusia* (Acanthaceae) in south and southeast Asia, *Baptisia tinctoria* (Fabaceae) in north America, and African indigo (*Philenoptera cyanescens*, Fabaceae) in west Africa.

The true indigo species, *Indigofera tinctoria*, is a shrub-like plant which grows to a height of 1.5 m and has delicately colored pink blossoms for three months in a year. This species will yield 1.5 to 2.0 kg of raw colorant from 100 kg of plant and has a twenty percent content of pure dye.

The subcontinent of India is not only the origin of the indigo plant proper (*Indigofera tinctoria*) but also the oldest centre of indigo dyeing in the Old World, though it also flourished among many races in Indo-China and Indonesia. In Africa indigo can be traced from the Mediterranean coastline to the Sudan and to the Guinea coast, even among primitive tribes, although it did not extend to the southern parts of the continent except for Madagascar. Even before Columbus there was evidence of indigo dyeing among the people of the Andes and Mexico. However the latter species may have been the perennial *Indigofera suffruticosa*.

In Europe first entries of indigo in weighing bills and accounts were made as early as 1140 in Genoa, 1194 in Bologna, 1228 in Marseille, and 1276 in London. Indigo was shipped from India via Alexandria to Venice, which was a world trading center of that period. Marco Polo (1254 to 1324) was the first to report on the preparation of indigo and the chief production centers in India, although these findings were not published until 1477. Whether he introduced indigo to Europe is unlikely.

In 1577 it was described by the Frankfort Police Regulations as "the newly invented, harmful, and balefully devouring corrosive dye." A union was formed, called the Woadites, that was an international political group of woad producers, united to fight indigo. Laws were passed in England, France, and Germany to prohibit the importation of indigo. It was not until ca. 1600 that indigo was officially approved in Germany as a self-shading dyestuff. Prior to that it could only be used to dye the background color or as an addition to woad.

It was not until 1737 that indigo was legally permitted into the rest of Europe after edicts made by Henry IV and Queen Elizabeth I. However, when large quantities of indigo were imported, the authorities had grave misgivings about permitting its use by dyers. The reason for this was probably not so much the fear that indigo would mean the end of woad-growing but rather that the new dye was harmful to the fabric. This could be due to the inexpert use of the yellow arsenic sesquisulfide (orpiment) baths necessary to obtain greener shades.

Meanwhile in 1649 Europeans attempted to break the Indian monopoly by planting *Indigofera tinctoria* in the New World. The first successful crop was produced by Eliza Lucas Pinckney in South Carolina (ca. 1740) but was dropped in favor of rice cultivation during the War of American Independence (1775 to 1783).

The British dropped any obstacles to indigo trade when they occupied India and began the exploitative activities of the East India Company (1756 to 1763). After the loss by England of her North American colonies in 1783 and the riots in the French colonies in 1795, the New World indigo went into decline and the Indian trade was encouraged back to prosperity.

World consumption of indigo was very high during the 19th century, which led the German Chemist Adolph von Baeyer (1835 to 1917) to attempt chemical synthesis. In 1880 he patented a method starting with cinnamic acid. But to make this acid from cinnamon bark was out of the question as cinnamon was as expensive as the indigo itself, but he found it could be made from coal tar. Further improvements to the synthetic material were made so that by 1897 synthetic indigo was finally marketed. By the end of the 19th century Germany could produce cheaper indigotine to monopolize the synthetic production. Between 1899 and 1900 the imports of natural indigo to Germany had reduced by forty-five percent and to England by forty percent, while exports of synthetic indigo from Germany increased by seventy-five percent during the same trading period.

By 1914 the Indian cultivators and the natural dye producers were practically ruined although the First World War kept the final collapse at bay for a few more years. Since then the production of natural indigo has all but ceased.

The use of indigo is now mainly associated with the denim fabric first produced in France, where the warp is indigo-dyed, and the weft is left undyed giving a two-sided color effect to the woven cloth.

Madder, C.I. Natural Red 8 *Rubia tinctorum*
Rubiaceae

Madder is the root of the herbaceous perennial plant *Rubia tinctorum*. It has weak stalks that prevent the plant from attaining its full height, these stalks being commonly used for animal fodder, although this often imparted a reddish hue to milk and a yellow tinge to butter. The bulk of the coloring material is contained in the red mass between the outer skin and the woody heart of the root. The plant is left in the soil for eighteen to twenty-eight months to improve the yield of coloring matter within the roots, which can contain approximately 1.9 percent of dyestuff. The quantity of the dye is dependent on the nature of the soil rather than the species of plant, the finest quality coming from plants grown on calcareous soil.

The genus *Rubia* occurs in many countries of the tropical and temperate zones. *Rubia cordifolia*, also known as Bengal Madder of India, was found in India, China, and Japan and yields an inferior dye called Munjeet; *Rubia tinctorum* was principally cultivated in Holland, France, and Turkey and to a lesser extent in Belgium, Germany, Italy, and North and South America.

Madder was known to ancient civilizations. The earliest references are from the Indus civilization of around 3000 BC. It has been identified on cloths found in Egyptian tombs, and Herodotus (ca. 484 to 424 BC), the Greek historian, noticed its use there around 450 BC. The collection of Hebrew laws and precepts which formed the basis of the Talmud and the Mishna permitted the growing of madder, though only for domestic rather than commercial use.

Pliny the Elder (23 to 79 AD) tells of madder cultivation near Rome, and Dioscorides reported that dyers had distinguished between the higher pigment content of the cultivated species to that of the inferior wild madder.

It was stated that madder cultivation was to be found in France in the vicinity of St. Denis near Paris during the 7th century. A century later it was ordered to be cultivated on the estates of Charlemagne (768 to 814), along with woad.

Holland was the most advanced madder-producer by the end of the 15th century. The town of Goes in Holland issued detailed instructions governing the planting and tending of madder in 1494. In the district of the so-called Onrust Polder, one of the marshy areas reclaimed from the sea, the madder plant was only allowed to grow for two years to prevent the roots from penetrating the soil to the underground level of the sea, which might have caused flooding. After each crop a period of ten years was allowed to elapse before madder was again planted in the soil.

Madder was grown practically all over Germany and exported in large quantities from Magdeburg to Poland in the 14th century, and to Flanders, Italy, and England in the 15th century. Norwich

Original pen and ink drawing of the Maddermarket, Norwich, by H.J. Stone.

appears to have been an important English center for the trade, as a church (St. Johns), a theatre, and one of its streets was named the Maddermarket.

During the 16th and 17th centuries Holland was still the major producer. Madder was probably not cultivated in the United Kingdom before 1624 when a patent was granted to "use, exercise, practice, the sowing, setting and planting of the herbe, roote, or plant called madder." Although also grown in Bohemia, the Thirty Years' War (1618 to 1648) brought about its destruction, and the cultivation of madder could never be resumed there. The only other country that could compete with Holland was France. It did not become a major competitor until the 18th century, as the level of cultivation reached during the 7th and 8th centuries had declined, but it was encouraged to thrive again to prevent large revenues being spent on imported madder and to meet the demands of the French dyeing industry.

The warlike period of the French Revolution and the First Empire (1789 to 1815) did great harm to the industry, and the industry did not recover until Louis Philippe (1830 to 1848) ordered the trousers and caps of the French soldiers to be dyed red using madder. The government imported madder seeds from Cyprus and had them planted in Alsace. In 1868 French madder exports were valued at over £1,200,000. However in 1869 alizarin, its synthetic alternative, was first produced by Graebe and Liebermann, and this proved disastrous for madder-growing countries.

Turkey Red. This was the most famous shade produced from madder, giving great brilliance and fastness. The dye extracted in an aqueous solution from the madder root imparts no lasting color without subjecting the textile to a treatment before dyeing. The original Turkey Red process took many repeated steps and one month to complete. It was produced on cotton or flax with oil and using pure alum as a mordant. It was developed in India, causing a sensation when brought to Europe by French, English, and Dutch merchants. However, the Greeks knew the use of alum as a mordant, and two alum pits are known to have existed in Turkey in the 15th century: one in Constantinople and one in Smyrna (the former name of the Turkish port of Izmir).

It is said the Turks developed the growing of madder in the Balkans and began to produce fabrics themselves by a process that was then known as Adrianpole or Levant red (the Levant being the east Mediterranean region, especially Turkey, Syria, Lebanon, and Israel). In 1747 a French dyeing company introduced Greek workers to Rouen and they brought the process to Europe, dyeing the first Turkey Red in the west.

John Wilson of Ainsworth (Greater Manchester) was reported to be the first man in Britain to dye Turkey Red.

However, the French method of dyeing Turkey Red had already been given to the world in a publication of 1765 in what would appear to be the method but not the know-how of its application.

In Scotland Turkey Red production was introduced to the Vale of Levin in 1790. A whole group of allied services grew up to supply the industry including manufacturers marketing a so-called Turkey Red alum. The scientist John Mercer researched the production of Turkey Red oil, produced by sulfuric acid and castor oil giving a product known as soluble oil, which was used for many years. Turkey Red remained important into the 20th century but declined after the 1869 introduction of synthetic alizarin.

Logwood, C.I. Natural Black 1 *Haematoxylum campechianum*
Fabaceae

The logwood tree, also known as the campeachy wood, blackwood, and *bois bleu* or *blauholz*, grows throughout Central America and the Caribbean, but the finest wood occurred in Yucatan. It was first taken from Campeachy on the Gulf of Mexico and successfully propagated in the West Indies. The tree would grow to a height of 30 to 45 feet within about 10 years, growing very rapidly in marshy ground. The interior of the wood is yellow immediately after cutting but oxidizes rapidly to dark brown.

One commercial form of the dye is logwood extract, obtained by boiling logwood chips with water under pressure and concentrating the liquor to dryness to form haematin crystals.

The medieval and ancient dyers knew other great natural dyes, such as indigo and madder, but logwood was only introduced into Europe following contact with America. Before this the dyeing of black was a problem requiring re-dipping to a navy with either woad or indigo and topping up with either weld or madder, but this was an expensive process. Logwood is a mordant dye that gives a reasonable black when treated with copperas and, hence, is much simpler to apply. Using an alum mordant a blue was obtained that was much less fast to light than indigo navy blues.

This lack of fastness was probably the cause of its prohibition in 1580 by Queen Elizabeth I who passed an Act "abolishing certain deceitful stuffs employed in dyeing cloths." In fact logwood, together with indigo, when found in dyehouses, was under order to be destroyed. The Act was repealed in 1661, stating "that the ingenious industry of these times hath taught the dyers of England that art of fixing colors made of logwood: so that by experience they are found as lasting and serviceable as the color made with any other sort of dye-wood."

The importance of the logwood trade during the 18th and 19th centuries was not always realized. The importation of logwood during 1896 was recorded as from Jamaica, 22,495 tons; San Domingo, 13,068 tons; Honduras, 21,076 tons; Laguna, 7,176 tons; Cuba, Yucatan, and Campeachy among other locations, 9,413 tons. The British colony of Honduras (now Belize) owed both its origin as a colony and its retention to the logwood trade. Logwood continued to be used for dyeing blues in imitation of indigo blue where the mordanting was carried out using alum and for purples where it was mordanted with tin crystals. The use of logwood died out towards the end of the 19th century as the first synthetic competitor was introduced (Alizarin Black S, patented in 1887). A better competitor against cost was to be Diamond Black F which was introduced into the United Kingdom in 1892. It continued to be used for black on wool until the 1920s when it was almost entirely replaced by the synthetic chrome blacks.

Weld, C.I. Natural Yellow 2 *Reseda luteola*
Resedeaceae

Commonly known as Dyer's Rocket, this is a liquid or dried extract of the herbaceous plant found in many parts of central Europe and formerly cultivated in England, France, Germany, and Austria. Weld is supposedly the oldest yellow of reasonable fastness to have been used during ancient times.

It precedes both fustic and quercitron bark (a species of oak indigenous to North America, see following) until the discovery of the Americas, although it's tinctorial strength is less than that of the newer introductions. The plant was sold in the sheaf-like straw; however the coloring matter, Luteolin, was mainly obtained from the leaves.

The plant was referred to by Dioscorides during the 1st century AD, but the name "reseda" is related to "sedation" which suggests that the plant was used mainly as a sedative. Pliny also claimed it was used for exorcism of evil spirits. However, seeds found at Zurich, Switzerland prove that even the lake-dwellers had been cultivating the weld plant for dyeing during the 1st century.

In a study of natural dyes in textile applications by the York Archeological Trust, about five hundred textile samples of the period 1500 to 1850 were received from museums in Europe and America and the dyes examined. Of the yellow dyes about eighty percent of the samples were dyed with weld, similar to results of Dutch textiles of the 17th century.

Fustic, C.I. Natural Yellow 11 *Maclura tinctoria* (syn. *Morus tinctoria*)
Moraceae

Fustic is sometimes known as "old fustic" to distinguish it from another dye called "new fustic," or "young fustic." It is also known as Cuba wood, yellow wood, yellow brazilwood, dyer's mulberry, *Bois jaune* in French, and *das Gelbholz* in German. It was probably the best of the natural yellows. The dye is obtained from the trunk of a tree found in North, Central, and South America, the Cuban variety being the best quality. One kilogram of wood yields approximately 40 g of dyestuff.

The precise period of its introduction into Europe for use in dyeing is unknown, but it was used as a drug in Europe soon after the middle of the 16th century via Spanish settlers in South America.

The dye was mainly used on wool and gave an olive to old gold color when mordanted with chrome or a greenish olive when mordanted with copper and iron. Both colors were fast to light and washing. The main uses of this dye were in combination with logwood, madder, and indigo for olive browns and drab browns. When the dust-color khaki was introduced for uniforms for the British and native troops in India ca. 1850 the use of fustic was increased, as it was the ideal dye for the production of this shade.

Soluble Red Wood, C.I. Natural Red 24 *Caesalpinia spp.*
Fabaceae

A number of trees in the legume family were important sources of red dyes. Their wood is a light yellow colour, changing rapidly to deep red on exposure to air. The first of these dyes to be used in the Old World was sappanwood or brazilwood (*Caesalpinia sappan*), imported to Europe from the 14th century onwards, from south and southeast Asia. In the early 16th century, Portugese traders found an alternative source of red dye in several closely related species, primarily Pernambuco or brazilwood (*Caesalpinia echinata*), which grew in the Atlantic forests of Brazil. The country of Brazil is named after the dye exported from there in such large quantities.

Saffron, C.I. Natural Yellow 6 *Crocus sativus*
Iridaceae

The stigmas of the autumn crocus have been used as a textile coloring matter and as a foodstuff colorant since ancient times. It is a very costly material as it takes between 70,000 to 80,000 stigmas to yield 1 kg of saffron, which contains about 10 g of crocin and 60 g of crocein as actual dye components.

It was known as a ladies plant due to its use as an aromatic, an ornamental, and a medicinal plant. Saffron yellow clothes were typically feminine attire. Roman brides were allowed to be colorful by wearing the flammeum, a kerchief dyed reddish yellow with saffron, and a red-yellow ceremonial robe was worn by the priestesses.

Because of the expensive nature of the precious saffron, stories of adulteration occur wherever saffron is grown. Marigold strands were a favorite substitute and if caught in Nuremberg in 1444 the penalty for this offence was to be burnt alive. Around this time other offenders were also reported as being buried alive.

Picture of town arms, Saffron Walden.

In Semifonte, a town destroyed in 1202 by the Florentines, it was considered easier to raise money on a few pounds of saffron than by mortgaging real estate. The rise of Basle to the status of being a city was largely due to the saffron trade. Its coat of arms, and that of Florence, depicts the saffron lily. During 1374 a consignment of goods bound for Basle was ambushed, 800 pounds of saffron (the most valuable part of the booty) were stolen, valuable enough for the city to engage in a lengthy feud to recover them and started the Saffron War. The incident occurred near Innsbruck, which suggests that the goods had been brought from Venice across the Brenner Pass.

Saffron was introduced to Britain in the 15th century and was grown in northwest Essex and parts of Cambridgeshire. It found a center of trade in the town of Walden, later to be known as Saffron Walden. Saffron was an important trade for the town, harvested every autumn for over 250 years. The crocus is even depicted in the town arms and representations made on several buildings.

The whole flowers were picked from before sunrise until late morning and then carried by the "crokers" (growers) to a center to be prepared. The stigma or chives were plucked out of the flower and dried while all the petals and stalk were discarded. There were reports of the gutters of Walden streets clogged with saffron petals.

William Harrison, the rector of Radwinter, a few miles from Saffron Walden, wrote in the 16th century, "warme nights, sweet dewes, fat grounde, and misty mornings are very good for saffron; but frost and cold do kill and keep backe the flower or else shrinke up the chive." The English saffron trade flourished until 1768.

Safflower, C.I. Natural Red 26 *Carthamus tinctorius*
Asteraceae
Safflower is an annual plant, growing 2 or 3 feet high with a whitish stem, upright but branching near the top. It has oval, spiny, sharp-pointed leaves with bases clasping around the stem. The orange-yellow flower is similar to that of a thistle and is commonly known as the Dyer's thistle, bastard saffron, or false saffron, although not related to *Crocus sativus*. Carthamin (or carmine) is the red coloring matter of the safflower, which contains two useless yellow coloring matters (thirty to thirty-six percent), one of which is Safflower yellow [C.I. Natural Yellow 5], but has no dyeing properties, and which may be removed in cold water, but more importantly, leaving the red colorant. This is only present in quantities of about 0.2 to 0.3 percent and is recovered by extraction with dilute sodium carbonate solution after removal of the yellow component.

The plant is a native of Western Asia and naturalized in Egypt. The dyestuff has been found in Egyptian tombs dating back to 2500 BC on yellow- to reddish-colored ribbons. It is cultivated in the Levant and the southern parts of Europe. The Chinese have long used this dyestuff which they know as *Hung hua*, meaning red flower. The colorant was used by the Japanese ladies as a cosmetic for coloring lips. It has also been used to dye the tape that binds legal documents affectionately referred to as red tape.

See: Nuts, Seeds, and Pulses, pp. 150–1

Alkanet, C.I. Natural Red 20 *Alkanna tinctoria*
Boraginaceae

Several forms of anchusa (also known as alkanna, alcanea, and bugloss) grow wild in Europe, including Britain. A red dye from the root was popular with Egyptians in textile dyeing and also used for cosmetics in Greece and Rome.

It is a plant that grows to about 2 feet high with very hairy stems and large forget-me-not type flowers that bloom in June.

Anchusa was cultivated in Europe specifically as a dye plant, but North American settlers also used it for medicinal purposes. However madder was a much more suitable and fast red dyestuff and was used for dyeing by European settlers in preference to anchusa. However the latter was probably used by the native Indians.

Dyer's broom, C.I. Natural Yellow 2 *Genista tinctoria*
Fabaceae

Also known as Dyer's greenwood, Dyer's weed or woad-waxen, it is a small, tufted shrub bearing racemes of yellow flowers. Cattle eating this wild plant add bitterness to their milk and to the cheese and butter made from it. It is found in England, rarely in Scotland, is wild throughout Europe, and has been established in the eastern part of the United States. It has been purposely cultivated in the United States due to its profusion of yellow flowers. It has been used for several ailments including dropsy, gout, rheumatism, sciatica, and even rabies. During the 14th century it was used to make an ointment called *Unguentum geneste*, "goud for alle could goutes", et cetera. The seed was also reported to be used in the plaster to heal broken limbs.

All parts of the plant are used to yield a yellow dye, which was used by the ancient Greeks. It has been used in combination with woad to give a green color. In England it was a source of income to the poor who collected the plant and sold it to dyers.

The process of dyeing linen, wool cloth, or leather using this plant was described by Tournefort in 1708 when he saw it being used on the island of Samos. It is still used for the same purpose on some Greek islands. However, as a dye, this plant has been largely superseded by the use of *Reseda luteola*.

Turmeric, C.I. Natural Yellow 3 *Curcuma* spp.
Zingiberaceae

Native to both India and China turmeric is also known as "Indian saffron". This underground stem or root of the Chinese *Curcuma longa* is the commercial variety of turmeric plant that is used for dyeing, the coloring matter being curcumin. It is one of the few natural dyes that is used as a direct dye, that is, requiring no mordant to give it a reasonable degree of fastness to the textile. The dried rhizomes, about 1 inch or longer and ¼ inch in diameter are very hard and have an external yellow-ish-grey color. For dyeing purposes the rhizomes are simply ground to a fine powder. The fine dust that was prevalent during this operation gave the workers a strong yellow complexion and is still used as a skin dye during festivals and celebrations.

Since the introduction of synthetic direct yellow dyes turmeric is much less used, being much more sensitive to acids and alkalis and having a lower degree of fastness to light. Other varieties of

Curcuma that are grown in southern Asia are more suited to the important uses of food coloring or condiments, the color yield of these varieties being less.

Orchil, C.I. Natural Red 28

Also known as orseille, archil, orchella moss, or orchella weed, this coloring matter is produced from various lichens. These are to be found on rocky coasts such as the Azores, the Canaries, Madeira, Corsica, and the Cape of Good Hope from the species *Roccella* or found in Scotland from the species *Lecanora*.

The Papyrus Leiden, one of the preserved papyri that were written during the 3rd century BC and discovered near Thebes in 1828 contained, among other things, dyeing recipes that threw a light on the early dyeing methods. It tells of a red coloring matter that was used in those days that was orchil or "substitute for the purple shell" (the very important historical purple coloring, purpurin, that came from a shellfish). Such was the importance in status of the color purple that the use of an alternative colorant for this shade was of great interest.

Theophrastus (371 to 287), a Greek philosopher on the island of Lesbos and a pupil of Aristotle, wrote about a vegetable dye produced on the same Greek island where purple is said to be used for the first time: "An alga grows abundantly on Crete, on a rock near the shore, enabling dyeing not only of ribbons but also of wool and garment cloth. It is more beautiful than purple when the dye solution is fresh." This coloring matter was orcein, similar to litmus. Rather than being obtained from algae it was from lichen, in this case from *Variolaria rocella*.

The extraction is carried out with ammonia at 50°C for six to seven days. Blue orchil is formed first followed by red orchil. When the orchil paste that has formed is dried and ground the redder Cudbear is obtained. Litmus is obtained from both these species in a similar manner to the extraction of orseille but with prolonged boiling (forty days) in the presence of potassium carbonate.

It was in 1766 that Dr. Cuthbert Gordon patented Cudbear (the name deriving from his mother's name). It is one of only two natural dyes whose discovery has been credited to an individual, the other being Quercitron, credited to, and patented in 1775, by Bancroft.

Cutch, C.I. Natural Brown 3 *Acacia catechu* and *Areca catechu*
Fabaceae

Dyers and tanners employ numerous varieties of cutch. The extract is taken from the wood of khair trees (*Acacia catechu*) which are found in India and Burma, and it represents the commonest variety of cutch on the Asian market and is known as Bengal catechu. The color was also obtained from the Asian palm (*Areca catechu*) that yields the betel nut. However, there are several other catechu derivations: Gambier catechu is an extract of the leaves and twigs of a bush, *Uncaria gambir* (Rubiacea), cultivated in Malacca, Penang, and Singapore; New catechu is a European extract of coniferous trees and contains tannic acid.

Catechu was used extensively (ca.1890) on cotton for fawns, browns, and blacks. For blacks the cotton was first dyed with indigo and topped up with catechu and chrome. It was the only black available at that time to give reasonable fastness.

Huebner recommended catechu for heavy browns, which are remarkable for their fastness. It was used in the dyeing of ships' sails due to its preserving action, preventing seawater from rotting the cotton, and it was also less likely to be affected by mildew.

Quercitron, C.I. Natural Yellow 10 *Quercus velutina*
Fagaceae

This is one of the two natural dyes whose discoverer is known. Bancroft reputedly introduced its use in 1775, the Colour Index stating the year of discovery as 1784. It is the ground inner bark of the *Quercus velutina*, a North American oak indigenous to the eastern United States. Bark liquor is its aqueous extract. Patent bark, or Quercetin industrielle (which was first manufactured by Leeshing in 1855), is obtained by boiling Quercitron bark with dilute sulfuric acid and is three times as concentrated as bark liquor. Flavine yellow shade is prepared by extracting the bark under high-pressure steam and consists mainly of Quercitrin, and flavine red shade, consisting mainly of Quercetin, is similarly obtained by rapidly extracting the bark with dilute ammonia and then boiling the extract with sulfuric acid.

Goldenrod *Solidago canadensis* and *virgaurea*
Asteraceae

This is a 2 to 5 foot perennial with small yellow flowers that grows wild in many parts of North America and Europe. The flower heads are broken off, with some stem, just as they come into flower. Although the whole plant can be used it is the flower heads that contain the majority of the dyestuff. The plant also contains tannin.

The generic name comes from solidare, as the plant is known as a vulnerary, or one that "makes whole." As well as its use as a dyestuff and tanning agent it is also used to treat ailments, as a diuretic, an astringent, and, among other things, in the treatment of diphtheria. The species *virgaurea* is the only one of over eighty species that is native to the United Kingdom.

Pokeweed *Phytolacca americana*
Phytolaccaceae

This plant is also known as pokeberry, scoke, pigeon berry, coakum, inkberry, and many others. It is a perennial that grows from a large, poisonous root to between 3 to 10 feet tall. The leaves are 5 to 12 inches long, and it has white flowers growing in terminal racemes, each floret having 4 to 5 sepals. These develop into dark purple, ten-ribbed berries. They are a favorite food for birds (hence known as pigeon berry) but are poisonous to humans.

Pokeweed grows east from Ontario in Canada through to Maine and south to Texas, Florida, and Mexico. It has also been naturalized in many parts of Europe but whether accidentally or on purpose is not known. The Navajo Indians and other tribes used the root bark of pokeweed together with that of plum to obtain a purple-brown color. The reddish-tinged leaves and the stems can also be used as a dyestuff.

References and Further Reading

Plant fibers

Baines, P. 1985. *Flax and Linen*. Princes Risborough: Shire.
Barber, E.J.W. 1992. *Prehistoric Textiles: The Development of Cloth in the Neolithic and Bronze Ages with Special Reference to the Aegean*. Princeton, NJ: Princeton University Press.
Bayer Farben Revue (1964–1986). Leverkusen, Germany: Bayer AG.
Beckett, J.C. 1966. *The Making of Modern Ireland 1603–1923*. London: Faber & Faber.
Brink, M. and Escobin, R.P. 2003. *Fibre Plants. Plant Resources of South-East Asia 17*. Leiden: Backhuys Publishers.
Burton, Anthony. 1984. *The Rise and Fall of King Cotton*. London: André Deutsch.
CIBA Review (1937–1972). Basle, Switzerland: CIBA Limited.
Cook, J.G. 1968. *Handbook of Textile Fibres*. Watford: Merrow.
Dempsey, J.M. 1975. *Fiber Crops*. Gainsville: University of Florida Press.
Dewey, L.H. 1943. *Fiber Production in the Western Hemisphere*. Washington, DC: United States Government Printing Office.
Dewilde, B. 1987. *Flax in Flanders Throughout the Centuries: History, Technical Evolution, Folklore*. Tielt: Lannoo.
Dodge, C.R. 1897. *Useful Fiber Plants of the World*. Washington, DC: United States Government Printing Office.
Harris, J. 1993. *5000 Years of Textiles*. London: British Museum Press.
Hollen, N. 1988. *Textiles*. New York: Macmillan.
Jarman, C. 1998. *Small-scale Textiles: Plant Fibre Processing, a Handbook*. London: Intermediate Technology Publications.
Jenkins, D. 2003. *The Cambridge History of Western Textiles. Two Volumes*. Cambridge: Cambridge University Press.
Kirby, R.H. 1963. *Vegetable Fibers: Botany, Cultivation and Utilization*. New York: Leonard Hill.
Mauersberger (Ed.). 1954. *Matthews Textile Fibers*, (6th ed.). New York: Wiley.
Montinola, L. 1991. *Piña*. Manila: Amon Foundation.

Moore, Alfred S. 1922. *Linen* (vol. 3): *Staple Trades and Industries*. London: Constable.

Munro, J.M. 1987. *Cotton*. Harlow: Longman.

Royle, J.F. 1855. *The Fibrous Plants of India*. London: Smith, Elder, and Co.

Smith, F.W. 1876. *The Irish Linen Trade Handbook and Directory*. Belfast: W.H. Greer.

Smith, C.W. and Cothren, J.T. 1999. *Cotton: Origin, History, Technology and Production*. New York: John Wiley.

Stout, E.E. 1970. *Introduction to Textiles* (3rd ed.). New York: Wiley.

Trotman, S.R., and E.R. Trotman. 1925. *The Bleaching, Dyeing and Chemical Technology of Textile Fibres*. London: Charles Griffin.

Dye plants

Adrosko, R.J. and Furry, M.S. 1968. *Natural Dyes in the United States*. Washington, DC: Smithsonian Institution Press.

Balfour-Paul, J. 1990. *Indigo*. London: British Museum Press.

Balfour-Paul, J. 1997. *Indigo in the Arab World*. Richmond, Surrey: Curzon.

Bayer Farben Revue (1964–1986). Leverkusen, Germany: Bayer AG.

Bliss, A. 1981. *A Handbook of Dyes from Natural Materials*. New York: Scribner.

Bliss, A. 1993. *North American Dye Plants*. Loveland, Co: Interweave Press.

Böhmer, H. 2002. *Koekboya: Natural Dyes and Textiles: a Colour Journey from Turkey to India and Beyond*. Ganderkesee: Remhèob.

Buchanan, R. 1999. *A Weaver's Garden: Growing Plants for Natural Dyes and Fibers*. Mineola, NY: Dover Publications.

Cannon, J.F.M., Cannon, M.J. and Dalby-Quenet, G. 1994. *Dye Plants and Dyeing*. Portland, OR: Timber Press.

Cardon, D. 2003. *Le Monde des Teintures Naturelles*. Paris: Éditions Belin.

CIBA Review (1937–1972). Basle, Switzerland: CIBA Limited.

Chenciner, R. 2000. *Madder Red: a History of Luxury and Trade*. Richmond, Surrey: Curzon.

Dean, J. 1999. *Wild Color: the Complete Guide to Making and Using Natural Dyes*. New York: Watson-Guptill Publications.

Delamare, F. and Guineau, B. 2000. *Colors: the Story of Dyes and Pigments*. New York: Harry N. Abrams.

Dronsfield, A. and Edmonds, J. 2001. *The Transition from Natural to Synthetic Dyes: 1856–1920. Historic Dyes Series, no. 6* Little Chalfont: J. Edmonds.

Edmonds, J. 2003. *Medieval Textile Dyeing. Historic Dyes Series, no. 3* Little Chalfont: J. Edmonds.

Fereday, G. 2002. *Natural Dyes*. London: British Museum.

Fraser, J. 1985. *Traditional Scottish Dyes: and How to Make Them*. Edinburgh: Canongate.

Gibbs, P.J. and Seddon, K.R. 1998. *Berberine and Huangbo: Ancient Colorants and Dyes*. London: British Library.

Green, C.L. 1995. *Natural Colourants and Dyestuffs: a Review of Production, Markets and Development Potential. Non-Wood Forest Products, 4* Rome: Food and Agriculture Organization of the United Nations.

Grieve, M. *A Modern Herbal*, www.botanical.com

Herzog, I. and Spanier, E. 1987. *The Royal Purple and the Biblical Blue: Argaman and Tekhelet: the Study of Chief Rabbi Dr. Isaac Herzog on the Dye Industries in Ancient Israel and Recent Scientific Contributions*. Jerusalem: Keter.

Hurry, J.B. and Dawson, W.R. 1930. *The Woad Plant and its Dye*. London: Oxford University Press.

Lemmens, R.H.M.J. and Wulijarni-Soetjipto, N. 1991. *Dye and Tannin-producing Plants. Plant Resources of South-East Asia 3*. Wageningen: Pudoc.

Nieto-Galan, A. 2001. *Colouring Textiles: a History of Natural Dyestuffs in Industrial Europe*. Dordrecht: Kluwer.

Noble, E. 1998. *Dyes & Paints: a Hands-on Guide to Coloring Fabric*. Bothell, WA: Fiber Studio Press.

Polakoff, C. 1980. *Into Indigo: African Textiles and Dyeing Techniques*. Garden City, NY: Anchor Press/Doubleday.

Ponting, K.G. 1980. *Dictionary of Dyes & Dyeing*. London: Mills & Boon.

Rawson, G., Gardner, W.M. and Laycock, W.F. 1918. *A Dictionary of Dyes, Mordants, and other Compounds used in Dyeing and Calico Printing*. London: C. Griffin.

Rieske, W.M., Colton, M.-R.F., Bryan, N.G. and Young, S. 1994. *Navajo and Hopi Dyes*. Salt Lake City, UT: Historic Indian Publishers.

Robertson, Seonaid M. 1973. *Dyes from Plants*. New York: Van Nostrand Reinhold Company.

Rowe, F.M. (Ed.). 1924. Section B—Natural Organic Dyestuffs. In *Colour Index* (1st ed.). Bradford: The Society of Dyers & Colourists, 292–302.

Sandberg, G. 1989. *Indigo Textiles: Technique and History*. Asheville, NC: Lark Books.

Sandberg, G. 1997. *The Red Dyes: Cochineal, Madder, and Murex Purple: a World Tour of Textile Techniques*. Asheville, NC: Lark Books.

The Society of Dyers & Colourists. (Ed.). *Colour Index* (3rd ed.). Natural Dyes, Vol. 3. Bradford: The Society of Dyers & Colourists and North Carolina: The American Association of Textile Chemists and Colorists, 1971.

Weigle, P. 1974. *Ancient Dyes for Modern Weavers*. New York: Watson-Guptill Publications.

16
Wood

TONY RUSSELL

Wood is strong but flexible and easily worked, and is one of the most widely used natural products. Archaeological evidence for its use as fuel goes back at least 700,000 years, to the first use of fire by *Homo erectus*. Evidence for the use of wood in tools can be inferred from Neanderthal spear points, dating to about 160,000 years ago, which must have had wooden shafts. A wide range of wooden artifacts, including a paddle, and a wooden platform made of poplar, were found at the Mesolithic site of Star Carr, in northern England (ca. 7500 ^{14}C years BC). Wooden artifacts are common at waterlogged sites from this period onwards.

Natural regeneration and the establishment of plantations mean that timber is potentially a highly renewable resource. However, the strong demand for tropical hardwoods has outstripped the supply. Typically, as one species declines beyond economic viability, trade will switch to another species. There are estimated to be 100,000 tree species of which some 7,300, mainly in tropical areas, are threatened with extinction. Clear-felling (the complete removal of trees from an area) of temperate forests (i.e., in Russia, Canada, and the United States) can also be highly damaging to ecosystems, particularly where old-growth forests are being cut. Trees are not only an important part of biodiversity in themselves, but also form a very important component of forest ecosystems. For example, the great apes of west and central Africa are now severely threatened by the illegal logging of their forest habitat.

Tree conservation now focuses on promoting "wise use," in which harvesting of wild species is not banned, but instead managed so as to ensure sustainability. Conservation practices that involve local communities are more likely to be successful while concurrently promoting economic development. Over and beyond providing timber, forest environments are often an economically important source of other plant and animal resources, known as Non Timber Forest Products (NTFP). Harvesting of NTFP, combined with controlled cutting of trees, can offer long-term economic benefits that outweigh those from uncontrolled logging. Research centers, such as the International Center for Research in Agroforestry in Kenya and the Center for International Forestry Research, are also active in developing the cultivation of threatened species. There is particular interest in agroforestry, in which the cultivation of trees is carried out on-farm, as an integral part of crop cultivation.

By far the greatest threat to tropical hardwoods is demand from consumers in developed countries, mainly Japan, Europe, and North America. Even where forest protection laws exist in source

315

Ulmus Americana. From *The American Woods,* vol. 2, published between 1888 and 1910 by Romeyn Beck Hough.

countries, lack of import controls means that illegally-logged timber can be freely traded. However, in the 1990s, increasing demand from consumers for sustainably traded products led to the establishment of non-governmental certification schemes. Led by the Forest Stewardship Council (FSC), such schemes are increasingly popular. However, a significant change in forestry practices is unlikely to come about unless western governments tackle illegal imports and adopt sustainable sourcing themselves.

Woods are conventionally divided into two groups: hardwoods, derived from angiosperms (flowering plants, usually deciduous), and softwoods from gymnosperms (conifers and other related groups, usually evergreen). These terms are sometimes misleading, as some angiosperm woods are soft and some gymnosperm woods hard.

Hardwoods

Aboudikro *Entandrophragma* spp.
Meliaceae

All eleven species of this large deciduous tree are used for timber. The most common are *E. angolense* and *E. cylindricum*. They grow in the lowland forests of tropical Africa from Angola and Ivory Coast in the west to Uganda in the east. Common in west Africa, the much smaller populations in east Africa are now threatened.

Aboudikro is a mahogany-type wood with a pinkish-yellow sapwood and light reddish-brown heartwood. It has an interlocked grain which, when quarter sawn, has striped or occasionally 'fiddleback' figuring. This interlocked grain can—and does—cause distortion when drying. It works reasonably well with hand and machine tools, although the interlocked grain does have a blunting effect.

Aboudikro has an attractive appearance and is used extensively for furniture and cabinetmaking, high class interior and external joinery, window frames, and doors. It is also used for boat and vehicle building and for piano cases. Selected logs are sliced for decorative veneers.

African blackwood, Mpingo *Dalbergia melanoxylon*
Fabaceae

This small, often multi-stemmed, misshapen tree from Mozambique produces the darkest timber of any tree, including ebony. Large trees are becoming rare due to heavy exploitation, though there is no imminent threat of extinction.

The grain of this wood is straight to irregular with a fine even texture which feels oily to the touch. It is an extremely dense, heavy wood, weighing almost as much as Lignum-vitae (*Guaiacum officinale*) and must be dried with great care to avoid splitting and checking. Once dry, however, it is very stable. Blackwood is hard to work (it splinters easily and quickly blunts cutting tools), but once smoothly finished, it will take a high polish.

The natural oils within Blackwood give it excellent resonance qualities and this is why it is often used in the production of woodwind instruments including clarinets, oboes, flutes, recorders, and the chanters for bagpipes. In Africa, it is extensively used to carve animal and human figures and chess pieces. The wood was imported into ancient Egypt and used for statues. Many other *Dalbergia* species are logged in Africa (including Madagascar) for export purposes and most are severely threatened.

Afzelia *Afzelia spp.*
Fabaceae

Afzelia occurs across the central band of Africa from Sierra Leone in the west to the Sudan in the east. It is in fact five species of tree sold as a single commercial timber. Although several of the species are common, they are now under pressure from heavy exploitation. All species produce a highly distinctive wood with a pale straw-colored sapwood and rich red-brown heartwood with a coarse texture. The grain is sometimes irregular and can contain yellow or white calcium deposits which cause staining in damp conditions.

This dense timber dries slowly but is as stable as teak, stronger than oak, and extremely durable. It is fairly hard to work and has a blunting effect on tools. *Afzelia* has a certain resistance to acid and has therefore been widely used in chemical works and laboratory benches. The many uses of this wood include internal and external joinery including window and door frames, staircases, office, and garden furniture.

African mahogany *Khaya* spp.
Meliaceae

The name "African Mahogany" relates to five different species of *Khaya* which grow throughout the equatorial evergreen forests of Central Africa. The timber is pale pink to reddish-brown with an interlocked grain which produces distinctive striped figuring. It is a fairly light wood which dries rapidly with little degradation and is very stable in use. African mahogany works well and can be stained and polished to an excellent finish.

The wood has long been a favored timber for furniture, office desks, cabinets, shop fittings, and high quality joinery for staircases, banisters, handrails, and paneling. It is a popular timber for boat manufacture, being light and moderately durable. Today some logs are rotary-cut to make high-quality plywood. This and other species in the genus are severely threatened by overharvesting. African mahoganies became popular in the 19th century as supplies of genuine, American mahogany declined.

Ash *Fraxinus* spp.
Oleaceae

Ash thrives throughout Europe, North America, and Japan and as a timber is commercially important in all these regions.

European Ash (*F. excelsior*) is white with a pink tint when first cut. Occasionally, dark brown heartwood may occur; this is marketed separately as "Olive Ash." Ash is a ring-porous wood with a conspicuous grain pattern on plain-sawn surfaces. It is straight-grained with a coarse texture. Ash dries rapidly with little degradation or shrinkage and is a strong wood with excellent shock resistance qualities. It works well with hand and machine tools, bends well under steam and can be brought to a good finish. It is not naturally durable, so is unsuitable for outside use unless treated.

Ash is extensively used for chair and cabinet making, but its ability to absorb shock loads has made it a firm favorite for sports equipment. Hockey sticks, baseball bats, cricket stumps, and gymnasium equipment are all made of ash. It has been used to make carpenter's tools since classical times, and is used today to make handles for striking tools such as axes, picks, hammers, and garden tools such as spades and rakes. Like oak and yew, ash has long had strong supernatural and ritual associations. For instance, ash is the wood preferred by the Iroquois Indians of North America when cutting their "false face" masks from trees.

Australian blackwood *Acacia melanoxylon*
Fabaceae

There are about 1,200 species of *Acacia,* mainly growing in the southern hemisphere. They produce a range of timbers varying in density and color. One of the most attractive of these is Australian blackwood which originates from southeast Australia and Tasmania where it can attain heights in excess of 100 feet (30 m), with a bole diameter of 3 feet (1 m).

Although called 'blackwood', the sapwood is a golden color giving way to a dark brown heartwood. It is generally straight-grained but, where it becomes wavy, it produces a fine 'fiddleback' figure. It is even-textured with a high natural luster, giving it a decorative appearance.

Australian blackwood dries evenly, saws easily, and is a fairly dense, heavy timber which bends well under steam. This decorative timber is in great demand for quality paneling, furniture, and cabinet making. 'Fiddleback' logs are often sliced for veneer and used to face interior doors and office and shop fitments.

Balsa *Ochroma pyramidale*
Bombacaceae

Balsa wood is probably best known for its lightweight properties; it is in fact the lightest wood in general use. "Balsa" is Spanish for "raft," referring to its excellent floatation qualities. The balsa tree occurs naturally throughout Central America from Mexico to Brazil, with the best wood coming from Ecuador. It has also been grown throughout Asia as a commercial timber crop.

Balsa is a very fast growing tree, easily reaching a height of 65 feet (20 m) in 15 years. Commercial balsa wood is the sapwood rather than the heartwood. This sapwood is a creamy-white color, often with a pinkish tinge. It is a very difficult timber to dry as it splits and warps easily, but once dry, it can be worked well to a smooth finish.

Probably the most famous use of balsa wood was for the Kon Tiki raft on which Thor Heyerdahl sailed from Peru to Polynesia. Its ability to float better than any other wood has meant that, for centuries, it has been used for lightweight boats, life-belts, floats, and buoys. It is also a firm favorite for model making and theatrical props.

Beech *Fagus sylvatica*
Fagaceae

Beech grows extensively in northern temperate woods across North America, Europe, and western Asia. In Britain, beech timber is used in larger quantities than any other hardwood. The timber is white to pale pinkish-fawn in coloring. It is straight-grained, with a fine, even texture and an attractive brown fleck radiating out from the center of the tree to the bark when quarter-sawn. Beech dries quickly, but tends to distort in the process. Once dry, it is not particularly stable, moving quite substantially in changing humidity levels. It works easily with both hand and machine tools and can be steam-bent to great effect.

The heartwood is perishable, making it unsuitable for outdoor use, however it does readily absorb preservative. Beech is an excellent general purpose, interior furniture timber and is used for school desks, chairs, cabinets, and bookcases. It is also a good turnery wood, being made into tool and brush handles, toys, brush backs, bobbins, kitchen utensils, and cutting blocks. It makes hard-wearing flooring for domestic use.

See: Nuts, Seeds, and Pulses, p. 134

Birch *Betula spp.*
Betulaceae

Of the thirty-five species of birch, three are favored for their timber: *B. alleghaniensis* (yellow birch), *B. papyrifera* (paper birch), and *B. pendula* (European birch).

The first two are confined to North America, the third to Europe. All three produce fine-textured, straight-grained wood which, in the main, has little figuring. Occasionally however, logs are found with a curly figure known as 'Flame Birch'.

The predominant coloring of both paper birch and European birch wood is creamy-white to light fawn. Yellow birch has a distinctive yellow sapwood and reddish-brown heartwood. All three

Paper birch (*Betula papyrifera*). USDA-NRCS PLANTS Database/Herman, D.E., et al., 1996. *North Dakota tree handbook*. USDA NRCS ND State Soil Conservation Committee; NDSU Extension and Western Area Power Admin., Bismarck, ND.

are fairly dense, with good strength properties comparing well with ash for toughness. The wood works well with both hand and machine tools and is excellent for turnery.

Birch has traditionally been used to produce bobbins, shuttles, dowels and spools for the textiles industry. Because of its high strength properties it makes excellent quality, structural plywood. The British Mosquito fighter bomber planes of World War II were built of birch ply. Selected logs are peeled for decorative veneers and small roundwood is predominantly used as pulp for paper-making.

See: Materials, p. 347

Boxwood *Buxus sempervirens*
Buxaceae

This much prized wood comes from a small evergreen, broadleaved tree which is native in Europe, from Britain (where its native status is uncertain) to Turkey and Iran. It has bright creamy-yellow coloring with a very fine even texture. The grain is more often than not irregular, especially from the curved misshapen stems which are characteristic of this tree's growth in Britain. It is a very dense, heavy wood which dries extremely slowly, and is prone to splitting. Once dried, however, it is an excellent turning wood and can be carved to give superb detail.

Boxwood is often used for carving chess pieces, for corkscrews, for shuttles for the silk industry, and for skittles and croquet mallets due to its dense, hard durability. It has long been a popular wood for wood-engraving blocks. Boxwood was prized by ancient Greek and Roman cabinet-makers, particularly for inlay.

Brazilwood *Caesalpinia echinata*
Fabaceae

The name 'brazil' has been used since the Middle Ages to describe plants that produce red dye. When the Portuguese colonized South America in the 16th century, they discovered a small tree which also produced red dye. They called this tree 'brazilwood' and named the land where they found it 'Brazil'. Harvesting from 1501 to the 1870s severely affected populations, and only a few natural stands survive, on the coastal plain. Brazilwood is a small tree with white sapwood and heartwood that turns from rich orange when freshly cut, to deep red as it matures. It has a fine, even texture with natural luster and a distinctive figuring of dark red-brown variegated stripes.

Brazilwood is a hard, heavy wood which needs to be dried very slowly to avoid splitting. Because of its weight, flexibility, and strength, it is regarded as the finest timber for violin bows and is also used for high quality gun butts, rifle stocks, and parquet flooring.

Cherry *Prunus* spp.
Rosaceae

There are almost one hundred true cherry species, growing mostly in the northern temperate regions of the world and countless cultivars and garden varieties. However, only two species are widely recognized for their timber potential. These are *P. avium* the European cherry or gean, and *P. serotina* the American cherry.

Cherry wood is a straight-grained wood with a fine, even texture. Initially pinkish-brown when first cut, the wood matures to a rich red-brown. It is quite difficult to dry, tending to warp and shrink, but once dry is reasonably stable in use. It works well with both hand and machine tools and can be polished to an excellent finish.

This attractive timber has long been favored as a decorative wood for special items of furniture and cabinets. It is also used to make tobacco pipes, detailed wooden toys, and figurines. Cherry wood peels well and makes an attractive veneer for cabinets and paneling.

See: Fruits, pp. 85–6, 86–7, and 87–8; Ornamentals, pp. 269–70

Common alder, Black alder *Alnus glutinosa*
Betulaceae

The thirty-five species of the *Alnus* genus have a wide distribution throughout the northern hemisphere. Common alder grows in Europe and the Near East and is a small- to medium-sized tree that does best in damp, moist growing conditions.

Freshly cut alder wood is bright orange-brown, which matures to a dull reddish-brown with a straight grain and fine texture but little figuring (the pattern of the grain). This medium dense timber dries quickly with little shrinkage. It machines easily, provided that tools are kept sharp. It is not particularly strong or durable but can be stained and polished to a good finish.

Alder makes good plywood, but the small size of the trees means the wood is mainly used in small-scale industry and for domestic purposes such as broom handles, brush backs and wooden toys. It has been used since medieval times for clog making, and charcoal for gunpowder-making. It has also been used for the manufacture of artificial limbs. The wood becomes very durable when immersed in water, and has been widely used as water-front piles and as foundations, including in Venice and under many medieval cathedrals.

Ebony *Diospyros spp.*
Ebenaceae

The name ebony encompasses all *Diospyros* species with black heartwood. They occur from tropical Africa and Madagascar through to India and Sri Lanka. The tree itself tends to be rather small, seldom achieving diameters in excess of 2 feet (0.6 m). Most larger trees have been logged, and the species is threatened in several African countries. *Diospyros crassiflora*, a lowland rainforest species, has the darkest heartwood, contrasting starkly with its creamy-white sapwood which is referred to as 'white ebony'. Timber from all species is usually straight-grained with a fine, even texture and extremely heavy. All ebony timber has to be dried slowly to avoid surface checking. Once dry, the wood is very stable, with good strength properties.

Black ebony has been in great demand since ancient Egyptian times. This distinctive wood has long been used for sculpture and carving and as well as for small, decorative items, i.e., handles for cutlery and pocket knives, butts for billiard cues, and door knobs. Traditionally, it was used for the small fittings on violins, organ stops, and the black keys of pianos and organs.

Elm *Ulmus* spp.
Ulmaceae

Elm occurs naturally throughout northern temperate regions of Europe, North America, continental Asia, and Japan. Eighteen species grow in different regions, but the characteristics of the wood are broadly similar. The heartwood is dull brown, sometimes with a reddish tinge, and has prominent, irregular growth rings and crossed grain which gives the wood a distinctive and attractive figuring.

Elm dries readily, but care needs to be taken because it is prone to severe distortion because of the irregular growth pattern. Once dry, it works reasonably well. Although not a particularly strong wood, it is very durable in waterlogged conditions. As far back as Roman times, it has been used as a conduit for water (the heartwood is bored out to create a basic drainpipe). The 'Rialto' in Venice stands on elm piles. Elm is used for ship building, weather-boarding, ladder rungs, and coffins. Its decorative appearance has meant that it has long been used for cabinet making, the seats of Windsor chairs, flooring, and as a turnery wood for bowls. Parts of the chariot in Tutankhamun's tomb (Egypt, ca. 1320 BC) were made from elm, probably *U. minor*.

Sadly, due to Dutch Elm Disease, elm timber is in short supply across much of Europe. In England, over ninety percent of mature elm trees were killed by the fungus *Ceratocystus ulmi* during the 1960s and 1970s.

Gaboon mahogany, Okoumé *Aucoumea klaineana*
Burseraceae.

The English name "Gaboon" derives from the West African state of Gabon where *A. klaineana* (the only species in the genus) is still found in abundance. However, plantation trials are underway in order to meet the heavy demand for this wood. It also grows in the Congo Republic and Equatorial Guinea. More Gaboon is exported than any other African wood.

Gaboon mahogany is a pink-brown even textured wood, with a straight grain and little character. It is a light timber which dries quickly with little degradation or movement but has little strength and poor durability. As gaboon contains grains of silica, which quickly blunts machine tools and saws, it is rarely sawn. Instead, it is rotary-cut—the first tropical hardwood to have been worked in this way—for veneer for the plywood industry. It is also used for the manufacture of blockboard and laminoboard. Gaboon ply is used for door skins, cabinets, paneling, and partitions.

Gmelina *Gmelina arborea*
Lamiaceae

Gmelina is native to India and Burma, where wild populations are declining. Over the last thirty years this medium-sized tree has been planted as a commercial timber species in many parts of the world, including Malaysia, West Africa, the West Indies, and Brazil. It grows very quickly, producing saw-mill sized logs, up to 15 inches (40 cm) diameter, within 10 to 15 years.

The timber is pale creamy-pink in coloring with a high natural luster. It has an interlocked grain and a rather coarse texture. Gmelina dries well, with little degradation and is stable in use. It saws easily and can be polished to a good finish. Despite its rapid growth rate, gmelina is strong and dense enough to be used for construction, general joinery, and furniture. As more plantations reach harvestable size, gmelina is being used more and more for pulp for paper-making and as a plywood timber.

Greenheart *Chlorocardium rodiei*
Lauraceae

Greenheart is the major commercial timber of Guyana in South America. It is an evergreen tree which can grow to a height of 130 feet (40 m). It has a pale cream sapwood which gradually changes to a yellow-green or olive-brown heartwood, sometimes with black markings. It is a fine, uniformed, textured wood with a straight or sometimes interlocked grain. The tree is considered to be vulnerable as regeneration is very slow.

Greenheart needs to be dried slowly; otherwise it will split. It is exceptionally strong and consequently difficult to work. It splinters easily and toxins in the wood may cause septic wounds to those working it. Greenheart is very durable, particularly in sea-water. For centuries it has been used for ship-building and marine construction such as docks, lock gates, pier decking, wharves, jetties, and bridges.

Hickory *Carya* spp.
Juglandaceae
There are over twenty species of *Carya*, all native to North America, but only four of these are harvested commercially for their timber, namely *C. glabra*, *C. laciniosa*, *C. ovata*, and *C. tomentosa*. All are medium to large trees with dense, coarse-textured, creamy-gray sapwood and red-brown heartwood. Although difficult to dry, hickory is very stable in service. Its main attribute is its ability to withstand crushing and impact forces. This makes hickory ideal for the handles of striking tools such as picks and axe handles and sledgehammer handles for which it is the white sapwood which is favored. It is also extensively used for sports equipment such as tennis rackets, skis, golf clubs, baseball bats, and lacrosse sticks.

Traditionally hickory chips have been used for smoking food such as hams and fish. There has been renewed interest in hickory for this purpose in recent years.

See: Nuts, Seeds, and Pulses, p. 138

Hornbeam *Carpinus betulus*
Betulaceae
A native of Britain and Europe through to Turkey and Iran, the hornbeam is a medium-sized tree similar to beech in appearance.

It has a uniform, dull white sapwood and heartwood with an irregular grain which is normally due to its fluted and frequently misshapen trunk. Hornbeam dries quickly and well, with little degradation, but it moves considerably under conditions of changing humidity. It is a heavy, dense wood, resistant to splitting but fairly difficult to work, having a blunting effect on both machine and hand tools. When used in turnery, hornbeam can produce a smooth, fine finish. Before environmental health requirements stipulated that food preparation had to take place on impervious surfaces, hornbeam was used extensively for butcher's blocks. Today it is used for drum sticks, snooker cues, skittles, mallets, and wooden pegs. Prior to the introduction of lignum vitae (*Guaiacum officinale*) in the 16th century, hornbeam was used for cogs, pulleys and bowls.

Horse chestnut *Aesculus hippocastanum*
Hippocastanaceae
Originally from the Balkan peninsula or Anatolia, this tree has been widely planted across Europe and North America as an ornamental in the last three hundred years; as a commercial timber it is of minor importance.

The wood is creamy-white in coloring (not dissimilar to holly) with an irregular grain. It dries well with little degradation or movement and can be worked easily with both hand and machine tools. It is however a structurally weak timber and fairly brittle. It has a low resistance to fungal attack. It is used for general turnery, producing such things as brush backs and handles, kitchen utensils, storage racks, trays, and boxes. The glossy-brown nuts are widely used in Britain for the children's game of conkers, popular since the mid-19th century. (A nut [conker] is threaded onto a string, and the two players, each using their own conker, take alternate strikes at the other's conker until one is broken by the other.)

See: Plants as Medicines, p. 215

Iroko *Milicia excelsa*
Moraceae

This important African tree grows throughout tropical Africa and is a popular substitute for teak, despite being not as strong as teak. West Africa is now the major source, as over-exploitation has greatly reduced populations in east Africa. It is a large tree, producing good quality, cylindrical logs, which are exported throughout the world.

The wood has pale yellow sapwood and deep brown heartwood with light fawn flecks radiating out from the center of the wood to the bark, when quarter-sawn. The grain is quite often inter-locked and irregular. Generally, the wood works well with both machine and hand tools, although occasional calcium deposits known as 'stone' can damage the cutting teeth of saws. Iroko is extremely durable and is widely used in ship building, piling, marine work, and garden furniture. It is also a favorite wood for sculpture, wood-carving, and parquet flooring.

Katsura *Cercidiphyllum japonicum*
Cercidiphyllaceae

This extremely attractive tree originates from Japan, China, and Korea and produces highly prized timber. It has warm nut-brown coloring with fine even texture, straight grain and a high natural luster. It is a rare tree, now threatened by over-exploitation.

Katsura dries easily without distortion or degradation and is stable in use. Although not strong, it is a delight to work with both hand and machine tools and finishes to a very smooth surface.

This is the ideal wood where detail is required such as in engraving, molding, carving, and foundry patterns. It is widely used for high-class cabinet making, interior joinery, and in the manufacture of pencils, cigar boxes, and the Japanese shoes called 'Geta'.

Keruing, Gurjun, Yang *Dipterocarpus* spp.
Dipterocarpaceae

There are some seventy different species of *Dipterocarpus* producing timber in south and southeast Asia. They are commercially sold under a group name depending on the country of origin; keruing in Malaysia and Indonesia, gurjun in India and Burma and yang in Thailand. Many species are critically endangered by habitat loss, and research into cultivation is urgently needed.

Depending on the species, the heartwood varies in color from light to dark brown. They are all generally straight-grained with a coarse but open texture and rather plain appearance. Several species exude a sticky resin which is a source of an essential oil, gurjun balsam, used in perfumery as a substitute for patchouli oil from *Pogostemon cablin*. The density of the wood varies with species. They are all quite difficult to dry without causing splitting or cracking and even when dry, are not very stable in use. Strength qualities are also not exceptional but because of their commercial availability and reasonable cost, keruing, gurjun, and yang are used extensively for building work. They are particularly useful for cladding, window frames, and sills. Selected logs are rotary-cut for plywood and veneers.

Lignum-vitae *Guaiacum officinale*
Zygophyllaceae

The literal translation of Lignum-vitae is "The Wood of Life." It was given this name in the 16th century when it was first discovered in the West Indies because its resin was believed to cure many diseases. Lignum-vitae is the collective timber name for three different species of *Guaicum* which occur from Florida and the Bahamas southwards to Columbia and Venezuela. The tree is extremely slow-growing, and poor regeneration has led to extinction on some Caribbean islands.

The timber is a very distinctive greenish-black color with a closely interlocked grain and a fine, even texture. It has a high resin content and is one of the heaviest of all timbers weighing some eighty percent more than oak. Lignum-vitae is a very strong, exceptionally hard and durable wood, with a high resistance to abrasion. It dries very slowly and is very difficult to work because of its hardness. However, its self-lubricating properties make it ideal for bearings for ships' propeller shafts. It has long been used to make flat green and crown green bowls. At one time it was the favored timber for truncheons carried by the British police force.

Mangrove *Rhizophora* spp.
Rhizophoraceae

Mangrove trees occur throughout tropical coastlines, but only three species grow large enough to be commercially harvested for their timber. Mangrove is a reddish-brown, heavy, hard, fine-textured wood. It is difficult to dry as it shrinks considerably and tends to split readily. Once dry, it is hard to work, blunting cutting tools quickly.

Over the centuries it has been a useful timber for local use in the countries where it is found. The wood has been used to make charcoal, and for fuel and building poles. The bark is an excellent source of tannin. Today it is quite often chipped at source and then transported to Japan for paper-making.

Maple, Sycamore *Acer* spp.
Aceraceae

This is a deciduous genus of 110 species that grows across North America, Europe, and Asia. In North America there are two main types producing timber; rock maple (*A. saccharum*, also the source of maple syrup) and red maple (*A. rubrum*). In Europe the main species is sycamore (*A. pseudoplatanus*).

The timber of all three is creamy/white in coloring (rock maple being the darkest), with a fine, even texture and natural luster. Although usually straight-grained, both rock maple and sycamore may sometimes produce wavy grain which in turn produces a very attractive 'fiddleback' appearance.

All three timbers are of medium density and dry well with little movement or degradation (i. e., splitting, cracking or shrinking during drying process). They are not durable and hence not suitable for use outdoors unless treated. Sycamore has been traditionally used for musical instruments including lutes and violins, and for joinery and polo mallets. 'Fiddleback' sycamore is still used for violin-backs today. Rock maple is very hard and has long been used for heavy industrial flooring, dance halls, squash courts, and bowling alleys. Soft maple is eminently suitable for furniture, interior joinery, and wooden utensils.

Acer campestre (field maple) grows wild throughout Europe and is the only species native to Britain. As the wood does not stain or taint food, it was widely used for culinary utensils, such as Neolithic ladles found in the Swiss lake villages.

Maranti, Balau *Shorea* spp.
Dipterocarpaceae

Maranti is a generic name for the wood of many species of *Shorea* which occur in Indonesia, Malaysia, and Sarawak. Some species are highly endangered, in part by logging, and in part by habitat loss.

The wood is pale pink to red in coloring with coarse texture and an interlocked grain which produces darker stripes on quarter-sawn surfaces. It is a light wood which dries rapidly without serious degradation. It can be worked easily with both hand and machine tools and produces a good finish. Maranti is used for light structural work, interior joinery, domestic flooring, and basic furniture. It is also sliced for plywood and peeled for decorative veneers for paneling.

Muhuhu *Brachylaena huillensis*
Asteraceae

This medium-sized east African tree grows along the coastal belt of Tanzania and Kenya. The wood is very hard, heavy, and dense and is yellow-brown in color. The texture of the wood is fine and even, and quite often there is an irregular or interlocked grain which gives an attractive delicate appearance. The timber needs to be dried slowly to avoid surface cracks and splitting, but once dry, it is very stable and extremely resistant to abrasion. It is difficult to work and quickly blunts machine and hand tools. The tree contains aromatic oil and when cut, gives off a pleasant, spicy aroma. This oil is distilled and sold as an alternative to sandalwood oil.

Due to its characteristic twisting and forking (the tree seldom has a straight trunk), muhuhu wood is rarely available in long lengths which limits its use. It is widely used for flooring blocks where footfall is likely to be heavy. Over-harvesting is seriously threatening this tree, particularly in Kenya where it is used for carving wooden animals for the tourist market. A UNESCO-sponsored project is helping wood-carvers switch to sustainably harvested softwood species.

Oak *Quercus* spp.
Fagaceae

There are more than four hundred different species of oak, mostly occurring within the temperate regions of the world. Collectively, the most important group for timber production is known as the "white oaks" and includes *Q. petraea* and *Q. robur* (English Oaks), *Q. alba* (American White Oak), and *Q. mongolica* (Japanese or Mongolian Oak).

White oak typically has a creamy-fawn sapwood and yellow-brown heartwood. It is a coarse-textured wood, normally straight-grained with a characteristic silvery-gray figuring when quarter-sawn. White oak is without doubt one of the world's most popular timbers and has been so for centuries. In Britain, it was used to build the ships of the British Royal Navy for over four hundred years, hence the saying and song "Hearts of oak." Because it is strong but easy to cleave, oak was used as the main structural building material for barns, dwellings, and cathedrals throughout Europe.

Oak has always been an important wood for furniture and cabinet making, sculpture, and carving. Because the wood is relatively impermeable, oak has been used for whisky, sherry, and brandy casks. Oak has also been used since at least classical times for boat construction.

Bog oak results from the natural burial of oak trees during the formation of peat bogs, particularly in Ireland. The jet black, heavy wood can be up to seven thousand years old and has been the source of popular Irish souvenirs since the early 19th century. However, the main importance of bog oak lies in the use of measurements of tree ring width in developing dendrochronology, widely used for dating ancient timbers and as an indicator of past climates.

See: Nuts, Seeds, and Pulses, pp. 133–4; Materials, pp. 347–8

Pau Marfim *Balfourodendron riedelianum*
Rutaceae

This South American tree is the only species of this genus, and occurs in southern Brazil, Paraguay, and northern Argentina. It has only recently begun to be exported to North America and Europe. Common along the banks of the Paraná and Uruguay River systems, the tree is still abundant.

The timber has a creamy-yellow appearance with few distinctive features and little to distinguish between the sapwood and heartwood. It normally has a straight grain, even texture, and a natural luster.

Pau Marfim is a dense, heavy wood similar to hickory. It dries well, is stable, and can be easily sawn and worked. This wood is extremely strong and particularly resistant to impact, making it a favorite timber to be used in the manufacture of handles for striking tools such as sledge-hammers. In South America, it is used for construction flooring, furniture making, and turnery.

Poplar, Cottonwood, Aspen *Populus* spp.
Salicaceae

There are scores of different poplar species and cultivated varieties growing throughout northern temperate regions of the world. The timber from all thirty-five species varies little, although locally, they may be known by a distinctive name. In North America, poplar is known as cotton-wood or aspen.

Poplar is a fast-growing deciduous tree, in some cases reaching 100 feet (33 m) tall. The heart-wood is creamy-white to pale straw-colored. It is generally straight-grained with a fine, even tex-ture. It is a light wood which dries readily with little cracking or splitting and is worked easily with hand or machine tools. Although not strong, poplar is tough for its weight and has been an impor-tant commercial timber for many years.

Selected grades are used for interior joinery, furniture frames, toys, and turnery. At one time, vast quantities of poplar wood were used to make matches and match-boxes. This market has declined dramatically over the last thirty years mainly due to the demise of domestic open fires. Today, poplar is widely used to make crates, boxes, and pallets. It also peels to give a good veneer used for fruit baskets and general purpose plywood. In the 15th–16th centuries, poplar was used in England for arrow shafts, as well as wagon and cart bottoms. In the Near East today, the long, straight trunks of poplar are widely used as beams for the flat roofs of mud-brick houses.

Ramin *Gonystylus bancanus*
Thymelaeaceae

This medium-sized tree grows widely in freshwater swamps in Sarawak, Malaysia, and Southeast Asia. The low rate of regeneration in over-exploited forests is a cause for concern. Both sapwood and heartwood are of a creamy-white to pale yellow color. It is generally straight-grained with a fine, even texture, but discolors quickly and needs to be treated immediately after sawing. It dries easily, but tends to split if dried too quickly. It works well with machine and hand tools and finishes well.

Ramin is widely used as a substitute for beech in furniture making. Although a little lighter than beech, its strength compares favorably and, like beech, it is not suitable for external use. Ramin is also extensively used for picture frames, toys, and louver doors. It is used for light construction work and domestic flooring. Selected logs are rotary cut for plywood.

Rosewood *Dalbergia* spp.
Fabaceae

About one hundred species of *Dalbergia* are widely distributed throughout tropical regions. The best-known species, originating from Brazil, is rosewood. This has long been regarded as one of the world's finest timbers. Sadly, centuries of commercial logging means indigenous rosewood timber is becoming increasingly rare and *D. nigra* is now CITES I (Convention on Trade in Endangered Species) listed, banning international trade in the wood. However it has been cultivated in com-mercial plantations in India, Madagascar, and Honduras.

Rosewood has creamy-white sapwood which contrasts dramatically with deep chocolate-brown to purple-black heartwood. True Brazilian rosewood (*D. nigra*) is strongly figured, not so with rose-wood from India and Honduras (Indian rosewood *D. latifolia)* which has a pinkish hue and much finer figuring. The grain in both wild and cultivated supplies is straight to wavy with a coarse tex-ture which is oily to the touch.

Rosewood dries slowly with little problem and is not unduly difficult to work. When sawn, it gives off a pleasant rose-like fragrance. For more than two hundred years it has been treasured for high quality furniture, cabinet making, and as inlay for jewellery and music boxes. It is also an excellent carving and turning wood. Specialized uses include cutlery handles, keyboards for marimbas, and xylophones.

Sandalwood *Santalum album*
Santalaceae

True sandalwood originates from southern India where it has been highly prized for centuries both for its wood and its oil content. Although the export of sandalwood from India is banned, smuggling is large-scale, and the species is considered vulnerable. Several other species of *Santalum* occur in Australasia, all of which produce an aromatic wood which may be used as a substitute for Sandalwood oil.

Santalum album is a small parasitic tree, which lives on the roots of other trees. When it is harvested, every piece is removed, including that growing below ground. The wood, which is a pale yellow-brown color, has a very fine, even texture and a straight or irregular grain. It is a dense, heavy wood and, once cut, gives off a strong aroma and has a slightly oily feel. Sandalwood dries slowly but does not degrade. Once dry it can be sawn and worked easily. Heartwood chips and shavings are distilled to produce sandalwood oil which is used throughout the world for perfume. The timber itself is an excellent carving wood and can be worked to very fine detail. Decorative knife-handles, picture frames, and carvings made from sandalwood are sold to tourists throughout Asia.

See: Fragrant Plants, p. 252

Sweet chestnut *Castanea sativa*
Fagaceae

This magnificent tree is a native of southern Europe and was probably introduced into Britain by the Romans. Since then it has become naturalized across much of the country.

The heartwood is pale brown, resembling oak, and has prominent growth rings and fine rays. Chestnut frequently has a spiral grain and, as a large tree, has a tendency to be badly shaken, causing it to split when felled. Although lighter than oak, it does not dry well, tending to collapse and warp, but its natural durability makes it an ideal timber for outside use. For centuries, chestnut has been grown as coppice, the resulting poles being cleaved and made into fencing pales. Furniture and coffin boards are frequently made out of chestnut as a substitute for oak.

See: Nuts, Seeds, and Pulses, p. 136

Tasmanian oak *Eucalyptus delagatensis*
Myrtaceae

Tasmanian oak grows in south eastern Australia from New South Wales to Tasmania. The timber bears a slight resemblance to European oak with a heartwood color which varies from pale fawn to light pinkish-brown. However, it lacks any silvery grain and does not have the tiny pinhole-like hollows in the growth ring, as does true oak. The wood is usually straight-grained, coarse-textured with a slight natural luster. It dries readily but quite often develops surface checks and suffers some shrinkage and distortion.

"Tasmanian Oak" works satisfactorily with both hand and machine tools and can be polished to an excellent finish. It is moderately dense and is popular for interior and exterior joinery, building construction, cladding, and weatherboards. Other uses include making furniture, sports equipment, and domestic flooring. It is favored for pulp by the Australian paper-making industry.

Teak *Tectona grandis*
Lamiaceae

Teak grows naturally in Burma, India, Thailand, Indonesia, and Java. It is widely cultivated there, and has also been introduced into Central America and tropical Africa as a commercial crop. It is a large tree which can reach 150 feet (45 m) in height and a diameter of 8 feet (2.4 m). The sapwood is a pale yellow color and the heartwood a rich dark brown. It has straight grain with a rough, uneven texture and feels oily to the touch. Trade in teak is large-scale and dominated by Burma, India, Thailand, and Indonesia. Teak harvesting from the wild is restricted in many countries and this, combined with cultivation, means that teak is not endangered.

Teak is a medium weight timber that is heavier than mahogany but lighter than oak, very strong, and naturally durable. It needs to be dried slowly to avoid degradation, but once dry is very stable. Teak works fairly well with hand and machine tools but blunts cutting edges rather quickly.

Teak has been used for a multitude of purposes over the years. Traditionally, it has been used within the maritime industry for shipbuilding, decking, oars, and masts. It has also been long favored for furniture and cabinet making. It is a high-class joinery timber which is extensively used for window and door frames, staircases, and paneling. More recently, it has become widely used to make garden and patio furniture. It has a resistance to acid and therefore is quite often used for laboratory benching. Because the wood is brittle, it is not used for tool handles and other applications requiring high resilience.

Walnut *Juglans* spp.
Juglandaceae

Of the fifteen species of walnut, two are important timber-producing trees, *J. nigra* (the American walnut) and *J. regia* (the European walnut). European walnut is native to the Black Sea and Caspian region, but spread early to the Mediterranean. Although grown in Britain by the Romans, it appears not to have become widespread there until the Medieval period. Both species are big trees attaining heights of up to 100 feet (33 m), and have rich brown heartwood with a natural wavy grain which produces very decorative figuring. The wood needs to be dried carefully to avoid splitting and degradation, but once dry there is little movement in service. It is easy to work and can be polished to a superb finish.

This beautiful timber is one of the world's outstanding decorative woods and has been used to make high-class cabinets and furniture since the Queen Anne period (in the early 1700s). It has also long been popular for attractive rifle butts and gun-stocks. Today it is used extensively as a decorative veneer for doors, paneling, reproduction desks, and fascias on expensive cars.

See: Nuts, Seeds, and Pulses, p. 141

Zambezi redwood *Baikiaea plurijuga*
Fabaceae

Sometimes called Rhodesian teak, this medium-sized tree grows in Zambia and Zimbabwe, in lowland tropical forest on the Kalahari sands. It is not a true teak, nor does it work like teak, but it does share similar properties. Although the tree is abundant, logging is threatening the overall forest ecosystem which is being converted to grassland.

Zambezi redwood is an attractive rich red-brown wood with black flecks or lines. The grain is usually straight or interlocked, however the variation of coloring within the heartwood gives the appearance of figuring. It has a fine, even texture and a smooth lustrous surface. Like oak, the tannin content in Zambezi redwood is high, making it liable to stain if in contact with iron in damp conditions.

This dense, heavy timber, (it is about forty percent heavier than true teak), needs to be dried slowly to avoid warping and splitting. Once dry, it is stable, durable, and resistant to abrasion, which also makes it very difficult to work. It is extremely hard wearing, making it ideal for heavy-duty and commercial flooring. In Africa, it is regularly used for wagon building and railway sleepers.

Softwoods

Cedar *Cedrus* spp.
Pinaceae

There are three true cedars: *C. libani*, the Cedar of Lebanon from the Near East; *C. atlantica*, the Atlas cedar from North Africa; and *C. deodara*, the Deodar cedar from the Western Himalayas. The wood of all three is similar in appearance, quality, and, particularly, aroma.

The Cedar of Lebanon is probably the best-known species. King Solomon is said to have built his temple with cedar from Mount Lebanon. The ancient Egyptians used it within the tombs of Egyptian Kings, particularly for coffins. It was at one time a favorite tree for ornamental planting in parks and gardens across the temperate world. Natural stands are abundant in southern Turkey, but the species is highly endangered in Lebanon.

When cut, all cedarwood gives off a strong, attractive fragrance and is distilled for both aromatherapy and air freshener products. The wood is light brown with prominent growth ring figuring. Both Lebanon and Atlas cedars tend to be knotty, but Deodar has a straight, even grain. Cedarwood dries easily, but with some distortion. It is not a particularly strong wood, but is incredibly durable and can be polished to a good finish. The aromatic components give excellent protection against insect and fungal attack. Selected logs are used for furniture, interior joinery, and door frames; otherwise, its uses tend to be limited to more decorative work including veneers for cabinets and paneling.

See: Fragrant Plants, p. 242

Douglas fir *Pseudotsuga menziesii*
Pinaceae

This softwood is not a true fir, nor a true pine, although in the USA it is known as 'Columbian' or 'Oregon' pine. It is closely related to spruce and occurs naturally in northwest America but has been introduced as a timber species to many other countries including Britain, Australia, and New Zealand. It is a large tree, regularly attaining heights in excess of 150 feet (50 m). The heartwood is a rich, reddish-brown with slightly paler sapwood. There is a prominent growth ring figure on plain-sawn surfaces and rotary-cut veneers. It dries well with little movement and is very stable in service. Because of its strength, it is favored for large-scale construction work, roof trusses, beams, interior and exterior joinery, flooring, and decking. Other uses include shipbuilding, railway sleepers, and vats and tanks for breweries, distilleries and chemical plants. Douglas fir is one of the largest sources for plywood in the world.

English yew *Taxus baccata*
Taxaceae

Yew grows naturally from Britain, through Europe and North Africa, to the Himalayas and Burma. It is a small to medium-sized evergreen tree with a short, fluted trunk. The heartwood is one of the most attractive of all timbers. It varies in color from orange-brown to purple-brown and has an irregular growth pattern which helps to produce a superb decorative appearance. Yew is one of the heaviest, strongest, and naturally durable softwood timbers. It dries quickly and well, with little distortion and can be steam bent. For many centuries, yew was favored for archer's bows, especially the English Long Bow. The bow found with the Ötztal Ice Man in the Tyrolean Alps, dating to 3200 ^{14}C years BC, was made of yew.

Yew is an excellent turning and carving wood and is widely used for reproduction furniture making. Traditionally, it was used for the bent wood parts of Windsor chairs. It is quite often sliced for decorative veneers.

See: Plants as Medicines, p. 231

Japanese cedar, Sugi *Cryptomeria japonica*
Taxodiaceae

This attractive tree is native to the mountains of Chekiang and Fukien provences in China and in the Honshu, Shikoku, and Kyushu regions of Japan. It has, however, been extensively planted as a commercial forest tree throughout both countries and many other areas of the temperate world.

It has pinkish-brown colored wood with straight grain, distinctive growth ring figuring, and even texture. When sawn, it does have a fragrance, although not as powerful or as attractive as the true cedars. *Cryptomeria* dries easily with little degradation and, once dry, is very stable in use. It works well with both machine and hand tools and can be polished to a good finish. It has become one of the most important commercial softwood timbers within both Japan and China, where it is used extensively for construction work, joinery, and flooring. It is also used as a decorative wood for interior paneling and furniture.

Larch *Larix* spp.
Pinaceae

Nine different species of larch grow naturally from North America and Canada, through parts of Europe to Siberia and Japan. Elsewhere it has been widely planted within commercial softwood plantations. Larch is unusual amongst softwoods as it is a deciduous conifer.

The heartwood color varies from pale pink to red-brown with very distinctive darker growth rings. It has straight grain and fine texture. Larch dries well with little distortion and is stable in service. It has good natural durability, takes preservatives well and hence is used extensively for boat building, transmission poles, bridge construction, fencing panels, and other exterior construction. Selected logs are also rotary cut to provide ornamental veneers. The Romans used larch for bridges and boat-building.

Parana pine *Araucaria angustifolia*
Araucariaceae

Parana pine is not a true pine, but is, in fact, closely related to the monkey puzzle tree, *A. araucana*. Parana pine grows in Brazil, Paraguay, and Argentina and takes its name from the Brazilian state of Parana. It is a large tree, reaching 130 feet (40 m) in height and a diameter of 4 feet (1.2 m), with a long, branch-free trunk. Although abundant, some eighty percent of Brazilian *Araucaria* forest has been lost to logging in the last hundred years, and the tree must now be considered vulnerable.

Parana pine is a very attractive, honey-colored wood. It is straight-grained with a fine, uniform texture and little annual growth-ring distinction. Its weight varies considerably, depending on where it has been growing (perhaps due to differing water balances). However, regardless of weight, it is a very difficult wood to dry without splitting. If it is not well dried, movement in service can be considerable. Parana pine is Brazil's major timber export. It has become one of the most common woods available from "Do It Yourself" stores across the world and is an excellent timber for interior joinery, especially staircases, because of the large sizes available and lack of knots. It is also used for a good quality plywood.

Scots pine *Pinus sylvestris*
Pinaceae

Pinus sylvestris occurs right across Europe from Britain to Siberia. It is also known as "European redwood" because of its pale red-brown heartwood which contrasts markedly with its creamy-white sapwood. It is one of the heaviest softwoods and is a fairly coarse, knotty wood with clearly marked annual growth rings. It dries rapidly and well and is stable in use; however, it is prone to a fungal disease called 'bluestain' in the drying process.

The wood is graded for specific uses, the best grade being used for interior joinery and furniture and the other grades used for general building work, boxes, pallets, and crates. It will take preservative treatment and is then suitable for railway sleepers and telegraph poles. Selected logs are used for plywood and decorative veneers. Demand for Scots pine was so great in medieval Britain that trees were being imported from the 13th century onwards. Other species of pine were widely used for shipbuilding and construction in the ancient Mediterranean.

Spruce *Picea* spp.
Pinaceae

A group of over thirty evergreen conifers found growing naturally in most of the cool temperate regions of the northern hemisphere. Only two are commercially important for their timber; *P. abies* (Norway spruce) and *P. sitchensis* (Sitka spruce). Although both species are sustainably harvested, large-scale plantations of Sitka spruce have been regarded as a threat to biodiversity in northern Europe. Spruce trees cast deep shade that inhibits growth of other plants.

Norway spruce, sometimes known as "whitewood," occurs throughout most of Europe. Sitka spruce is from the Pacific coast of North America. Both are straight-grained with pale yellow to pale pink coloring and fairly well defined growth rings. They have a high natural luster. They dry rapidly but with a slight risk of distortion. Once dry, they are easily worked with hand and machine tools.

Spruce wood is not durable and it is only the sapwood which takes preservatives well. Norway spruce was introduced to Britain ca. 1500, and since the mid-19th century, has been its traditional Christmas tree; Sitka is far too sharp-needled for this purpose. Both are widely used for interior building work, general joinery, boxes, and crates and provide excellent "tone-woods" for keyboard instrument soundboards and for guitar and violin casings. Sitka produces the vast majority of the world's virgin pulp supply for newsprint.

Western hemlock, Alaskan pine *Tsuga heterophylla*
Pinaceae

This North American tree has been widely planted as a timber-producing species in many northern hemisphere countries, including Britain. It has a cream colored wood with occasional pale brown banding and prominent growth rings. It is a straight-grained, strong wood with an even texture. It requires careful drying to avoid splitting and checking and there is a small degree of movement when in use. It works well with both machine and hand tools and can be worked to a good finish.

Due to its strength qualities, it is an important commercial timber and a valuable export for North America. It is used for general building construction, joists, rafters, and both interior and exterior joinery. Western hemlock is widely used for packing cases, crates, and pallets. It is also sliced for decorative veneers for plywood and paneling.

Western red cedar *Thuja plicata*
Cupressaceae

This large tree, native to North America, is one of the most durable softwoods of all. It has a creamy white sapwood which contrasts dramatically with the heartwood which can vary from dark chocolate-brown to shrimp-pink. It is straight-grained with prominent growth rings. Western red cedar is a very light wood and, as such, is quite weak and brittle. It dries easily when cut into thin boards, but thicker planks may collapse and distort. Once dry, it is very stable in service.

Its natural durability means it is extensively used for lightweight exterior work, particularly weather-boarding and roofing shingles. It is also favored for wooden greenhouses and sheds. The honey frames within beehives are also made from this wood.

References and Further Reading

Desch, H.E. and J.M. Dinwoodie. 1981. *Timber, Its Structure, Properties and Utilization.* London: Macmillan.

Dudley, N. Jeanrenaud, P.-P. and Sullivan, F. 1995. Bad Harvest? The Timber Trade and the Degradation of the World's Forests. London: Earthscan.

Edlin, H.L. 1973. Woodland Crafts in Britain: An Account of the Traditional Uses of Trees and Timbers in the British Countryside. Second Edition. Newton Abbot: David and Charles.

Flynn, J.H. and Holder, C.D. 2001. *A Guide to Useful Woods of the World.* Second Edition. Madison, WI: Forest Products Society.

Gale, R. and D. Cutler. 2000. *Plants in Archaeology: Identification Manual of Vegetative Plant Materials Used in Europe and the Southern Mediterranean to c. 1500.* Otley, UK: Westbury.

Glastra, R. 1999. *Cut and Run: Illegal Logging and Timber Trade in the Tropics.* Ottawa: International Development Research Centre.

Hora, B. 1981. *The Oxford Encyclopedia of Trees of the World.* Oxford: Oxford University Press.

Hough, R.B. 2002. The Wood Book: Reprint of "The American Woods" (1888–1913), 1928. London: Taschen.

Johnson, H. 1976. *The International Book of Wood.* London: Mitchell Beazley and New York: Simon and Schuster.

Latham, B. 1957. Timber: A Historical Survey of Its Development and Distribution. London: Harrap.

Meiggs, R. 1982. *Trees and Timber in the Ancient Mediterranean World.* Oxford, UK: Clarendon Press.

Sentence, B. 2003. *Wood: The World of Woodworking and Carving.* London: Thames and Hudson.

Soerianegara, I. and R.H.M.J. Lemmens. 1993–1998. *Plant Resources of South-east Asia. No 5. Timber Trees.* 1–3. Wageningen, The Netherlands: Pudoc.

Tsoumis, G. 1991. Science and Technology of Wood: Structure, Properties, Utilization. New York: Van Nostrand Reinhold.

UNEP World Conservation Monitoring Center Website, *Tree Conservation Database,* www.unep-wcmc.org/trees/

17
Materials

DAPHNE HAKUNO

Humanity's quest for material goods created the global economy. What began with the manipulation of locally available plant resources led to commerce between groups spanning vast distances. As knowledge of materials spread to new areas, demand for these items fueled the cultivation of formerly wild-gathered plants. Innovations in plant processing techniques and the introduction of new plants changed the structure of societies and environments. Skill and craftsmanship in working with plant materials developed to surpass making objects just to meet basic needs, to that of creating works of art and ornamentation. Materials such as latexes, fibers, dyes, oils, waxes, poisons, and others have been obtained from plants and have been utilized for a variety of purposes ranging from basic building materials, buttons, and containers to products for the operation of the machines of the industrial age.

Resins, Gums, and Exudates

Chicle *Manilkara zapota*
Sapotaceae

Chicle is the latex harvested from *Manilkara zapota*, or sapodilla trees (sapodilla is the fruit), originally from Central America and Mexico. The wild trees, which grow up to 70 feet (20 m) tall, were tapped laterally by harvesters called chicleros. The edible fruits can also be used to obtain a latex by pricking or slicing and letting the latex flow out without squeezing. The heavy sapodilla wood polishes well and is also valuable. The latex is primarily used as a base in chewing gums, but has also had applications in plasters and insulations. Chicle was used as a chewing gum by the Mayans and others in Central America, but this application was not what initially brought it to the attention of industry in the 1800s. It was first experimented on by the inventor Thomas Adams around 1876 after Antonio López de Santa Anna, a notorious Mexican politician, had shipped two tons of it to New York, hoping it would prove to be a viable alternative to *Hevea* latex for rubber manufacture. The experiments proved it lacked the qualities needed in rubber, but led to its use as a popular chewing gum. Limited amounts of chicle chewing gum are still produced much as the original product was, by heating and blending the latex with sugar.

Yokuts baskets, 1924. Library of Congress, Prints & Photographs Division, Edward S. Curtis Collection, LC-USZ62-118769. Usually, willow was used as the foundation, sedge roots for the weft, and redbud for the designs.

Dragon's blood *Dracaena draco* and *Daemomorops draco*
Agavaceae and Arecaceae

Dracaena draco, a tree native to the Canaries and East Indies, and *Daemomorops draco* (formerly known as *Calamus draco*) and other *Daemomorops* species, a native tree of Southeast Asia, both produce resins known as dragon's blood. *Dracaena draco* produces an exudate from its trunk. In India, this red exudate has been used medicinally as a powerful styptic (to stop bleeding), and as a colorant for caste marks. The Guanaches, early inhabitants of the Canary Islands, considered it a sacred tree. Dragon's blood was used in Medieval Europe as an artist's pigment and in the 18th century, when *Dracaena* resin was used by the great violin makers of Italy in their varnishes. *Daemomorops draco* produces fruits with a resinous coating, which is used in Sumatra as a coloring agent. The resin is also burned as incense.

Guar gum *Cyamopsis tetragonoloba*
Fabaceae

An herbaceous perennial native to tropical Africa, the guar or cluster bean, *Cyamopsis tetragonoloba*, produces guar gum, which gained popular use as a replacement for locust bean gum during World War II. (Locust bean gum, extracted from the seeds of the carob tree *Ceratonia siliqua*, became in short supply during the war as the tree is only found in the Mediterranean region.) The gum is extracted from the ground endosperm of the seed. Guar has been cultivated in India for centuries as a cattle fodder crop, and was introduced into the United States from India in 1903. Its superiority to starch as an additive for the papermaking industry led to research into further industrial applications. Aside from its use as a stabilizer in food products, guar gum is used as a sand- suspending agent in oil well drilling, a binding agent in explosives, a settling agent in mining, and an additive for water in fire hoses to reduce friction.

See: Nuts, Seeds, and Pulses, p. 142

Guayule *Parthenium argentatum*
Asteraceae

Guayule is a latex-producing shrub native to arid regions in Mexico and the southwestern United States. It was used as a minor alternative rubber source until World War II, when cultivation was begun in an attempt to internally satisfy the natural rubber needs of the United States. Even as early as 1910, guayule was the source of nearly ten percent of all natural rubber used worldwide.

This production was dependent mostly on wild stands of guayule in Mexico which were eventually severely depleted due to overharvesting: the entire plant was collected to obtain the latex contained within the vacuoles of root and stem cells. The Mexican revolution (1910–1920), in combination with overharvesting, led to the cultivation of guayule across the border in the United States. Production in the United States increased through the 1920s and slowed during the depression with the world slump in rubber prices until guayule was again viewed as an important alternative rubber source during wartime scarcity. Large amounts were cultivated until the mid-1950s when it was no longer deemed profitable and further scarcities in overseas supplies seemed unlikely. Though harvesting of wild stands still occurs in Mexico, attempts at cultivation in other arid regions have the potential of creating a cash crop where most other conventional crops are neither feasible nor sustainable.

Gum arabic *Acacia senegal*
Fabaceae

Gum arabic is an exudate produced from the *Acacia senegal* trees native to northeast Africa. In use since at least the time of early Egyptian civilization (ca. 5,000 years ago), it became known as gum arabic in Europe in the Middle Ages because supplies were shipped from Arabia. Largely collected by hand, the exudate is allowed to dry on the wounded trees, where it forms into droplets. As a colorless, odorless, and hot or cold water soluble exudate, it is commercially desirable. Industrially significant, it has a history of use as an adhesive, candy glaze, and sugar crystallization preventer, stabilizer for the foam in beer, fat emulsifier for hand soaps and lotions, waterproofing agent, and in watercolor paints.

Gutta percha *Palaquium gutta*
Sapotaceae

Gutta percha is a resin obtained from *Palaquium gutta* trees and used in a similar fashion as rubber. Indigenous to Malaysia and Indonesia, gutta trees were cultivated in plantations from the 1890s, following introduction of samples to Europe in the 1840s. Unlike most gums, gutta percha did not become brittle when cooled. This property made the resin quite popular, and groups of gum gathers began roaming the forests in search of the tall, straight trees. When the demand for gutta was noticed by a Sultan in Ligua during this time, he began confiscating some of the harvest by declaring it a royalty. The malleable substance gained widespread use in handles and other parts of surgical implements, but was most important as a non-conducting insulator for subterranean and submarine telegraph wires, enabling the first wires to be laid across the Atlantic in the 1850s. Gutta percha can also be vulcanized like rubber and used for shoe soles, or extracted by another method, and was used in the manufacture of golf balls from 1848 to 1900.

Rubber *Hevea brasiliensis*
Euphorbiaceae

Hevea brasiliensis, a large latex-producing forest tree native to the Amazon region of Brazil, played a dramatic role in the history of industrial development. Even with the introduction of petroleum-derived rubber substitutes, natural rubber is still an important world crop today, providing about thirty percent of all commercial rubber.

Rubber remained in relative obscurity for several centuries after one of the earliest written records of its use appeared in 1536 in *Historia general y natural de las Indias* by Oviedo y Valdés. Some early reports of rubber came from several early European explorers of the New World who recorded seeing games played with bouncy rubber balls. Other early uses of rubber in South America included waterproofing and making figurines, syringes, and molded containers. Many of these reports most likely referred to another rubber-producing plant, *Castilla*, which inhabits Mexico and Central America as well as the Amazon basin. *Hevea* and several other latex-producing

Rubber tapping in Brazilian Amazonian forest, Acre, Brazil. Copyright Edward Parker, used with permission.

plants were utilized by the Aztecs and Mayans for a variety of purposes, including ceremonial. In certain rituals, the latex was symbolic of blood, represented by the latex flows from plant wounds.

The term rubber was given to *Hevea* latex in 1770 by Sir Joseph Priestly, the chemist who discovered oxygen. In a preface to one of his books he mentioned that the material could be used to remove lead pencil marks from paper by 'rubbing' them out. Rubber was first used for erasers, and later for waterproofing fabric such as mackintoshes and shoes. Natural rubber had its limitations, though: natural rubber products would soften and become sticky in heat and crack and split in colder temperatures. Then the process of vulcanization was developed in 1839, which made the manufacture of all-weather products such as the automobile tire possible.

The process of vulcanization was accidentally discovered during experiments on rubber by Charles Goodyear in 1839. Goodyear patented the process of stabilization of rubber with sulfur and lead oxide. The term vulcanization was later coined by Thomas Hancock, who patented it in Europe under his own name. Attempts by Goodyear to sue Hancock for his theft of the process were in vain; Goodyear was never able to profit from his invention and died in poverty.

Demand for rubber production greatly expanded with the discovery of vulcanization. *Hevea* became the main source of rubber when the wild *Castilla* populations became depleted from the common practice of felling entire trees to collect larger amounts of latex than could be collected by tapping living trees. *Hevea* trees proved to be a more sustainable source of latex, as they could continue to be tapped over many years. Although with lower initial per tree yields than *Castilla*, *Hevea* yields gradually increase over time with continued tapping. Only available from wild trees in Brazil, alternative growing regions for *Hevea* were sought to provide a way into the lucrative rubber market. Attempts at plantations within its native Amazonian range had failed due to the presence of disease. Manaus continued to flourish as the capital of the Brazilian rubber industry until 1912. Vast quantities of rubber were hand-collected by a huge labor force, with many of the laborers essentially working as slaves for the large rubber companies. Yellow fever, malaria, and the difficult working conditions caused the deaths of thousands of these workers.

Hevea cultivation in other humid tropical locations fared somewhat better. Seeds which were long reported as being smuggled out of Brazil by Henry Wickham in 1876 were in fact legitimately collected; they were later successfully germinated at Kew in England and planted as seedlings in Southeast Asia, bringing about an end to the Brazilian monopoly on the world's rubber supply. These plantings finally began to exceed Brazil's rubber production in 1913. By 1918, the amount of rubber supplied by Brazil was small in comparison to that from British plantations. Production restrictions imposed by the British during the 1920s in order to maintain high prices led the Dutch and American rubber industries to seek alternative supplies. The Dutch worked on developing higher-yielding plants using disease resistant rootstocks while the Americans invested in plantations in Liberia, Ghana, and the Philippines. By the time restrictions were lifted in 1927, the Dutch had gained control of a majority of the world's rubber trade.

Rubber harvesting techniques improved during the late 1890s, which also helped increase production. The Brazilian method of multiple tapping incisions damaged the trees so much that it eventually made them untappable. H.N. Ridley from Kew replaced the old technique with that of sloping incisions that removed a thin strip of bark below the cut. If sufficient rest periods were allowed, the tree would be able to heal smoothly and be tapped again in the same area once the bark was renewed. A modified version of this method is still used for harvesting.

The rubber plantations in Asia were supplying almost all of the world's rubber by World War II, and fighting eventually cut the United States, the main rubber consumer, off from this supply. Synthetic substitutes were created in the United States using petroleum, and accounted for the majority of the rubber on the market by the end of the war.

The need for rubber remains strong, especially for use in transportation for the manufacture of tires. Although the majority of automobile tires are now composed of both natural and synthetic rubber, there is still a great demand for natural rubber, due to its superior properties. The continued demand for tires and other products requiring natural rubber has encouraged research on other plant sources of rubber such as guayule.

Lacquer Toxicodendron vernix (syn. *Rhus vernicifera*)
Anacardiaceae

Lacquer is obtained from the lacquer tree, native to China but long cultivated in Japan. The lacquer resin has been used as a varnish since at least the 13th century BC in China, and in Japan the earliest records are from the 4th century. The art of lacquering objects was perfected in Japan during the Ming Dynasty (1368–1644 AD), reaching its peak during the 17th century. Rashes were accepted as a common occupational hazard for those who worked with lacquer, for it is a rather toxic substance. The resin remains quite stable over time, as Chinese lacquer from thousand-year-old tombs has still caused rashes when handled.

Turpentine *Abies balsamea, Pinus* spp.
Pinaceae

Turpentine is a resin collected from several different *Pinus* species native to North America and Europe, and from *Abies balsamea*, or Canada balsam, also native to North America. Today, turpentine is primarily used as a solvent for oil-based paints, but Canada balsam is still used for sealing microscope slides. Turpentine is also used to produce other chemicals such as limonene, a lemon flavoring, and in making synthetic pine oil for insecticides. From around 1830 to the early decades of the 20th century, farmers in the American South who collected resin from trees on their land often operated small stills, but these were replaced by larger processing plants from about 1950.

Oils and Waxes

Candlenut *Aleurites moluccanus*
Euphorbiaceae

Aleurites moluccanus, the candlenut or kukui tree, is native to Polynesia, the South Sea islands, Hawaii, the Malay Peninsula, and the Philippines. Rich in oil, the kernel inside the hard shell of the fruit was commonly used for lighting from ancient times in Polynesian history until kerosene was introduced to the islands. Several kernels would be strung together on a bamboo or palm midrib and burned to produce a weak light. Stone containers of oil extracted from the nuts and burned with a wick provided even better illumination. Candlenuts were used to start fires, and the soot from burnt candlenuts was used as a pigment for tattooing.

See: Spices, p. 157

Carnauba *Copernicia prunifera*
Arecaceae

Carnauba is a wax harvested from *Copernicia prunifera*, a palm native to northeastern Brazil. The wax began to be more widely used in Brazil around 1810, and the first carnauba plantations may have been planted as early as the 1890s. Growing in a semi-arid region, the palms are harvested twice a year, the wax separating from the leaves as they dry, when it can then be collected, melted down, and packaged. Carnauba wax is harder than beeswax and many synthetic waxes, and is used primarily as a car wax and shoe polish. It is also used in wax varnishes, lipstick, soaps, crayons, inks, and as a coating for candies and pills.

Jojoba *Simmondsia chinensis*
Simmondsiaceae

The *Simmondsia chinensis* tree, known as jojoba, is native to the Sonoran and Great Basin deserts in North America. The first written mention of the tree is from the early 1700s by a Jesuit exploring the Sonora desert. Southwestern tribes used jojoba extensively as medicine, food, an appetite suppressant to alleviate hunger, a hair growth promoter, and as a hair dressing.

Industry remained largely uninterested in jojoba until the 1930s, when it was discovered that the oil from the seeds is in fact a liquid wax. Similar in properties to sperm whale oil, jojoba was explored for industrial needs after the United States banned the use of sperm whale oil in 1970. Requiring less processing than whale oil, jojoba is used as a machine lubricant, and has even been marketed as an automotive oil. Its primary use is as an ingredient in cosmetics, as jojoba yields are not yet capable of supplying the industrial demand for machine lubricant.

See: Nuts, Seeds, and Pulses, p. 148

Palm oil *Elaeis guineensis*
Arecaceae

The oil palm, *Elaeis guineensis,* originated in Africa, where there is evidence of its early use in west Africa dating to 5000 BC. Palm oil and other palm products did not become widely known in Europe until the European involvement in slave trade, and by the 16th century it was known as a medicine in Europe. Palm oil was closely associated with the slave trade because traders used it to feed the slaves and to oil their bodies before sale to make them look more attractive to buyers. The market in palm oil did, however, eventually help the cause of abolition, providing traders with an alternative interest in Africa after the abandonment of the slave trade in England. Introduced into northeastern Brazil trade during the 17th century slave trade, *Elaeis* palms now occur in there in semi-wild groves. Extensively used in soaps and candles, palm oil was being traded internationally by the end of the 18th century. Demand for palm oil increased in the Industrial Revolution when its use as a machine lubricant and in antifrictional railway compounds began.

Palm oil is similar to olive oil, which is also commonly used in soaps, in that both oils are extracted from the fruit pulp, not just from seed like many oil crops. Plantation production in Africa increased supplies after 1920. This helped meet industry demand as the use of palm oil in manufacture expanded beyond soaps and lubricants to include glycerine, a by-product of palm oil processing, for use in munitions manufacture during World War I and later, in the making of margarine.

See: Nuts, Seeds, and Pulses, pp. 149–50; Fruits, p. 77

Beneficial Poisons

Curare

The varied accounts of the source of curare poison were most likely due to the fact that several plants were used in similar fashion, or were used together, by different Amazonian Indian groups. In South America where the plants are native, extracts from parts of *Chondrodendron tomentosum* (Menispermaceae), *Strychnos nux-vomica* (Loganiaceae), and other less commonly used plants were long known to be useful arrow and fish poisons. Some of the early printed references to curare are found in *De Orbe Novo* from the early 1500s and Sir Walter Raleigh's 1596 book, *The Discovery of the Large, Rich and Beautiful Empire of Guiana*. These poisons were feared by the early European explorers who attempted to find antidotes using camphor, tobacco, coffee, sulfur, garlic, and sugar. The first physician to write about tobacco as an arrow poison antidote was Nicholas Monardes, whose book was translated from Spanish into English in 1577. Later, in the 1700s, investigations on arrow poisons by scientists and explorers invalidated many of these antidote claims.

Knowledge about the preparation and uses of curare and similar poisons increased within the Amazon groups after contact with Europeans in the 1600s, although ingredients in mixtures were rarely revealed to foreigners. Different methods of hunting with curare continued to evolve, including the use of arrows, blowguns, and spears to administer the poison. As a hunting poison, curare does not cause death immediately. It blocks nerve impulses, causing the prey to lose muscle control, and eventually leading to asphyxiation. This muscle-relaxing quality is what led to the first medicinal application of curare in 1938 when curare was administered to psychiatric patients undergoing convulsive therapy to relax their muscles. It was later used during surgery to help relax muscles that remain contracted under anesthesia. Strychnine, a popular rat and mouse poison, was also used medicinally as a central nervous system stimulant, but like curare was later replaced in medicine by synthetic compounds that had less risk of respiratory complications.

Neem, Margosa *Azadirachta indica*
Meliaceae

Neem has an ancient history in its native region of India. Leaves from *Azadirachta indica* trees were found during excavation of a 2000 BC site in what is today a part of Pakistan. Neem is also encountered in Hindu mythology, which refers to it as being of divine origin. Called margosa by the European colonizers in the 16th century, it was primarily used in medicine, and young twigs were used as chewsticks, to clean teeth and relieve toothache. Neem's most economically important application was learned by Indian farmers observing that neem trees were left uneaten during locust swarms. This indication of insecticidal properties was not widely known until the 1960s, after which time research on its pest-repellent characteristics began in earnest. Neem was found to be an effective feeding inhibitor and growth regulator for insect pests without producing toxic environmental effects.

See: Plants as Medicines, pp. 227–8

Pyrethrum *Tanacetum cinerariifolium* (syn. *Chrysanthemum cinerariifolium*)
Asteraceae

Tanacetum cinerariifolium flowers provide a natural insecticide, pyrethrum. A perennial native to Asia, the history of its specific use is somewhat unclear, but similar uses of two other species of Chrysanthemum likely originated in Persia. One story of the discovery of its insecticidal effects is that of a German woman noticing dead insects near a discarded bouquet of the flowers and then being inspired to go into business manufacturing powdered pyrethrum. Regardless of its discoverer, pyrethrum from this plant was produced in Europe after 1820, and early commercial production occurred in Dalmatia and later in Japan, after being introduced to Japan in 1881. Centers of production shifted during times of war. Japan surpassed Dalmatia in production after World War I and remained the principal source of pyrethrum until Japanese production became insignificant during World War II. Introduced into eastern Africa in the 1920s, *Tanacetum cinerariifolium* thrived, increasing in quality and generating yields superior to other regions. By the time of World War II, Kenya's production was already greater than Japan's.

Now pyrethrum is commercially grown primarily in eastern Africa in Kenya and Tanzania, and in some parts of northeastern South America as well. The main use of pyrethrum in the early 1900s was as a household and horticultural insecticide. Another major use arose later with the discovery that pyrethrum was effective at blocking neurosensors in adult female mosquitoes, one type of which can transmit malaria to humans. Long-burning mosquito coils manufactured with pyrethrum came into use to help prevent nighttime bites from mosquitoes.

Rotenone

The natural insecticide rotenone is found in the roots of many plants in the Fabaceae family. Used as a fish poison in Asia, and in the Amazon prior to European contact, rotenone-containing plants were not greatly used as pesticides in industry until the discovery in the 1950s of DDT's toxicity and bio-accumulation in animals and humans. The South American plant barbasco, or *Deguelia rufescens* var. *urucu* (syn. *Lonchocarpus urucu*), was used by South American Indians to stupefy fish in rivers and streams so they could be gathered easily. A relatively safe insecticide is made with rotenone, which has been found to leave no residue when used in the amounts needed to make it effective.

Botanical Jewelry

Plant materials are often used in making ornaments to wear on the body. Seeds and other plant parts have been used throughout the world in the making of necklaces and other jewelry as well as decorative objects. Unlike gems, seeds used for ornamentation come from plants which are often utilized for purposes other than decoration. One material for botanical jewelry, amber, was actually believed by some to be a mineral for much of the history of its use.

Amber

Amber was collected and used primarily in jewelry for many centuries before it was discovered to be plant-derived in origin. A fossilized resin, it is believed to come from several different sources, generally *Pinus* or pinelike trees. A recent hypothesis suggests the origin of the famous Baltic amber may be from ancient precursors of modern *Pseudolarix* trees.

The oldest amber artifacts known indicate that as long ago as 11,000 to 9000 BC, amber was prized and used as a mineral for jewelry and other decorative objects. Trade in amber items such as beads and carvings expanded during the Neolithic period, and amber products were later traded by the Romans when they gained control of many amber source locations along the shores of the Baltic Sea and other regions. By 1312, a monopoly in amber trade was held by the Prussian knights, who imposed strict collecting regulations. Control of the market passed through several hands before again being held by the Prussians around 1642, but the industry truly grew only after large-scale mining of amber began in the 1800s. The amber found washed upon the shores of the Baltic Sea was only a fraction of what was present in deposits beneath the sea. Dredging for these deposits began in 1865, and mining of the nearby land deposits began a few years later. One early mine in particular, the Palmnicken mine at Kaliningrad, has remained the world's most productive amber mine.

The majority of the amber mined does not contain inclusions of preserved prehistoric organisms which are highly valued and have been studied since the 1800s, nor is it suitable for ornamental objects. Instead, the lower quality ambers are used industrially for making varnish, amber oil, and distilled acids. Processing techniques were developed over time to improve the appearance of these ambers by clarifying them. Most of the techniques involved heating the amber either in sand to expand air bubbles or in oils to replace the air with oil. A technique for processing the large amounts of small amber chips not otherwise usable produces what is called ambroid, which was popularly used for buttons and pipe mouthpieces in the early 1900s. Amber and amber products continue to be utilized for items of beauty and utility and for giving scientists an opportunity to study in detail ancient organisms not available in most fossils.

Job's tears *Coix lacryma-jobi*
Poaceae

The seeds of *Coix lacryma-jobi*, also called Job's tears, are edible and ornamental. Native to Southeast Asia, the grass was in cultivation in India as early as 1000–2000 BC. Job's tears were known to the Greeks and Romans as well, although they were not widely recognized in Europe until the 17th century

Vegetable Ivory (palm fruits), Namibia. Copyright Edward Parker, used with permission.

after introduction by Arab traders. The seeds have a porcelain-like exterior and are used as ornamental beads, primarily for necklaces.

See: Grains, p. 53

Vegetable ivory, Tagua, Corozo
Arecaceae

Fruits from a palm native to the New World tropics, *Phytelephas macrocarpa*, or tagua, were once used as a replacement for small ivory objects, and thus became known as vegetable ivory or ivory nut. Seeds from other palm genera *Ammandra* and *Aphandra* are also produced as vegetable ivory. Sir William Hooker may have first introduced vegetable ivory to England around 1826. The hardened albumen of the tagua nut could be cut only with special saws and was used for making buttons, knobs, handles, chess pieces, and other small carved objects. Plastics largely replaced the vegetable ivory products, but some carvings are still made today to supply tourism and other markets.

Dyes and Pigments

Dyeing with plant materials is an ancient practice for decorating objects and even the human body. During some periods of history, civilizations have viewed colors as magical or capable of denoting the wearer as someone with different rank or privilege. Trade in botanical dyes was important since certain colors were difficult to obtain and variety was greatly desired. Plant-based dyes are found in many plant parts, such as leaves, roots, and seeds, and have been used for coloring many items such as foodstuffs, fibers, and hair.

See: Natural Fibers and Dyes, pp. 303–14

Annatto *Bixa orellana*
Bixaceae

Bixa orellana, known as annatto, the lipstick plant, or achiote, is native to South America and has been used throughout South and Central America and the West Indies since pre-Columbian times. Certain tribes in the Amazon basin and other areas of the Americas used the red dye obtained from the seeds as a body paint, which is how its use was first encountered by European explorers. Seen as a potential commercial dye source, *B. orellana* seeds were exported to Europe in the 1700s and 1800s, with commercial production first beginning in India by the late 1700s for use as an orange-red

Colorado Indian with annatto paste in hair, Santo Domingo. Copyright Edward Parker, used with permission.

dye for textiles, although the dye was found to fade quickly. Traditionally used in food coloring in Central America and the West Indies, research on the properties of annatto found it was well absorbed by oils and fats. This led to its use in Europe and the United States as a dye for cheeses, eggs, and as a coloring to make margarine look more like butter.

Henna *Lawsonia inermis*
Lythraceae

A perennial shrub native to North Africa, Asia, and the Middle East, *Lawsonia inermis*, or henna, was used as a hedge plant for thousands of years. Henna leaves were used as a hair dye beginning around 3200 BC, and continue to be used for that purpose today. Egyptian mummies have been found with henna-dyed nails, and perhaps henna was used as the first natural fingernail paint. Henna also has a long history of use as a body paint. Elaborate and intricate designs, applied primarily to the hands and feet, gradually fade with washing and can then be reapplied. Henna has also been used as an orange and yellow dye for wools and silks.

Beneficial Insect Host Plants

Certain insects have evolved unique relationships with the plants they depend on for food and other needs. Although there is often a struggle against insects that feed on agricultural and horticultural crops, for those insects that are useful to humans, these plant-insect interactions can be exploited to our advantage. Two of the most notable plant-insect relationships are that of the *Opuntia* cactus and the cochineal insect, used to produce cochineal dye, and the mulberry and the silkworm, which is raised to produce silk.

Prickly pear cactus *Opuntia* spp.

Cactaceae insects, a parasite of the *Opuntia* cactus, were exploited for producing a rich red dye centuries before the Spanish invasion of their native range in Central America. Based on Spanish accounts from the 1500s, cultivation of *Opuntia* and cochineal may have first centered around the Mixteca area of Mexico around that time. Other Spanish accounts related how cochineal was collected as a tax under Montezuma in Oaxaca, its primary production site until Spanish rule increased the area of its cultivation. Cochineal production was under Spanish monopoly beginning in the 1600s. Traditionally used in paints and dyes, cochineal replaced the use of kermes as a scarlet dye in Europe. Most of the production came from smallholdings where *Opuntia ficus-indica*, the prickly pear cactus, and other species would be seeded and cultivated by farmers known as napoleros. Cochineal use waned in the mid-1800s and was for the most part replaced by synthetic red dyes by the late 1800s.

See: Invasives, pp. 382–3

White mulberry *Morus alba*
Moraceae

The beginning of the use of the white mulberry tree, *Morus alba*, as the host plant for silkworms is told in several legends that are over 4000 years old. One is about an Empress known as Hsi Ling-Shi (also known as Xi Ling Shi) who was closely observing the Emperor's prized mulberry trees to discover what had been eating the leaves. It was then that silkworm caterpillars were seen spinning their cocoons; later, each cocoon was found to be composed of a single silken strand much stronger than thread. The potential of a fabric made from these strands was quickly seen as reason enough to keep the source of silk a secret in China by Emperor's decree. Eventually, however, the secret was taken out of China. One story involves four Chinese maidens who were kidnapped and forced to reveal the secret to their Japanese captors. In another version, silk is introduced into India as a hidden bridal dowry. Yet the monopoly of the silk industry continued to be held by China until Emperor Justinian (435–565 AD) sent two monks to bring the secret back to Constantinople. The monks smuggled the silkworm eggs out of China inside their bamboo walking canes. Once hatched, these silkworms were used to start a new silk industry outside of China.

Italy became a major silk producer during the Renaissance, ranking with the Asian producers. Attempts at English production were made by King James I after 1603. Mulberry trees were sent to the southern colonies in America and by 1619 landowners could incur penalties if they had not planted them. To produce twelve pounds of raw silk, almost one ton of mulberry leaves is needed to feed between thirty and thirty-five thousand silkworms. Production of silkworms couldn't compete with cotton and tobacco in the American colonies and was largely abandoned, the focus shifting instead to weaving raw silk imported from the main producers.

Basketry and Containers

The need for storage of foods, beverages, and other goods led to creative use of many plant materials. Baskets and other containers have been made from barks, fibers, fruits, and other plant parts. These materials have also had applications in making musical instruments, toys, and utensils, among others.

Bamboo *Bambusa* spp.
Poaceae

Bamboo, a native plant of Asia, has the distinction of being the most rapid-growing of all plants, with some bamboos growing up to 3 feet (1 m) or more in only a single day. The common bamboo, *Bambusa bambos*, is also amazing in its array of uses. In Asia, bamboo is more important than timber both as construction material and as a raw material for scaffolding, walls, cloth, and a variety of decorative objects, especially bamboo baskets.

The artistic history of bamboo baskets dates back to the Jomon period (10,000–300 BC) in Japan, when they were being used for everyday needs. Crafted in many shapes and sizes, baskets made for more purely aesthetic reasons became quite popular during the Edo period (1615–1868). Individual artisans rarely took credit for their basketry until after this period, and since then many have achieved fame for their unique designs.

Beyond the needs of containers and decoration, bamboo also has a practical history of use as writing material. Before pulp paper, bamboo stems were split and the interior was used as a writing surface, the narrow shape being well suited to Asian writing styles. Use of bamboo leaves for a more conventional paper is evidenced in the archeological records from possibly as early as 105 BC. After soaking in water, the leaves would be beaten into a paste which was then poured out onto mats to dry. Bamboo paper tends to be fibrous and coarse, but its strength makes it desirable for certain applications. Even the pens used to write on bamboo have been made of cut bamboo stems.

See: Herbs and Vegetables, p. 114

Birch *Betula* spp.
Betulaceae

Birch trees, species of *Betula*, are native to northern Asia and Europe. The paper birch was a primary source of paper in Russia from 800 to 1700 AD, and continued to be used after this time for certain documents requiring its durability. Paper birch and other birches were long used as containers and even clothing. Several birch bark containers dating from about 5000 BC were found at the Starr Carr site in North Yorkshire. In northern Europe, birch bark shoes were worn during the Middle Ages. Many indigenous tribes in North America used birch to make canoes, maps, buckets, and baskets. They considered birch bark containers superior for food storage, believing they prevented stored food from spoiling for a longer period of time than other containers.

See: Wood, pp. 319–20

Cork *Quercus suber*
Fagaceae

Cork, the outer bark of *Quercus suber*, has long been used to seal bottles and other containers in the Mediterranean countries where it is native. The early Greeks and Romans used cork harvested from local forests in continuous production for many centuries. One ancient use of cork was in the manufacturing of beehives, cork being a good insulating material to maintain hive temperatures within a desirable range. The insulating properties of cork were put to use in human dwellings as well, lining walls and ceilings as early as medieval times, when they were mentioned as being used in monasteries. Slabs of cork were used for roofing and flooring of more simple housing.

The cork industry truly began to grow during the 17th century with the more widespread use of glass bottles that required cork stoppers. The cultivation of cork trees began in the mid-1700s in Spain, moving into nearby countries and northern Africa by the early 1800s. New plantings of cork trees are not harvestable until the tree is about twenty years old; the trees remain productive for around 150 years, and can be harvested for cork every eight to ten years. The invention of corkboard

Harvested cork oak, Andalucia, Spain. Copyright Edward Parker, used with permission.

insulation in 1891 by John Smith, crown cork in 1892 by William Painter, and composition cork (which utilized the large amounts of cork scraps being generated) in 1909 by Charles E. MacManus expanded the potential uses of cork and increased demand for cork products. Attempts were made to cultivate *Q. suber* in America as early as 1859, but they failed and the production center for cork remained in the Mediterranean region.

See: Wood, p. 326

Gourds *Lagenaria* spp., *Cucurbita* spp., and *Luffa* spp.
Cucurbitaceae

Gourds and calabashes have been used as containers for thousands of years. Several different plants produce fruits used as gourds, among them many *Cucurbita* species: the tree gourd or calabash tree, *Crescentia cujete* (Bignoniaceae), native to tropical America, and the bottle gourd, *Lagenaria siceraria*, native to Africa. Bottle gourds, which were already present in Asia and the Americas in prehistoric times, have been uncovered in archeological sites dating from 13,000–11,000 BC in Peru, and from 10,000 to 6000 BC in Thailand. Gourds were believed to be a symbol of water in many arid regions, and are often mentioned in early writings and in creation myths around the world. An inventive use of gourds reported in early Guatemalan history is that of being used for defense when the Quiche'Cakchiquel were about to be invaded. The leaders filled gourds with wasps and hornets, releasing them to attack the invaders all at once.

An important crop in South America, gourds continue to provide many useful items such as food, containers (especially for water and other liquids), utensils, toys, and musical instruments. Gourds have long been used as a medium and model in art and music. Beginning around 1000 AD in China, ornate cricket houses were created from gourds by applying carved molds around the growing fruit; as the gourds matured and enlarged, the carvings made a decorative impression on the gourd's surface which remained after the mold was removed. Gourds form a natural rattle and have been hollowed out and used in constructing marimbas and stringed instruments, and in ancient Mayan times were placed on the ends of long wooden trumpets to amplify the sound. The early Caribs viewed their gourd rattles reverently, and some groups believed they could only be used by shamans.

Lagenaria, the bottle gourd, had been introduced to Europe prior to 1555, when they were first mentioned in writings, and were being cultivated in England by 1597. The most popular of all the gourds, the bottle gourd is utilized throughout the world. Among its many applications, this gourd is well suited for storage of liquids and dry goods because of its shape (which is why it is referred to as the bottle gourd), and is also useful for making floats for fish nets and swimmers. Gourds have even been turned into clothing in Africa, South America, and the southwest Pacific, where they were used as penis sheaths, a practice largely abandoned in most areas but still ongoing in Papua New Guinea.

Willow, Osier *Salix* spp.
Salicaceae

The flexible qualities of willow made it a popular fiber for weaving baskets and other containers very early in European history. With a worldwide distribution, there are several *Salix* species that have been commonly used for weaving. The Romans may have learned to weave baskets from the Britons. In Europe, willows for osier production have been propagated from cuttings with little variation in methods since Roman times. *Salix viminalis*, used in furniture making, was later introduced to America to meet the demand for wicker production.

See: Ornamentals, p. 284; Plants as Medicines, p. 217

Coconut harvesting. Copyright Edward Parker, used with permission.

Fibers

Coir *Cocos nucifera*
Arecaceae

Coir, a rough fiber from the coconut palm *Cocos nucifera*, is a pantropical native. The first written mention of coir is in Arabic, and dates from the 11th century. The fiber is obtained from the outer husk of coconuts after they have been retted (soaked) in seawater for several months. Because of its ability to withstand salt water, it was early put to use as cable, rigging, and for other purposes on ships and other ocean vessels. Arab traders most likely spread the use of coir to India and Sri Lanka, which are the main producers of coir. Introduced to England around 1832, coir was a minor import until it was displayed at the Exhibition of 1851 and use of it became more widespread, especially for cordage.

The highest quality coir fibers come from immature coconuts that are less than a year old. This unfortunately means that the copra used for dried coconut is not the best source. The majority of coir comes instead as a by-product of copra production, and is a coarser, lower quality fiber. It is used primarily for ropes, matting, and stuffing, but has also had applications as a soundproofing material and in filtration systems.

See: Natural Fibers and Dyes, pp. 292–3

Kapok *Ceiba pentandra*
Bombacaceae

The traditional use of *Ceiba pentandra* for fiber was first observed by Dutch colonists of Java early in the 17th century. Referred to as kapok, the seed fibers were used as a stuffing for bedding and other items. A native of East Asia, kapok was cultivated in Java since before the tenth century, perhaps having been earlier introduced from Bengal. It was not introduced into Europe until the middle of the 19th century, and was not well known until after being exhibited at the 1893 World's Colombian Exposition in Chicago.

Characterized by a sheen similar to silk, it was referred to as silk cotton and tree cotton by those who hoped it might be a replacement for silk or cotton. But kapok fibers proved impractical for commercial fabric production because their slippery nature required hand spinning and weaving. However, western manufacturers found its light weight and its insulating ability perfect for use in stuffing upholstery, bedding and cushions, life preservers, and cold-weather garments as well as for soundproofing insulation and padded safety gear. Indonesia was the main producer of kapok prior to World War II, after which the main center of production moved to Thailand.

See: Natural Fibers and Dyes, pp. 293–4

Paper mulberry *Broussonetia papyrifera*
Moraceae

The Polynesjans and Mayans both invented use of the bark of paper mulberry, *Broussonetia papyrifera*, native to Asia and Polynesia, pounded into thin sheets, as a paper and cloth. The inner bark fibers are still used for making specialty papers and tapa cloth in the South Sea Islands. Papyrus, cloth, and tree bark fibers were the common writing surfaces until pulp paper manufacturing was invented. The beginning of pulp paper manufacture occurred in China possibly as early as the first century BC, but some claim it was not until at least the second century AD. Knowledge of the pulping process spread westward from China, reaching Morocco around 1100 and then Europe in the late 1100s. The bark of the paper mulberry was handbeaten to separate the fibers, which were then soaked and boiled; afterwards, the pulp was poured into molds for making sheets of paper. Later, the process was refined to dipping the molds into the pulp, which made for smoother sheets. Paper for printing has been made on a large scale using a blend of paper mulberry and bamboo pulp.

Papyrus *Cyperus papyrus*
Cyperaceae

Papyrus may have been in use as a writing surface in Egypt as early as 2500 bc, remaining the primary source of "paper" until the 8th century AD. Used not just for writing, papyrus was for centuries used to make baskets, sandals, rope, and other items, and was also burned as a fuel where wood was scarce. The Greeks and Romans were also familiar with the fiber: in the 5th century BC, Herodotus mentioned the use of papyrus for boat sails. *Cyperus papyrus*, a reed native to Africa, could no longer be found growing in the Nile region when French explorers went there in 1798–1801. Its use had been largely replaced by the invention of pulp paper which could be made from a variety of plants.

Toquilla *Carludovica palmata*
Cyclanthaceae

The most popular straw hat fiber comes from the *Carludovica palmate*, or toquilla plant, native to South America. The true toquilla hat is manufactured from fibers grown and harvested in Ecuador. Hats woven from this plant became known as "Panama hats" because so many were distributed and sold through the ports of Panama, especially during the 1800s. Toquilla was used in South America for weaving head coverings and other items long before its discovery by European explorers. The first written record of an artisan in the Panama hat trade- Francisco Delgado- was not made until 1630. During the17th and 18th centuries, returning explorers introduced the hats to Europe, stimulating an industry of collectors and weavers in Ecuador as the hat gained wider distribution and popularity. Even Napoleon wore one of the finest quality "Montecristi" hats during his long exile on Saint Helena in the early 1800s. Later, thousands of hats were purchased by California gold seekers passing through Panama during the mid-1800s.

The best quality Panama hats are produced from fiber originating in the limited area of the province of Guayas (Ecuador), where the plant is harvested only during the five days after the waning quarter moon in every lunar cycle, when it is easiest to cut. Leaves 2–3 m (6–9 feet) long are cut and stripped for fiber, and the fibers are boiled for up to an hour to soften them. The few highly skilled master weavers for these specialty hats are still located in Montecristi, with the main center of the hat production market being based in Cuenca since the mid-1900s. The demand for Panama hats led to the export of toquilla straw from the provinces of Guayas and Manabí during the 1800s. Fearing they would lose their hold on this lucrative market if the hats were produced elsewhere, Ecuadorian weavers lobbied their national government to ban the export of raw toquilla straw. The ban was implemented in 1835 and was not lifted until 1843.

Loofah *Luffa aegyptiaca*
Cucurbitaceae

The loofah gourd *Luffa aegyptiaca*, unlike most gourds, is prized for its interior fibers. The tough, wear-resistant loofah has been utilized by many societies for cleaning and scrubbing. It is also used to make sandals, baskets, and matting, and was industrially useful as a filter for steam and diesel motors and as insulating material. Native to Asia, the plant was first commercially grown in Japan in 1890. Introduced to the American tropics after European contact, New World loofah production was started to make up for wartime shortages, but lessened after the war because Japan produced the highest commercial quality loofah.

Other applications for the gourd have been investigated. Even the seeds have been used in Brazil to extract an odorless, clear oil. Because loofah exfoliates and cleanses the skin easily, its most popular widespread modern use is as an alternative to wash cloths.

References and Further Reading

General

Barrett, O.W. 1928. *The Tropical Crops*. New York: MacMillan.
De Candolle, A. 1885. *Origin of Cultivated Plants*. New York: D. Appleton and Co..
Erichsen-Brown, C. 1979. *Use of Plants for the Past 500 Years*. Ontario: Breezy Creeks Press.
Freethy, R. 1985. *From Agar to Zenry: A Book of Plant Uses, Names and Folklore*. Dover: Tanager Books.
Harlan, J.R. 1975. *Crops and Man*. Madison: Crop Science Society of America.
Hill, A.F. 1937. *Economic Botany: A Textbook of Useful Plants and Plant Products*. New York: McGraw-Hill.
Jackson, J.R. 1890. *Commercial Botany of the Nineteenth Century*. London: Cassell.
Kochhar, S.L. 1981. *Tropical Crops: A textbook of Economic Botany*. Dehli: Macmillan India.
Lee, R. 1854. *Trees, Plants, and Flowers: Their Beauties, Uses and Influences*. London: Grant and Griffith.
Roecklein, J.C., and Leung, P. 1987. *A Profile of Economic Plants*. New Brunswick: Transaction.
Simpson, B. 1986. *Economic Botany: Plants in Our World*. New York: McGraw-Hill.
Smith, N.J.H., et al. 1992. *Tropical Forests and Their Crops*. New York: Cornell University Press.
Wickens, G.E. 2001. *Economic Botany: Principles and Practices*. Dordrecht: Kluwer Academic.
Wilson, C.M. (Ed.). 1945. *New Crops for the New World*. Westport, CT: Greenwood Press.

Cork

Armstrong Cork Company. 1909. *Cork: Being the Story of the Origin of Cork, the Processes Employed in Its Manufacture & Its Various Uses in the World Today*. Pittsburgh, PA: Armstrong Cork Co..
Cooke, G.B. 1961. *Cork and the Cork Tree. International Series of Monographs on Pure and Applied Biology*. New York: Pergamon Press.
Faubel, A.L. 1941. *Cork and the American Cork Industry*. Rahway: Quinn & Boden.

Containers

Browne, E.A. 1924. *Peeps at Industries—Vegetable Oils*. London: A. & C. Black.
Cotsen, L., Johnson, H., and Graham, P.J. 1999. *Japanese Bamboo Baskets: Masterworks of Form and Texture*. Chigaco: Art Media Resources.
Dunkelberg, K. 1985. *Bambus-Bamboo*. Stuttgart: Institute fur leichte Flachentragwerke.
Gourd Society of America. 1966. *Gourds, Their Culture and Craft*. Bridgewater: Dorr's Print Shop.
Heiser, C.B. 1979. *The Gourd Book*. Norman: University of Oklahoma Press.
Hunter, D. 1930. *Papermaking through Eighteen Centuries*. New York: B. Franklin.
Krishnaswamy, V.S. 1958. *A Note on* Broussonetia papyrifera *(Paper Mulberry)*. Dehra Dun: The Northern Circle, Survey of India.

Peyton, J.L. 1994. *The Birch: Bright Tree of Life and Legend*. Blacksburg: McDonald & Woodward.
Sentence, B. 2001. *Basketry: A World Guide to Traditional Techniques*. London: Thames and Hudson.
Tilton, J.K. 1947. *The History of Silk and Sericulture*. New York: Scalamandre' Silks, Inc..
Warren-Wren, S.C. 1972. *The Complete Book of Willows*. New York: A.S. Barnes & Company.
Wilson, E.W. 1947. *The Gourd in Folk Literature*. Boston: The Gourd Society of America.

Resins, Gums, Exudates

Coppen, J.J.W. 1995. *Gums, Resins and Latexes of Plant Origin*. Non-Wood Forest Products 6. Rome: Food and Agriculture Organization.
Howes, F. N. 1949. *Vegetable Gums and Resins*. Waltham, MA: Chronica Botanica.
Langenheim, J.H. 2003. *Plant Resins: Chemistry, Evolution, Ecology, and Ethnobotany*. Portland, OR: Timber Press.
Pearson, H.C. 1909. *Crude Rubber and Compounding Ingredients*. New York: India Rubber Publishing Company.
Piper, J.M. 1992. *Bamboo and Rattan: Traditional Uses and Beliefs*. Oxford: Oxford University Press.
Ruskin, F.R. (Ed.). 1977. *Guayule: An Alternative Source of Natural Rubber*. Washington, DC: National Academy of Sciences.
Verdoorn, F. 1949. *Vegetable Gums and Resin*. Waltham: Chronica Botanica Company.
Whistler, R.L. and Hymnowitz, T. 1979. *Guar: Agronomy, Production, Industrial Use and Nutrition*. West Lafayette: Purdue University Press.

Rubber

Brown, H. 1913. *Rubber; Its Sources, Cultivation and Preparation*. New York: D.Van Nostrand.
Collier, R. 1968. *The River That God Forgot; The Story of the Amazon Rubber Boom*. New York: E.P. Dutton.
Dean, W. 1987. *Brazil and the Struggle for Rubber: A Study in Environmental History*. Cambridge: Cambridge University Press.
Drabble, J.H. 1973. *Rubber in Malaya, 1876–1922: The Genesis of the Industry*. Kuala Lumpur: Oxford University Press.
Firestone, H.S. 1932. *The Romance and Drama of the Rubber Industry*. Akron: Firestone Tire & Rubber Company.
Fisher, H. 1941. *Rubber and Its Use*. New York: Chemical Publishing Co..
Morris, D. 1898. *Cantor Lectures on the Plants Yielding Commercial India-Rubber: with Special Reference to the Rubber Industries Connected with Her Majesty's Colonial and Indian Possessions*. London: Society for the Encouragement of Arts, Manufactures, & Commerce.
Pearson, H.C. 1909. *Crude Rubber and Compounding Ingredients*. New York: India Rubber Publishing Company.
Polhamus, L.G. 1962. *Rubber: Botany, Production and Utilization*. London: Leonard Hill.
Sethuraj, M.R. and Mathew, N. M. 1992. *Natural Rubber: Biology, Cultivation and Technology*. Amsterdam: Elsevier.
Stanfield, M.E. 1998. *Red Rubber, Bleeding Trees*: *Violence, Slavery and Empire in Northwest Amazonia, 1850–1933*. Albuquerque: University of New Mexico Press.
Stevens, H.P., and Stevens, W.H. 1940. *Rubber Latex*. New York: Chemical Publishing Co..
Webster, C.C. and Baulwill, W.J. 1989. *Rubber*. Harlow: Longman.
Webster, C.C., and Baukill, W.J. 1989. *Rubber*. New York: Longman Scientific & Technical.

Dyes

Buchanan, R. 1987. *A Weaver's Garden*. Loveland: Interweave Press.
Cannon, J., and Cannon, M. 1994. *Dye Plants and Dyeing*. Portland: Timber Press.
Donkin, R.A. 1977. *Spanish Red. An Ethnographical Study of Cochineal and the Opuntia Cactus*. Independence Square: American Philosophical Society.
Gre, I. 1974. *Nature's Colors: Dyes from Plants*. New York: Collier Books.
Krochmal, A. 1974. *The Complete Illustrated Book of Dyes from Natural Sources*. New York: Doubleday.
Leggett, W.F. 1944. *Ancient and Medieval Dyes*. New York: Chemical Publishing.
Lemmens, R.H.M.J. and Wulijarni-Soetjipto, N. 1991. *Dye and Tannin-Producing Plants. Plant Resources of South-East Asia 3*. Wageningen: Pudoc.
Manniche, L. 1989. *An Ancient Egyptian Herbal*. Austin: University of Texas Press.

Poisons

Bovet, D., Bovet-Nitti,F., and Marini-Bettolo,G.B. (Eds.). 1959. *International Symposium on Curare and Curare-like Agents*. Amsterdam: Elsevier.
Casida, J.E., and Quistad,G.B. (Eds.). 1995. *Pyrethrum Flowers: Production, Chemistry, Toxicology, and Uses*. New York: Oxford University Press.
Gnadinger, C.B. 1933. *Pyrethrum Flowers*. Minneapolis: McLaughlin Gormley King.
Gupta, B.N., and Sharma, K.K. (Eds.). 1998. *Neem, a Wonder Tree*. Dehra Dun: Indian Council of Forestry Research and Education.
Higbee, E.C. 1949. *Lonchocarpo, Derris y Piretro*. Washington, D.C.: Union Panamericana.
Murty, A.J.S., Samba,S., and Subrahmanyam, N.S. 1989. *A Textbook of Economic Botany*. New Dehli: Wiley Eastern.

Puri, H.S. 1999. *Neem, the Divine Tree:* Azadirachta indica. London: Dunitz Martin.

Thomas, K.B. 1963. *Curare, Its History and Usage.* Philadelphia: Lippincott.

Fibers

Barrett, T. 1983. *Japanese Papermaking: Traditions, Tools and Techniques.* New York: Weatherhill.

Bell, L.A. 1981. *Plant Fibers for Papermaking.* McMinnville. OR: Liliaceae Press.

Bell, L.A. 1985. *Papyrus, Tapa, Amate & Rice Paper: Papermaking in Africa, the Pacific, Latin America & Southeast Asia.* McMinnville, OR: Liliaceae Press.

Brink, M. and Escobin, R.P. 2003. *Fibre Plants. Plant Resources of South-East Asia 17.* Leiden: Backhuys Publishers.

Buchet, M. 1995. *Panama: A Legendary Hat.* Quito: Ediciones Libri Mundi.

Child, R. 1964. *Coconuts.* London: Longmans Green.

Dewey, L.H. 1943. *Fiber Production in the Western Hemisphere.* Washington, DC: United States Government Printing Office.

Dodge, C.R. 1897. *Useful Fiber Plants of the World.* Washington, DC: Government Printing Office.

Gaur, A. 1979. *Writing Materials of the East.* London: British Library.

Hughes, S. 1978. *Washi: The World of Japanese Paper.* Tokyo: Kodansha International.

Hunter, D. 1971. *Papermaking through Eighteen Centuries.* New York: Burt Franklin.

Kirby, R.H. 1963. *Vegetable Fibers: Botany, Cultivation and Utilization.* New York: Leonard Hill.

Miller, T. 1986. *The Panama Hat Trail: A Journey from South America.* New York: Morrow.

Parkinson, R., and Quirke, S. 1995. *Papyrus.* Austin: University of Texas Press.

Zand, S.J. 1941. *Kapok: A Survey of Its History, Cultivation and Uses.* New York: Trade Commissioner for the Netherlands.

Oils and Waxes

Abbott, I.A. 1992. *La'au Hawaii: Traditional Hawaiin Uses of Plants.* Honolulu, HI: Bishop Museum Press.

Bloomfield, F. 1985. *Jojoba and Yucca.* London: Century.

Duffus, C.M., and Slaughter, J.C. 1980. *Seeds and Their Uses.* Chichester: Wiley.

Grimaldi, D.A. 1996. *Amber: Window to the Past.* New York: Harry N. Abrams.

Linam, D. 1981. *Jojoba Fever.* Burbank: Burbank Books.

Martin, S.M. 1988. *Palm Oil and Protest.* Cambridge: Cambridge University Press.

Scarlett, P.L. 1978. *Jojoba in a Nutshell.* Summerland: Scarlett-Trotter.

Van der Vossen, H.A.M., and Umali, B.E. (Eds.). 2001. *Vegetable Oils and Fats. Plant Resources of South-East Asia 14.* Leiden: Backhuys, 2001.

Zeven, A.C. 1967. *The Semi-wild Oil Palm and Its Industry in Africa.* Wageningen, The Netherlands: Centrum voor Landbouwpublikaties en Landbouwdocumentatie.

Part III
Today and Tomorrow

GHILLEAN T. PRANCE

This final section of this volume brings together the results of all the human activities that have been described in the previous sections and puts then into the context of today's environment.

The way in which many plants were used changed radically with the advent of the industrial revolution. The history of the industrialization of various crops is given in the first chapter. This led to a much greater demand for the raw material of many crops, which in turn led to the conversion of many more natural areas to use for commercial agriculture and forestry. It also led to a much more intense and specialized breeding of crops that tended to select for specific qualities without maintaining the genetic diversity of either the wild relatives or the many landraces and varieties developed by local peoples around the world. There was both erosion of genetic diversity of the plants most useful to humans and the loss of wild species through habitat destruction.

Some useful plants when moved to other places escaped from gardens and farms and became weeds of an enormous proportion. For example, guava, a most useful fruit crop, is one of the most serious invasive species in Hawaii and elsewhere. People transported plants both deliberately and accidentally, and the invasive species that are described here were introduced for many different reasons, some as crops and some as accidental weeds. Many were also taken as ornamentals, but instead of beautifying gardens they became pests. The forests of Hawaii are full of such things as ornamental gingers from tropical Asia, Mexican daisies, and nasturtiums from South America. Even the Mediterranean Sea ecosystem is threatened by the introduction of a tropical marine alga, *Caulerpa*. The lakes and ditches of Great Britain are becoming clogged by an invasive species of *Crassula* from Australia and New Zealand that was discarded by aquarists and is now taking over and replacing native pondweeds. The result is that invasive species have become one of the biggest threats to natural ecosystems and to native species around the world. This is particularly true on tropical islands rich in endemic species and in sensitive ecosystems such as the species-diverse Cape flora of South Africa, as the well-chosen examples cited in the chapter on invasive species will show.

It is in this section that we see the downside of the cultural relationship between people and plants. Even without modern techniques of molecular biology, humans have been ingenious in breeding crops, but this has been at considerable cost to the environment. In an article in *Science*, the botanists Nigel Pitman and Peter Jorgensen estimated that between thirty and forty-seven percent of all plant species are threatened with extinction and could disappear within the next fifty years. This would be a serious loss to our natural heritage, for it is both interfering with natural processes that maintain the equilibrium of our planet and reducing our options for the future use of plants. Some good—actually, unfortunate—examples of species threatened with extinction are given in the final chapter. This represents a challenge to conservationists, scientists, and politicians around the globe and especially to botanic gardens, most of which have now placed conservation high on their agendas.

If greater action is not taken to preserve the plants upon which our cultural history and our cultural future are based, then humankind will face a grave crisis that could even lead to its own extinction. The greatest challenge for the future is use the scientific data we now posses to avoid this massive wave of biological extinction. We cannot exist without plants. This book has demonstrated well that our culture is intricately interwoven with plants, and so we need to return to fostering the respect for plants that is shown by many primitive societies but is rapidly becoming lost in our globalized, industrial, and largely urban society.

18

Age of Industrialization and Agro-industry

ANDREW B. JACOBSON

The Industrial Revolution began in England during the mid-18th century and quickly spread through the rest of Europe and the world. Industrialization meant changes in all forms and walks of life. The 19th century was the hey-day of the age of industrialization. This chapter focuses on how the cultivation of some of the major crops of the world was affected by those advances.

Corn, Maize *Zea mays*
Poaceae

Corn, or maize, is perhaps one of the most versatile plants in the world. Native to the Americas, where it had been grown for thousands of years, it was introduced to the Europeans and the rest of the world in the 15th century. After European colonists were introduced to maize, over time they found it to be the ultimate cash crop and it became firmly entrenched in the newly developing Euro-American farm culture. In a few hundred years maize became the most common and valuable crop grown for animal products, direct human consumption, and industrial product development, in what was fast becoming the United States of America. The most common use of corn (i.e., the non-sweet variety) during the 1800s was for animal feed (or fodder). Feed corn had a rougher texture than the sweet corn eaten by humans, and was better suited for livestock. In the early 1900s, thanks to the campaign for health by John H. and William K. Kellogg, and their famous Corn is Flakes cereal, sweet corn soon became more important for human consumption. Corn is used to make other products, including industrial materials and pharmaceuticals. In the late 1800s and early 1900s, these products included industrial packing materials, and later, laundry chemicals (such as collar starch). However, during the 1800s many other uses for corn were also being researched.

In 1840, the Englishman Orlando Jones developed and patented a method of vegetable starch extraction which used an alkaline as a catalyst. Four years later, in 1844, Colgate & Company applied Jones's patent to sweet corn in its New Jersey wheat-starch factory. Thomas Kingsford, employed by Colgate & Company, was impressed by their cornstarch processing system and set up his own cornstarch factory in Oswego, New York. In 1854, Wright Duryea, a former millwright, opened his own cornstarch factory and coined the name Maizena for his starch. By 1891 Duryea had the largest starch factory in the United States. His plant was capable of grinding up to 7,000 bushels of corn in one day. His company was still producing cornstarch into the 1930s (Fussell 1992, 265–9).

The most common use of cornstarch during this mid to late 19th century era was as a form of "flour," mostly used to make Johnny Cakes, a simple corn-based food.

By the mid to late 1800s, due to declining sugar exports from the United States and increasing sugar imports from places like China, plenty of American sugar processing mills and apparatii were being left to deteriorate. With some modification, it was believed, they could be used to process sugar from cornstarch. Sugar cane required many types of equipment and multiple steps in processing, and the same would apply to maize processing; although, as it later turned out, the steps used in processing corn sugar (syrup), were not identical to those used in processing sugar cane.

On July 1, 1877, upon becoming Commissioner of Agriculture, the Hon. W.G. Le Duc began intensive research into the problems afflicting sugar production and the possible solutions (Le Duc 1877, 228–37).

Sugar beet cultivation as a replacement for sugar cane was considered, but not found to be completely satisfactory. France was then one of the few countries to successfully produce and utilize beet-sugar. However, in the United States, the use of the sugar beet was not totally ruled out from future consideration. Le Duc also carried out his own study into the use of corn sugar (syrup) as a substitute for cane sugar, and he felt very positive about the results (Le Duc 1877, 228–37).

In 1877, an average of 45 million acres of corn was planted annually. If this abundance could be used to extract sugars from the corn, the U.S. sugar production problems could well have been solved. After much research, it was proclaimed by Le Duc that corn was suitable for sugar production, with only a slightly lower yield than typical sugar cane. He concluded that a realistic breakdown of products made from an acre of corn was: 70 gallons of syrup, 5 pounds of stem, 3,000 pounds of sugar, and 66 gallons of molasses (Le Duc 1877, 247–8).

In modern commercial corn syrup manufacturing, after the corn starch is separated from the gluteus corn material using methods similar to those used in late 1800s (only now used on much larger scale) and converted to corn syrup, some factories now take corn syrup production one stage further. Today, when glucose in corn syrup is converted to fructose enzymatically (derived from *Streptomyces* bacteria), the resulting corn syrup is much sweeter. This refined product is known as High Fructose Corn Syrup or HFCS. All types of sugar are ranked on a sweetness scale by comparison to cane sugar sweetness. HFCS is considered to have twice the sweetness of cane sugar. The Clinton Corn Processing Co of Clinton, Iowa first commercialized the production of HFCS in 1967. This company was also noted for its development and patent of *Isomerose*, an enzyme that converts glucose to fructose. By 1972, the company had increased the sweetness of their regular fructose corn syrup by 14 to 42 percent, making it equivalent to cane sugar (sucrose) in sweetness. Sugar prices rose once again during the 1970s, and food and beverage manufacturers began to replace cane sugar sucrose more extensively with HFCS. By 1976, HFCS production jumped from two hundred thousand pounds a year to two and a half billion pounds a year. By 1992, HFCS had become a major component of all major soft drinks (Fussell 1992, 269–74).

Other modern corn products made during and after cornstarch conversion to corn syrup are laundry "starches" (made from dextrin), refined corn oil (made from germ separation), molasses or hydrol (made at the same time as corn syrup), lactic acid (made from glucose), sorbitol and mannitol (made from glucose/dextrose), and methyl glucoside (also made from glucose/dextrose).

Today, well over one hundred years since Kirchhof, Le Duc, and Jones performed their experiments, corn syrup may not have totally replaced imported cane sugar; nonetheless, if you look on any food label (and on some non-food labels) at your local supermarket, you will notice that corn syrup, if not the first ingredient, is a close second or third. Today, Le Duc's predictions would not be considered too far off the mark. Corn has come a long way from an indigenous plant and has become a primary source of human and animal nutrition, sugar, and non-edible products, all of which are made and used throughout the world.

However, the twentieth and twenty-first centuries have also seen an increase in the genetic modification of crop plants. In an effort to increase crop yields and crop strength (resistance to diseases, pests, and weeds), agricultural scientists have been directly manipulating the genetic makeup of corn and soybean plants in particular; for example, to develop maize that is herbicide-resistant or resistant to pests. Bt corn is corn modified by the addition of a gene from the bacterium *Bacillus thuringiensis* to produce a toxin in the pollen and corn tissues that makes the hybrid resistant to the European corn borer and other pests. The resulting products, collectively known as transgenic plants or GMOs, Genetically Modified Organisms, have led to controversy (see the section on soybeans below).

Cotton *Gossypium*
Malvaceae

Cotton must be spun into thread or yarn before it can be woven into cloth. Cotton threads must be woven, a very time-consuming task, by way of a shuttle moved in and out of the warp threads. In 1733, John Kay invented a "fly shuttle," which only needed one groove to move back and forth through the warp. This device made weaving quicker and easier. In 1764, James Hargreaves invented a hand powered "Spinning Jenny," patented in about 1770, which could spin eight strands of thread at one time, whereas the traditional spinning wheel could only spin one at a time. Cotton fibers also needed to be separated from their seeds, and up until 1793 this too had been done by hand. In 1793, during the Industrial Revolution, Eli Whitney invented the famous "cotton gin." While a human worker could only clean about one pound of cotton a day, the "gin" cleaned about fifty pounds of cotton a day. The machine used a system of revolving wire spikes to comb the seeds from the fiber, while a human had to pick out small amounts of seed with his or her fingers.

Cotton on the right has been ginned and is ready for conversion into yarn at a textile mill. ARS/USDA.

By the mid to late 1800s, these machines were fairly common and even more advanced. Soon their use revolutionized the cotton-processing industry, particularly in the United States. Between 1800 and 1809, the average production of U.S. cotton was 136,000 bales. By the 1850s the average production was 307,100 bales, and by the 1890s the average production was 8,905,000 bales (Nickerson 1954, 140–1). As these figures show, in about one hundred years the amount of U.S. cotton bales produced for sale in the world market increased by 6,500 percent.

However, cotton sales and production could have become much more lucrative for Eli Whitney and the United States much earlier, if not for a twist of fate. By the late 1790s Whitney's company was not doing well financially. Competitors' petty and professional jealousies had made Whitney's life a living hell, for they constantly tried to drive him out of business, and some were producing their own versions of Whitney's gin to further make his and his partner's life and profit difficult (Green, 1956, 67–89).

Had Whitney been more business savvy and less concerned with immediate profit, and had indeed earlier gone to England to secure a cotton trade agreement as urged by his partner, Phineas Miller, things might have turned out differently. In the mid to late 1700s and early 1800s England was very interested in cotton processed by the United States, although the United States mainly processed long staple cotton. In England, however, there was a special need for well-processed short staple cotton. Whitney's machines with only minor adjustments could have filled this need. Unfortunately, Whitney kept putting off his trip to England, and it was only a matter of time before, spurned on by his detractors, word spread that Whitney's machines were inferior rip-offs. The English began to believe the rumors and thus began purchasing processed cotton from Whitney's detractors and competitors, further putting the squeeze on Miller and Whitney (Green 1956, 67–79). The company would never recover, and would later force Whitney's factory into making firearms in order to make ends meet. Adding insult to injury, worldwide demand for cotton would increase manifold in the decades and centuries to come. Today, cotton is the most extensively used clothing material.

In the Americas, agricultural production of cotton and other crops required many human laborers and some animal labor to till fields, plant crops, harvest crops, process crops, and to take the resulting products, usually by horse-drawn wagon, to market. So, as a way to cut overhead and increase profits at minimal cost to the plantation owners, "free" labor was determined a necessity, as was a way to control it; thus was born the New World institution of slavery.

In America, Senator James Henry Hammond of South Carolina decided that the southern United States plantations should become more involved in the cotton industry, and more African slaves should do more of the laborious work. This opinion was expressed in his infamous speech to the Senate in 1858:

> In all social systems there must be a class to do the mean duties, to perform the drudgeries of life; that is, a class requiring but a low order of intellect and but little skill. Its requisites are vigor, docility, [and] fidelity. . . . Fortunately for the South, she found a race adapted to that purpose to her hand—a race inferior to herself, but eminently qualified . . . to answer all her purposes. We use them for the purpose and call them slaves . . . (Dodge 1984, 114–5).

Apparently Hammond had a lot of kindred spirits; in the United States, since the late 1600s, slaves were used to pick cotton, and by the time of the cotton "gin," slavery and cotton production had become synonymous. Slavery had also been associated with the production of indigo, tea, tobacco, corn, coffee, and sugar cane. This U.S. slave labor state was to be seriously changed with the outbreak of the Civil War and its subsequent conclusion, of course; the total collapse of the entire economic foundation of the South, of which slavery was only one part.

However, cotton fiber production still remained and, in some cases, was revived. Only this time, it was planted, harvested, processed, and spun by low paid human labor, otherwise known as share-croppers. These farmers were allowed to grow crops on sub-divided land owned by someone else in exchange for a percentage or share (usually one third or one half) of their harvest going to the land-owner. The remaining percentage of their crops served as a default wage for these farmers (Hurt, 1994, 166–70). Sharecroppers were often former slaves who after being freed needed to earn a liv-ing. So they went back to what they knew, the skills they had acquired while under the yoke of sla-very. The sharecroppers could sell their crops at market, but due to a depressed Southern economy they barely made any profit. Thus, they could not afford most items of necessity, such as coffee, cal-ico, and tools. Inevitably, they had to look toward the landowners for help. The landowners would pay for these "furnishings" in return for a greater percentage of the crop yields, or reimbursement on credit. In many cases, landowners would only take cotton product as reimbursement (Hurt 1994, 166–70.). The end result of this unbalanced economic system was former slaves living in per-petual debt to landowners. Although physically free, these sharecroppers lived in virtual economic and geographical slavery.

The sharecropping industry for cotton and other plants lasted well into the 20th century, although most cotton-producing tasks by this time were mechanized, depending on plantation finances. In the United States and around the world, however, cotton picking itself still remained a task mostly performed by humans. Up through the 20th century, sharecroppers also handpicked tea, coffee, tobacco, and orchard fruits. Many to this day still do.

Over the last two decades, the demand for cotton, especially for the long staple variety, has increased dramatically. Long staple cotton is better suited for more modern means of mechanical harvesting and processing, and its long fibers are the easiest cotton material for spinning. The world market, especially the United States and Europe, demands the best machine cloth (i.e., that with the highest thread count). Nowhere is this developing need for long staple cotton more noticeable than in the Near East. In 2001 a federal committee in Islamabad asked the government to produce at least ten percent cotton of long staple variety by 2005. Up until this point, the demand by the local textile industry for higher count yarn was met through imports. These textile companies now want Islamabad to produce more of its raw cotton locally, thus decreasing dependence on outside sources and at the same time increasing exports of processed and raw cotton.

Hemp *Cannabis sativa*
Cannabaceae

In the 1890s, hemp was produced all over the world. There were many cordage plants referred to as Russian hemp, Manila hemp, and sisal hemp, but true hemp *Cannabis sativa* was the standard to which all other cordage fibers were compared.

In 1896, while other countries were using machines to plant, harvest, and process hemp, Japanese farmers still carried out most of these tasks by hand. Also, the process of retting (i.e., the separation of soft unusable plant tissue from the tough usable plant fibers) in Japan and some other countries was still performed by hand: first, the plants were manually pulled out of the ground, then the leaves and roots were cut off with a sickle. The remaining stems were sorted into short, medium, and long lengths and bound into bundles. These bundles were then steamed in a specially built bath and then laid out in piles on straw mats to dry in the sun. They were then dipped in water again and left out to dry, then stored for future use. This method of moistening and drying caused the fiber casings to rot, allowing for more effective removal of the usable fibers. Finally, the dried bundles were then unbound and "broken" using a device similar to a flax break. (Dodge 1895, 215–21).

Up until the 1870s, most hemp plants were processed by hand in a way similar to the Japanese method discussed above. However, inventors from different countries began developing machinery to achieve more efficient and cost-effective hemp planting, harvesting, processing, and packaging. Between 1875 and 1880, about forty-five patents for hemp and other fiber processing machines were issued. These inventions sped up processing and improved trade to a significant extent.

On May 8, 1877, Norbert de Landtsheer of Paris patented a device that separated workable fibers from hemp and flax plants, the scutching machine. It was able to clean about 1,000 pounds of hemp or flax straw in about ten hours. The de Landtsheer design was quite complex, but easily operated. The material to be cleaned was passed through a combination of fluted rollers. Then it was passed into a scutching drum composed of two cast-metal rings covered with sheet metal. Attached to the plaited rings' circumference were between eight and sixteen blades, sometimes more, depending on the model. These blades heated the material after the fluted rollers had broken it up. This additional step helped to clean the hemp during the scutching process. The resulting long-stapled lengths of fiber ("line") were nearly perfectly scutched, as were the leftover materials ("tow") (Le Duc 1879, 600–609.). Faster processing meant better profits.

In the United States, up to the 1930s hemp was used to make paper, rope, twine, clothing, and sail canvas. However, in the 1930s its role was threatened by petroleum companies wanting to develop and produce synthetic fibers for rope and textiles. These petroleum companies launched negative advertising campaigns against the hemp industry. It is no secret that hemp (cannabis) leaves and seeds have narcotic effects, but the rope and textile fibers produced from the stems have no narcotic content. Even today, the stigma associated with hemp as a narcotic in general is a by-product of these campaigns. The large lumber and newspaper conglomerates also launched similar anti-hemp advertising because they felt that hemp was the greatest threat to the wood-fiber based newspaper production industry (Bielby 2002, 1). Due to environmental concerns in the 1990s in Europe, new agricultural initiatives were proposed as sustainable alternatives to over-produced food crops. England successfully lobbied the EC for standardized legislation regarding alternative crops to cover all of Europe. The result was that in 1992/1993, England and other countries received the first licenses granting farms permission to grow low-narcotic hemp plants, provided it was grown for "special purposes" or "in the public interest" (Bielby 2002, 1–2). The hemp industry was on the rise again.

At the forefront of this positive awareness and economic hemp revival in the United Kingdom was, and still is, The Cornish Hemp Company Ltd. Set up in 1995 to develop and research "new" uses for hemp, the Company has so far produced a hemp crop which yields 5–9 tons of biomass, two tons of which are long useable fibers for textiles. The remaining short fibers and "waste products" can be used to make Oakum (i.e., a material used in caulking seams of ships and in pipe joints), or used as upholstery stuffing. However, more importantly, the Cornish Hemp Company Ltd. is currently experimenting with short fiber spinning machines, and new hemp processing methods are now able to produce hemp fibers that are comparable to cotton (Bielby 2002, 2). In short, hemp products may one day dominate the textile and fiber industries again. Much depends on the attitude of governments towards allowing the legalized use of hemp for factory purposes and the creation of demand from consumers.

See: Psychoactive Plants, pp. 198–9, Natural Fibers and Dyes, p. 297 and 302

Sisal, Henequen *Agave sisalana, A. fourcroydes*
Agavaceae

Sisal was a plant used as cordage to make things such as nets, rope, mats, brushes, wall hangings, and basketry. Native to Mexico, sisal was used by pre-Columbian people to make these very same items long before the arrival of the Europeans. Sisal, known as "maguey" to the indigenous peoples of Mexico, came from the long spiked leaves of the *Agave* plant. Sisal fibers are too coarse to spin into yarn, so they were made into more utilitarian items. The average sizes of cordage lengths made

Sisal extraction with a small portable machine.

from sisal were 40–50 inches (120–150 cm) long, and these lengths were very strong (Hochberg 1980, 55). Up until the 1800s, the harvesting and processing of sisal was mostly done by hand.

Sisal farming utilized the latest in technology for planting, cultivating, harvesting, and processing. From 1875 on, special machinery was used in the Yucatan to work the sisal fibers. Previous to this, crude tools were used to extract the fibers from the sisal plant. From the 1870s to the early 20th century, steam-powered machinery was used to process sisal fibers.

By the second half of the 19th century, mechanized American farm equipment such as hay balers and reapers were designed to use twine to bundle the harvested crops. Sisal was one of the major types of fiber used and was one of the most important exports from Yucatan, Mexico to the United States. As the demand for more self-binding equipment increased, so did the demand for binding fibers. By between the 1870s and 1915, some sisal farms in Yucatan were becoming large-scale commercial agricultural businesses, or plantations, although they only grew sisal for export. Major American manufacturers of self-binding harvesting machinery such as International Harvester were looking to Yucatan to supply their sisal fibers, which was then made into binder-twine (Wells 1985, 30–50).

Sisal fibers were grown in northwestern Yucatan, and required a great deal of labor to plant, harvest, and process. Family-run, large-scale sisal farms (haciendas and later plantations) were run with the same iron rule, discipline, and minimal wages as the sugar or tobacco plantations elsewhere. As the profitability of sisal increased, the plantation owners pushed the laborers harder. The field laborers were mostly the indigenous peoples, in this case the Mayans (campesinos), who were treated with little respect by those in authority, the intermingled Mayan/Spanish peoples (ladinos) representing the owners and managers (Wells 1985, 8–25).

By 1880, Yucatan manufacturers had invested $5 million in the sisal industry, and by 1879, Yucatan had 18,000,000 plants under cultivation. Processing this number of plants required over 420 scraping-wheels, run by 229 steam engines with a force of 1,732 horsepower, and thirty wheels moved by animal power. Each scraping-wheel extracted an average of 300 pounds of fiber a day. The scraping-wheels and animal wheels were only run about 202 days out of the year (Dodge 1879, 546–7). These figures indicate that by the 1880s, sisal was no longer just a small farm crop; it was now a major economic agricultural force.

However, this boom in the sisal trade was soon to change. By the late 1890s and early 1900s, the sisal industry was heading for difficulties, due to competition from the hemp and cotton markets and price fluctuations, all controlled by the importing foreign companies and not Yucatan sisal producers themselves. This, plus over-extension of sisal company finances, would lead to the collapse of many major sisal merchants, planters, and manufacturers. The investment of all of a company's resources into the production and processing of a single major crop in the Yucatan was not the basis for a stable economy. Having no alternative crop in case of economic depression would inevitably lead to the ruin of major sisal producers. In addition, the terrible plantation conditions and treatment of workers led to major unrest and violence from the Mayan workers directed against the sisal industry and its proponents, the Ladinos. This violent rebellion came to be known as the "Caste War," and the resulting government military action decimated the labor population, thus damaging sisal production (Wells 1985, 144–82). In addition, the failure of the Yucatan's railroad network to effectively bolster the sisal trade only added fuel to the fire. All of these problems exacerbated the ongoing economic decline of the Yucatan sisal trade. By 1915, the Yucatan's 'gilded age' (time of economic prosperity), developed by the sisal trade, was ending. Even now, the Yucatan has never been able to resurrect this 'gilded age'.

Today, sisal is primarily used to make rope, string, matting, rugs, brushes, sacking (burlap), cable insulation, and for general industrial use. It is still produced in Yucatan, on a small scale. Currently, Kenya and Tanzania together produce 50 percent of the world's total sisal supply. Sisal fibers are no longer used to make binder-twine; most crop harvesting, binding, and bailing machines now use wire, synthetic twine, or plastic strapping to bundle the crop.

See: Natural Fibers and Dyes, pp. 301–3

Soybeans, Soya, Soja *Glycine max*
Fabaceae

In the Far East, soybean meal and other soybean products have been eaten on a regular basis for centuries. This activity over time spread west to the United States. By 1917, soybeans were used in the United States for a variety of purposes including in cooking oils, animal feed, fertilizer, paint medium, and explosives. By the 1940s, the United States was gradually dominating the soybean trade, a trend that was going to continue through the 1960s and into the 21st century. Today, soybeans in their many forms are consumed and utilized throughout the United States, with a major boost from the health food movement. This same trend is also occurring in other countries throughout the world. In fact, the soy industry has become one of the major players in world trade. Many of the soy products mentioned above are still used today. Soybeans are now even being considered as an alternative fuel source to petroleum. Further research in this particular area is still needed, and the cooperation of major oil (fuel) companies has yet to be secured.

During the late 1800s and early 1900s, the importance of the soybean in the world market increased quite noticeably. It is believed that in 1770, Benjamin Franklin introduced soybeans to the American colonists. There is an archived letter stating that Franklin sent soybeans home from England (Liu 1997, 1–3). Other records indicate that soybeans were also grown in the United States in 1804. Soybeans were initially thought of as botanical curiosities. In 1875, Professor Haberlandt of Austria made extensive studies into the use of soybeans for human and animal food. Although his experiments were not as successful as he had hoped they would be, they did create interest in soybeans as a farm crop (Morse 1918, 101–102).

In the early 1900s, northern U.S. soybeans were used as silage. Soybean silage kept well, and livestock fed on soybean silage "showed good gains in flesh and milk production" (Morse 1918, 104). At this time, as a supplement to corn feed, the use of soybeans as a forage crop for hog consumption was very profitable. In 1910, the United States imported soybeans from Manchuria which were then used to make oil and meal. The success of these products can be noted by the ever-increasing imports of Manchurian soybeans. America first began growing its own soybeans for oil in about

1915. At this time, due to a shortage of cottonseed and a surplus of soybean seed, the established cottonseed-oil mills in places like North Carolina switched to the processing of soybean seed oil. However, during the 1916–1917 season, the high price of domestic-grown soybeans meant that it was not economical to process these, so once again the Southern cottonseed-oil mills looked to imported Manchurian soybeans for a solution. Many mills throughout the cotton belt then used these beans for the production of oil and meal.

The use of soybean seed in place of cottonseed for oil production did not require any extensive re-design of the cottonseed processing machinery. In order to make the soybean seed oil, soybean seed, like cottonseed, had to be crushed. The resulting pressure produced the oil and cake (i.e., the fibrous residue left when the seeds were crushed). According to 1917 records of various cottonseed-turned-soybean-seed mills, one ton of soybean seed yielded from 28 to 31 gallons of oil and about 1,600 pounds of meal (Morse 1918, 105). The extracted soybean oil resembled cottonseed oil in several ways. It had a good color, a faint odor, and a pleasant taste. In 1917, soybean oil was fast becoming a major competitor with other vegetable cooking oils such as corn oil or peanut oil.

Soybean oil also began to be used far more extensively in the production of non-food items. In the search for new oils to replace the linseed oil (a flax product) being used in the production of paint, soybean seed oil was found quite suitable. Soon companies were using large quantities of this soybean oil in the manufacture of types of paint and varnishes. The oil was also used to manufacture soft soaps, linoleum, and explosives. As mentioned above, a by-product of soybean seed processing was the cake. This material was also used in the soybean industry, ground into meal and used to manufacture foodstuffs for cattle feed and fertilizer for field crops. Yellow American-grown soybean meal varieties when consumed fresh have a sweet nutty flavor and are high in protein; samples of meal were found to have from 46 to 52 percent protein (Morse 1918, 106).

Perhaps the most commonly known form of soybean food product in Asia is tofu (bean curd). In 1917, tofu was made in a similar way to cheese or yogurt. Dried soybeans were soaked in water for a few hours and then finely crushed. They then were boiled in a large quantity of water. About thirty minutes later, a milk emulsion was obtained which had the appearance of cow's milk and had similar properties (kept in a warm environment the milk became sour and could be processed into buttermilk or sour cream).

However useful and multipurpose soybeans are as a food source, non-food source, or energy source, it has always been difficult to harvest massive quantities of soybeans in a short period of time. Their fragile nature has always made it necessary to hand-cut, gather, and thresh soybeans in a way so as not to shatter the soybeans from the pods. This painstaking and time-consuming process prevented the further advancement for the commercialization of soybeans. Until an effective mechanical means of harvesting and processing large quantities could be developed, the crop's future as a 'mass-produced' plant product was in serious doubt.

This all changed, however, in 1924 when a combine was successfully adapted for use in Illinois soybean harvesting. It was the statements and reflections of Elmer J. Baker, Jr. that made this technological change possible. His pushing and prodding of farm equipment manufacturers led to the development of the combine floating cutter-bar and, later, the 1934 Hume-Love Company pickup reel (Quick and Buchele 1978, 225–32). Today, more advanced versions of these harvesting machines are used on a scale beyond even Mr. Baker's dreams.

During the late 1940s, China's erratic exporting of soybeans led to more lucrative markets for other growers. China's main trading partner had been Japan, but after WWII, Japan was no longer economically in a position to maintain trade with China or other countries. So it was the United States that took up the soybean "slack." Soon, with the help of, and easy access to, modern agricultural technology such as mechanized planters, harvesters, and processing machinery (Liu 1997, 3), the United States began to dominate the soybean trade. Also, during and after WWII, international trade in fats and oils was controlled so that supplies were allocated to countries with a critical deficit of soybeans and other products.

By the 1960s, the United States grew 72 percent of the soybeans grown in the main soybean producing nations. It exported 89 percent of the total world trade in soybeans. By 1967, China was second to the United States both in soybean production and exportation (Houck et al. 1972, 21–49). Brazil has an increasingly important role in world trade, accounting for just under 30% of world soybean exports in 2002. This increase is controversial because it is thought to be linked to increased cultivation in the Amazon rainforest, and the introduction of GM (Genetically Modified) soya.

Since the 1980s, commercial agriculture has been beset by many controversies. However, one major issue stands out more than the others. It is the question of Genetically Modified Organisms, or GMOs. Some scientists are against the use of Genetically Modified Organisms because they feel that they pose a threat to some animal species, such as Monarch butterflies (through exposure to pollen from Bt corn, modified to express insecticide), or to non-modified crops through cross-fertilization and seed dispersal (horizontal gene transfer) with hybridizing relatives. GMO crops could encourage the emergence of "superweeds" or "superpests"—virulent races of weeds or insects that are resistant to the pesticide or herbicide used on the GMO crops. It is also feared that, indirectly (through large-scale application of weedkiller), herbicide-resistant crops will lead to a reduction in wild plant diversity and damage to species high up in the food chain; for example, to the invertebrates that feed on the weeds and their predators. Other scientists feel GMOs are dangerous to humans (Hurt 2003, 159–66); for example, there may be allergic reactions to new proteins expressed by the introduced genes.

On the other hand, some agricultural scientists feel GMOs are not harmful, but beneficial, to growers and consumers. These scientists feel that GMOs such as corn and soybean seeds will lead to increased crop yields, reductions and changes in pesticide usage, and hardier herbicide-resistant and insect-resistant plants. Thus, stronger, larger yield crops mean more profits and more efficient ways to feed the hungry (Hurt 2003, 159–66).

At this point, whether GMOs are more efficient or more harmful remains to be seen.

Sugar cane *Saccharum officinarum*
Poaceae

Sugar derived from sugar cane was once considered to be a luxury item only available to the wealthy or in very small quantities to the poor. Sugar cane only grows in tropical conditions; it is highly susceptible to low temperatures. This, coupled with a very time-consuming production process, made it relatively difficult to mass-produce, distribute, and sell cheaply. Until the mid to late 1800s, it had to be transported by horse-drawn wagon or by boat. The invention of the steam engine, and steam powered tractors, helped overcome these problems. Steam engines allowed farmers to plant, plow, and harvest larger crop acreage in less time, allowing for greater crop yield. Steam engines also allowed trains carrying sugar cane to get to markets more quickly. They also allowed much greater amounts of sugar to be transported to these same markets.

Prior to the Civil War era, in most producing countries sugar cane had be cultivated, harvested, and processed by hand (usually by slave labor), then transported and distributed by the same laborers. In Martinique and Guadeloupe, for instance, the Industrial Revolution led to faster and more efficient ways of producing, processing, and transporting sugar. Previously, the farmers of these two islands grew and processed sugar cane in the same vicinity as the plantation. In the late 1800s, there was a move to separate cane agriculture and sugar manufacturing into two distinct processes and geographic locations. Each area was governed separately, but worked cooperatively. This allowed the farmers to focus on planting, growing, and harvesting, while factory workers performed processing and distributing tasks (Le Duc 1879, 23–33.).

The sugar cane was grown on the plantations and transported by tram or railroad to a central factory (usine). At the factory, the cane was put into mills where the juice was squeezed out, falling through strainers into a tank heated by steam. The juice then was treated with small amounts of bi-sulphite of lime and then sent through a system of heating pipes and charcoal filters which clarified the sugar. This liquid was then boiled off, resulting in the crystallization of the sugar in a large pan. The sugar was then dropped into boxes which were passed over centrifugals where the sugar was cured (usually physically broken down). This became the first-quality sugar. The remaining molasses from the boiling was in turn re-boiled into a jelly-like substance, then dropped into boxes and left to granulate for a couple of days. This became the second-quality sugar. The resulting molasses left over from this process was combined with other skimmings and residues of clarifiers to make rum. By the 1870s, in Martinique and Guadeloupe most of these processes were powered by steam engines, as was transportation of the raw and final product. In addition, most of the mechanical functions in the factories used hydraulic power (Le Duc 1879, 46–7).

During the 1870s, sugar cane was a major crop in Cuba, Martinique, and Guadeloupe. Up until the late 1870s, sugar cane was mainly exported to the United States and the rest of the world from countries like China or India. However, by 1877, it was being developed as a cash crop in Louisiana.

In the early Colonial period and through the 19th century, sugar cane was grown from stem cuttings. Each cutting contained three or four nodes and each node had a least one bud. The cuttings were planted horizontally in shallow furrows and then covered with soil, leaving the buds exposed. Each of the buried nodes would then grow new primary root systems (rhizomes). New stems would also grow from the buds. Once the new stems developed their own root systems, the original primary root systems died away. It took about twelve to eighteen months of growing before the sugar cane was ready to harvest (Galloway 1989, 12–13). Once it was ready, the sugar cane was cut and then taken to the processing factories, where the various forms of table sugar were produced as described above.

The nature of sugar cane was such that a plantation could often yield one or more additional crops from the same original root systems once the first (plant cane) crop had been harvested, as the remaining root system would later sprout new stems. This process of growth and re-growth could allow for several harvest seasons before replanting was necessary (Levetin and McMahon 1996, 57). Thanks to the renewable nature of sugar cane, a plantation owner could produce several crops from one planting, thus reducing labor costs and increasing profits. The downside of this was that subsequent crops, depending on soil fertility, lost their profitability over time (Galloway 1989, 13). Even now, sugar cane planting follows the same basic procedure only on a much larger scale.

After the sugar extraction process, the left over stems, known as mill trash or baggase, could be used for several purposes. For instance, in Brazil in the 1800s there was a shortage of firewood, so baggase was used as fuel for the newly developed factory processing furnaces (Galloway 1989, 98–9). Over the years, baggase has also been used commercially to make heavy-duty board and card (Galloway 1989, 235). Baggase could also be used to make paper or even compost (Levetin and McMahon 1996, 57). Thus, baggase became a profitable by-product of the sugar cane industry.

Towards the end of the 19th century, researchers learned how to breed high-yield sugar cane, and their hybrid plants soon replaced those of the Java type and others like it around the world (Galloway 1989, 12). By 1914, despite competition from sugar beet, sugar cane accounted for about 50 percent of the world's total sugar output. Much of World War I was fought in European sugar beet fields, devastating untold numbers of sugar beet crops. Indeed, by 1919, sugar cane was supplying 78 percent of the world's sugar, a percentage that to this day has never been surpassed (Galloway 1989, 235).

Today, cane sugar faces great competition. High fructose corn syrup (produced from maize (see previous entry for Corn) is used as a sweetener in numerous food products. With an increasingly health- and weight-conscious society, synthetic low calorie, non-tooth decaying sugar substitutes such as aspartame, saccharin, or phenylketonurics have come to prominence. Yet, despite this competition, cane sugar still manages to hold its own.

See: Caffeine, Alcohol, and Sweeteners, p. 186

Sugar beet *Beta vulgaris*
Chenopodiaceae

The sugar beet, a member of the parsnip family, had been thought of as a possible source of sugar since the 16th century. It was known for hundreds of years that the beets accumulated some form of sugar in their thickened roots, however agriculturalists did not know how to extract this sugar efficiently, in a controlled environment. In fact, it was not until some two hundred years later that this changed. In 1747 a Berlin professor, Andreas Marggraf, discovered a process for extracting sugar from red and white beets. By this process, Marggraf was able to obtain half an ounce of white sugar from half a pound of dried beet material. Although, this result was not as positive as he might have hoped, it was definite proof that the extraction of sugar was possible (Galloway 1989, 130).

Other European countries soon took notice of this potential new industry, but because of the low cost of imported cane sugar, they did not enter the sugar beet industry or trade. In order for the sugar beet industry to thrive, a proprietary market was needed and the Napoleonic Wars provided such a market. During these wars England set up a blockade of Europe, stopping the supply of cane sugar to the French and their allies. Napoleon's government had to turn to another source of sugar, mainly the sugar beet. However, due to various difficulties and with the resumption of cane sugar trade in 1815, the beet industry collapsed, if only temporarily (Galloway 1989, 130–131).

Due to the abolition of slavery in British and French colonies in 1833 and 1848 respectively, sugar cane was no longer as accessible to England and France so other sources of sugar were sought. Beets once again become Europe's main source of sugar. This reversal was also bolstered by the fact that sugar beet cultivation, as part of a crop rotation system, allowed other crops like wheat to grow more effectively. A further incentive was that until the 1840s Britain as well as other European nations did not tax beet sugar. By the 1840s beet sugar was taxed, but at this point the sugar beet industry was well established, and trade was booming with the sugar beet industry supplying eight percent of the world's sugar production. In 1845, Britain reduced beet sugar taxes in the interests of free trade and by 1845 cane sugar had lost most of its continental markets (Hobhouse 1985, 74–5). Between 1874 and 1901 Britain did not levy taxes on either beet sugar or cane sugar. By the end of the 19th century two-thirds of the sugar entering Britain was beet sugar (Galloway 1989, 130–4).

By the 1870s beet sugar had become the main food sweetener for England, France, Russia, Germany, and Canada. Sugar cane could not be grown in any of these countries, as sugar cane was strictly a tropical or semi-tropical plant (see previous Sugar cane entry). Although sugar beets could never truly replace the yield of sugar cane sugar, in terms of volume in Europe sugar beets were considered suitable substitutes. If compared to corn and sorghum (a plant similar to sugar cane) sugar beets lack sugar content. They tended to have about two percent less of the desirable sugars, and about five percent less sugar content than sugar cane. One of the biggest problems involved with beet manufacturing in 1870s' America was that it required more work and energy in planting, cultivating, harvesting, and processing, than was required for sugar cane or corn. The costs of processing sugar beets in America were much too high for them to be considered any further as a replacement for sugar cane.

By 1885, beet sugar overtook cane sugar in world trade. So easy was it to come by and cultivate sugar beets in the 1880s to the late twentieth century in Europe particularly, that even today if one of these countries so chooses they could become sugar beet self-sufficient (Hobhouse 1985, p. 74–5). Over the next hundred years the sugar beet trade industry was to fluctuate back and forth between gains and loses, but by 1996 sugar beets provided about 30% of the world's sugar, with the largest producers being France, Germany and the United States.

In the late twentieth century, sugar beet crops joined maize and soybeans as genetically modified crops. Genetically modified maize has been commercially grown in the United States and Canada for both animal and human feed since approval by the U.S. Environmental Protection Agency (EPA) in 1995, but acceptance of GM crops in Europe has been slower. In the 1990s, trials of herbicide-resistant sugar beet crops were carried out in Belgium, the Netherlands, Germany, England, and Ireland.

Though GM sugar beet is approved for sale in the United States, GM beet is not yet cleared for food use in Europe and acceptance of GMOs in the European market continues to face environmental and consumer concern. However, following a moratorium on the approval of new GM foods by a science advisory committee of the European Commission in February 1998, EU farm ministers in July 2003 approved new labeling regulations for GM food that would eventually open up the European market to GM food from the United States and other GM crop-growing nations. If this trend continues, there is a potential for hundreds of millions of dollars in profits for growers and producers of GM seeds. Only the future will show us if this economic boom will occur.

Of greater immediate significance are the ethical issues surrounding European Union subsidy of sugar beet production. High tariff barriers in Europe discourage import of cane sugar from developing countries, while export subsidies for sugar beet result in instabilities in world markets. Oxfam (2004) estimate that the European Union spent €3.30 in subsidies for every €1 worth of sugar exported. Oxfam argue that such subsidies are immensely damaging to sugar-beet growers in the poorest parts of the world.

See: Caffeine, Alcohol, and Sweeteners, p. 186

Tea *Camellia sinensis*
Theaceae

The tea plant has been grown in Asia for thousands of years, originating in China. The Chinese have drunk tea for over two thousand years, and dominated the tea trade up until the mid-19th century. Until the early 1800s, Canton was the only port in China through which trade with the rest of the world could pass. It was the primary supplier of tea to Europe, and the British Empire in particular. The British East India Company dominated the tea trade out of China throughout the British Empire and colonies. This trade route ran from Canton to London and then to the East Indies. From the East Indies, the tea would then travel to the "American" British Colony of Boston, and later to New York. However, this trade route was not straightforward and involved trading opium from Calcutta with the Cantonese in exchange for their tea. All parties would exchange goods offshore under cover of darkness. If any money exchanged hands, it was always in silver.

During the early 1800s, the political climate in the British Empire changed. Those in power wanted to reform these illicit trade practices and to open up trade with China to others besides the East India Company. Conversely, the Chinese authorities were anxious to curtail the ever-increasing Western influence on China, including the import of opium to China. Matters came to a head in 1839 when the Chinese seized all British-owned opium in Canton and prohibited trade with Britain. This led to the opium war of 1840–1842 between China and Britain, eventually won by the British forces. As a result, the British acquired Hong Kong and the tea trade was opened up.

The new tea traders designed and developed new ships to transport the tea directly from China to New York or London. These ships, "clippers", could store more goods and sail faster than the East India Company ships, completing the same journey in 90–120 days, half the time required previously. These improvements allowed traders to get tea and other goods to the various ports in less time, and in a "fresher" or "new season" state.

During the early 1800s, tea traders wanted to develop tea plantations in countries other than China in order to expand the tea trade. Tea needs a moist and warm climate with deep friable soil, high in humus, a pH level of 5.0–5.5, and plenty of available cheap labor. Based upon these criteria, countries such as India, Indonesia, and Sri Lanka (formerly Ceylon) seemed perfect candidates for tea cultivation. Tea seeds were taken and transplanted in various European "colonies" first by the East India Company in the late 18th and early 19th centuries, followed by the Dutch in 1827–1833, and then by other dominant tea-trading nations. Robert Fortune, the plant hunter, covertly took Chinese tea plants and seeds out of China between 1842 and the 1850s. The tea plant was extensively studied over many years in order that tea could be successfully cultivated in these other regions. As a result, tea plantations were established in 1860 in India, followed by Sri Lanka and Java in 1890 (Hobhouse 1985).

These new plantations used processing machinery to carry out various tasks that had been carried out entirely by laborers. However, the picking of the leaves was still done by hand. The workers at these plantations were "imported" from other countries by European trade barons and were paid very low salaries. The plantations, in particular those located in Assam, were managed by European (British) ex-civil servicemen, who ran them like military compounds. This autocratic socioeconomic method of agriculture became know as the "planter raj" system. So fierce became the tea garden/factory reputations that it led to the saying "better to go to the Andamans (a notorious penal colony) than to the tea garden" (Gardella 1994, 124–8). However inhumane or unprofessional the tea garden practices were in Assam, by the 1840s they had made Assam the center of tea production in British-controlled India and its surrounding regions. By 1859, Assam's tea output was over 1.2 million pounds (Gardella 1994, 124–5).

By the late 1800s, East Indian plantations already had advanced tea-processing machines. These machines, although still muscle-powered, were much more effective then the older, more traditional methods of tea production. In East India, a special winnowing machine (i.e., a fanning mill) was developed which separated the tea into five grades simultaneously. Tea was placed in the hopper at the top of the machine. One person would then regulate the flow of the tea into the machine, at the same time turning the fan crank. The tea would be blown across five or six troughs in the machine. The lightest tea and dust particles would be blown the furthest; they would hit the end of the machine and fall through the trough into a chute to be collected. The next lightest tea (the Broken Pekoe) was not blown as far, and would fall through another trough and chute. Thus, the Pekoe, Souchong, Coarse Souchong, and Gunpowder grades of tea were separated. So rather than laborers using one out of four winnowing sieves at a time and having to spend time changing sieves, this machinery meant that at least five grades of tea could be sorted extremely quickly. To speed up this process even further, a steam engine or horse treadmill could be belted to the tea-winnowing machine (Le Duc 1877, 349–67).

Most tea manufacturing factories/plantations in China, India, Sri Lanka, and Indonesia used similar processing technology to that described above. In addition, the more successful plantations used special tea rolling machines. Rolling of the tea sped up the process known as fermentation, which gave the tea its characteristic flavor and smell. Of all the technological innovations in tea processing, the development of rolling machines was by far the most important. Today, one machine can perform the work of up to one hundred laborers (Hobhouse 1985).

In the 1930s, the "cut-tear-curl" machine (CTC machine) became popular. This machine mass-produced a kind of "standardized" tea. In using this machine, plantation owners strived for the fastest tea production to bring in the quickest profits. This type of processing did eliminate the undesirable lowest quality tea; unfortunately, it also eliminated the most desirable highest quality tea as well (Hobhouse 1985, 127).

Cutting Burley tobacco and putting it on sticks to wilt before taking it into the curing and drying barn. Russell Spears' farm, near Lexington, Ky. Library of Congress, Prints & Photographs Division, FSA-OWI Collection LC-USF34-055375-D DLC.

Since the 1960s, tea-producing companies have tried many new and different ways to package and market their product. Developments of the late 20th and early 21st centuries included tea bags and dehydrated instant tea.

See: Caffeine, Alcohol, and Sweeteners, pp. 175–6

Tobacco *Nicotiana tabacum*
Solanaceae

In the 17th century, tobacco was one of the first crops grown strictly for commercial profit in what is now the southern United States. It was taken in the form of snuff (i.e., inhaled as a powder through the nose) or smoked in a pipe. John Rolfe was the first Virginia colonist to develop tobacco for export from the American Colonies in 1612. Over time, Virginia became one of the primary snuff and pipe tobacco producers in the world, and a major player in the world tobacco market. The formal name of this product was Dark Fire-Cured Tobacco of Virginia. As of 1905, Virginia Tobacco was still dominating the market even though the technology used in planting, cultivating, harvesting, processing, and packaging of tobacco was essentially the same as that used in the 1800s (McNess and Mathewson 1905, 222–7).

From the late 18th century to the early 19th century and onwards, tobacco was a major part of life in many European countries. It was mostly grown in Spanish Caribbean colonies such as Cuba and Costa Rica, which exported most of their tobacco to Spain; Spain could then regulate the Cuban tobacco trade with other countries. Since the late 18th century, Cuba's main tobacco export was in the form of its famous cigar. Its main market for the cigar was the northeast United States.

Havana cigars became the mainstay of the American wealthy and powerful. In order to maximize the potential of this important market, in 1817, Spain relaxed some of its monopoly restrictions on the sale and production of tobacco in Cuba. This was also an attempt by Spain to bolster Cuba's loyalty to the Spanish empire and government (Gately 2001,171–3).

With the arrival of this freer trade system, Cuban tobacco production increased significantly, allowing Cuba to open more trade routes. By 1845, tobacco had replaced sugar (cane) as Cuba's principal export (Gately 2001, 171–3.). By 1855, Cuba had 9,500 tobacco plantations, about 2,000 cigar manufacturers, and employed over 15,000 workers in both of these industries (Gately 2001, 172). The importance of the Cuban exports to the United States continued until Fidel Castro's Communist take-over of Cuba in 1959, which led to United States trade restrictions against all Cuban imports.

The most significant impact of industrialization on the Virginia tobacco plantations was in the area of field fertilization. Up until the mid 1800s, tobacco farmers used basic livestock manure or compost as fertilizer for their fields, spreading the manure over the tobacco seedbeds by hand or using shovels or special manure spreader wagons pulled by horse or oxen. Broadcast spraying of nitrate of soda (soda nitrate) over the seedbeds was also found to be beneficial to the growing process, giving a more stable, rapid, and abundant growth. This way, about fifteen pounds of nitrate of soda were applied to every 100 square yards of surface. In 1904, Virginia Dark Fire-Cured Tobacco plantations yielded about 800 pounds of tobacco per acre, slightly more than in the late 1800s. By 1904, however, commercial fertilizers were thought to have increased tobacco yield from about 800 pounds to 1,400 pounds per acre (McNess and Mathewson 1905, 222–3).

In 1839, a plantation slave named Stephen in Caswell County, North Carolina discovered what was to become the most widely consumed form of tobacco, that used in cigarettes. This plantation produced a lighter, non-intoxicating, smooth burning flavor called Piedmont. When by chance Stephen cured the tobacco with charcoal instead of logwood, the new golden colored Piedmont leaf, or 'yallacure,' was born. This type of easy-to-inhale tobacco soon set a precedent for all future cigarettes to come. This easy inhaling concept soon spread to the rest of the World (Gately 2001, 183–5).

It was not until the 1850s that cigarettes were manufactured or sold in Britain. One of the soldiers returning from the Crimean War, named Robert Peacock Gloag, set up the first British cigarette factory in 1856. His factory produced cigarettes in the Russian style; the tobacco used was in a dust format and the paper used to hand-roll the cigarettes was yellow tissue, while attached mouthpieces were made of cane. Yet at this point cigarettes did not catch on, and Gloag turned to making cigars. At about the same time as Gloag's business, the Bond Street manufacturing company Philip Morris also began to produce hand-rolled cigarettes (Gately 2001, 184).

Although cigarettes may not have caught on in Victorian Britain, tobacco in other forms certainly did. Cigars, mainly smoked by royalty and the rich, were a definite symbol of the upper classes, and the pipe became the symbol of the new and growing middle classes. As the middle class developed in financial scope and area of influence, so did their pipe smoking. Soon the middle classes, reflecting their growing financial success, were using more decorative and expensive pipes. The famous Meerschaum pipe bowls, made from the hydrated silicate of magnesia found mainly in Eskişehir, Turkey, were considered the pinnacle of pipes in Victorian middle class society (Gately 2001, 185–90).

By the 1880s, the largest producer of tobacco for the world market had become the United States, and much of this tobacco production was in the form of cigarettes. It was during this period that the tobacco industry was becoming mechanized. James A. Bonsack invented a machine that increased the number of cigarettes manufactured in one day from 3,000 to 120,000 cigarettes per factory (Licht 1995, 126–7). This increase in production and sale of cigarettes was further developed by James B. Duke, another American tobacco company owner and producer. He purchased two of Bonsack's cigarette machines, later acquiring the sole rights to the machine, and increased his company's production of cigarettes by tenfold. This allowed Duke to far out-produce his competitors. Through sales and advertising teams he actively promoted his product and very successfully increased his customer base. Thus, in 1890, Duke's company earned $400,000 in profits from the $4.5 million sale of 834 million cigarettes. An even more profitable development that same year was the merger by Duke with his four biggest competitors, to form the conglomerate American Tobacco Company (Licht 1995, 145).

By the turn of the 20th century, tobacco was being enjoyed in ever increasing quantities throughout the world. Cigarettes were being smoked in nations as far away from Europe as Asia. For example, in 1902, China alone consumed 1.25 billion cigarettes (Gately 2001, 220). Despite research into the health risk of cigarettes in the early 1900s, cigarettes continued to be sold in great amounts.

When World War I broke out in 1914, tobacco production got another boost. Cigarettes and cigars were considered to be important needs for the fighting soldier. In fact, cigarettes and cigars were distributed by the Red Cross to Austro-Hungarian soldiers in the form of rations. Cigarette rations were also distributed to British and Prussian soldiers. The mild narcotic attributes of cigarettes and their appetite-suppressing properties seemed to help ease the various physical and mental stresses of horrible trench warfare conditions. It became a common ritual to smoke a cigarette before a potential fatal assault.

In the 1920s, cigarette companies decided to pursue the previously un-tapped market of the female consumer. The companies began to manufacture slender and "prettier" cigarettes for women. Philip Morris led this new marketing and manufacturing scheme with its revival of the Marlboro cigarette. In 1924, its new brand slogan became "Mild as May." These cigarettes were "specially designed for a woman's needs." The cigarettes had ivory tips and were rolled with greaseproof papers, so as not to stick to lipstick. Originally a British tobacco company, by this time Philip Morris had developed trade and factories overseas in the United States, based out of New York. Thus, Philip Morris had truly become an international company and could afford to take a few sales risks (Gately 2001, 243–5).

The popularity of cigarettes and other tobacco products with both women and men (and, covertly, children) increased through the next four decades; even the Great Depression did not hinder cigarette sales. However, by the 1970s, all this was about to change. Advances in medical research and technology finally concluded without a doubt that tobacco was unhealthy and addictive for both animals and humans, and later it was determined to cause that 20th century pre-AIDS plague of the world, cancer. The results of these findings were government restrictions on tobacco smoking, advertising, production, and sales. In the United States, the vocal protests of nonsmokers against second-hand smoke inhalation—or passive smoking—led to the establishment of "No Smoking" zones in various public areas or, the reverse of this, segregated smoking sections. Cigarette ads were banned from television and tobacco companies were forced to put health-warning labels on cigarette packages.

European countries, such as Great Britain, did not totally escape this purging of their pleasure either. In 1971, the Royal College of Physicians produced the film, "Smoking and Health Now." The film described smoking as 'this present holocaust,' and led to British cigarette companies voluntarily putting warning labels on their packages. However, the British government considered smoking too financially profitable and not enough of a financial burden on medical treatment to restrict it. To ease its collective conscience, the government decided to raise tobacco company taxes (Gately 2001, 306–15). In the late 1970s, the disco craze led to a renewed interest in cigarettes. On the other hand, health concerns were still there, leading to the birth of a new industry, the "Quit Smoking" industry, which produced seminars and nicotine patches. By this time cigarette tar was isolated as one of the chemicals in tobacco that caused cancer, and nicotine as the chemical that caused addiction.

The period from the 1980s to the beginning of the 21st century saw a major boost in anti-tobacco sentiment around the world. But tobacco companies did not give up and they led successful campaigns of "incidental" product placement in various Hollywood movies like "Bladerunner." Another way that tobacco companies protected their assets, and increased profits, was to purchase food and beverage manufacturers. At the forefront of this purchasing boom once again were tobacco conglomerates like Philip Morris and R.J. Reynolds. By 1988, Philip Morris owned Miller Breweries, Kraft, and Philadelphia Cream Cheese. R.J. Reynolds followed suit by purchasing Del Monte and Nabisco. In the UK, British American Tobacco's diversification fell into the realm of insurance with the purchase of Eagle Star, Allied Dunbar, and Farmers Insurance.

Despite political, social, and economic adversity towards tobacco and smoking from the late 20th–early 21st centuries, many die-hard smokers still refuse to give up smoking. This will always be the case unless governments ban smoking all together. To this day, there are 1.2 billion smokers in the world.

See: Psychoactive Plants, pp. 201–2

Conclusion

Based upon all of the above, it is clear that industrialization had a significant impact on the development of worldwide agriculture from the late 18th and 19th centuries. Major strides in technology, as well as other factors, brought about this development. Developments such as winnowing machines, de-pulping machines, extracting machines, fertilizer, cordage making, scutching machines, and fiber "gins," together with financial incentives, meant that some previously insignificant countries in terms of world trade became major players. The use of all of these innovations as they became available impacted the average farmer, for better or for worse, often transforming small farms or plantations into commercially viable agricultural 'factories'. Those country farms and plantations which did not change with the times by and large fell by the wayside. The exception to this rule is Colombia, which to this day still carries out most of its agricultural tasks with human labor. New technology was responsible for transforming some countries from a state of poverty into new financially viable economic powers, a case in point being Guatemala. Yet one could also say that the United States of America was a country born of agricultural technology and industry.

Agricultural industrialization also led to breakthroughs in various chemical and scientific processes and to the development of entirely new ones, such as the genetic engineering of plants. These new areas of science led to the establishment of major business conglomerates, which provided a boost to global economy and advertising. Agricultural industrialization also effected social and political change. Most importantly it opened up the world to new trading opportunities and possibilities previously inaccessible. But along with these new processes and horizons came new controversies.

Perhaps the most negative or controversial aspect of the industrialization of agriculture was the slave trade. People were forcibly removed from their native countries or regions to work on plantations and in factories for complete strangers. These strangers in turn mistreated the laborers, and in some cases tortured them. The slave owners carried out these atrocities in the name of making the greatest profit in the shortest time. Yet, ironically, it was agricultural and industrial technology that ultimately made slavery commercially unnecessary. Industrialization of agriculture also led to the decline of many independent small-scale farmers who were not prepared to adapt to modern practices. In short, the industrialization of agriculture around the world had mixed results, some very positive and some not so positive.

References and Further Reading

Bielby, Derek. 2002. *The History of Hemp*. Website, www.hemp-union.karoo.net/main/info/history/book1.htm

Biggar, H.H. 1918. *The Old and the New in Corn Culture*. Office of Corn Investigations, Bureau of Plant Industry. Yearbook of the Dept. of Agriculture: U.S.D.A.

Daniel, P. 1985. *Breaking the Land: The Transformation of Cotton, Tobacco and Rice Cultures since 1880*. Chicago: University of Illinois Press.

Devine, T.M. 1975. *The Tobacco Lords: A Study of the Tobacco Merchants of Glasgow and their Trading Activities c. 1740–90*. Edinburgh: John Donald Publishers.

Dodge, B.S. 1984. *Cotton: The Plant That Would Be King*. Austin: University of Texas Press.

Dodge, C.R. 1879. *Vegetable Fibers in the Collection of the Department of Agriculture*. Yearbook of the U.S. Dept. of Agriculture: U.S.D.A.

Dodge, C.R. 1895. *Hemp Culture*. Yearbook of the U.S. Dept. of Agriculture: U.S.D.A.

Edwards, H.T. *Sisal and Henequen as Binder-Twine Fibers*. Yearbook of the Dept. of Agriculture: U.S.D.A.

Fussell, B. 1992. *The Story of Corn: The Myths and History, the Culture and Agriculture, the Art and Science of America's Quintessential Crop*. New York: Knopf.

Galloway, J.H. 1989. *The Sugar Cane Industry: An Historical Geography from Its Origins to 1914*. Cambridge: Cambridge University Press.

Gardella, R. 1994. *Harvesting Mountains: Fujiian and the China Tea Trade, 1757–1937*. Berkeley: University of California Press.

Gately, I. 2001. *Tobacco: A Cultural History of How an Exotic Plant Seduced Civilization*. New York: Grove Press.

Green, Constance McL. 1956. *Eli Whitney and the Birth of American Technology*. Boston: Little, Brown and Company.

Hobhouse, H. 1985. *Seeds of Change: Five Plants that Transformed Mankind*. London: Sidgwick & Jackson.

Hochberg, B. 1980. *Handspinner's Handbook*. Santa Cruz: Bette Hochberg.

Houck, J.P., Ryan, M.E., and Subotnik, A. 1972. *Soybeans and Their Products: Markets, Models, and Policy*. Minneapolis: University of Minnesota Press.

Hurt, R.D. 1994. *American Agriculture: A Brief History*. Ames: Iowa State University Press.

Hurt, R.D. 2003. *Problems of Plenty: The American Farmer in the Twentieth Century*. Chicago: The American Ways Series.

Le Duc, Hon. W.G. 1877. *The Chinese Tea-Plant*. Report of the Commissioner of Agriculture: U.S.D.A..

Le Duc, Hon. W.G. 1878. *Maize and Sorghum as Sugar Plants: Prefatory Remarks by the Commissioner*. Report of the Commissioner of Agriculture: U.S.D.A..

Le Duc, Hon. W.G. 1879. *Vegetable Fibers in the Collection of the Department of Agriculture*. Report of the Commissioner of Agriculture: U.S.D.A.

Levetin, E., and McMahon, K. 1996. *Plants and Society*. Dubuque: Wm.C. Brown.

Licht, W. 1995. *Industrializing America: The Twentieth Century*. Baltimore: Johns Hopkins University Press.

Liu, K. 1997. *Soybeans: Chemistry, Technology, and Utilization*. New York: Chapman & Hall.

McNess, G.T., and Mathewson, E.H. 1906. *Dark Fire-Cured Tobacco of Virginia and the Possibilities for its Improvement*. Yearbook of the Dept. of Agriculture: U.S.D.A..

Morse, W.J. 1918. *The Soy-Bean Industry in the United States*. Yearbook of the Dept. of Agriculture: U.S.D.A..

Nickerson, R.F. 1954. Cotton: history, growth and statistics. In *Matthews' Textile Fibers*, edited by H.R. Mauersberger. New York: Wiley.

Oxfam. 2004. Dumping on the world: how EU sugar policies hurt poor countries. Oxford: Oxfam International.

Quick, G., and Buchele, W. 1978. *The Grain Harvesters*. St. Joseph, MI: American Society of Agricultural Engineers.

Rakosky, J. 1989. *Protein Additives in Food Service Preparations*. New York: Van Nostrand Reinhold.

Wells, A. 1985. *Yucatan's Gilded Age: Haciendas, Henequen, and International Harvester, 1860–1915*. Albuquerque: University of New Mexico Press.

19
Invasives

VERNON HEYWOOD

Humans' exploitation of the plant kingdom through the ages can be described as a reshuffling of species around the globe in a giant game of botanical chess. The vast increase in global trade, travel, and communications in the last two hundred years has contributed to this mixing and exchange of plants around the world (McNeely 2001). While some of these introductions are accidental, agriculture and horticulture in most parts of the world are mainly dependent on species that have been deliberately introduced from other regions. Some of these plants that are brought into cultivation, whether for agriculture, forestry, or ornament, escape from where they are grown and become naturalized, replacing native vegetation. Such plants are known as *invasives* or *plant invaders* (Cronk and Fuller 1995) and form part of the group called *Invasive alien species* (IAS)—non-native organisms that cause, or have the potential to cause, harm to the environment, economy, or human health. They have been described as one of the most significant drivers of environmental change worldwide.

Invasive plants have had a very serious impact in many parts of the world. When they cause significant habitat transformation, leading to loss of biodiversity and reduction in ecosystem services, they are often known as *transformers* or *transformer species*. In Hawaii, for example, of the 20 percent of the land area not occupied by cattle pastures, plantations, and urban development, nearly half of the flowering plant flora is made up of naturalized species that dominate large areas of the vegetation (Wagner, Herbst, and Sohmer 1990). The abandoned sugar cane fields of Hawaii are turning into a strange synthetic ecosystem dominated by weed trees, including *Leucaena* from Mexico, *Spathodea* (African tulip tree) from tropical Africa, *Cecropia* from South America, and *Schleffera* and *Macaranga* from Asia. Sometimes a single invasive species may spread over large areas, such as the strawberry guava (*Psidium cattleianum*), a native of Brazil that forms dense, uniform stands in disturbed mesic and wet forest on all the main Hawaiian islands. This plant has had a similarly catastrophic effect on Mauritius and other islands in Polynesia and Australasia. In less than one hundred years, the hurricane sourgrass (*Bothriochloa pertusa*), a native of the Old World tropics, became the most common grass in lowland Jamaica, and is also widespread in other Caribbean countries and parts of North and South America. Even common European species, such as gorse (*Ulex europaeus*) and broom (*Cytisus scoparius*), have become invasive as far afield as India, Australia, South Africa, and North America; in the latter it has invaded more than one million acres in the states of Oregon and Washington.

In South Africa, many species that are today invasive aliens were introduced for food, forage, and shelter. Within days of landing at the Cape of Good Hope in 1652, Jan van Riebeck, chief merchant and governor, established the Cape Town garden on the lower slopes of today's city. Seeds brought from Europe were sown, and soon a fruit and vegetable garden was supplying fresh produce for the scurvy-ridden sailors. Van Riebeck sent an urgent plea to his masters in Amsterdam saying "send us anything that will grow", and a remarkable number of plant were introduced in the first few decades of the settlement— fifty or so crop plants in the first few years. Two centuries later, overgrazing on the Cape Flats led to serious erosion; to tackle this, many species of *Acacia* and *Hakea* trees were introduced for dune stabilization. These genera are now known to contain some of the most serious invasives in South Africa today. The Cape region is recognized as one of the world's great hot spots for plant life, and the fynbos—shrub and heath vegetation on the nutrient-poor soils of the Western Cape—is not only exceptionally rich in species, but highly susceptible to invasion. Indeed, much of the lowland fynbos has been seriously invaded by alien species such as *Acacia cyclops* and *A. saligna*, introduced in the mid-1800s from Australia; these are believed to be the primary cause of the extinction of fifty-eight plant species. In the mountain fynbos, species of *Hakea* (notably *H. sericea*, the silky wattle or needlebush) introduced in 1830, and pines and acacias have spread over wide areas. Alien trees such as *Pinus pinaster* were planted on the recommendation of the Colonial Botanist in the 1860s, and by foresters to increase the water supply in the mountain catchment areas (Huntley 1996).

In the New World tropics, subtropics, and warm temperate regions of the Americas, the deliberate introduction of African grasses, following the clearing of native forests and grasslands to create pasture for livestock grazing, had an extensive and significant effect on the appearance of the landscape (Williams and Baruch 2000). The introduced African C4 grasses (grasses with the C4 metabolism have a particular leaf structure that is well adapted to hot dry environments) escaped from cultivated pastures and invaded areas of natural vegetation at alarming rates. It is estimated that more than 100 million hectares of the humid and subhumid neotropics have been converted to grassland dominated by African C4 grasses, such as serrated tussock (*Nassella trichotoma*) in Australia.

Invasive plants are perhaps the most serious threat to biodiversity after habitat loss. So great is the concern that many countries have introduced regulations to combat the threat from invasive plants and animals. Article 8h of the Convention on Biological Diversity (CBD) calls for action by countries to "prevent the introduction of, control or eradicate those alien species which threaten ecosystems, habitats, or species", and the Conference of the Parties to the CBD has adopted interim guidelines for the prevention, introduction, and mitigation of impacts of invasive species.

The economic and social costs of invasive plants have seldom been properly assessed. In terms of direct costs, they represent billions of dollars annually. Only in the United States are detailed figures available: one study estimates that the total costs of invasive species amount to more than $100 billion each year (Pimental et al. 2000).

The number of invasive plant species runs into thousands, and individually they affect most areas of the world to a greater or lesser degree. A number of representative examples will now be described.

Black wattle *Acacia mearnsii*
Fabaceae

Several species of *Acacia* are serious invaders in various parts of the world. The black wattle is also listed by the Global Invasive Species Database one of the one hundred "world's worst invaders." It is a small- to medium-sized, evergreen leguminous tree, native to Southeast Australia and Tasmania. It is grown in eastern and South Africa, Zimbabwe, India, and South America (Rio Grande do Sul) for tannin, fuel, nitrogen fixation, erosion control; *Acacia* is also a good timber tree, producing wood that polishes well. It has become invasive in Hawaii, New Zealand, South Africa, and the Indian Ocean islands of Réunion and Seychelles.

It was introduced into the Cape Town Botanic Garden, South Africa, in the 1850s and has subsequently invaded shrublands, grasslands, and savannas in the Cape region, especially along water courses, and now occupies 6 million acres (2.5 million ha). On the island of Réunion, it was introduced in 1887 and subsequently planted in the 1950s by the Forestry Service, mainly on high slopes. It became invasive in 1962 when the *Pelargonium* production industry declined, and it occupied the lands relinquished by the growers colonizing ravines, river valleys, and gullies. Today it also occurs in ravines at low altitudes, probably with the help of hurricanes that spread the seeds. The main invasion patches consist of long and narrow bands limited by the boundaries of abandoned parcels of land (Balent and Tassin 1999). The plant has become so invasive partly due to its ability to produce large quantities of seeds. Germination of the seeds and resprouting by basal shoots are stimulated by fire, leading to the formation of dense, impenetrable thickets. *Acacia is* controlled is by mechanical means, such as uprooting, or by cutting the stems and treating them with herbicides to prevent resprouting. Glyphosate is used to control seedlings and saplings.

Another highly invasive species, golden wreath wattle or Port Jackson willow (*Acacia saligna*), was a native of Australia introduced into South Africa in about 1850 to stabilize the sands of the Cape Flats and elsewhere in the fynbos. It is now the most troublesome weed in the lowlands. A third species, rooikrans (*Acacia cyclops)* was also introduced for the same purpose at this time and is also highly invasive (Cowling and Richardson 1995).

In southern Europe, silver wattle (*Acacia dealbata*) is widely naturalized, notably in the Côte d'Azur. Although it is sometimes planted for timber, soil stabilization, or a source of tannin, it is mainly grown as an ornamental mimosa, providing considerable quantities of cut flowers to the European flower trade.

Gorse, Whin, Furze *Ulex europaeus*
Fabaceae

Gorse is a spiny, evergreen shrub with an extensive woody root system. It produces masses of bright yellow, pea-like flowers that form small hairy pods (legumes) containing three to eight seeds. The pods burst open explosively and expel the seeds, which possess elaiosomes (lipid-rich, fleshy appendages) and are distributed by ants. It is native to Atlantic parts of western Europe and has been introduced in many parts of the world, where it is cultivated as an ornamental, and used as a hedge or for fodder. Gorse is a serious pest, naturalized in scattered localities throughout central Europe; despite its temperate origins, it has also been introduced to many montane regions of the tropics, including Peru, Réunion, Sri Lanka (where it is now naturalized), St Helena, and Hawaii. It is invasive in many temperate areas and has been recorded as an invader in Australia, Canada, China, Indonesia, New Zealand, the United States, Mauritius, Réunion, and South Africa. On the island of Réunion, it has extensively invaded heathlands, particularly in areas disturbed by cultivation and grazing. It forms monotypic stands in South America, including in previously forested areas. It is a major pest in parts of South Australia, much of Tasmania, and the greater Melbourne area of Victoria. It is also a major environmental weed of the Blue Mountains in New South Wales and parts of western Australia. It is often a troublesome weed in old gold mining areas and cooler grazing areas. In Hawaii it was introduced as a hedge plant, and now invades pasture and roadsides. In mainland North America it is primarily restricted to counties bordering the Pacific Ocean, where it has replaced much of the vegetation in salt spray meadows along the coast. Gorse is highly flammable, and the town of Bandon, OR, burned down in a 1936 wildfire that spread rapidly through the gorse-infested area. The plant regenerates rapidly after fire, by resprouting and from the soil seed bank. Gorse is difficult to control once established. Chemical methods are not very effective or feasible, while biological control using a species of weevil has been tried in Australia, Chile, and New Zealand. Experiments using native and alien trees to suppress gorse are underway on Maui and Hawaii Islands in the Hawaii chain. Native *Acacia koa* appears to be effective within five years. In the long term these methods may be more effective than herbicides, which provide quick suppression but are not a permanent solution (Tulang 1992).

Killer alga *Caulerpa taxifolia*
Algae

Caulerpa taxifolia, the killer alga, is a handsome tropical seaweed that was introduced into aquaria as an ornamental. Cuttings were acquired by the Oceanographic Museum of Monaco, where it was grown in 1982; two years later it was found in the ocean beneath the building, probably having been thrown out as waste. There is speculation that the species released into the Mediterranean was a hardier clone of the original tropical seaweed. It adapted well to colder waters and has spread throughout the northern Mediterranean, where it is a serious threat to the native marine flora and fauna. New colonies are able to start from small segments of this plant and, being an opportunistic hitchhiker, it is a threat to the whole Mediterranean. Wherever it has established itself, *Caulerpa* has smothered such habitats as the beds of native sea grass that serve as nurseries for many species. On June 12, 2000, divers in a lagoon near San Diego discovered a patch of *Caulerpa* measuring 65 feet by 30 feet (20 m by 10 m). It is thought that the infestation occurred after somebody emptied a fish tank into a storm-water drain. Luckily this invasion was discovered at an early stage and measures were taken to eradicate it (IUCN 2001).

Koster's curse *Clidemia hirta*
Melastomataceae

Koster's curse is named for the man believed to have introduced the noxious plant to Fiji, where it is a serious pest. It is an erect, hairy, densely-branched shrub up to 16 feet (5 m) tall, with ovate to oblong-ovate leaves 1.5–6 inches (4–15 cm) long, and white or pinkish flower axillary (growing between the leaf and stem) clusters that produce blue-black berries containing over 100 seeds. A mature plant can produce 500 to 1,000 of these berries each year. It is native to the Caribbean and tropical South and Central America. Once established, it forms dense thickets and suppresses most ground vegetation; even small patches are virtually impossible to eradicate. It is now an invasive pest on islands such as Fiji, Seychelles, Hawaii, Java, Samoa, and India and East Africa. The seeds are dispersed by birds and also form a very large soil seedbank. In Hawaii, seeds are locally spread by birds and feral pigs, with long-distance dispersal carried out by humans. According to Binggeli, Hall, and Healey (1998), the date of introduction to Hawaiian Islands is unknown, and it was first observed in 1941. *C. hirta* was grown in the Wahiawa Botanic Garden, and thought to be "very promising because it won't be spread by birds." It was first noted as having escaped in 1949 on the island of Oahu, and by 1952 covered at least 247 acres (100 ha). By the late 1990s it had invaded all suitable habitats, covering over 247,000 acres (100,000 ha). In the 1970s and 1980s it was accidentally introduced by humans to five other Hawaiian islands. On Fiji, *C. hirta* was probably accidentally introduced prior to 1890 with coffee plants imported from British Guiana, and by 1920 had become invasive. It has affected large areas of grazing land and plantations of cocoa and rubber. It was introduced into Java towards the end of the 19th century. The success of *Clidemia hirta* as an invasive is due to several reasons: its ability as a primary colonizer of habitats subjected to regular disturbance, such as pastures, roadsides, landslides, and riverbanks; its prolific seed production and effective dispersal; high germination rates and rapid growth rate; and its ability to form dense thickets that crowd out and displace native species. Biological control has so far met with limited success. The thrips *Liothrips urichi*, a sucking insect that attacks the terminal shoots, was released into Fijian fields in 1930, and then in 1953 in Hawaii. The thrips are effective in limiting the plant's growth in open sunny areas and reduce its competitive ability, although they are less effective under shady conditions. They have not been effective in controlling infestations in Fiji. Other biological control agents are undergoing trials. Physical means of control, such as removal of seedlings, are not feasible because of the scale of the infestations.

Lead tree, Wild tamarind *Leucaena leucocephala*
Fabaceae

The lead tree is a small, fast-growing, deciduous tree or shrub, up to 60 feet (18 m) tall, forming dense stands, with evergreen leaves, 4–10 inches (10–25 cm) long, which are fetid when crushed. The numerous white flowers occur in dense global heads (1–2 cm) on long axillary stalks. Pods (legumes) 3–7 inches (8–18 cm) long are produced, containing huge quantities of glossy brown seeds. It flowers and fruits throughout most of the year. *L. leucocephala* is native to Central America, although its status in some countries is uncertain. It is often called the "miracle tree" because of its remarkably fast growth and its many uses: fuelwood, fiber, fodder, erosion control, contour planting, water conservation, reforestation, soil improvement, as a shade plant in coffee, rubber, cacao, pepper, and cinchona plantations, windbreaks and firebreaks, and as a good cover and green manure crop. In the Philippine Islands, young pods are cooked as a vegetable, and seeds are used as a substitute for coffee. The seeds are also used as jumbie beads in necklaces. On the other hand, both the leaves and the seeds contain the amino acid mimosine, and can be toxic in large quantities and cause loss of hair in non-ruminant animals, such as horses. It has been widely cultivated in most tropical countries, covering up to 12 million acres (5 million ha); it regenerates readily and often escapes from cultivation and becomes naturalized and, in many cases, invasive. The origin of *Leucaena* is uncertain and the taxonomy of *L. leucocephala* is complex: it contains three subspecies, subsp. *leucocephala*, subsp. *glabrata,* and subsp. *ixtahuacana* (ILDIS World Database), and a large number of varieties and cultivars contribute to its gene pool. The commonest form in the tropics is the so-called "Hawaiian" type, possibly native to coastal Mexico, while the "Salvador" type is taller, tree-like, and less aggressive, and the "Peru" type, which is also tree-like, contains several cultivars that are good for forage. Lead tree was brought to Philippines by the Spanish about 400 years ago, and in the 19th century was widely introduced throughout the tropics as a fodder plant. It is invasive in the West Indies, parts of South America, the Pacific, Japan, India, north Australia, and Africa (Kenya, Tanzania). It was first collected in Hawaii in 1837, where today it sometimes forms dense thickets and is the dominant element of the vegetation at low elevations in disturbed habitats (Wagner, Herbst, and Sohmer 1990), and to the Marquesas Islands prior to 1893, where it is regarded today as an agricultural weed and is also replacing native vegetation. It is now common in the lowlands of many Pacific islands, often forming monotypic stands. In Australia it has been declared as a noxious weed in twenty counties, and in some municipalities in northern Australia, although it is not reported to invade undisturbed vegetation. Control of the Lead tree is made difficult because of the conflict between its enormous economic importance (which would be compromised by control measures) and the economic and environmental damage it causes as a weed and invasive. Biological control through the use of the psyllid insect *Heteropsylla cubana*, which causes defoliation, can be successful, while mechanical control is difficult because it is nearly impossible to uproot completely from cultivated land.

Melaleuca, Niaouli, Punk tree *Melaleuca quinquenervia*
Myrtaceae

Melaleuca is an evergreen tree up to 80 feet (25 m) tall, with spongy, whitish or pale-brown, paper-like bark, and leathery leaves 2–8 inches (5–20 cm) long, with a characteristic aromatic odor when crushed. The flowers are white or cream-colored, cylindrical, bottlebrush-like spikes, and the fruits are small, woody, short, cylindrical, button-like capsules, containing numerous minute seeds. It is native to the coastal regions of eastern Australia, New Caledonia, Papua, New Guinea, and Irian Jaya.

Melaleuca quinquenervia is the source of niaouli oil and, like tea tree and cajuput oils from other *Melaleuca* species, is widely used in industry as a solvent, cleaning agent, in cosmetics, as a flavoring, as an antiseptic and medication, and the wood is employed for a variety of purposes. It has been widely introduced into Africa, Central America, Florida, Hawaii, India, the Philippines, Puerto Rico, South America, and the West Indies. It is an aggressive invader in the United States. It is confined to southern Florida, where it was introduced as an ornamental in 1906, occupying several million acres, primarily within the Everglades and surrounding areas, where the trees grow into immense impenetrable monospecific forests, virtually eliminating all other vegetation. Seeding from airplanes was carried out in the 1930s, and in the 1940s trees were planted inland (Binggeli, Hall, and Healey 1998). It was first cultivated in Hawaii in 1920, and subsequently over 1.7 million trees have been planted, but natural regeneration is limited to the island of Maui, where it is now naturalized in undisturbed mesic forest.

Oxford ragwort *Senecio squalidus*
Asteraceae

The Oxford ragwort is so-called because it originated in the University of Oxford Botanic Garden from seeds sown by Jacob Brobart, the head gardener, of plants collected in Sicily. Recent studies have shown that Oxford ragwort is not a true species, but actually a hybrid derived from two other *Senecio* species—*S. aethnensis* and *S. chrysanthemifolius*—that grow on the slopes of Mount Etna in an altitudinal zone of 3,200–3,600 feet (1,000–1,800 m) between those of the two parental species (Abbott et al. 2000, 2002). It is a branched, straggling, annual to perennial herb, somewhat woody at the base, up to 20 inches (50 cm) tall, with dark green leaves. The large yellow flower heads, 0.6–1.3 inches (1.5–3.5 cm) in diameter, are in loose, flat-topped clusters with conspicuous black-tipped bracts; fruits have silky hair facilitating wind dispersal. All parts of the plant, especially the leaves, contain alkaloids (jacobine, seneciphylline) that cause irreversible liver damage in horses, cattle, goats, and humans. Oxford ragwort escaped from the Oxford Botanic Garden in 1794, when it was first recorded growing on the walls of the city. It then spread along the newly created railway system, whose limestone ballast provided a well-drained medium, to colonize much of the country, where it is found in waste ground, railway banks, and other disturbed habitats. It is one of the most rapidly spreading alien weeds in Europe and occurs in Albania, Austria, Bulgaria, the Czech Republic, Germany, Greece, Crete, Italy (including its native Sicily), Romania, Switzerland, and Yugoslavia. It has been declared a noxious weed in California.

Prickly pear, Barbary or Indian fig, Mission cactus *Opuntia ficus-indica* and Prickly pear, Erect prickly pear, Southern spineless cactus *Opuntia stricta*
Cactaceae

Prickly pear cacti (*Opuntia* spp.) are native to North, Central, and South America. Prickly pears can reach a height of 9–16 feet (3–5 m), with flattened green, wax-covered, usually flat, oval joints known as pads (technically cladodes), 1–2 feet (30–60 cm) long, 8–15 inches (20–40 cm) wide, and 0.7–1 inches (19–28 mm) thick. The fruit is a purple or red berry, 1.5–3.5 inches (4–9 cm) long, with reddish pulp that contains many small seeds. Several species (platyopuntias) have been introduced in many parts of the world, where they are grown for ornament, for their spiny edible fruits (hence prickly pears), edible pads called nopales or nopalitos, as forage plants, as living fences to mark territories, or for the production of a red dye, cochineal, that is obtained from a scale insect that some species host. A number of species have become naturalized and, in some cases, invasive.

One of these is *Opuntia ficus-indica*, which was introduced around 1500 from Mexico (its exact native distribution is not known) into Spain and, later, throughout the Mediterranean basin by sailors who used it as a vegetable to prevent scurvy. Later changes in land use have led it to become naturalized and invade natural and semi-natural habitats in the Mediterranean region, where *Opuntia* can be considered a noxious weed. The plants spread vegetatively by the pads although seedlings are also produced and play a role in establishment and spread of the species.

Cochineal is a traditional and highly prized red dye of pre-Hispanic Mexico, produced by a scale insect that feeds on prickly pears. Over a dozen *Opuntia* species, including *O. ficus-indica*, have been used for cochineal production. Following colonization by Spain, cochineal production became a Spanish monopoly for 250 years. This monopoly was broken in 1777, when the French botanist Thierry de Menonville smuggled some cactus pads with scale insects out of Mexico to Haiti, where he unsuccessfully tried to introduce it into cultivation. Later, material was taken to Guatemala, South America, India, the Canary Islands, and Portugal. Cochineal production became the leading export from the Canary Islands between the 1820s and the 1850s, especially in Lanzarote. However, following the gradual introduction in the 1870s of the synthetic red aniline dyes that replaced cochineal, the cultivation of *Opuntia* species was greatly reduced, and today it is a relict crop in the Canaries; prickly pears have escaped into the wild to become a serious invasive.

Opuntia stricta is a prickly pear native to the southeastern United States. It is invasive in Australia and South Africa and is listed by the Global Invasive Species Database as one of the one hundred "world's worst invaders." It was probably introduced into South Africa by collectors of succulents, and competes vigorously with more valuable pasture species. It forms dense spiny thickets that harm livestock and prevent them from getting to grazing land. It has invaded thousands of hectares, including large areas of the Kruger National Park in South Africa. It is also invasive in Yemen and Eritrea, and occurs in Ethiopia and Somalia. It is difficult to eradicate prickly pears, as fragments of the leaf pads regenerate readily, but biological control has been successful in Australia and Hawaii by the introduction of two cactus-eating insects and is being evaluated in South Africa.

Common rhododendron *Rhododendron ponticum*
Ericaceae

Another invasive species in Britain of hybrid origin is *Rhododendron ponticum*, an evergreen shrub introduced to Britain from southern Spain at the end of the 18th century, and later from Turkey, as an ornamental. It was widely planted and used as cover for game in large estates. *R. ponticum* subsequently escaped and became naturalized (and often invasive), replacing native woodlands in the British Isles and Ireland. Recent research indicates that it is at least partly, possibly largely, hybrid, formed in Britain between Spanish *R. ponticum*, American *R. catawbiense*, and other species (Milne and Abbott 2000).

Strawberry guava, Purple guava, Cattley guava *Psidium cattleianum*
Myrtaceae

The strawberry guava is a shrub or small tree, 6–18 feet (2–6 m), with a smooth trunk and dark green, aromatic, leathery leaves. The white flowers produce red to purplish-red, occasionally yellow, berries that can be eaten fresh or made into jams or other preserves. It is native to the Atlantic Forest of Brazil, from Espirito Santo south to northern Uruguay, but is widely cultivated and naturalized in many parts of the tropics and subtropics. It is one of the most serious invaders of montane tropical rainforest (Cronk and Fuller 1995) in a number of island ecosystems, including Hawaii, Fiji, Tahiti and the Cook Islands (Rarotonga and Mangaia), as well as Réunion and Mauritius in the Indian Ocean. In Hawaii it is found in all the main islands and is thought to have been introduced there for its fruit on the voyage of the *Blonde* in 1835; it often forms dense monotypic thickets in moist, lowland, and submontane forests and poses a major threat to the native biodiversity (Wagner, Herbst, and Sohmer 1990). It was introduced around 1822 in Mauritius, where it is invading the upland humid forests. It shows very aggressive behavior in indigenous forests of Norfolk Island and elsewhere in the Pacific, and there is concern that it could become a serious problem in New Zealand. It is also invasive in parts of peninsular Florida. The features that make it such a serious invader are its rapid, shade-casting, vegetative growth with a dense system of surface feeder roots; its prolific fruiting with the seeds being produced all year round and dispersed by birds and feral pigs; and its tolerance of shade. Hundreds of germinating guava seedlings can be found in a single pig dropping, so feral pig management is a necessary first step in control management. Manual, mechanical, and chemical methods of control are practiced, and experiments on biological control are underway.

Another South American species, common guava (*P. guajava*), is a serious invader in the Galapagos, Hawaii, Africa, and New Zealand.

Water hyacinth *Eichhornia crassipes*
Pontederiaceae

The water hyacinth is an attractive plant, with thick, fleshy, horizontal roots, and clusters of rounded or oblong shiny green leaves 9 inches (25 cm) in diameter, with bulbously inflated petioles. It has showy, pale violet flowers with a bright yellow spot on the large upper lobe. A native to the Amazon basin, the water hyacinth has become a pan-tropical invasive species, and is reported from Central America, the United States (California and southern states), Africa, India, Asia, Australia, New Zealand, and various Pacific islands. It is one of the most noxious freshwater aquatic invasive species known, and listed by the Global Invasive Species Database as one of the 100 "world's worst invaders." In many parts of Africa and the Middle East, infestation of lakes and rivers by the water

Water hyacinth (*Eichhornia crassipes*). U.S. Fish and Wildlife Service.

hyacinth has reached crisis proportions, with the cumulative environmental, economic, and social damage to date estimated to run into billions of dollars. It blocks waterways such as stretches of the Suez Canal, limiting boat traffic, swimming, and fishing, and has especially serious effects on communities that live adjacent to river-bank areas. In Lake Victoria, it clogs the ferry terminals in Uganda, Tanzania, and Kenya, and its effects on agriculture, power generation, fisheries, transport, recreation, water supply, and trade have been devastating. The water hyacinth multiplies vegetatively by means of stolons (a stem or runner that forms roots at the nodes), which, together with solitary plants or drifting mats, are readily distributed by water currents, wind, boats, and rafts. The plant also produces vast quantities of long-lived seed that can also be a major factor in the spread and persistence of the weed. Because it grows so rapidly, it forms dense mats that prevent the growth of native submersed and floating-leaved plants and so reduces biological diversity in aquatic ecosystems. The dense floating mats reduce the levels of oxygen in the water and impede water flow; this creates good breeding conditions for mosquitoes. In the United States, it is believed to have been first introduced at the World's Industrial and Cotton Centennial Exposition of 1884–1885 in Louisiana. A Florida visitor to the Exposition apparently returned home with water hyacinth plants and subsequently released them into the St. John's River. Water hyacinth may be controlled by chemical means, using herbicides, but this has not proved effective for large-scale infestations such as occur in the Sudan; by mechanical means, such as harvesting with specially designed machines, hand pulling or dragline; or by biological control, using two different species of weevil and a moth, but with limited success.

White sage *Lantana camara*
Verbenaceae

White sage is one of the most widely recognized shrubs of tropical and subtropical regions. It is a highly variable shrub, with dense axillary heads of flowers that open from the outside inwards, with yellow or sometimes orange corollas, turning pink, red, orange, rarely white, purple or blue. The fruits are fleshy drupes, 0.1 to 0.2 inches (3 to 6 mm) in diameter, containing one to two seeds, and mature rapidly, changing from dark green to purple or black. The seeds are dispersed by birds. It is native to Central and South America and the West Indies, although its exact native distribution is difficult to determine, as non-native forms have been introduced. It has been cultivated for over 300 years with numerous cultivars and hybrids having been produced, and it is widely introduced as an ornamental in tropical and sub-tropical regions, where it has also been used for hedges and erosion control. White sage has escaped and become a weed or naturalized and invasive, especially in disturbed areas, forming extensive impenetrable thickets in forest plantations, orchards, pasture land, waste land, and areas of natural vegetation. It is highly toxic to grazing animals and may cause death in children if a quantity of unripe berries are eaten. It is a major invasive species in the Galapagos Islands (where it is seen as a threat to bird breeding populations and plant communities containing rare endemics), Tonga, Cook Islands, New Caledonia, Hawaii, and other islands in Oceania, and in the Indian Ocean islands of Madagascar, Réunion, and Rodrigues. It is a serious pest in South Africa, where it occurs in 5.5 million acres (2.2 million ha) of forest and plantation margins, water courses, savanna and covers some 1 million acres (400 000 ha) in Natal alone. White sage has also appeared in Australia, India, Southeast Asia, China, Thailand, Cambodia, Vietnam, Malaysia, Indonesia, Philippines and North America. It has been placed on the noxious weed list of New Zealand and is banned from propagation, sale, and distribution there. In Tanzania it may be considered a serious health threat, as its dense thickets provide breeding grounds for Tsetse flies infected with trypanosomes of domestic animals (Binggeli, Hall, and Healey 1998). The success of *Lantana camara* as an invasive species is due to a range of different factors: bird dispersion allows long-distance dispersal and colonization of new disturbed areas; its production of allelopathic substances in the roots and shoots increases its competitive ability; it spreads readily by vegetative means; it has long periods of flowering and prolific seed production; is toxicity to many mammals; and is resistant to herbivory. Biological control has been attempted in many countries where *Lantana camara* is a serious pest, with as many as thirty-three species of insects being tried out. Success is variable, partly depending on the climatic conditions, and partly on the particular cultivars concerned. In Hawaii, despite a successful biological control program that eventually eliminated *Lantana camara* from large areas, some were subsequently invaded by *Schinus terebinthifolius*, the Brazilian pepper tree (Andres 1977). Physical control is labor intensive but can be effective in some areas, and needs to be followed up with removal of roots and seedlings. Chemical control is expensive and not very effective because of the risk of reinfestation by seedlings.

References and Further Reading

Abbott, R.J., James, J.K., Irwin, J.A., and Comes, H.P. 2000. Hybrid origin of the Oxford ragwort, *Senecio squalidus* L. *Watsonia* 23, 123–138.

Abbott, R.J., James, J.K., Forbes, D.G., and Comes, H.P. 2002. Hybrid origin of the Oxford ragwort, *Senecio squalidus* L.: morphological and allozyme differences between *S. squalidus* and *S. rupestris* Waldst. and Kit. *Watsonia 24*, 17–29.

Andres, L.A., 1977. *The Economics of the Biological Control of Weeds*. Aquatic Botany 3, 111–123

Balent, G., and Tassin, J. 1999. Landscape level to assess *Acacia mearnsii* invasion in the Reunion island (Indian Ocean). *5th International Conference Ecology of Invasive Alien Plants*, 13–16 October 1999, La Maddalena, Sardinia, Italy. Abstracts.

Binggeli, P., Hall, J.B. and Healey, J.R. 1998. An overview of invasive woody plants in the tropics. Bangor: University of Wales, School of Agricultural and Forest Sciences Publication Number 13.

Cowling, R., and Richardson, D. 1995. *Fynbos. South Africa's Unique Floral Kingdom*. Vlaeberg, South Africa: Fernwood Press.

Cronk, Q.C.B., and Fuller, J.R. 1995. *Plant Invaders. The Threat to Natural Ecosystems*. London: Chapman & Hall.

Goodland, T.C.R., Healey, J.R., and Binggeli, P. 1998. *Control and management of invasive alien woody plants in the tropics*. UK School of Agricultural and Forest Sciences Publication Number 14. University of Wales, Bangor, UK.

ILDIS World Database. http://www.ildis.org/

Herfjord, T., Østhagen, H., and Sælthun, N.R. 1994. *The Water Hyacinth: With Focus on Distribution, Control and Evapotranspiration. Publication No 1*. Oslo: Norwegian Water Resources and Energy Administration, Office of International Cooperation.

Huntley, B.J. 1996. South Africa's experience regarding alien species: impacts and controls. In *Proceedings Norway/UN Conference on Alien Species*, edited by O.T. Sandlund, O.T. Schei, and Å Viken. The Trondheim Conferences on Biodiversity, 11–15 July, 1996. Trondheim, Norway: Directorate for Nature Management (DN) and Norwegian Institute for Nature Research (NINA).

The Invasive Species Specialist Group of the Species Survival Commission of The World Conservation Union (IUCN). 2001. *100 of the World's Worst Invasive Alien Species. A selection from the global invasive species database*. Auckland: Invasive Species Specialist Group (ISSG).

McNeely, J.A. (Ed.). 2001. *The Great Reshuffling: Human Dimensions of Invasive Alien Species*. Gland, Switzerland and Cambridge: IUCN.

Milne, R.I., and Abbott, R.J. 2000. Origin and evolution of invasive naturalized material of *Rhododendron ponticum* L. in the British Isles. *Molecular Ecology* 9, 541–556.

Myers, J.H., Bazely, D., Usher, M., Saunders, D., Dobson, A. and Peet, R. (Eds.) 2003. Ecology and Control of Introduced Plants. Cambridge: Cambridge University Press.

Perrings, C., Williamson, M., and Dalmazzone, S. (Eds.). 2000. *The Economics of Biological Invasions*. Cheltenham, UK: Edward Elgar.

Pimental, D., Lach, L., Zuniga, R., and Morrison, D. 2000. Environmental and economic costs associated with introduced non-native species in the United States. *Bioscience 50*, 53–65.

Plant Talk. 2001. The cost of invasive plants. *Plant Talk 24*.

Tulang, M. 1992. The U.S. Department of Agriculture's rural development approach to alien plant control in Hawaii: a case study. In *Alien Plant Invasions in Native Ecosystems of Hawaii: Management and Research*, edited by C.P. Stone, C.W. Smithand J.T. Tunison. Honolulu: University of Hawaii Press, 577–583.

Wagner, W.L., Herbst, D.R., and Sohmer, S.H. 1990. *Manual of the Flowering Plants of Hawai'i 1*. Honolulu: University of Hawaii Press and Bishop Museum Press.

Williams, D.G., and Baruch, Z. 2000. African grass invasion in the Americas: ecosystem consequences and the role of ecophysiology. *Biological Invasions 2*, 123–140.

20
Conservation of Wild Plants

DAVID R. GIVEN AND NIGEL MAXTED

David Brackett, chair of the International Conservation Union's Species Survival Commission, asserted that, "If you like to breathe and you like to eat, you should care more about plants" (Given 1998). We share the world with perhaps as many as 30 million types of organisms, of which most are insects, but at least 235,000 are flowering plants, and 325,000 are the non-flowering algae, lichens, mosses, and fungi. Through photosynthesis, the conversion of the sun's energy to other forms of energy, plants provide the vital link that allows life as we know it to exist on Earth. In a real sense, "all flesh is grass."

Distribution of Wild Plant Diversity

Plant biodiversity is not evenly distributed across the surface of the Earth; in fact, relatively few plant genera or families are found worldwide. Most species have very discrete geographic and ecological distributions. As a result, there are uneven concentrations of diversity, with some regions being very rich in species and genera while others are relatively poor. Table 20.1 shows that plant species' diversity increases as one moves away from the poles towards the equator. Furthermore, even at similar latitudes, some ecosystems have higher natural levels of species diversity than others.

In the 1980s, ecologist Norman Myers identified ten globally important centers of diversity (mostly tropical moist forest) and later a further eight that are mostly Mediterranean-climate regions (Myers 1988, 1990). A recent monumental analysis of biological hot-spots identifies twenty-five areas on Earth with an unusually high concentration of species, many endemic to (only naturally occurring in) that area (see Table 20.2). These twenty-five regions have the greatest number of species, distinctive ecosystems, and unusual combinations of ecological characteristics. But, in many instances, they are also the places under intense pressure from development. Although not strictly hot-spots, the Galapagos and Juan Fernandez Islands are particularly important in terms of island-based plant diversity and should be regarded as mini hot-spots. These twenty-seven areas alone support about 131,339 endemic plant species, about 44 percent of the world total, which together with the non-endemic species means that about 70 percent of all vascular plants are contained in 1.44 percent of the world's surface.

It might be assumed that within-species genetic diversity would have a distribution pattern similar to that of species, but this is often not the case. Wide-scale study of genetic variation patterns is

TABLE 20.1   Regional Distribution of Higher Plants (World Conservation Monitoring Centre 1992)

Region	Sub-region	Number of Species Sub-region	Number of Species Continent	Endemics Number	Endemics %
Europe			12,500	3500	28
Americas			133–138,000		
	North America	20,000		4198	21
	Middle America	30–35,000		14–19,000	46–54
	South America	70,000		55,000	78.5
	Caribbean Islands	13,000		6,555	50
Africa			40–45,000	35,000	77–87.5
	North Africa	10,000			
	Tropical Africa	21,000			
	Southern Africa	21,000			
Asia					
	Southwest Asia & Middle East		23,000	7,100	31
	Central & North	17,500		2,500	14
	Indian Subcontinent	25,000		12,000	48
	Southeast Asia (Malesia)	42–50,000		29–40,000	70–80
	China & East Asia	45,000		18,650	41.5
Australasia	Australia & New Zealand		17,580	16,202	90
Oceania	Pacific islands		11–12,000	7,000	58–63
Totals			380–399,000	216–234,000	

in its infancy, and early indications are that although some areas of species richness also have matching genetic diversity, some unexpected patterns are emerging. The issue is partly clouded by the impact of centuries of migrating human groups taking favored plants with them when colonizing distant lands.

See: Conservation of Crop Genetic Resources, pp. 413–429

The Reality of Extinction

A recent declaration by leading botanical scientists drawing attention to the need for concerted global effort to conserve plants considers that "as many as two-thirds of the world's plant species are in danger of extinction in nature during the course of the 21st century, threatened by population growth, deforestation, habitat loss, destructive development, over-consumption of resources, the spread of alien invasive species and agricultural expansion" (BGCI 2000).

Extinction is not just a future phenomenon and not just the highly visible loss of larger animals: plant extinctions are being recorded at the present time. Thomas Givnish noted the demise of a species of *Cyanea* (*Lobelioideae* family) on the Hawaiian Islands, and Arnoldo Santos-Guerra documented the extinction through animal grazing of the last wild population of a plant species on La Palma in the Canary Islands. Fortunately, Santos-Guerra was able to salvage ripe seeds and introduce the species into a botanic garden (Josephson 2000). However, while salvaging a few viable seeds may literally save a species from extinction at the eleventh hour, there remains the problem of how often it is feasible to restore viable wild species back from such a bottleneck without enormous expenditure. We need to prevent human-induced extinction factors from operating in the first place.

TABLE 20.2 The Twenty-five Hot-spots, Their Characteristics, and Biodiversity (after Mittermeier, Myers, and Mittermeier, 1999)

| Hot Spots | Biome(s) | Original Extent | | Geographic Area | | | | Indicative Biodiversity | |
| | | | | Remaining Intact | | Area Protected | | Vascular Plants | |
		(km²)	(%)	(km²)	(%)	(km²)	(%)	Diversity	Endemism
Mediterranean Basin	Mediterranean type	2,362,000	1.59	110,000	4.7	42,123	1.8	25,000	13,000
Indo-Burma	Tropical rain forest/Tropical dry forest	2,060,000	1.40	100,000	4.9	160,000	7.8	13,500	7,000
Brazilian Cerrado	Tropical dry forest Woodland savannah Open savannah	1,783,169	1.20	356,634	20.0	22,000	1.2	10,000	4,400
Sundaland (Indonesia)	Tropical rain forest Tropical dry forest	1,600,000	1.08	125,000	7.8	90,000	5.6	25,000	15,000
Guinean forest	Tropical rain forest	1,265,000	0.85	126,500	10.0	20,224	1.6	9000	2,250
Tropical Andes	Tropical rain forest Tropical dry forest High-altitude grass	1,258,000	0.84	314,500	25.0	79,686	6.3	45,000	20,000
Atlantic forest region	Tropical rain forest Sub-tropical rain forest	1,227,600	0.82	91,930	7.5	33,084	2.7	20,000	6,000
Mesoamerica	Tropical rain forest Tropical dry forest	1,154,912	0.77	230,982	20.0	138,437	12.0	24,000	5,000
South Central China	Temperate forest Grassland	800,000	0.53	64,000	8.0	16,562	2.1	12,000	3,500
Madagascar and Islands	Tropical rain forest Tropical dry forest Xerophytic vegetation	594,221	0.40	59,038	9.9	11,546	1.9	12,000	9,704
Caucasus	Temperate forest Grassland	500,000	0.34	50,000	10.0	14,050	2.8	6300	1,600
Wallacea (Indonesia)	Tropical rain forest Tropical dry forest Xerophytic vegetation	346,782	0.23	52,017	15.0	20,415	5.9	10,000	1,500
California Floristic Province	Mediterranean type	324,000	0.21	80,000	24.7	31,443	9.7	4426	2,125
Philippines	Tropical rain forest	300,780	0.20	24,062	8.0	3910	1.3	7620	5,832

(Continued)

TABLE 20.2 (*Continued*)

| Hot Spots | Biome(s) | Geographic Area | | | | | | | Indicative Biodiversity | |
| | | Original Extent | | Remaining Intact | | Area Protected | | Vascular Plants | |
		(km²)	(%)	(km²)	(%)	(km²)	(%)	Diversity	Endemism
Central Chile	Mediterranean type	300,000	0.20	90,000	30.0	9167	3.1	3429	1,605
New Zealand	Temperate forest Grassland	270,534	0.18	59,400	22.0	52,068	19.2	2300	1,865
Caribbean	Tropical rain forest Tropical dry forest Xerophytic vegetation	263,535	0.17	29,840	11.3	41,000	15.6	12,000	7,000
Choco-Darien/ Western Ecuador	Tropical rain forest Tropical dry forest	260,595	0.17	63,000	24.2	16,471	6.3	9000	2,250
Western Ghats and Sri Lanka	Tropical rain forest	182,500	0.12	12,445	6.8	18,962	10.4	4780	2,180
Succulent Karoo (Africa)	Xerophytic vegetation	112,000	0.08	30,000	27.0	2352	2.1	4849	1,940
Cape Floristic Province	Mediterranean type	74,000	0.05	18,000	24.3	14,060	19.0	8200	5,682
Polynesia/Micronesia	Tropical rain forest Tropical dry forest	46,012	0.03	10,024	21.8	4913	10.7	6557	3,334
Eastern Arc Mountains	Tropical rain forest	30,000	0.02	2000	6.7	5083	16.9	4000	1,400
New Caledonia	Tropical rain forest Tropical dry forest Maquis shrubland	18,567	0.01	5200	28.0	527	2.8	3332	2,551
Galapagos	Xerophytic shrubland	7882	0.005	4931	62.6	7278	92.3	541	224
Juan Fernandez Island	Temperate forest	100	—	—	—	91	91.0	209	126
Totals		17,452,038	11.76	2,142,839	1.44	888,789	0.60		
Total Endemics									131,399
% Global Diversity									43.8

Despite our considerable knowledge of plant distribution and abundance, it can be extremely difficult to assess global extinction rates, especially when they are projected into the future. A critical review of the extinction problem using three contrasting analyses concluded that impending extinction rates are at least four orders-of-magnitude faster than the background rates seen in the fossil record (May et al. 1995, 1–24). This figure does not take into account the assertion that the extinction of an obvious, large keystone organism (such as a forest tree) probably results in the loss of up to ten or twenty other smaller organisms dependent on or confined to that single species.

Recent analyses of possible extinction rates by Stuart Pimm and Peter Raven paint a bleak scenario for the future of biodiversity unless markedly increased steps are taken to protect the remaining species-rich regions and habitats (Pimm and Raven 2000). This model suggests an accelerating rate of extinction, peaking in the mid-21st century at nearly 50,000 per million species (of both plants and animals), then decreasing into the 22nd century.

As expected, the most critical parts of the world are the biological hot-spots that are rich in species found nowhere else. A disturbing feature of this analysis is that in the seventeen tropical regions designated as biodiversity hot-spots, only 12 percent of the original primary vegetation remains. Applying the Pimm and Raven model, one arrives at the prediction that even if the entire remaining habitat in these seventeen areas were protected immediately, there would still be an 18 percent loss of their species, and if habitat depletion continued at the present rate for a further decade, the species loss would rise to 40 percent.

Urban areas where the species are visible to a greater number of people can also be major extinction sites. An example is *Thismia americana*, only ever known from one locality in Cook County, Chicago, where it was last seen in 1913. The site, originally described as being "the bottom prairie swale on the east side of Calumet Lake, between Torrence Avenue and Nickel Plate Railroad, between the Ford factory and the Solway Coke factory" is now under heavy industry. The species has been searched for repeatedly in the original area and in similar sites, but without success. It is not in cultivation and is now believed to be extinct (Lucas and Synge 1978).

Main Extinction Threats: Successes, Failures

What factors lead to extinction? It is important to appreciate that both species extinction and loss of genetic diversity, or "genetic erosion", can be natural events, just as some forms of evolution are. Nature is, and it seems has always been, dynamic—there are natural background rates of extinction. However, the contemporary situation concerning species' extinction and genetic erosion is quite different from that which existed in the past. Humankind now has the ability to alter drastically the world environment in ways not previously possible. Therefore, loss of plant biodiversity is mostly caused by humans, and these causes may be broadly grouped under the general headings of:

- Destruction, degradation, and fragmentation of natural habitats (e.g., by road and reservoir building)
- Overexploitation (e.g., by medicinal plant extraction from the wild, fuel wood gathering, overgrazing)
- Introduction of exotic species which compete with, prey on, or hybridize with native species (e.g., human introduction of alien species for exploitation that subsequently escape and become novel weeds or pests) (see Invasives)
- Human socio-economic changes and upheaval (e.g., extinction of tribal cultures, urban sprawl, land clearances, food shortages)
- Changes in agricultural practices and land use (e.g., displacement of land races or traditional cultivars by modern varieties or shift to monoculture)
- Calamities, both natural and man-made (e.g., floods, landslides, or wars)

Threats to survival can be direct (as with harvesting) or indirect (as with habitat removal or extirpation of an essential pollinator or seed disperser). Also, threats can be extrinsic—generated outside the plant and its population—or intrinsic—a factor in the biology of the species that makes it especially vulnerable to an environmental change. An example of the former is the destruction of individuals or predation on fruits or flowers; an example of the latter is relying on fertilization by a specific insect pollinator that can be readily eliminated from the ecosystem, leading indirectly to the demise of the plant species.

At the ecosystem level, climatic modeling studies suggest that since the beginning of agriculture, some 15 percent of forests had been converted to cultivation by 1970. A study by FAO and UNEP of eighty-seven countries found that, between 1980 and 1990, about 169 million hectares (418 million acres) of forest were lost, or about one percent per year. It is estimated that western Ecuador, one of the areas of the world most rich in biodiversity, has lost 90 percent of its forest since 1950, and possibly up to half of its complement of indigenous plant species. Over 90 percent of natural wetlands in New Zealand have been lost since European settlement in the 17th century (Given 1981). More than 90 percent of other major ecosystems are either disappearing or being fragmented, and so losing their self-sustaining viability. Loss of mangrove forests along tropical coasts is rapid, with figures of 40 to 80 percent in the last 20 years being quoted for countries such as Thailand, Philippines, and Gambia. A UNEP study suggested that some 69.5 percent of the world's drylands have, to some extent, been degraded as a result of adverse human impacts. It is more difficult to measure genetic erosion, either within or between species, and also the extent of genetic pollution of wild species through contact with cultivars and imported species, whether in the form of crops or amenity plantings.

Species do not have to entirely disappear to become functionally extinct. An increasingly common phenomenon is what has been sometimes called the "living dead." These are species that are still present as adults, but whose reproduction has virtually ceased, and so are functionally "dead". New Zealand provides a good example. Shrubby pohuehue (*Muehlenbeckia astonii*, of the dry scrub of the northeastern South Island of New Zealand) is a strange plant, shaped like a bouncy igloo of interlacing branches. About 2000 plants are known, but these are all probably at least 100 years old, and only a handful of seedlings and juveniles have been seen in recent years. Reproduction is virtually unknown, and few individuals of the species play an effective ecosystem role. Shrubby pohuehue is perhaps the most extreme example of functional extinction in this region, but close examination of a number of its companion shrubs suggests that many are doing little better. The intricate and fascinating mosaic of divaricating shrublands in this part of the world, unimaginatively called "grey scrub" in the textbooks, may join the dodo, the moa, and the passenger pigeon as museum pieces (Given 2001). There are strong arguments for a prophylactic approach to conservation, rather than waiting for a crisis of extreme endangerment and expending limited conservation funds on species that are close to extinction and may not be successfully saved for posterity.

Threatened Species and Priorities

Increasingly, data on plant species' abundance, biology, and distribution is being used to determine the *conservation status*, or degree-of-risk of extinction. There has been much debate about the categorization process and how to best list species. An important distinction is between categorizing species in terms of extinction risk (for example, the process of Red Listing—see later discussion) and setting priorities for conservation action that involves other types of information, such as cultural importance, significance for commercial use, and likely cost and success of a recovery program.

There is a general assumption that about 12.5 percent of the world's flora is globally threatened. This is probably an underestimate for two reasons. First, it can be difficult to prove that a species is extinct, even when there have been repeated searches of likely sites. Second, although threat assessment

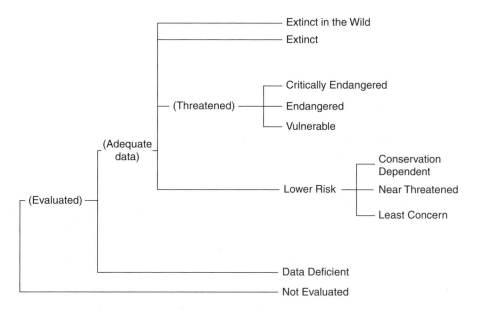

Figure 20.1. Structure of the IUCN categories of threat (http://www.redlist.org/).

systems have been in use both at the country level and globally for a number of years, it is only recently that an integrated and replicatable global system has been phased in to cope with large numbers of species from very diverse taxonomic groups.

The International Union for Conservation of Nature and Natural Resources (IUCN) has developed a global system of categories of conservation status, which ideally requires knowledge of distribution, population size, and population dynamics of the species concerned, although approximations can be made if precise data is not available (IUCN 2001). This is the IUCN Red Data System or Red List (Fig. 20.1). Inventories in 1997 and 1998 suggest that approximately 34,000 plant species are threatened to some degree worldwide, with some 6,500 in the Endangered category, meaning they are likely to become extinct in the near future (Oldfield, Lusty, and MacKinven 1998, Walter and Gillett 1997). The 1997 study lists 751 described species from 192 families that are thought to be extinct in the wild. There are numerous estimates of predicted extinction rates, but a widely held view is that 15 to 20 percent of all plant species could become extinct in the next 20 years, even though at present many countries recognize only about 12 percent of their flora as being at risk.

An important rationale for identifying hot-spots of biodiversity is that priorities must be set for use of the limited resources available for conservation. Norman Myers and others have warned that conservationists may not be able to assist all species under threat, if only because of lack of funding. Therefore, we must place a premium on setting priorities. This is part of the rationale of identifying biodiversity hot-spots. A concerted effort to achieve substantial and immediate protection of these twenty-five regions, which comprise only 1.4 percent of the Earth's land surface, would protect as many as 44 percent of all plant species. As noted in the chapter on conservation of crop genetic resources, there are good arguments for giving priority to those species (and ecosystems and geographic regions) that have multiple gene pools within the "gene sea" of interrelated gene pools, because they would better represent the breadth of the plant genetic diversity in the limited number of populations or accessions likely to be actively conserved.

Another aspect of prioritization useful when detailed information is not available is to estimate proneness to extinction, based on inherent biological characteristics of species (or genera). Frequently little is known about species that may be at risk since often time and resources do not allow in-depth study before decisions on priority have to be made. This has led to the development of various methodologies for pre-assessing extinction risk (Fiedler and Ahouse 1992, Given 1994, Mace et al. 2001).

Protected Areas and *in situ* Conservation

Conserving plants in their natural habitat is often regarded as the mainstay of plant species conservation, but this involves more than the simple expedient of putting a fence around a few individuals. Most species, except those that are recent immigrants, have evolved in concert with other species and environmental factors such as climate, soils, and moisture. Even the simplest of ecosystems are rarely random selections of animals and plants thrown together, but rather complex networks of mutual interactions and interdependencies. Thus, conservation of individual species needs to take into account the communities within which species occur, its genetic makeup, the particular role they play in maintaining other parts of the system, and the species on which they themselves depend.

Communities include many feedback loops with a constant interplay of animals, plants, and physical factors, which in various ways reinforce processes that change through time. This raises important questions for management of communities and ecosystems: What happens when communities are stressed in various ways? What species losses can be expected on the basis of community structure? Are some kinds of communities more susceptible to catastrophes and species invasions than others? What kinds of species invade and what is their effect? Will the loss of one species cause other losses to cascade through the community, and if so, how far? Are there critical key species in the food web? If substantial numbers of species are lost, will they be recovered when original conditions are restored?

Comprehensive species conservation requires preservation of species throughout their complete geographic and habitat range and a consideration of patterns of variation. Conservation of plants must also take animals into account. Continued evolution will involve the interaction of both internal and external factors at a range of levels right down to the individual, and even the genome. As a result, there are a number of fundamental questions to consider when conserving species or special vegetation types in protected areas:

- Should protected areas be selected on the basis of strict biogeographic criteria?
- Can existing protected areas be upgraded to improve their conservation effectiveness?
- Are several small reserves better than one large one?
- The minimum viable population (MVP) debate—how large does the population have to be to be viable, and does MVP differ widely for different species?
- How is protection integrated with traditional uses of plants?
- How can plant conservation best be integrated with local people's development aspirations and wealth generation?
- How can we involve local stakeholders in the management of protected areas?
- How do we protect a mosaic of many habitat types which may be necessary for life history stages, or to conserve the full habitat needs of key pollinators and dispersers?
- Should priority be given to protection of representative areas, or of areas with high levels of species richness or unique vegetation types?
- What strategies will best protect genetic diversity within populations?
- If we simply protect representative ecosystems, are all species adequately protected along with them?
- What management and monitoring regimes should be implemented?
- Should there be routine provision for harvesting target plants in the protected area for sustainable use or for germplasm?

Most people accept the validity of these questions, but justifiably argue that one usually has to be pragmatic. Social and economic issues may be sensitive, especially where dense human habitation or questions of traditional use exist. Final boundary placement may have to be a compromise of financial and political factors, existing patterns of land ownership and use, and the practical availability of sites for preserves.

When people think of protected areas, the first thing that often comes to mind is the National Park System. These have wide international recognition as "jewels in the conservation crown." They enjoy a high public profile and are often large, preserving extensive ecosystems. But they may be too large to facilitate management of highly endangered species. They are also public areas, in that traditionally people have reasonably unconstrained access. On the other hand, strict nature reserves are well suited for conservation of individual species and smaller ecosystems. Yet, neither may meet the needs of relict stands of vegetation growing in highly modified, urban habitats, or the requirements of genetic resource conservation.

The UNESCO Man and the Biosphere programme have drawn attention in many countries to the need for a range of habitat types, from core areas that are close to a natural state, through transition zones, seres, and ecotones, to human-influenced lands. Particularly valuable is the concept of buffer zones bounding protected areas, and the range of mechanisms that can be used to protect nature values on production lands, including land management agreements and conservation covenants.

Sites not specifically protected for nature conservation may have high conservation value and reasonable permaneance. Examples include sacred sites such as churchyards and temples, railway or motorway embankments, lighthouse reserves, and military sites. In the United Kingdom burial grounds and churchyards are sometimes extremely valuable for protection of wildflowers that have all but disappeared from the countryside. There is an increasing need to think laterally when considering sites for conservation of natural vegetation and native species. Even sports and recreation grounds and racecourses are known to provide protection for populations of threatened plants. Management may not be simple on such sites. Competing uses have to be carefully meshed with conservation needs and there are likely to be conflicts that require compromise solutions.

Preserving biotic communities is not the same as preserving genes. Since communities are classified according to vegetation structure, and the dominant plant and animal keystone species, it is quite possible to preserve a community-type and still lose many species. It is also possible to preserve a species and lose genetically distinct populations. Although on-site conservation requires that biological diversity be considered as a whole—that is, in the form of intact communities—the type of conservation strategies employed and their outcome will depend on the particular focus. A protected area may concentrate on conserving a particular ecosystem, such as the Peat Swamp or Heath Forests of Malaysia, Terra Firme forest in Amazonia, or *Scalesia* forest on the Galapagos Islands. It may be focused on a particular and important tree species of one of those communities, such as *Gonostylus bancanus* found in the Peat Swamp Forests or *Lippia* growing in *Scalesia* forest. A protected area may also be established to conserve genetically distinct populations of such species, rather than the whole range of variation.

It is often assumed that conventional reserves to protect species and habitats are sufficient to conserve all aspects of biological diversity. This ignores the particular needs of genetic resource conservation concerned with crop relatives and wild species with economic value. The aim of genetic resource conservation is to maintain as many as possible of the genes or groups of genes found in these species in a representative array of combinations (Prescott-Allen and Prescott-Allen 1986).

A special kind of protected area is an *in situ* (on-site) genetic reserve, which is a location where wild genetic resources are conserved by maintaining gene pools of species in their natural habitat.

The emphasis is on species of known or potential economic value, and a necessary function is to provide for use of the gene pool. There are two important distinctions between management for gene pool conservation and management for other types of conservation. First, the unit of management is the gene pool rather than the species, community, or ecosystem. Second, provisions must be made for collection and sustainable use of plants from the reserve by *bona fide* breeders and research workers, and for supply of germplasm to *ex situ* gene banks.

These are the first steps for effective genetic resource conservation in natural habitats:

1. Select target taxa or region and undertake an ecogeographic survey to establish foci of genetic diversity for the taxa. This will determine where the reserve could be located within the region.
2. Survey potential sites for the genetic reserve and select the reserve site(s) based on scientific, economic and practical factors. Sites already under some form of protection, or in common ownership, are likely to be most easily converted to genetic reserves.
3. Establish working relationships between protected area managers, major users (such as plant breeders, geneticists, population biologists, and ethnobotanists), local traditional resource users, and the local community. This provides a range of expertise to carry out steps 2 and 4.
4. Complete a more detailed survey of population status, ecology, and life history of the target taxa within the genetic reserves.
5. Establish the management and monitoring regime for the genetic reserve that promotes the retention of genetic diversity within the reserve.
6. Use the routine genetic reserve monitoring to feedback into the reserve management regime.
7. Promote use of the reserve and the value resource by professional and traditional users, while ensuring use does not adversely impact genetic diversity of the target taxa.
8. Establish a link with *ex situ* conservation to ensure any unique genetic diversity is duplicated to ensure security.

An *in situ* gene bank may be a formally designated protected area such as a nature reserve. It may be a zone designated within a protected area that has other objectives, such as a national park. It may be a protected area specifically set aside with genetic resource conservation as its only purpose. Where there is competition among different users of land wanted for a gene bank, there may have to be careful selection so that adequate gene pools are protected in relatively few sites. Once organized, a national system of genetic reserve will include an array of different types of protected areas (Prescott-Allen and Prescott-Allen 1986):

1. Protected areas with many objectives, and some parts zoned for gene pool conservation. It is logical that existing protected areas are candidates for consideration as pilot genetic reserves.
2. Protected areas whose objectives do not permit artificial maintenance of seres (a developmental series of communities following in succession to a climax stage) or subclimaxes, but instead conserve climax species through protection of ecosystems in their natural state.
3. Protected areas whose primary objective is gene pool conservation.

The distribution of protected areas should be such that they conserve at least one viable population of each major genetic variant of the target species. Protected populations must be sufficiently viable as to be self-sustaining, and so minimize the loss of genes that only occur in low frequency.

Those sites which have the highest numbers of target populations must be *a priori* for more detailed investigation and monitoring. It will be necessary to ensure that plant numbers are sufficient to maintain long-term populations, critical habitats are identified, essential associates such as pollinators and fruit dispersers are present, and other activities in the area do not jeopardize *ex situ* functions (MacKinnon et al. 1986). This is where it becomes essential to determine key biological parameters, such as minimum viable population size, natural, and induced fluctuations in the population, and interactions with co-existing species.

Between two areas of very dissimilar habitat or land use, there will be a transition zone where the habitats or land use interact as *edge effects*. Edge effects have two main consequences. In very small reserves, there is a risk of the whole reserve being subject to edge effects so that no undisturbed core remains; it becomes important to determine not only preferences of target species for core or non-core habitats, but also the preferences of inter-related organisms, such as pollinators. The concept has been particularly developed for the boundary between land grossly disturbed by people and predominantly natural habitat.

A widely adopted technique for minimizing edge effects and maintaining core areas is to establish buffer zones around reserves. Buffer zones are areas peripheral to national parks or reserves which have restrictions placed on their use to give an added layer of protection to the nature reserve itself, and to compensate villagers for the loss of access to strict reserve areas (MacKinnon et al. 1986).

There is considerable scope for protected areas and buffer zones for plant conservation to also help local people in a very practical way. Economic poverty of many tropical countries is aggravated by the adoption of new and inappropriate agro-silvicultural systems. Many traditional systems of agriculture (for example, multicrop and agroforestry systems) have evolved to minimize environmental impact. Their displacement can be disastrous. Protected areas are not only living laboratories for science and benchmarks against which to assess change; they can also offer valuable facilities, especially in buffer zones, for education and awareness programs, and for research into low-impact systems of land-use.

Local people may not have a profound understanding of modern conservation objectives, but there are many examples of traditional cultural practices that have themselves led to the protection of ecosystems and species. Where local practices such as type, timing, and intensity of harvesting are compatible with long-term conservation, management should aim to incorporate them. Other uses of plants by local people may have to be modified, especially where there are increasing demands on grazing land, fuelwood, or food plants. Necessary changes should be introduced at a pace that allows people to adapt and understand. Examples include progressing from wild harvest to sustainable cultivation of such crops as African ginger in South Africa and bulbs in Turkey. Common concerns provide a basis for dialogue to develop a system of management that will conserve plants and their environment, yet not alienate people. Local preference in employment is also an excellent way to integrate protected areas into the community and benefit people in the immediate vicinity. A high proportion of locally-generated wealth should be shown to directly benefit communities immediately adjacent to, and perhaps otherwise disadvantaged by, a protected area.

Managing protected areas does not come free. It can usually be assumed that costs of management are more likely to escalate than decrease. In too many instances, management has come to a standstill because it was assumed that necessary resources would become available or would not cost. Similarly, it is tempting to keep adding new protected areas into a system, but each time this happens it is necessary to budget not only for setting up a reserve (land purchase, fences, signs, etc.), but also for continued management. Cuts in research funding too often occur in those areas that are of most value to the conservation of plants, including population biology and monitoring. These are not seen as revenue earning, and monitoring in particular is open to criticism as being "open-ended". Funding must be sufficient to carry out the minimum necessary activities to maintain a viable conservation program.

Translocation and Re-establishment of Threatened Species

The subject of translocations (transfers of species of animals and plants to other areas for conservation purposes by humans) generates strong opinions and feelings. Some conservationists and protected area managers regard translocations as unreasonable or not necessary, while others see it as a routine protection mechanism fundamental to the survival of some critically threatened species. One point to consider is that translocations are not new, but have been occurring for centuries. Many prehistoric translocations took place as peoples moved across continents and oceans, taking with them such plants as the coconut and sweet potato.

Translocations for conservation purposes are essential when habitat is about to be destroyed in its entirety, and the only choice is either to allow plants to be destroyed or to shift them. Sometimes it is possible to shift plants from a site into a holding area on a temporary basis while rehabilitation is being done. In other instances, the shift may have to be permanent. Translocation may also be justified where uncontrollable disease or predation threatens a population, or where for species that are unable to breed sexually (usually through loss of one sexual state). By translocating plants from two or more sites to combine elements of them at one location, a new breeding population can be established.

Several precautions are necessary, especially where translocations involve long distances and different climates or soils. Here there is a risk that the target species become weeds at the new site. Accidental disease transfer must be avoided, especially where isolated target sites such as oceanic islands are involved. Such places may not have the normal complement of pests and diseases, especially microrganisms. Careful documentation and follow-up is essential with a monitoring program that can track any failures and set-backs and allow them to be rectified. Genetic considerations suggest that the variability of a population established at a new site should not be lower than that of the source site. Properly done, translocations are time-consuming and can also be expensive, so they should only be used where absolutely necessary.

The Role of *ex situ* Conservation

The general goal of conservation is the maintenance of viable populations of all species, preferably in the wild, where they can engage in a full range of interactions with other species and the abiotic environment, as well as continuing to evolve. However, conservation managers and decision-makers have to adopt a realistic approach to what is feasible as the threats to species *in situ* continue to grow. Already much biodiversity lives in human-modified environments. Many threats, which include habitat loss, climate change, unsustainable use, political instability, and invasive and pathogenic organisms, are difficult to control. The present reality is that we shall be unable to ensure the survival of as many species as possible without significant use of *ex situ* conservation, both as a primary conservation strategy and for back-up of *in situ* conservation.

If the decision to bring a species under *ex situ* management is left to the last minute, it is frequently too late to implement effectively, and risks permanent loss of the species. However, *ex situ* conservation should *only* be considered as an alternative to *in situ* conservation in the most exceptional circumstances (such as complete and sudden destruction of all wild habitat for a species), and effective integration between *in situ* and *ex situ* approaches should be sought wherever possible. What is also needed is to match good theory with effective practice, adopting the best methods and philosophy for the particular situation. *Ex situ* conservation should not be adopted simply because it is a last-minute act of desperation.

Botanic gardens have played a significant role in conservation, particularly education and awareness-raising, from the time when they were a driving force in man's exploration of the globe for both potentially economic plants and the discovery of scientific novelties. In recent decades, botanic gardens have assumed a new importance as centers for conservation. In some countries, especially the developing tropics, they are the primary institutions involved in research, collection, maintenance, and conservation of plants. The traditional roles of botanic gardens have been augmented through the evolution of a range of actions: holding comprehensive collections of threatened species, cultivating such species in sufficient numbers to arrest genetic erosion, having material available for research and assessment for economic use, propagating plants for reintroduction into the wild, and displaying plants and simulated habitats for education and awareness programs.

Gardens have long played a crucial role in preserving genotypes. Many of the earliest tropical botanic gardens were primarily established for the purpose of trialing wild species that could be grown commercially. This role continues, but has been augmented by growing plants that are rare, and also by maintaining general collections of both cultivars and interesting genetic variants, especially of ornamental plants.

Another important role for gardens is investigation into germination, silviculture, and optimum growing conditions for plants. A recent example has been the investigation by several botanic gardens in Australia, South Africa, and the United Kingdom of the benefits of exposing the seeds of a wide range of dryland species to smoke. Within smoke there is a very complex assemblage of chemicals, and some of these promote germination. Another has been exposing seeds that germinate only with difficulty to thaw-freeze cycles, acid-etch treatment, or scarification to promote germination. Some botanic gardens are now using novel techniques, such as pollen storage and cryopreservation in liquid nitrogen, to preserve genotypes on a long-term basis.

Botanic gardens are increasingly involved in propagating plants for reintroduction to the wild. The decision to implement a propagation program as part of a formalized recovery plan, and the appropriate design of such a program (away from present or historical habitat), will depend on the species' specializations and circumstances. A species-specific propagation plan may involve a range of objectives in reproduction, research, reinforcement, reintroduction, and so forth, but must be clearly stated and agreed among organizations participating in the program. Especially in developing countries, botanic gardens frequently have the ability to bring partners in such a plan together, and to seek funding to facilitate a conservation program.

Production Land—The Gray Area for Conservation

A tragedy for many people today is dislocation from nature. Just one example, but an important one, is that perturbation—floods, storms, and ecological succession—is seen as a negative to be overcome, rather than being something we live with, and even foster as necessary. This dislocation is challenged by proponents of a new ecological and conservation paradigm in which perturbation, change, and landscape heterogeneity are part of the normal dynamics of nature (Pickett et al. 1997). We either learn to live with and through perturbation, or struggle to maintain a static and declining world through precarious human intervention. Land-use planning and management needs to incorporate risk and uncertainty as a fundamental component of ecosystems, rather than ignoring or minimizing it. The discipline of landscape ecology teaches this as a fundamental. It also teaches that utilization of biodiversity at one scale inevitably relates to management of the landscape at other scales, and that over longer time periods, it is management of the landscape matrix that determines the future of patches such as small nature reserves within the matrix.

Long-term sustainable use of land by people must take into account ecosystems, natural landscape heterogeneity, and species diversity. Sustainability thus becomes far more than just a matter of economics, or maintaining protected areas as samples of what once existed as natural ecosystems.

Within production lands, what are some of the options for preserving plants and natural or semi-natural ecosystems? Plantations and hedgerows provide one option that has long been used in Europe. Road margins also provide another situation that can be used innovatively for conservation purposes. Australia, South Africa, and parts of the United States extensively utilize road margins for replanting vegetation types and rare or declining species. Railways have also been used in the same way. In Wellington, New Zealand, threatened plant species have been planted in large numbers in median strips and on road junction rotaries to provide seed sources that can be harvested for translocation into the wild. In Tasmania, shelter belts of scarce *Eucalyptus* are used as seed orchards.

Integrated Conservation

Practical conservation and sustainable use means utilizing as much varied information as possible; many disciplines contribute to understanding and managing biodiversity. Conservation and sustainable use is not just about individual plant species or populations, but concerns ecosystems, communities of species, species themselves, and the populations they consist of—right down to the genetic diversity within species and populations. Two useful distinctions are *ecological conservation,* which focuses on the conservation and sustainable use of whole communities, and *genetic conservation,* which focuses more explicitly on particular taxa (usually species) and attempts to conserve the full range of genetic (allelic) variation within those taxa.

In situ and *ex situ* conservation strategies should not be seen as alternatives or in opposition to one another, but rather as being complementary. Along with habitat restoration, they are the three legs supporting integrated conservation strategies. Each approach acts as a backup to the others, the precise mix of techniques being unique to the particular situation. Integrated holistic conservation also brings together a range of agencies as partners, including government agencies, nongovernment organizations, and the general public. Partnerships must also be forged with groups of people who have traditional use of particular plant resources.

Why Wild Plant Diversity Is Important

There are many and varied reasons why people regard plant diversity as important. Since the dawn of history, people have drawn heavily on the plants around them to satisfy the needs of daily life—food, shelter, clothing, fuel, and medicines, as well as recreational and cultural needs. Humans' selection of particular plants has led to the development of major crops, such as rice, barley, maize, wheat, and potato, and has also led to utilization of many thousands of plants for medicinal use. Sources of fuel are fundamentally of plant origin, whether directly as wood or charcoal, or indirectly as coal, petroleum, or animal dung.

If we take medicine, almost every tropical plant now important in modern medicine was "discovered" in the course of ethnobotanical studies. An outstanding example is provided by *Pilocarpus jaborandi*, a shrub native to the Brazilian Amazon. This yields pilocarpine, used for treatment of glaucoma. Another example is rosy periwinkle (*Catharanthus roseus*) of Madagascar. This was first investigated because of its local use as an oral hypoglycaemic agent. But it has become better known as a source of anti-cancer alkaloids.

See: Plants as Medicines, pp. 205–38

In highly industrialized societies of the North, it is all too easy to forget the dependence on plants for the basics of life, and how many everyday products are made from biological raw materials. Shops are just the end of a processing and distribution chain that starts with the natural world of plants and animals. Loss of this perspective in many Western nations has been a conspicuous feature of the last 100 years.

At an even more fundamental level, ultimately, all life (including both humans and other animals) is derived from the photosynthetic factories of plants. The term *life-support species* has been used in two distinct senses: for those plants that contribute most fundamentally to the basic welfare of human society, and for plants that provide a backstop in times of emergency, famine, and when crops fail. Increasing efforts are being made to identify life-support species as priorities for conservation. Life-support species include the major crops of the world. Only about 150 flowering plants have become important enough to enter into world commerce, with several hundred more being grown on a local scale. A much smaller number provide most of the nutritional needs for the world's people, and a mere eight provide three-quarters of their nutrition. Other major life-support plants include fuel-wood trees and shrubs, vital medicinal species, fiber plants, and plants that provide important poisons for hunting, such as several plants including a species of *Strychnos* from which western Amazonian Indians derive curare.

See: Materials, pp. 341–2

The palm family of plants (*Arecaceae*) provides an outstanding example of life-support species. One of the most versatile groups of plants used by humanity, palms provide all the basic necessities of life: food, shelter, fuel, fiber, and medicines. They also provide numerous oils, spices, waxes, gums, and poisons. It is little wonder that for many people they are considered "trees of life", or that they are repeatedly symbolized in major religions.

From ethnobotany we learn not only that people relate to plants in a practical sense, but also culturally. A particular plant can have high value to local people, even if it is not constantly used for food, fuel, or medicine. The cultural value of plants can outweigh immediate tangible benefits. Thus, amaranth (*Amaranthus*) in South America had religious value for the Aztecs, quite apart from its use as a food plant. Other plants are valued for dyestuffs, decorative weaving, and basket-making.

See: Natural Fibers and Dyes, pp. 287–314.

It is easy to neglect powerful arguments for conservation of plants that involve recreational, aesthetic, and moral justification for wild species conservation. An important question for many people is: *How much do we value the plants and the plant associations that contribute to our landscapes, along with those unique elements of biodiversity which give the places where we live familiarity, and make us feel at home?* A tragedy for many people today is dislocation from nature. Averting this means bringing people closer to nature—for example, through botanic gardens, school programs, and visits to wild habitats. There needs to be a blurring of the demarcation between urban and rural lands, and greater recognition of the "footprint" effect of urban areas on remnants of natural habitat. We all need to realize that beyond the supermarket, there is a world of nature that not only provides the basics of life, but also provides sense of home and inspiration.

Yet, of necessity, an anthropocentric and utilitarian view has to be taken when deciding conservation priorities, particularly in developing countries. When financial resources are scarce and people are starving, there is a compelling imperative to work initially to conserve those species that are of most direct use to people.

The CBD Global Strategy for Plant Conservation

Recent emphasis on protection and wise use of biodiversity today has its foundations in the Convention on Biological Diversity (CBD), agreed at the 1992 Earth Summit in Rio de Janeiro. The CBD is the only comprehensive global treaty for the conservation of biodiversity and sustainable utilization of the Earth's biological resources.

See: Conservation of Crop Genetic Resources, pp. 419–20.

It encompasses a very wide range of issues, including conservation of biodiversity and associated indigenous knowledge, property rights for biodiversity, access to genetic resources, the sharing of benefits arising from exploitation of biodiversity, facilitating support for conservation action in the biologically rich developing world, and development of sustainable systems for both economic and traditional production from plants. The CBD is the primary guiding framework for biodiversity conservation and its sustainable use.

In April, 2000, a small ad hoc group of plant biodiversity experts came together in Gran Canaria, Spain, to consider a global initiative on plant conservation that had been suggested in 1999 at the XVI International Botanical Congress. The Gran Canaria group developed a framework (BGCI 2000), subsequently discussed and supported at a number of regional meetings, such as the June, 2001, Planta Europa meeting in Czech Republic, for what has now become the first Global Strategy for Plant Conservation, approved in April, 2002, for implementation through the CBD.

The strategy is novel in including sixteen global targets that integrate social, economic, and biological approaches to plant conservation, so that all appropriate and available resources, technologies, techniques, and sectors are brought together. The targets also incorporate, link, and reinforce the individual strategies of major partners such as Botanic Gardens Conservation International, the IUCN Species Survival Commission, and the International Plant Genetic Resources Institute (IPGRI). There are fifteen targets for year 2010:

- A widely accessible working list of known plant species
- Assessment of the conservation status of all known plants
- Action models based on good conservation biology research
- Conservation of ten percent of the world's ecological regions
- Protection of fifty percent of the key areas for plant diversity
- Thirty percent of production land management, consistent with the conservation of plant diversity
- Sixty percent of threatened species conserved *in situ* and *ex situ,* with ten percent in recovery programs
- Seventy percent of genetic diversity of crops conserved, with indigenous knowledge maintained
- Management plans for one hundred major alien species that threaten plant diversity
- No species of wild flora endangered by international trade
- Thirty percent of plant-based products derived from sustainably managed sources
- To halt decline in local and indigenous knowledge, innovations, and practices supporting livelihoods, food security, and health
- The importance of plant diversity and conservation in communication, education, and public awareness
- An increase in trained people working in plant conservation
- Established and strengthened plant conservation networks

To achieve these targets will require an unprecedented level of cooperative involvement of private landowners, producer organizations, conservation organizations, and governments until 2010. The targets also require consideration of paradigms that are not all as yet widely accepted, such as relationships between extinction, conservation, and economics, and alleviation of such social issues as poverty.

Long-Term Prospects

Conservation activities will always tend to be limited by the financial, temporal, and technical resources available; conservation has a cost, and the effort expended is directly related to how much society values that species and is therefore willing to pay. It is impossible to actively conserve or monitor all species, so it is important to make the most efficient and effective selection of species to focus conservation efforts.

Two crucial steps are halting perverse governmental incentives that endanger biodiversity and the empowering and training of local communities to take responsibility for conservation and wise use. States can facilitate and guide, but should not preempt the cost of local community benefits. Empowering is most effective where local people are convinced that biodiversity is worth preserving, can see both immediate proximate and distant effects of actions, and observe the three Rs of restraint, respect, and resolution of conflict. When people are aware, they start to value. What people value, they will keep, preserve, and cherish.

Case Studies

The case studies below include examples submitted from individual specialists, and examples selected from the literature.

Teak conservation project in the Philippines

The Philippine Plant Specialist Group of the IUCN Species Survival Commission (SSC) is undertaking conservation of the Philippine teak (*Tectona philippinensis*), an endemic and threatened tree species found along highly disturbed forest edges on limestone along the seashores of Batangas and in Iling Island, Mindoro. This tree produces hardwood used by the local people for house construction, furniture, and high-quality firewood. The species is currently threatened by habitat destruction, overcollection, and by occasional fires. Moreover, immature teak trees are preferred for building material, thus threatening the reproductive capacity of the population. This project, funded partly by Flora and Fauna International under the Flagship Species project, and the Philippines National Museum, started in 2001 and will continue for two years. A conservation program is being proposed to ensure the reestablishment of a stable natural population, and includes setting up long-term ecological research, establishment of a recovery and management program, public education, community consultation, and resource stewardship, and policy initiatives. (Thanks to Domingo Madulid, chair of the IUCN/SSC Philippines Specialist Group.)

Conservation of the lady's slipper

Cypripedium calceolus, commonly named lady's slipper, is the most spectacular of all Swiss orchids and also the most endangered, with marked decrease in the size and number of populations. Habitat destruction, habitat alteration, and overcollection are the main causes of the decline. People now seem to be more aware of the damage they can do by picking plants, or even only flowers, since they will then not produce any seeds. But man is not the only enemy: the chamois deer, for example, find orchids a delicacy and therefore are capable of exterminating all the lady's slippers in a particular habitat in just a few years. Chamois and *Cypripedium calceolus* are both protected by law . . . so is there a problem? Encouraging results elsewhere led to the creation of Fondation Orchidée, supporting a project to reintroduce *C. calceolus* to le Chasseral in the west of Switzerland, where environmental damage had led to its disappearance. After studying the inventory of existing and extinct populations of *C. calceolus* in Cantons of Bern, Fribourg, and Neuchatel and making an ecological study of the habitats, the project leaders selected healthy plants in a habitat very similar to and close to le Chasseral. Plants were hand-pollinated and seeds collected for germination in the United Kingdom. After germination, plants were sent back to Switzerland, where they were refrigerated at 4°C for about twelve weeks to simulate winter, then potted in soil and covered with pine needles to protect them from excess heat or cold. Plants are repotted each year and will bloom after 5 to 7 years. When they flower for the first time they will be reintroduced to le Chasseral, where the ecological conditions are most suitable. With the help of the Swiss conservation organization Pro Natura and forest guards, the new habitats will be protected by a fence, and the chamois will be kept away. The new populations will be monitored for evaluation. (Thanks to Vinciane Dumont, Fondation Orchidée.)

Xanthocyparis vietnamensis, a new but threatened genus of conifer

It rarely happens that a new conifer discovery turns out to be sufficiently distinct to be a new genus. Since the addition of *Metasequoia* in the 1940s, there have been only two: *Cathaya* in the 1950s, and *Wollemia* in 1994. In October, 1999, a botanical survey to the Karst Mountains in the far north of Vietnam, on the border with Yunnan (China), came upon an unknown conifer, which could only be identified to the family *Cupressaceae*. At first, not much material was collected, but a second visit to the site, high on a steep limestone ridge, yielded good herbarium samples; botanists then set to work to determine its identity. Other information was gathered by some of the botanists involved in the field work. As was the case with the other newly discovered genera, by the time botanists had found this Vietnamese conifer, it was virtually on the brink of extinction. Similar to the Chinese *Thuja*, only small trees seem to remain high on steep rocks, while evidence of a similar type of forest on lower slopes strongly suggests it could have grown there as well in the past. Villagers know of the tree and its valuable wood, and confirm that good trees are now almost completely gone. The forest is disappearing now even from the steeper slopes. It is not known how many trees are left, but it cannot be many. They seem to be restricted to a single mountain ridge. Therefore, this conifer has been classified as Critically Endangered (CR) in the IUCN Red List system. (Thanks to Aljos Farjon, chair IUCN/SSC Conifer Specialist Group.)

Rediscovery and protection of an "extinct" plant in California

Some 26 miles from downtown Los Angeles are some of the last remaining remnants of southern California's ancient oak woodlands and native grasslands. In May, 1999, a team of botanists investigating these discovered a species thought to be extinct, the San Fernando Valley spineflower (*Chorizanthe parryi* var. *fernandina*). The site is on a ranch that was scheduled for residential and commercial development from 2001. Five plant communities had been identified as "sensitive", according to the California Natural Heritage Database, a statewide inventory of the locations and condition of the state's rarest species and natural communities. A number of other rare species were found during the survey including valley oak, blue oak, and Californian melica. The spineflower was formerly known from only eleven sites, and prior to this new sighting had not been seen since 1929. In August, 2001, the developers, Ahmanson Land Company, announced that they had agreed to set aside a 330-acre preserve for permanent protection of this plant. The new preserve would be adjacent to 400 acres maintained in a natural state, and would protect almost 1.6 million plants—more than 90 percent of the occupied habitat and population as mapped [*see reports in Plant Talk 19 (1999) and 26 (2001)*].

Rediscovery and protection of a lost Mauritius endemic species

The flora of Mauritius is one of the most endangered in the world, with nearly three hundred species threatened. The major threat is invasive weeds that have taken over almost all natural habitats. *Trochetia parviflora* (family *Sterculiaceae*) is a small tree that is endemic to Mauritius and had not been seen since 1863. In the course of a botanical survey in early 2001, a plant of *T. parviflora* was discovered growing on steep cliffs in a nature reserve on an isolated mountain. Fortunately, the plant is not facing strong competition from invasive plants, as it is isolated on a rocky ledge. But only a few meters away are dense thickets of invasive *Schinus* and *Hiptage*. Subsequent searches found a further fifty-three wild individuals, but of the fifteen adult plants found only three were producing seedlings. The main reasons for its lack of reproduction appear to be invasive alien weeds, monkeys, and rats. A grant from the Chicago Zoological Society will help save the plant by allowing weeding of the immediate vicinity, but there is need for greater funding to ensure *in situ* conservation and survival of the species [*see report by Vincent Florens in Plant Talk 24 and 25, (2001)*].

Parque Nacional Cordillera Azul, Peru—a plant-diverse park

The Cordillera Azul contains a diversity of habitats rich in unique plant and animal species, and on account of this has been identified as an *a priori* area for conservation. In August/September, 2000, a team of scientists carried out rapid inventory for three weeks, and recorded 1,600 of the estimated 4,000 to 6,000 species thought to occur in the region. Following recommendations from this survey, in May, 2001, the government of Peru established the Parque Nacional Cordillera Azul. The mountain complex in the park encompasses a wide range of habitats, including jagged peaks, tall lowland forests, elfin forest shrublands and meadowlands, and high altitude wetlands. Many of these ecosystems are very rare and unprotected elsewhere. A remarkable feature is the diversity of palms with forty-five of the 105 palm species known in Peru being found in the area. A number of new species were discovered, including a small daisy tree, a large-stemmed *Zamia* (cycad), and three palms. A small fern (*Schizaea poepigiana*) that was found during the survey had not been seen since 1929. Threats from logging have been averted, since the region was declared a national park. The creation of this park shows that there are still opportunities to act before habitat fragmentation and degradation transform the landscape [*see report by Jane Villa-Lobos in Plant Talk 25 (2001)*].

Garden genes conserve threatened plants

Garden plants often exhibit limited genetic variation, and this is often cited as evidence for the limited role of gardens in conserving wild species. However, a study of genetic variation among cultivated plants of the threatened Chilean climber *Berberidopsis corallina* has revealed a surprising amount of genetic variation. Analysis of the DNA of thirty-five cultivated samples revealed sixteen different genotypes, suggesting that the original introductions by Veitch and Sons Nursery in 1862 was through seed, rather than cuttings. In the wild, the species has disappeared through much of its range, and conifers introduced for forestry have replaced much of its wild habitat. Garden plants therefore comprise a significant off-site gene pool, and a good potential source of material for reintroduction into the wild [*see article by Martin Gardner in The New Plantsman 7 (2000), pp. 174–177*].

Conserving furtive flowers—the Hydnoraceae

The family *Hydnoraceae* has only two genera and a small number of species, mostly plants of arid lands. These plants spend their whole life underground as root parasites; only the flower emerges from the soil. One of the most remarkable is *Hydnora triceps*. The dominant vegetation in its native habitat in Namaqualand, South Africa, is shrubby *Euphorbia*, one species of which is a host for the parasite. *H. triceps* had been rarely collected since its discovery in the early 19th century, and although thought to be extinct, was collected during a survey in 1999. One of the amazing things about this plant is that it flowers underground, punching up through the soil so that it can be pollinated by flying insects. But it also has an underground "chamber flower" that may be pollinated by weevils and nitulids, as has been shown to occur in the similar species *Prosopache americana*. No fruits have ever been described, although fruits are known to local shepherds, who eat them roasted. No way is known yet of growing this species in gardens, so its preservation depends on maintaining its wild habitat. Apart from the problem of knowing little of its ecology and reproduction, the region is highly threatened by diamond mining. Also, poisoning of predators of livestock could impact the small mammals that feed on the fruits and scatter the seeds. "Out of sight, out of mind" is not a good policy for ensuring the future of one the planet's most curious plants [*see report by L.J. Musselman and P. Vorster in Plant Talk 21 (2000)*].

Mutinondo (Zambia)—combining conservation and low impact tourism

The Mutinondo Wilderness Area is a 10,000 ha rare treat for the botanist. It is a concession being developed for low-impact tourism in Zambia's northern province. The developers have restricted development to a single road in and out of the area, and to camping facilities. Perhaps the most striking feature are the inselbergs— granite peaks where plants must either find a way to the soil below, or survive in dewfall in the dry season. The miombo woodland of this remote part of Zambia is largely free from human disturbance, and is rich in plant species that are less common elsewhere. The area also includes riparian forest and numerous "dambos" that are seepage zones of edaphic grassland. The uniqueness of this site stems partly from the habitats it protects, but also from its history of isolation—despite deep granite soils, this area has been uninhabited for many generations. The result is a pristine miombo wilderness unrivalled in this part of Africa, and now available for low-impact tourism visits [*see report in Plant Talk 20 (2000)*].

Preserving the Badgeworth buttercup in Britain

A classic example of setting up a species reserve to protect a rare wild flower concerns Adderstongue spearwort (*Ranunculus ophioglossifolius*). This occurs at Badgeworth in Gloucestershire, U.K., famous as "the world's smallest nature reserve" in the *Guinness Book of Records*. The reserve was created in 1932 and fenced to exclude grazing, but the buttercup rapidly declined. In five of the years between 1934 and 1962, no plants at all occurred in the reserve; in only two of the years were there one hundred or more plants. In 1962, it was realized that young plants could not develop in the dense tangle of vegetation that had developed. Ecological research also established that the plant likes warm, long summers, and that there is an essential role played by standing water. The plant is adapted to shallow pools that flood in late autumn, but partly dry out in late summer. It also needs a sufficient depth of water in winter to protect growth points from frost. It is this kind of autecological (single species) research that can often greatly help scientists understand what is needed to look after species in nature reserves. The local villagers have taken the buttercup to their hearts, and for some years a Buttercup Queen was crowned at the village fete with a garland of flowers [*see report by Peter Marrin in Plant Talk 19 (1999)*].

Protecting the "living dead": Muehlenbeckia astonii (Polygonaceae) in the South Island, New Zealand pastoral farmlands

M. astonii, or shrubby tororaro, is a medium- to large-sized shrub found in four discrete areas: coastal Wellington, northeastern Marlborough, north Canterbury, and Kaitorete Spit, just south of Christchurch. In total, about 2600 plants have been found, of which over 90 percent are known from one site. Shrubby tororaro was not included on threatened plant lists until the late 1980s, a reflection of both its wide distribution and, until recently, the few times it had been seen in the wild. In North Canterbury and Marlborough provinces, it has a particularly fragmentary range that reflects profound land use modification and loss of habitat. One of the remarkable features of this species is that virtually no site preserves the original habitat. *Muehlenbeckia astonii* is a valuable horticultural plant with great potential as a shrub for hedges and stock shelter in drier parts of the South Island. During severe droughts of the 2000–2001 summer it showed no signs of obvious drought stress, when many exotic species and even some indigenous species showed considerable drought stress. This provides a potential selling point for its reestablishment on both private and public land. A concern regarding the future of this species is that reproduction is practically unknown. Flowers are either females or inconstant "males," and seeds are produced where there is opportunity for cross-pollination. It is assumed at the present time that existing wild plants largely represent an aging cohort of currently unknown longevity. (Given 2001)

Brighamia in the Hawaiian Islands

Brighamia, an endemic genus with two critically endangered species in the Campanulaceae, survives on the high sea cliffs of Hawaii. *Brighamia insignis*, which was last seen on Ni`ihau in 1947, is currently known from two separate populations on Kaua`i. Those populations occur along the Na Pali coast of northwest Kaua`i, and include five plants on the Hoolulu cliffs and two plants on the headland cliffs of Waiahuakua. A third population, on the southeast side of Kaua`i along the basalt cliffs of the Haupu Range and above the Huleia River, has recently gone extinct. That population consisted of twelve individuals in 1990. Recent inventories of *Brighamia insignis* by National Tropical Botanical Garden (NTBG) staff estimate a total of only seven wild plants remaining. This is a catastrophic drop in numbers from the 1990 inventory that estimated ca. 142 individuals, and totals a loss of 95 percent (135 individuals) of the wild *Brighamia insignis* population. *Brighamia rockii* is now extinct on Lana`i and Maui, and is presently known only from the northern sea-cliffs of Moloka`i, which are considered the tallest in the world. A population inventory made in 1990 estimated approximately 173 individuals of *Brighamia rockii* within six naturally occurring populations, which included the basalt cliffs of Haupu Bay, East & West Wailau, Anapuhi, Waiehu, and Huelo Islet. A recent survey found only eight plants on Huelo Islet and seventy plants at Haupu Bay. The wild populations of both species are setting very little seed, it is hypothesized that the endemic pollinator, thought to be a hawk moth, has declined to such a point that pollination is not occurring. Conservationists have been hand-pollinating the surviving populations. Abundant seed has been generated, and managed populations of *B. insignis* have been successfully established at coastal sites on Kauai, most notably the Limahuli Botanic Garden and Preserve and the Kilauea National Wildlife Reserve. Two more populations are planned in the near future. (Thanks to Mike Maunder, Royal Botanic Gardens, Kew.)

Hyophorbe lagenicaulis: a Round Island endemic, Indian Ocean

Endemic to Round Island, off the north-east coast of Mauritius, this extraordinary palm may have originally occurred on mainland Mauritius and some of the small offshore islets, such as Ile aux Aigrettes. The wild population on Round Island has decreased dramatically over the last 150 years following the introduction of goats and rabbits. Travelers' reports and photographs from the 1920s suggest that this species was an abundant component of the island's palm woodland. Following conservation concerns expressed by the Government of Mauritius and the Jersey Wildlife Preservation Trust (now Durrell Wildlife Conservation Trust), the island was cleared of goats and rabbits; this resulted in a dramatic recovery in both plant cover and the endemic reptilian fauna. About fifteen individuals were recorded in 1978, dropping in 1986 to only eight adults and twenty-seven seedlings. Subsequently, through natural regeneration the population has increased, with about 350 seedlings observed in 1999. In addition, seeds have been planted and distributed around the island during management trips. With only one wild population, it was regarded as prudent to establish new populations on rat-free offshore islands. Accordingly, this species has been introduced onto Ile aux Aigrettes as part of a long-term restoration program. The older plantings at the Pamplemousse Botanic Garden are derived from nineteenth and early twentieth century wild collections made on Round Island prior to the population bottleneck. An initial genetic screening of wild and cultivated stocks is being used to plan the supplementation of wild populations with old cultivated sources. (Thanks to Mike Maunder, Royal Botanic Gardens, Kew.)

Maintenance of Lythrum hyssopifolia in agricultural landscapes in Britain

Lythrum hyssopifolia (hyssop loosestrife) is a widespread annual species, but in Britain it is sufficiently rare to be included in the British Red Data Book. Like many species found in pioneer vegetation on exposed mud, it has diminished in frequency. In recent years it has been confined to only one site in Cambridgeshire and in the Channel Islands, where it occurs locally in shallow depressions in arable fields. It is dependent on both winter flooding and regular ploughing of these sites for its survival. *Lythrum* is able to persist because the combination of local topography and agricultural practice favors an annual plant of low long-term competitive ability. Annuals predominate in this highly disturbed habitat. They germinate mostly in spring, most are self-pollinating, and have potential longevity of seeds (an advantage, as sites may not be flooded or disturbed every year). *Lythrum* was absent from one site for three years because winter flooding did not occur. The habitat is highly specialized, and its survival (along with the species found there) depends on the peculiar combination of flooding and disturbance. (Preston and Whitehouse, *Biological Conservation 35* (1986), 41–62.)

Reestablishment of the extinct native plant Filago gallica L. (Asteraceae), narrow-leaved cudweed, in Britain

Reintroduction was used for the rare annual narrow-leaved cudweed (*Filago gallica* L.), under Plantlife UK's Back from the Brink project. The species had been recorded in about thirty confirmed sites, mainly from sands and gravels in southeast England. It became extinct in England in 1955, but survived in one site on Sark, Channel Islands. Native mainland material of *F. gallica* had been maintained in cultivation since 1948, and provided an opportunity to reestablish the plant. Historical records were used to help plan the reintroduction; these indicated that it used to be found on light sandy/gravelly soils, in areas with a high summer temperature, in open vegetation usually with other *Filago* species. Pot-grown plants and seed were reintroduced to the wild in 1994 at the last known site, and by 1998 the species was successfully reestablished and continues to do well (*Rich, Gibson, and Marsden 1999*).

Rediscovering obscure members of the carrot family: The role of amateur botanists

Flora Europaea describes 423 species of the carrot family (*Apiaceae*) in 110 genera. Many are well known, but between forty and fifty of these species are very poorly known, and some have not been seen in the wild for many years. One example is *Peucedanum achaicum* from southern Greece. The original site is described as being on cliffs in the Vouraikos Gorge below Zachlarou, but in a country where detailed maps were hard to obtain until recently, finding such a site may not be easy. Zachlarou is a monastery with a station on a cog railway, but penetration of the gorge below the monastery revealed a talus slope with many plants of this species. The genus *Seseli* (moon-carrot) is rich in obscure species, with one which had even been thought to be extinct. But assiduous exploration revealed 600 plants of this species in the Sierra de Gador of Spain. One of the problems is that rare and obscure species may be accompanied by much more common species, adding to the confusion. This is the case with *Seseli tomentosum*, which occurs in Croatia and is easily confused with the much more common *S. montanum* subsp *tommasinii*. The lesson is to have good knowledge of what such plants look like, and to undertake fieldwork [*see report by Mervyn Southam in Plant Talk 11 (1997)*].

Cloning the Mascarene café marron

The islands of Mauritius, Rodrigues, and Réunion (Mascarene Isles) are famous for their endemic plants. One of the very rarest is the café marron (*Ramosmania rodriguesii*) in the family *Rubiaceae*. It is reduced to just a single wild tree, which was discovered by a schoolboy in 1980. The Rodrigues Forestry Service put a fence around the tree to save it from being hacked at or cut down. Continuing damage meant that eventually the tree was protected by three more concentric fences and two pad-locked gates. In 1986, a cutting was taken to the Royal Botanic Gardens, Kew, from which a number of plants were propagated. An attempt to repatriate the plant in 1989 failed. In 1996, further cuttings were taken and successfully rooted on the island. This will now provide the basis for a new genera-tion of plants to be placed back into the wild. Other good news from the Mascarenes is that several other plants feared to be extinct have been found in the wild and will now be propagated. One of these is *Trochetia parviflora*, a small tree found on Mauritius that had not been seen since 1863 and was feared extinct [*see reports in Plant Talk 12 (1998) and Plant Talk 24 (2001)*].

Collapse of ant-plant mutualisms through competition

Sometimes invading species can have a devastating effect on the reproduction and eventual survival of plants. An example comes from the unique southern African fynbos flora. Indigenous ants are important seed dispersers, taking seeds for food and storing them in their nests, where a proportion eventually germinates. Invasion by the Argentine ant (*Iridomyrmex humilis*) has resulted in a dra-matic drop in indigenous ant numbers. Whereas seedling emergence of *Mimetes cucullatus* (family *Proteaceae*) was 35 percent in sites not infested by Argentine ant, where the ant has invaded this has dropped to 0.7 percent. It is predicted that continued invasion of fynbos by these ants will lead to eventual extinction of a number of fynbos endemic shrubs by slow and subtle attrition of seed reserves.

Creating a new population of the desert endemic annual: Amsinkia grandiflora

Amsinkia grandiflora (family *Boraginaceae*) is an endangered annual plant known only from three natural populations in the dry grasslands east of Livermore, California. Current threats are both intrinsic (low genetic variability, low seed production, and specific habitat require-ments) and extrinsic (exotic grasses, livestock grazing, and intensive land use). Using existing data on ecology and distribution, the natural range of the species was evaluated and twelve can-didate sites selected, with one being chosen for the first experimental introduction. A total of 3460 nutlets were sown into twenty plots that were given a variety of treatments (burnt, hand-clipped, grass-specific herbicide) to keep down weed competition. Sowing of nutlets was done using frames that allowed the fate of individuals to be followed. Success was evaluated by mea-suring germination, stress, survivorship, nutlet production, and demography. Overall, the new population was successful. Of the 3460 nutlets, 1774 produced germinules in the first season and 1101 survived to reproduce. From these, over 35,000 nutlets were produced. Prior to the second growing season, the new population was divided into two treatment blocks—one being burnt in the fall and sprayed with grass herbicide in late winter, while the other was herbicide-sprayed only. Nutlet production grew by 44 percent. By next spring there were 1640 reproductive plants in the population. This demonstrates that even for precarious species, sites for new populations can be located and planted, evaluated, experimentally treated, and managed to a recovery phase (Pavlik 1994).

Saving the "Wooden Rhino": making the Kenyan woodcarving industry sustainable

The Kenyan woodcarving industry has been a rural development success. Ironically, the economic success of this industry has severely undermined the natural resource on which it is based: slow-growing hardwoods such as ebony (*Dalbergia melanoxylon*) and "muhugu" or mahogany (*Brachylaena huillensis*). As a result, the livelihoods of 60,000 carvers and their families, as well as globally-important forest biodiversity, are under threat. The felling of over 50,000 trees per year for carving alone poses a major conservation problem, because it degrades forest habitat. Carvers are aware of the threat to their livelihoods and have identified fast-growing tree species that can be sustainably harvested, such as neem (*Azadirachta indica*), as alternatives. The People and Plants Initiative of WWF-UK, UNESCO, and the Royal Botanic Gardens, Kew, has been promoting the use of such "Good Woods" for a number of years. Stumbling blocks exist, though. For example, carvers are used to carving hardwoods and, at present, have few incentives to make the switch. Carvers can still buy hardwood, illegally-cut elsewhere, for little extra cost compared to Good Woods. In contrast to Good Woods, hardwoods do not have to be cured before carving. However, a strong, market-led demand for Good Wood carvings could change their practices. The carvers are very responsive to trends in demand. People and Plants is working with the carvers and farmers' group towards Forest Stewardship Council certification, which could prove a pivotal tool, allowing the economic benefits of carving to continue to accrue to the carvers and farmers, while helping to conserve the environment (*see People and Plants project, www.kew.org/peopleplants*).

References and Further Reading

Adams, W.M. 2004. *Against Extinction: the Story of Conservation*. London: Earthscan.

Akerele, O., Heywood, V. and Synge, H. (Eds.) 1991. *Conservation of Medicinal Plants*. Cambridge: Cambridge University Press.

Bond, W., and Slingsby, P. 1984. Collapse of an ant-plant mutalism: The Argentine ant (*Iridomyrmex humilis*) and myrmecochorous Proteaceae. Ecology 65, 1031–1037.

Botanic Gardens Conservation International. 2000. *The Gran Canaria Declaration*. Kew, Surrey BGCI.

Cunningham, A. 2001. *Applied Ethnobotany: People, Wild Plant Use and Conservation*. London: Earthscan.

Fiedler, P., and Ahouse, J.J. 1992. Hierarchies of cause: towards an understanding of rarity in vascular plant species. In *Conservation Biology*, edited by P.L. Fiedler and S.K. Jain. . New York: Chapman and Hall, 23–48.

Given, D.R. 2001. The wiggy-wig bush—orphan or geriatric? *Muehlenbeckia astonii* (Polygonaceae) in North Canterbury and Marlborough. *Canterbury Botanical Society Journal 35*, 48–51.

Given, D.R. 1998. All flesh is grass. *World Conservation 2*(98), 3–4.

Given, D.R. 1994. *Principles and Practice of Plant Conservation*. Portland, OR: Timber Press.

Given, D.R. 1981. *Rare and Endangered Plants of New Zealand*. Wellington, New Zealand: Reed.

Guerrant, E.O., Havens, K. Maunder, M. and Raven, P.H. (eds.) 2004. Ex Situ Plant Conservation: Supporting Species Survival in the Wild. Washington, DC: Island Press.

IUCN. 2001. *IUCN Red List Categories and Criteria: Version 3.1*. Cambridge: IUCN Species Survival Commission.

Josephson, J. 2000. Going, going, gone. Plant species extinction in the 21st century. *Environmental Science and Technology 35*(5), 130A–135A.

Howard, P.L. (Ed.) 2003. *Women & Plants: Gender Relations in Biodiversity Management and Conservation*. London: Zed Books.

Lucas, G., and Synge, H. 1978. *The IUCN Plant Red Data Book*. Morges: IUCN.

Mace, G.M., Baillie, J.E.M., Beissinger, S.R., and Redford, K.H. 2001. Assessment and management of species a risk. In *Conservation Biology. Research priorities for the next decade*, edited by M.E. Soulé and G.H. Orians. Washington, DC: Island Press, 11–30.

MacKinnon, J., MacKinnon, K., Child, G., and Thorsell, J. 1986. *Managing Protected Areas in the Tropics*. Gland, Switzerland: IUCN/UNEP.

Maunder, M., Clubbe, C., Hankamer, C., and Groves, M. (Eds). 2002. *Plant Conservation in the Tropics: Perspectives and Practice*. Kew: Royal Botanic Gardens.

May, R.M., Lawton, J.H., and Stork, N.E. 1995. Assessing extinction rates. In *Extinction Rates*, edited by J.H. Lawton and R.M. May. Oxford: Oxford University Press, 1–24.

Mittermeier, R.A., Myers, N., and Mittermeier, C.G. 1999. *Hotspots: Earth's Biological Richest and Most Endangered Terrestrial Eco-regions*. Washington, DC: CEMEX, Conservation International.

Myers, N. 1988. Threatened biota: hot spots in tropical forests. *The Environmentalist 8*, 1–20.

Myers, N. 1990. The biodiversity challenge: expanded hot-spots analysis. *The Environmentalist 10*, 243–56.

Myers, N. et al. 2000. Biodiversity hotspots for conservation priorities. *Nature 403*, 853–858.

Oldfield, S., Lusty, C., and MacKinven, A. 1998. *The World List of Threatened Trees*. Cambridge: World Conservation Press.

Pickett, S.T.A., Ostfeld, R.S., Shachak, M., and Likens, G.E. 1997. *The Ecological Basis of Conservation*. London: Chapman and Hall.

Pimm, S.L., and Raven, P.R. 2000. Extinction by numbers. *Nature 403*, 843–845.

Prescott-Allen, C. and Prescott-Allen, R. 1986. *The First Resource: Wild Species in the North American Economy*. New Haven: Yale University Press.

Rich, T.C.G., Gibson, C. and Marsden, M. 1999. Re-establishment of the extinct native plant *Filago gallica* L. (Asteraceae), narrow-leaved curdweed, in Britain. *Biological Conservation 91*, 1–8.

Tuxhill, J. and Nabhan, G.P. 2001. *People, Plants and Protected Areas: A Guide to in Situ Management*. London: Earthscan.

Walter, K.S., and Gillett, H.J. 1997. *1997 IUCN Red List of Threatened Plants*. Cambridge: World Conservation Monitoring Centre.

21

Conservation of Crop Genetic Resources

NIGEL MAXTED AND DAVID R. GIVEN

Plants: A National and International Resource

At a time of exponential human population growth—there will be nine billion humans on Earth by 2025 (Lutz, Prinz, and Langgassmer 1993)—our plant genetic resources should be as valuable to us as gold and oil, if not more so. Unfortunately, however, history shows us that far too often, humans have repeatedly mismanaged nature and shown a thoughtless disregard for non-human biodiversity.

Loss of plant biodiversity has a fundamentally detrimental impact on humankind because of the direct and indirect benefits humans gain from the existence and use of plants. The ways in which plants and plant products are used by humans are very diverse, including food, food additives, animal foods, materials, fuels, poisons, medicines (human and veterinary), ornamentals, recreational use, windbreaks, soil improvers, erosion control, pollution indicator species, and gene donors. Of the approximate 300,000 plant species, some 5,000 provide food; however, thirty species provide 95 percent of the world's calorie intake, and three alone (corn, wheat, and rice) supply almost 50 percent of the calories and protein in the human diet (FAO 1998). There is clearly scope for diversification based on further domestication and the promotion of so-called neglected and under-utilized crops. It has also been reported that approximately 80 percent of the world's people rely on traditional medicine for their primary health care needs, and an estimated 85 percent of traditional remedies are derived from plant extracts (Lewington 1990). The value of plants, however, is not restricted to socio-economic criteria; they also have cultural value in religion, folklore, and art, and in themselves for their contribution to the structure of a community and its processes. In practice, the term "plant genetic resources" is applied to the plant diversity that is most immediately exploitable and, therefore, has greater socio-economic value.

In terms of use, it is not only diversity of species, but also diversity within the species that is important. Genetic diversity within species ensures the short-term viability of individuals and populations, and by allowing adaptation to natural selection pressures in a changing environment ensures long-term evolutionary survival of populations and species. Individuals vary in traits such as height, fecundity, disease resistance, or tolerance to such extreme environmental conditions as drought. For millennia, farmers have used genetic diversity, evaluating, selecting, and exchanging desired traits or genotypes to improve their local crops, just as contemporary plant breeders or

To increase the genetic diversity of U.S. corn, the Germplasm Enhancement for Maize (GEM) project seeks to combine exotic germplasm, such as this unusually colored and shaped maize from Latin America, with domestic corn lines. Photo by Keith Weller. ARS/USDA.

biotechnologists do when they introduce new genes into crop varieties. Both traditional farmers and plant breeders need continued access to genetic diversity if they are to make progress in their ongoing battle with evolving pests and diseases, and thus maintain food security.

The economic impact of breeding disease resistance in a crop can be high. For example, the use of Hessian fly resistance in wheat in the U.S.A. saved $17 million in a single year (Bouhssini et al. 1998). In contrast, however, the consequences of lack of genetic diversity in crops can be devastating, as shown by the often-quoted case of potato blight in Ireland in 1845. An infection of late potato blight (*Phytophthora infestans*) wiped out the potato crop in Ireland, leading to the Great Potato Famine of 1845–1849, and the starvation and emigration of millions of people. The existing varieties of potato at that time had no resistance to *P. infestans*. Resistance was subsequently found by the 1870s in the Chilean subspecies of *Solanum tuberosum* subsp. *andigena* and in several wild potato species, particularly *Solanum demissum* from Mexico (Hawkes, Maxted, and Ford-Lloyd 2000).

See: Roots & Tubers, pp. 63–5

No country is sufficiently rich in native genetic diversity to make it independent of other regions of the world. In Brazil, for example, which has an abundance of plant-life, two-thirds of its calorie consumption is based on crops originating from other continents (see Table 21.1).

TABLE 21.1 Source of Plant-Derived Calories Consumed in Brazil (Crucible Group 1994)

Crop	Share of Plant-Derived Calories (%)	Center of Origin
Sugar	20.38	Indochina
Rice (paddy)	17.64	Asia
Wheat	15.29	Southwest Asia
Corn	12.20	Central America
Soybean	8.84	China/Japan
Cassava	7.10	Brazil/Paraguay
Beans	6.40	Andes
Bananas	2.22	Indochina

Of course in addition to their direct economic use by humans, plants play a pivotal role in natural ecosystems and contribute to the functioning of the biosphere, as well as having aesthetic and recreational value; each is a significant impetus to conservation. The conservation of wild plants, with these important roles, is considered in the previous chapter.

Gene Pool Concepts

The history of human exploitation of plants is as long as human evolution itself, stretching from hunter-gatherers in pre-agricultural societies to targeted exploration and collection by European colonists in Asia and South America. But it was the Russian botanical geneticist Nikolai I. Vavilov who, in the early years of the 20th century, realized the importance of conserving the total genetic diversity contained within any crop complex (such as peas)—not only diversity within the crop species itself, but also that found in its wild relatives—to make the full range of genetic diversity available to plant breeders. This total range of genetic diversity is known as the *crop gene pool*. This concept was developed and formalized by Harlan and de Wet (1971) (see Figure 21.1).

- **Primary gene pool (GP-1):** all cultivated, wild, and weedy forms of a crop species. Hybrids among these taxa are fertile, and gene transfer to the crop is simple and direct. This pool of taxa is often referred to as a *biological species*.
- **Secondary gene pool (GP-2):** the group of species that can be artificially hybridized with the crop, but where gene transfer is difficult. Hybrids may be weak or partially sterile, or their chromosomes may pair poorly during meiosis.
- **Tertiary gene pool (GP-3):** including all species that can be crossed, though with some difficulty (e.g., requiring *in vitro* hybrid embryo culture), and where gene transfer is impossible or requires radical techniques (e.g., radiation-induced chromosome breakage).

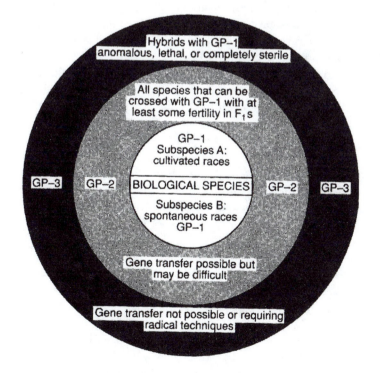

Figure 21.1. Schematic diagram of gene pool concept (Harlan and de Wet 1971).

TABLE 21.2 Common Features of Crops, Crop Progenitors and Wild Relatives.

Crop	Crop Progenitor	Wild Relatives
Large fruit, seed, or flower size	Fruit, seed, or flower size larger than average for wild species	Average fruit, seed, or flower size for wild species
Very flavorsome fruit or seed	Fruit or seed slightly more flavorsome than average wild species	Average flavor of fruit or seed for wild species
Very nutritious fruit or seed	Fruit or seed slightly more nutritious than average wild species	Average nutritional value of fruit or seed for wild species
Highest protein content	High protein content	Low protein content
Highest oil content	High oil content	Low oil content
Non-shattering fruit	Partially-shattering fruit	Shattering fruit
Aesthetically pleasing color combinations in flowers and fruits	Unusual color combinations in flowers and fruits	Drab color combinations in flowers and fruits
Examples		
Barley (*Hordeum vulgare*)	*Hordeum spontaneum*	Other *Hordeum* species
Faba bean (*Vicia faba*)	*Vicia faba* ssp. *paucijuga*, Narbon bean (*Vicia narbonensis*)	*Vicia bithynica* and other vetch species

If this concept is applied to barley as an example, *Hordeum vulgare* and its progenitor, *H. spontaneum*, would belong to GP-1, *H. bulbosum* to GP-2, and all the other species of the genus to GP-3. The highest value would be ascribed to the GP-1 species, then GP-2, and finally GP-3. This kind of priority setting can be refined, as some species within a given gene pool may be more likely to harbor desired traits. In general, the wild progenitors of crops are the most promising source of new genes, because they often share characteristics with the crop plant that are lacking from species in the secondary and tertiary gene pools (GP-2 and GP-3).

From his observations of crop gene pools, Vavilov also noted similar patterns of variation between crops and their wild relatives for each gene pool. In other words, wild relatives from another genus sometimes showed some features of domestication (large seed, lack of fruit shattering, nutritious fruits, etc.) that were absent from most related wild species (see Table 21.2). He noted this parallelism in domestication characteristics in many different crop complexes (for example, vetches, lentils, and peas) and this forms the basis of his Law of Homologous Series (Vavilov 1922). The importance of Vavilov's law is that it has predictive value, in that if one crop plant is found with a particular trait, then this same desirable trait is likely to be found in the gene pool of unrelated crop species. There is an obvious tie-in here with contemporary views of phylogenetic evolution and synteny.

The gene pool concept is, by definition, crop- or single species-centered, in that there is always one taxon in GP-1A. However, a single species may conceivably be in GP-1 for one crop, GP-2 for another and GP-3 for a third. Expanding on Harlan and de Wet's work, Maxted, Ford-Lloyd, and Hawkes (1997) have developed the concept of the "gene sea," where each species is at the centre of its own gene pool, but all individual gene pools are interrelated in one expanse of genetic diversity, as shown in Figure 21.2.

Thus, species that are present in multiple overlapping gene pools within the gene sea would be given the highest priority for conservation, because they would better represent the breadth of plant genetic diversity in the lowest number of populations or accessions.

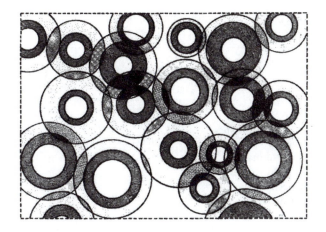

Figure 21.2. Schematic diagram of a segment of the gene sea (Maxted, Ford-Lloyd, and Hawkes 1997).

The Threat to Plant Genetic Resources

Plant genetic resources are increasingly threatened at the ecosystem, species, and genetic levels, largely as a result of human activities. For example, in the case of crop plants, the proportion of the wheat crop in Greece contributed by landraces or old, indigenous varieties declined from 80 percent in 1930 to less than 10 percent in 1970. In China, nearly 10,000 wheat varieties were in use in 1949, but only 1000 were still in use by the 1970s (FAO 1998). In Cambodia, unique rice varieties were lost in the 1970s when war disrupted agricultural production. Stored seed in the national gene bank was eaten or rotted, and numerous landraces would therefore have died out, were it not for the duplicates preserved in the International Rice Research Institute (IRRI) gene bank in the Philippines. In Mexico and Guatemala, urbanization has displaced some of the populations of teosinte (*Zea mexicana*), the closest relative of corn, and these populations have also suffered genetic pollution from genetically modified corn (Quist and Chapela 2001).

It is likely that virtually all plant species are currently suffering loss of genetic variation to varying degrees: it was estimated that 25 to 35 percent of plant genetic diversity could be lost over the next 20 years (Maxted, Ford-Lloyd, and Hawkes 1997).

National and international agencies must deal with a paradoxical confrontation between conservation and development. Plant breeders throughout the world are rightly engaged in developing better and higher-yielding cultivars of crop plants. This involves the replacement of genetically variable, lower-yielding landraces with products of modern agriculture, which are much more genetically uniform. Thus, genetic uniformity is replacing diversity. These same plant breeders are, however, dependent upon the availability of a pool of diverse genetic material for success in their work, and thus are unwittingly causing the genetic erosion of plant diversity that they themselves will need in the future—hence the paradox. Of course, replacement of traditional landraces by modern cultivars is not the only cause of genetic erosion or loss of genetic diversity; other changes in farming systems, the intensification of production systems, overexploitation, introduction of exotic cash crops, human socio-economic changes and upheaval (e.g., extinction of tribal cultures, urban sprawl, land clearances, food shortages), as well as both natural and man-made calamities (e.g., floods, landslides, or wars) have all acted against the retention of socio-economically important biodiversity.

TABLE 21.3 Estimated Annual Markets for Genetic Resources Products (ten Kate and Laird, 1999)

Sector	Lower Estimate (US $ billion)	Upper Estimate (US $ billion)
Pharmaceutical	75	150
Botanical Medicine	20	40
Major Crop	300+	450+
Horticultural	16	19
Crop Protection	0.6	3
Biotechnology	60	120
Cosmetics & Personal Care Products	2.8	2.8
Total	500	800

Plant genetic resources cost/benefit analysis

The economic benefit of plant genetic resources use has recently been reviewed by ten Kate and Laird (1999). Although it is very difficult to estimate precisely the annual global market value of plant genetic resources, they suggest a range of figures between US $500 to $800 billion; the break-down of figures is given in Table 21.3. The use of wild species by local communities should not be underestimated; for example in Tanzania in 1988, it was estimated that the value of all wild plant resources to rural communities, whether through subsistence consumption or sale, was more than US $120 million (8 percent of agricultural GDP) (FAO 1998).

Of the industries that depend on diversity, agriculture remains by far the largest. Phillips and Meilleur (1998) estimate that endangered food crop relatives have a worth of about US $10 billion annually in wholesale farm values. Various studies, mostly conducted on cereals, have estimated that more than 50 percent of the increase in crop production has been due to the improvement of crop cultivars, and such improvement is brought about by transferring desirable genes/traits to crops from landraces and other more distant germplasm sources. Thus, the transfer of dwarfing genes from Japanese semi-dwarf material to U.S. and Mexican wheat stocks led to the revolution of wheat production in the world during the 1960s and 1970s (see Grains, pp. 45, 53). This so-called "green revolution" helped food-deficient countries like India become food sufficient, and ultimately even net exporters within a short period of 10 years.

The transfer of genes for high sugar content to the tomato (*Lycopersicon esculentum*) from its wild relative (*L. chmielewskii*) has generated an additional income of US $5 to $8 million per year for the tomato industry (Iltis 1988). Although precise estimates of the global value associated with the use of plant genetic resources do vary, it is clear that plant genetic resources have a real and substantial value.

However, there is a cost involved in conserving this diversity. Using the FAO (1998) figure of 6.1 million accessions in world gene banks, and using the estimates of Smith and Linington (1997) for the cost of obtaining the material (US $597 each) and incorporation of the material into the gene bank (US $273 each), we then have a total cost of US $5.3 billion for collecting and conserving the world's germplasm in gene banks. Even the cost of maintaining existing *ex situ* gene bank accessions is not insignificant consid-ering the commitment to conserve is open-ended. Taking Smith and Linington's (1997) estimates of the annual cost of maintaining an accession of US $5 each, then for 6.1 million accessions the running cost is US $30.5 million per year. We have no estimate of *in situ* expenditure, but for the United States at least, it has been estimated that more than 98 percent of all conservation expenditure is spent on *in situ* activities related to wild species (Cohen et al. 1991). Although these figures are relatively high, they are small compared to the annual market for genetic resources use. Therefore simple cost/benefit analysis clearly indicates humans are very short-sighted to carelessly oversee loss and threat to plant genetic diversity.

International treaties and plant genetic resources

Recognition of the fundamental importance of these issues was highlighted at the United Nations Conference on Environment and Development (UNCED) held in Rio de Janeiro, Brazil in 1992, and has been enshrined in the resulting Convention on Biological Diversity (CBD). Its objectives are:

> … the conservation of biological diversity, the sustainable use of its components and the fair and equitable sharing of the benefits arising out of the utilization of genetic resources, including by appropriate access to genetic resources and by appropriate transfer of relevant technologies, taking into account all rights over those resources and to technologies, and by appropriate funding (www.biodiv.org).

The CBD was the first global treaty that linked the conservation of biodiversity to sustainable utilization. It represents a milestone in biodiversity conservation thinking, reflecting international acknowledgement of the loss of our biological resources, their role in human development and wealth creation, and the urgent need for conservation action and sustainable exploitation. The CBD is now recognized as the primary guiding framework for the conservation, management, and use of biodiversity.

More specific reference to plant genetic resources is made in the International Treaty on Plant Genetic Resources for Food and Agriculture (ITPGRFA), agreed by 116 countries in Rome in November, 2001. The International Treaty is in harmony with the CBD, and many of its articles make explicit propositions in the CBD, especially as regard Farmers' Rights—recognizing the contribution that the local and indigenous communities and farmers of all regions of the world have made to the conservation and development of plant genetic resources. Farmers' rights are equated with plant breeders' rights. It is a legally binding international agreement, which will come into force when ratified by at least 40 states. The objective of the ITPGRFA was expressed in Article 1.

> The objectives of this Treaty are the conservation and sustainable use of plant genetic resources for food and agriculture and the fair and equitable sharing of the benefits arising out of their use, in harmony with the Convention on Biological Diversity, for sustainable agriculture and food security (www.fao.org/ag/cgrfa/itpgr.htm#text).

It takes into consideration the particular needs of farmers and plant breeders, and aims to guarantee the future availability of the diversity of plant genetic resources for food and agriculture on which they depend, and the fair and equitable sharing of the benefits.

These Treaties both provide a broad framework for plant conservation linked to sustainable and equitable use of resources, but they lack any specific strategy for achieving their objectives. The CBD's Conference of the Parties, who are charged with implementing the CBD, adopted a Global Strategy for Plant Conservation (GSPC) at its seventh meeting in November, 2001. GSPC provides the necessary specific conservation targets that are to be achieved by 2010, several of which relate to plant genetic resources (the full list of targets is provided in Conservation of Wild Plants, p. 402):

- Thirty percent of production lands to be managed consistent with the conservation of plant diversity
- Seventy percent of the genetic diversity of crops and other major plant genetic resources to be conserved

- No species of wild flora is to be subject to unsustainable exploitation resulting from international trade
- Thirty percent of plant-based products to be derived from sources that are sustainably managed
- A reversal of the decline of plant resources that support sustainable livelihoods, local food security and health care
- Every child to be aware of the importance of, and the need to conserve, plant diversity
- The number of trained people working with adequate facilities in plant conservation and related activities to be doubled
- Networks for plant conservation activities established or strengthened at international, regional, and national levels

Even more detailed targets are being used by many national governments and regions—for example, the European Plant Conservation Strategy.

Where Are Plant Genetic Resources Found?

Plant biodiversity, as has been seen in the chapter on conservation of wild plants, is not evenly distributed across the surface of the Earth. A similar picture of uneven geographic distribution also emerges for that of plant genetic resources. Vavilov developed a concept of centers of diversity for crop plants, where crop gene pools were focused. Although his belief that all of these were centers of crop domestication is no longer accepted, all are still recognized as important centers of diversity. These were eight, generally mountainous areas, situated in tropical or sub-tropical regions:

I **Chinese Center:** Western and Central China—millets, beans, onion, radish, cabbage, fruit trees, as well as plants producing oils, spices, medicines, and fibers

II **Indian Center:** India, Indo-Malaya, Indo-China, Burma, and Assam—rice, chickpea, beans, many tropical fruits (including *Citrus* species *Musa, Mangifera,* etc.); oil-producing species, fibers, spices, stimulants, and dye plants; *Saccharum*

III **Inner-Asiatic Center:** Northwestern India, Afghanistan, Tadzhikistan, Uzbekistan, and western Tien-Shan—wheat, peas, cabbage, lettuce, sesame, cotton, various vegetables and melon species, spice crops; fruit and nut trees

IV **Asia Minor Center:** Transcaucasia, Iran, Turkmenistan, and Anatolia—wheat, rye, oats, chickpeas, lentil, vetches, peas; alfalfa, clover, and sainfoin; melons; vegetables; fruit crops, including *Malus, Pyrus, Punica, Ficus, Vitis, Pistacia*

V **Mediterranean Center:** Mediterranean countries—Forage and vegetable species; various oil-producing plants and spices; olive, beets, cabbages, onion, asparagus, lettuce, parsnip; ethereal oil species and spices

VI **Ethiopian Center:** Ethiopia—wheat, barley, peas and beans, lupins, teff, finger millet, coffee, banana, and sorghum

VII **South Mexican and Central American Center:** South Mexico and Central America—corn, beans, marrow, sweet potato, peppers, cotton, and tobacco

VIII **South American Andean Center:** Peru, Ecuador, and Bolivia—Potato and other tuberous crops, some fruit crops, lupins, beets, corn, and various beans

VIIIa **The Chilean Center:** Chile—Potato, oilseed, grasses, and strawberries

VIIIb **The Brazilian-Paraguayan Center:** Brazil and Paraguay—Manioc, peanut, cocoa, rubber plant, and maté

Discussion of the geographical distribution of diversity almost always focuses on the spatial distribution of species, but biodiversity shows patterns of distribution at all levels. If we consider genetic diversity, we might assume that infra-specific genetic diversity is evenly spread throughout the range of the species—but this is often not the case. Studies of wild lentils (Ferguson et al. 1998), for example, have clearly shown that the genetic diversity is not distributed evenly across its geographic range, but is concentrated in a relatively small region. Therefore, if we wish to conserve a gene pool, we must understand the distribution of genetic diversity in relation to species ecogeographic range. The general picture that is emerging for crop plants is also supported by studies of within- and between-population genetic variation in wild plant species.

Plant Conservation

In order to develop practical techniques to achieve conservation and sustainable use objectives, conservation managers must use their knowledge of genetics, ecology, geography, taxonomy, and many other disciplines to understand and manage the biodiversity they wish to conserve. Conservation and sustainable use is not just about individual plant species. Even if the conservation target is a population of a species, no population can survive in isolation; it exists within a community or ecosystem, interacting with other species and the abiotic environment. Examples of such interactions include pollinators, seed dispersers, microbial symbionts, herbivores (whether natural or introduced by humans), and pathogens. Thus, when applying genetic *in situ* conservation for a socio-economically important plant species, the maintenance of whole ecosystems is crucial. Plant genetic resource conservation acts as an essential link between the genetic diversity of a plant and its utilization or exploitation by humans (Figure 21.3).

Plant genetic diversity

The ultimate goal of genetic resources conservation is to ensure that the maximum possible genetic diversity of a taxon is maintained and available for utilization, and as such diversity tends to focus on specific target taxa. Faced with limited financial, temporal, and technical resources and the impossibility of actively conserving or monitoring all species, it is important to make the most efficient and effective selection of species to focus conservation efforts.

Selection of target taxa

This choice should be objective, based on logical, scientific, and economic principles. The factors that provide a species with "value" include:

- Current conservation status
- Potential economic use
- Threat of genetic erosion
- Genetic distinction
- Ecogeographic distinction
- National or conservation agency priorities
- Biologically important species
- Culturally important species
- Relative cost of conservation
- Conservation sustainability
- Ethical and aesthetic considerations

It is rare for any one factor alone to lead to a taxon being given conservation priority. More commonly, all or a range of these factors will be assessed for a particular taxon, and then it will be assigned a certain level of national, regional, or world conservation priority. If the overall score passes a threshold level or is higher than competing taxa, then proposals will be made to conserve it, either *in situ* in a genetic reserve or on farm, or collected for *ex situ* conservation.

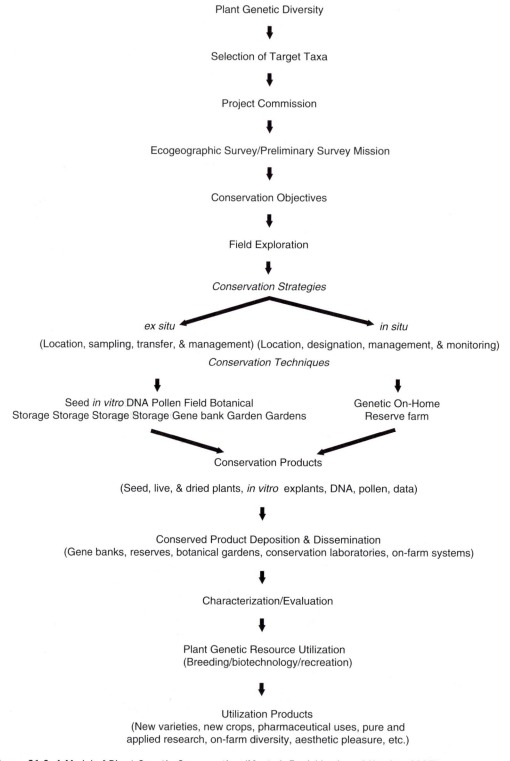

Figure 21.3. A Model of Plant Genetic Conservation (Maxted, Ford-Lloyd, and Hawkes 1997).

Project commission

Once a taxon is selected for conservation, a commission statement necessarily precedes the actual conservation activities. It establishes the objectives of the conservation proposal, and specifies the target taxa and target areas, how the material is to be utilized, and where the conserved material is to be duplicated. It will also give an indication of which conservation techniques are to be employed. The commission may vary in taxonomic and geographic coverage, such as onion (*Allium*) species of Central Asia, *Cymbidium* orchids worldwide, or chickpeas (*Cicer*) from the Western Tien Shen. In each case, however, a particular group of taxa from a defined geographical area must be considered currently insufficiently conserved (either *in situ* or *ex situ*), of sufficient actual or potential use, and/or endangered, to warrant active conservation.

Ecogeographic survey and preliminary survey mission

Once the target taxon or group of taxa is chosen, fundamental biological data is used to formulate the most appropriate conservation strategy. The process of collating and analyzing geographical, ecological, taxonomic, and genetic data for use in designing conservation strategies is referred to as ecogeography (Maxted, van Slageren, and J. Rihan 1995). Ecogeographic studies involve the use of large and complex data sets obtained from the literature and from the compilation of passport data associated with herbarium specimens and germplasm accessions. The synthesis of these data enables the conservationist to clearly identify the geographical regions and ecological niches that the taxon inhabits; therefore, not only can areas with high numbers of target taxa be identified, but also areas that contain high taxonomic or genotypic diversity, uniqueness of habitat, and economic or breeding importance.

If the available ecogeographic data for the target taxon are limited, there will not be sufficient background biological evidence to formulate an effective conservation strategy. In this case, it would be necessary to undertake an initial survey mission to gather the novel ecogeographic data required on which to base the conservation strategy. The survey mission may be in the form of "coarse grid sampling", which involves traveling throughout a likely target region and sampling sites at relatively wide intervals over the whole region. The precise size of the interval between sites depends on the level of environmental diversity across the region, but it may involve sampling every 1 to 50 km. (Hawkes, Maxted, and Ford-Lloyd 2000).

Conservation objectives

The products of the ecogeographic survey or survey mission provide a basis to formulate future conservation priorities and strategies for the target taxon. Within the target area, zones of particular interest may be identified—for instance, areas with high concentrations of diverse taxa, very low or very high rainfall, high frequency of saline soils, or extremes of altitude or exposure. In general, it can be assumed that areas with very distinctive ecogeographic characteristics are likely to contain plants with associated distinct genes or genotypes. If a taxon is found throughout a particular region, then the conservation manager can use the ecogeographic data to positively select a series of diverse habitats to designate as reserves. If a taxon has been found at one location, but not at another with similar ecogeographic conditions, then these similar locations should be searched.

Field exploration

The ecogeographic information provides the general locality of the plant populations, but will rarely be sufficient to precisely locate actual populations. Therefore, the preparatory element of conservation activities will be followed by field exploration, during which actual populations are located. Ideally, populations of the target taxon that contain the maximum amount of genetic diversity in the minimum number of populations will be identified. Commonly, there is too much diversity both in wild and domesticated species to conserve all their alleles; therefore, we must attempt to conserve the range of diversity that best reflects the total genetic diversity of the species. To identify how many population samples are required, the conservationist should ideally know the amount of genetic variation within and between populations, local population structure, breeding system, taxonomy, and ecogeographic requirements of the target taxon, as well as many other biological details. Some of this information will be supplied following the ecogeographic survey, but some will remain unavailable. Therefore, the practice of field exploration will be modified depending on the biological information on the target taxon and target area that is available.

Conservation strategies and techniques

There are two basic conservation strategies, *ex situ* and *in situ*, each composed of a range of techniques. There is an obvious fundamental difference between these two strategies: *ex situ* conservation involves the location, sampling, transfer, and storage of target taxa from the target area, whereas *in situ* conservation involves the location, designation, management, and monitoring of target taxa where they are currently encountered. The two basic conservation strategies may be further subdivided into several specific applications of the strategies or techniques (see Table 21.4). It is now generally agreed that a mix of both strategies incorporating *in situ* and *ex situ* techniques, possibly also incorporating an element of ecosystem restoration, provides the best practical approach to integrated conservation of a particular species (Falk and Holsinger 1991).

In situ conservation involves the maintenance of genetic variation at the location where it is encountered, either in traditional farming systems or in the wild. *In situ* conservation of wild plants, such as the wild relatives of crop plants, is covered in the section on genetic reserves in the chapter on conservation of wild plants. The process of domestication (the selection and adaptation of wild plants for use by humans) has taken place over thousands of years, and has led to the existence of an enormous number of different landraces. Each season the farmer keeps a proportion of harvested seed for resowing, and seed may be exchanged locally between villages. Thus, the landrace is highly adapted to the local environment, and is likely to contain locally-adapted alleles that

Botanist David Spooner (right) and Alberto Salas, plant genetic resources specialist with the International Potato Center, Lima, Peru, collect potato germplasm in Peru for deposition in national and international gene banks. Photo by Alejandro Balaguer. ARS/USDA.

may prove useful for specific breeding programs. The fact that the application of *in situ* conservation techniques permits this continued evolution, as opposed to freezing diversity in the *ex situ* context, is cited as the key benefit of *in situ* techniques. However, maintaining cultivation of landraces at a time when economic forces are in favor of new cultivars is difficult, and may involve provision of long-term incentives for farmers. Another approach is to develop specialist markets for landraces. This has been highly successful in the Tuscan region of Italy, where a long-term decline in cultivation of emmer wheat, *farro,* has been reversed following its promotion as a health food.

Ex situ conservation involves the maintenance of genetic variation away from its original location. Samples of populations are taken and conserved, either as living collections of plants, or samples of seed, tubers, tissue explants, pollen, or DNA maintained under special artificial conditions. In an ideal world, it would be preferable to conserve all biodiversity *in situ* in nature, rather than move it into an artificial environment to be conserved. However, due to the threat of genetic erosion in the original location and the need for easy access for evaluation, *ex situ* conservation techniques are desirable for many plant species. These techniques also provide a safety backup for *in situ* conservation techniques, where *in situ* conservation in a genetic reserve or on farm alone cannot guarantee long-term security for a particular domesticated or wild species. One significant advantage of *ex situ* conservation is that the genetic material is always available to researchers and to the breeder for evaluation of characteristics, such as resistance to a particular pest or disease.

The most convenient way of maintaining most plant germplasm is by storing seeds; however, this option is not open for all species. For the majority of species, the seed can be stored for longer periods by lowering the storage temperature, and also the moisture content to 2 to 5 percent, or even lower. These species are termed *orthodox.* The lowering of the storage temperature to zero or below does not generally damage the seeds of orthodox species, providing that the moisture content has been reduced previously to at least 15 percent. Thus, most gene banks for cereals, pulses and other species with small seeds are able to store their seeds at about 0 to 20°C after they have been dried to about 2 to 5 percent moisture content. In contrast, a second group of species lose seed viability if moisture content is reduced to between 12 to 13 percent, and are considered *recalcitrant.* This group includes many temperate and tropical tree species, including oil palm, rubber, cocoa, and coffee. There are also species with intermediate storage behavior, where the seeds can be dried to relatively low moisture levels (7 to 12 percent)—low enough to qualify as orthodox, but that are still sensitive to the low temperatures typically employed for storage of orthodox seeds. These may still be stored by cryopreservation (which involves dehydrating the plant material, then freezing it rapidly by direct immersion in liquid nitrogen). Fortunately, the majority of plant genetic resource species are orthodox, allowing storage for relatively long periods.

The two most important factors in promoting seed storage are seed moisture content and temperature. Roughly speaking, longevity is doubled for each 5°C fall in temperature or for each 1 percent fall in moisture content. This formulation, developed by Harrington (1970), is now known as "Harrington's rule of thumb". The general recommendation for orthodox seeds is that they should be dried to 5 to 6 percent moisture content, and placed in sealed containers stored at temperatures below −18°C. Although storage under optimal conditions will result in the maintenance of high viability over a long period, a certain decrease in viability is inevitable necessitating periodic regeneration of seed lots.

In practice, there may also be cases where conservation cannot be accurately termed *in situ* or *ex situ.* Take, for example, the conservation of the legume tree genus, *Leucaena,* where germplasm is often collected from native habitats and taken *ex situ* to more easily managed agroforestry areas for conservation by local communities (Hughes 1998). The trees are not conserved using standard applications of field gene bank or arboreta techniques, but within local communities; the transplanted trees are to all intents and purposes managed by local people using traditional silvicultural

TABLE 21.4 Strategies and Techniques of Genetic Conservation

Strategies	Techniques	Definition	Advantages	Disadvantages
In situ conservation	Genetic Reserve	The location, management, and monitoring of genetic diversity in natural wild populations within defined areas designated for active, long-term conservation	• Dynamic conservation in relation to environmental changes, pests, and diseases • Provides easy access for evolutionary and genetic studies • Appropriate for recalcitrant species	• Materials not easily available for utilization • Vulnerable to natural and man-directed disasters • Management regimes poorly understood • High level of supervision and monitoring • Limited genetic diversity in a reserve
	On-farm	The sustainable management of genetic diversity of locally-developed traditional crop varieties with associated wild and weedy species or forms by farmers within traditional agricultural, horticultural, or agri-silvicultural cultivation systems	• Dynamic conservation in relation to environmental changes and diseases • Ensures the conservation of landraces of field crops • Ensures the conservation of weedy crop relatives and ancestral forms	• Vulnerable to change in farming practices • Management regimes poorly understood • Possible payment of incentives to farmers • Restricted to field crops • Only limited diversity can be maintained on each farm
	Home Garden	The sustainable management of genetic diversity of locally-developed traditional crop varieties by householder within domestic cultivation systems	• Dynamic conservation in relation to environmental changes, pests and diseases • Ensures the conservation of landraces of minor crops, fruit and vegetables, medicinal plants, flavorings, herbs, fruit trees and bushes, and so forth	• Vulnerable to changes in management practices • Management regimes poorly understood • Requires maintenance of domestic cultivation systems
Ex situ conservation	Seed Storage	Collection of seed samples at one location and their transfer to a gene bank; samples are dried to low moisture content and then stored at sub-zero temperatures	• Efficient and reproducible • Medium and long-term storage • Easy access • Little maintenance cost	• Cannot store recalcitrant seeded species • Freezes evolutionary development • Genetic diversity lost with regeneration

(Continued)

TABLE 21.4 (*Continued*)

Strategies	Techniques	Definition	Advantages	Disadvantages
	in vitro Storage	Collection and maintenance of explants (tissue samples) in a sterile, pathogen-free environment	· Suitable for recalcitrant, sterile, or clonal species · Easy access for evaluation and utilization	· Risk of somaclonal variation · Need to develop individual maintenance protocols for most species · High technology and maintenance costs
	Field Gene Bank	Collection of seed or living material from one location and its transfer and planting at a second site; large numbers of accessions of a few species are conserved	· Suitable for recalcitrant species · Easy acces · Material can be evaluated while being conserved	· Susceptible to pests, disease, and vandalism · Involves large areas of land, but even then genetic diversity is likely to be restricted · High maintenance cost once conserved
	Botanic Garden/ Arboretum	Collection of seed or living material from one location and its transfer and maintenance as living plant collections in a garden or (for tree species) an arboretum small numbers of accessions of a large number of species are usually conserved.	· Freedom to focus on wild or non-economic plants · Easy public access for conservation education	· Space limits the number and genetic diversity of the species conserved · Involves large areas of land or glasshouses, so genetic diversity is likely to be restricted · High maintenance costs in glasshouse once conserved
	DNA/ Pollen Storage	The collecting of DNA or pollen and storage in appropriate, usually refrigerated, conditions.	· Relatively easy, low-cost conservation	· Regeneration of entire plants from DNA cannot be envisaged at present · Problems with subsequent gene isolation, cloning and transfer

techniques within an *in situ* on-farm system. In cases like this, the term *circa situ* has been used to describe this form of conservation (Hawkes, Maxted, and Ford-Lloyd. 2000).

Conservation products and dissemination

The products of conservation activities are conserved germplasm (seed, embryos), live plants, dried plants (herbarium specimens), cultures, and conservation data. Whether the conservation products are conserved *in situ* or *ex situ*, they should ideally be duplicated in more than one location to ensure their safety. The distribution of duplicate sets of material avoids accidental loss of the material due to fire, economic or political difficulties, warfare, or other unforeseen circumstances. Duplication of the "passport data" is relatively easy from the conservation database, and the commissioning agency, relevant host country institutes, and other interested parties should hold copies.

Complementarity and sustainability are fundamental

In situ and *ex situ* conservation strategies should not be seen as alternatives or in opposition to one another, but rather as complementary, with one conservation strategy or technique acting as a backup to another. The degree of emphasis placed on each depends on the conservation aims, the type of species being conserved, the resources available, and whether the species has utilization potential (Maxted, Ford-Lloyd, and Hawkes 1997). In terms of plant genetic resources, *in situ* conservation is still in its infancy (Hawkes 1991), and *ex situ* techniques are much more established. The handful of *in situ* genetic conservation projects that have been established are still largely experimental and, perhaps most worryingly, none has secured long-term funding. Sustainability in the sense of continuance is a fundamental concept for conservation. It is therefore vital that target populations in genetic reserves or on-farm programs are also sampled and duplicated *ex situ* to ensure long-term security.

Linking conservation to sustainable use

Lastly, it is important to stress that there should be an intimate link between conservation and utilization (Maxted, Ford-Lloyd, and Hawkes 1997; Hawkes, Maxted, and Ford-Lloyd 2000) and in fact this point is implicit in the CBD. The products of conservation, whether they are "living" or "suspended", should be made available for utilization. It is perhaps easiest to make this case for socioeconomically important plant species, but conservation of any plant resource can be seen as safe keeping for utilization and management at a future date. In certain cases, the material can be used directly, for instance, in the selection of forage accessions, where little breeding is undertaken, or specific alleles can be incorporated into cultivars. The conserved material may also be used in reintroduction programs where the primitive landrace has been lost locally.

Commonly, the first stage of utilization involves recording genetically-controlled characteristics (characterization). Genetic material may also be tested in diverse environments to evaluate and screen for tolerance to harsh conditions, or be deliberately infected with diseases or pests to screen for particular biotic resistance (evaluation). The biotechnologist screens for single genes, which once located may be transferred into a host organism, while the biochemist (bioprospector) screens for particular chemical products that may be of use to the pharmaceutical industry. The products of utilization are therefore numerous, including new varieties, new crops, improved breeds, and pharmaceuticals, as well as a "good" environment for recreational activities.

References and Further Reading

Brush, S.B. (Ed.) 2000. *Genes in the Field: On-Farm Conservation of Crop Diversity.* Boca Raton, FL: Lewis Publishers.

Bouhssini, M. et al. 1998. Identification in *Aegilops* species of resistant sources to Hessian fly (Diptera: Cecidomyiidae) in Morocco. *Genetic Resources and Crop Evolution 45*, 343–345.

Cohen, J.I., et al. 1991. *Ex situ* conservation of plant genetic resources: global development and environmental concerns. *Science 253*, 866–872.

Crucible Group. 1994. *People, Plants and Patents: The Impact of Intellectual Property on Trade, Plant Biodiversity and Rural Society.* Ottawa: IDRC.

Falk, D.A., and Holsinger, K.E. 1991. *Genetics and Conservation of Rare Plants.* Oxford: Oxford University Press.

Food and Agriculture Organization of the United Nations. 1998. *The State of the World's Plant Genetic Resources for Food and Agriculture.* Rome: FAO.

Ferguson, M.E., et al. 1998. Mapping the geographical distribution of genetic variation in the genus *Lens* for the enhanced conservation of plant genetic diversity. *Molecular Ecology 7*, 1743–1755.

IPGRI. 1999. *Diversity for Development.* Rome: International Plant Genetic Resources Institute.

Harlan, J.R., and de Wet, J.M.J. 1971. Towards a rational classification of cultivated plants. *Taxon 20*, 509–517.

Harrington, J.F. 1970. Seed and pollen storage for conservation of plant gene resources. In *Genetic Resources in Plants: Their Exploration and Conservation,* edited by O.H. Frankel and E. Bennett. Oxford: Blackwell, 501–521.

Hawkes, J.G. 1991. International workshop on dynamic *in situ* conservation of wild relatives of major cultivated plants: summary of final discussion and recommendations. *Israel Journal of Botany 40*, 529–536.

Hawkes, J.G., Maxted, N., and Ford-Lloyd, B.V. 2000. *The ex situ Conservation of Plant Genetic Resources.*-Dordrecht: Kluwer.

Hughes, C.E. 1998. *Leucaena. A Genetic Resources Handbook. Tropical Forestry Papers 37*. Oxford: Oxford Forestry Institute.

Iltis, H.H. 1988. Serendipity in the exploration of biodiversity: What good are weedy tomatoes? In *Biodiversity*, edited by E.O. Wilson. Washington, DC: National Academy Press, 98–105.

Lesser, W. 1998. Sustainable Use of Genetic Resources under the Convention on Biological Diversity : Exploring Access and Benefit Sharing Issues. Wallingford: CABI.

Lewington, A. 1990. *Plants for People.* London: Natural History Museum.

Lutz, W., Prinz, C., and Langgassmer, J. World population projections and possible ecological feedbacks. *POPNET 23*, 1–11.

Maxted, N., Ford-Lloyd, B.V., and Hawkes, J.G. 1997. *Plant Genetic Conservation: The* in situ *Approach*. London: Chapman & Hall.

Maxted, N., van Slageren, M.W., and Rihan, J. 1995. Ecogeographic surveys. In *Collecting Plant Genetic Diversity: Technical Guidelines*, edited by L. Guarino, V. Ramanatha Rao, and R. Reid. Wallingford: CAB International, 255–286.

Phillips, O.L., and Meilleur, B. 1998. Usefulness and economic potential of the rare plants of the United States: a status survey. *Economic Botany 52*, 57–67.

Quist, D., and Chapela, I.H. 2001. Transgenic DNA introgressed into traditional maize landraces in Oaxaca, Mexico. *Nature 414*, 541–543.

Smith, R.D., and Linington, S. 1997. The management of the Kew Seed Bank for the conservation of arid land and UK wild species. *Bocconea 7*, 273–280.

Smith, R.D., Dickie, J.B., Linington, S.H., Pritchard, H.W. and Probert, R.J. (Eds.) 2003. Seed Conservation: Turning Science into Practice. Kew, Surrey: Royal Botanic Gardens, Kew.

ten Kate, K., and Laird, S.A. 1999. *The commercial use of biodiversity: access to genetic resources and benefit sharing*. London: Earthscan.

Vavilov, N.I. 1922. The law of homologous series in variation. *Journal of Genetics 12*, 47–89.

Websites

Convention on Biological Diversity (CBD) website, www.biodiv.org

European Plant Conservation Strategy website, www.plantaeuropa.org/html/plant_conservation_strategy.htm

International Plant Genetic Resources Institute website, www.ipgri.cgiar.org

International Treaty on Plant Genetic Resources for Food and Agriculture (ITPGRFA) website, www.fao.org/ag/cgrfa/

Global Strategy for Plant Conservation (GSPC) website, www.biodiv.org/programmes/cross-cutting/plant/

Plant Names

For most of the plants in this volume, the botanical (Latin) name is unambiguous. However, in some cases the same species is known by more than one botanical name. For these, this book usually follows the accepted name used by the Germplasm Resources Information Network (GRIN) of the United States Department of Agriculture (Wiersema and León 1999) or, in some cases, that used by the Institute of Plant Genetics in Gatersleben (Hanelt and Institute of Plant Genetics and Crop Plant Research 2001). Both sources can be consulted for author names and for alternative or obsolete plant names omitted from this volume.

For the eight plant families for which both old and new names are in use, this book follows the new usage:

New Name	Old Name
Apiaceae	Umbelliferae
Arecaceae	Palmae
Asteraceae	Compositae
Brassicaceae	Cruciferae
Clusiaceae	Guttiferae
Fabaceae	Leguminosae
Lamiaceae	Labiatae
Poaceae	Gramineae

References

Hanelt, P., and Institute of Plant Genetics and Crop Plant Research 2001. *Mansfeld's Encyclopedia of Agricultural and Horticultural Crops (except Ornamentals)*. Berlin: Springer. Also available at http://mansfeld.ipk-gatersleben.de/
Wiersema, J.H., and León, B. 1999. *World Economic Plants: A Standard Reference*. Boca Raton: CRC Press. Also available at http://www.ars-grin.gov/npgs/

Radiocarbon Dating

Most dates given in this book are straightforward historical dates in calendar years. However, some dates for archaeological material are given in radiocarbon years, shown in this volume as ^{14}C years.

Radiocarbon dates tend to underestimate ages; for example, a seed that is 11,400 years old would have a radiocarbon date of 10,000 ^{14}C years. For younger dates the difference is less; for example, a seed that is 5000 years old will have a radiocarbon date of about 4400 ^{14}C years. Radiocarbon dating of tree rings has allowed approximate calibration of radiocarbon dates to calendar years. Calibration has only recently become possible for the older dates (>5000 years before present) relevant to the beginnings of farming, and many archaeologists still use uncalibrated dates for this early period.

General References on the History of Useful Plants

Brücher, H. 1989. *Useful Plants of Neotropical Origin and their Wild Relatives*. Berlin: Springer-Verlag.

Davidson, A. 1999. *The Oxford Companion to Food*. Oxford: Oxford University Press.

de Cleene, M. and Lejeune, M.C. 2003. Compendium of Symbolic and Ritual Plants in Europe: Botanical, Cultural, Uses. Ghent: Mans & Culture.

Hanelt, P. (ed.) 2001. *Mansfeld's Encyclopedia of Agricultural and Horticultural Crops. Six volumes*. Berlin: Springer.

Harris, D. R. (ed.) 1996. *The Origins and Spread of Agriculture and Pastoralism in Eurasia*. London: UCL Press.

Harris, D. R. and G. C. Hillman (Eds.) 1989. *Foraging and Farming. The Evolution of Plant Exploitation*. One World Archaeology 13. London: Unwin Hyman.

Kiple, K. F. and K. C. Ornelas (Eds.) 2000. *The Cambridge World History of Food. Two Volumes*. Cambridge: Cambridge University Press.

Lewington, A. 2003. *Plants For People*. London: Transworld.

Mabberley, D. J. 1997. *The Plant Book. A Portable Dictionary of the Vascular Plants. Second Edition*. Cambridge: Cambridge University Press.

Musgrave, T., and W. Musgrave. 2000. *An Empire of Plants: People and Plants that Changed the World*. London: Cassell.

Nicholson, P. T. and I. Shaw. (Eds.) 2000. *Ancient Egyptian Materials and Technology*. Cambridge: Cambridge University Press.

Price, T. D., and A. B. Gebauer. (Eds.) 1995. *Last Hunters-First Farmers*. Santa Fe, NM: School of American Research Press.

Sauer, J. D. 1993. *Historical Geography of Crop Plants. A Selective Roster*. Boca Raton, FL: CRC Press.

Smartt, J. and N. W. Simmonds (Eds.) 1995. *Evolution of Crop Plants. Second Edition*. Harlow: Longman.

Smith, B. D. 1998. *The Emergence of Agriculture. Paperback Edition*. New York, NY: Scientific American Library.

Vaughan, J. G. and C. A. Geissler. 1997. *The New Oxford Book of Food Plants*. Oxford: Oxford University Press.

Zohary, D. and M. Hopf. 2000. *Domestication of Plants in the Old World: the Origin and Spread of Cultivated Plants in West Asia, Europe, and the Nile Valley. Third Edition*. Oxford: Clarendon Press.

Index

435